T0142181

Lecture Notes in Networks and Systems 1035

The series "Lecture Notes in Networks and Systems" publishes the latest developments in Networks and Systems—quickly, informally and with high quality. Original research reported in proceedings and post-proceedings represents the core of LNNS.

Volumes published in LNNS embrace all aspects and subfields of, as well as new challenges in, Networks and Systems.

The series contains proceedings and edited volumes in systems and networks, spanning the areas of Cyber-Physical Systems, Autonomous Systems, Sensor Networks, Control Systems, Energy Systems, Automotive Systems, Biological Systems, Vehicular Networking and Connected Vehicles, Aerospace Systems, Automation, Manufacturing, Smart Grids, Nonlinear Systems, Power Systems, Robotics, Social Systems, Economic Systems and other. Of particular value to both the contributors and the readership are the short publication timeframe and the world-wide distribution and exposure which enable both a wide and rapid dissemination of research output.

The series covers the theory, applications, and perspectives on the state of the art and future developments relevant to systems and networks, decision making, control, complex processes and related areas, as embedded in the fields of interdisciplinary and applied sciences, engineering, computer science, physics, economics, social, and life sciences, as well as the paradigms and methodologies behind them.

Indexed by SCOPUS, INSPEC, WTI Frankfurt eG, zbMATH, SCImago.

All books published in the series are submitted for consideration in Web of Science.

For proposals from Asia please contact Aninda Bose (aninda.bose@springer.com).

Jawad Rasheed · Adnan M. Abu-Mahfouz ·
Muhammad Fahim
Editors

Forthcoming Networks and Sustainability in the AIoT Era

Second International Conference FoNeS-AIoT 2024 – Volume 1

Editors
Jawad Rasheed
Department of Computer Engineering
Istanbul Sabahattin Zaim University
Istanbul, Türkiye

Muhammad Fahim
School of Electronics, Electrical Engineering and Computer Science
Queen's University Belfast
Belfast, UK

Adnan M. Abu-Mahfouz
Council for Scientific and Industrial Research (CSIR)
Pretoria, South Africa

Department of Electrical and Electronic Engineering Science
University of Johannesburg
Johannesburg, South Africa

ISSN 2367-3370 ISSN 2367-3389 (electronic)
Lecture Notes in Networks and Systems
ISBN 978-3-031-62870-2 ISBN 978-3-031-62871-9 (eBook)
https://doi.org/10.1007/978-3-031-62871-9

This Springer imprint is published by the registered company Springer Nature Switzerland AG
The registered company address is: Gewerbestrasse 11, 6330 Cham, Switzerland

Contents

Determining the Digits of Turkish Sign Languages Using Deep Learning Techniques

Emine Karataş and Gökalp Çınarer(✉)

Yozgat Bozok University, Yozgat, Turkey
gokalp.cinarer@bozok.edu.tr

Abstract. Sign language is a physical language that enables people with disabilities to communicate with each other by using hand and facial movements as a whole to express themselves. It is very important that sign language is learned by everyone and used as a communication tool for the disabled to adapt to social life and to express themselves easily. For this reason, people's learning of sign languages, which are specific to the country's spoken language, will increase the quality of life of people with disabilities. In this study, 12981 images of the numbers 0–10 in Turkish Sign Language taken from different angles were used as a data set. In the last stage of the study, the detection of digits over images was carried out with CNN, Resnet-50, VGG-16, Densenet-201, and Inception-V3 deep learning architectures. In the study, an effective model of deep learning algorithms is proposed to determine which number an action corresponds to in sign language. Examining the models, VGG-16 and Densenet-201 were the architectures that gave the highest accuracy with 100% accuracy. After these architectures, Inception-V3 architecture comes with 99.91% success in determining the numbers. It has been seen that it is very successful in detecting numbers in Turkish Sign Language using deep learning models.

Keywords: Hand Gesture Recognition · Sign Language Translation · Deep Learning

1 Introduction

Sign language vital tool that deaf people make use of to express themselves using body language [1]. It is necessary to know sign language correctly and evaluate its importance accordingly. Even though deaf people can communicate with each other through sign language, people who go to public or private places have difficulty explaining themselves and understanding others. As stated by the findings of the research carried out by the World Health Organization in Europe in 2018, 34 million people with hearing impairment have been identified. It is estimated that this number will approach 46 million in 2050 [2].

Hearing-impaired people were cut off from social life until recently. It has been mentioned in the studies that hearing-impaired individuals from different communities have difficulty communicating even among themselves, despite participating in international sports-related events together [3, 4]. In addition, in Gondon's research, he stated

J. Rasheed et al. (Eds.): FoNeS-AIoT 2024, LNNS 1035, pp. 1–10, 2024.
https://doi.org/10.1007/978-3-031-62871-9_1

that there are more than 124 sign languages in the world and that individuals from different nationalities have problems in communicating even if they have similar aspects [5]. The World Report on the Disabled states that individuals with hearing impairments generally have difficulty interpreting sign language. According to a study involving 93 countries, translation services are not available in 31 countries and the number of translators authorized for translation services in 30 countries is 20 or less. With the developing technology, it is necessary to benefit from artificial intelligence technologies to eliminate such problems.

In this study, the success of deep learning architectures in detecting numbers from 0 to 10 for Turkish Sign Language has been examined.

2 Literature Review

In the literature research, deep learning is exploited in several fields such as software industry [6], image processing [7], and noise detection [8], however, it is seen that the studies on sign language are quite limited in Turkey. On the other hand, the number of studies in this field in the USA is quite high. Ravinder Ahuja proposed a model for performing gesture recognition of American Sign Language using a Convolutional Neural Network (CNN). They evaluated 24 hand signals in the study where they used the user's camera recordings. At the end of the study, they reached an accuracy rate of 99,7% [9]. Abdulwahap et al. used CNN deep learning architecture to classify American letters and got a result of 99.33%. They revealed that they got a higher result for CNN when compared with SVM and ANN algorithms [10].

In another study, Indian Sign Language alphabets and numbers were used. To train 36 static movements, a classification process was performed using 45,000 RGB and 45,000 depth images with CNN architecture. As a result of the training, 98.81% accuracy was obtained. [11].

Selvi and Kemaloğlu [12] classified Turkish sign language digits with the CNN and reached a test accuracy of 98.55%. Fernandez and Kwolek presented a CNN-based algorithm for hand gesture recognition from a color camera. They used 6000 tagged images. They were labeled in 10 classes. They did their work on CNN and completed their work with a high degree of accuracy [13].

Rahmat and his team have signed two separate studies. In the first study, 6 different hand movements were carried out, for the second study, 4 different hand movements were performed. The most successful test result from the study was obtained from the test performed for four different hand movement types. In the study 300 and 50 hidden neurons were used for each layer, in two hidden layers [14].

In another study [15], 99.90% training accuracy was achieved in the classification made with 35,000 images of facial static signs. In a study conducted with the Arabic sign language recognition method, in a study that recognized 32 hand gestures, the VGG16 and ResNet152 models reached 99% accuracy [16].

A comparative analysis of similar studies in the literature is shown in Table 1. On the other side sample images of numbers used in Turkish sign language are given in Table 2.

Table 1. Studies examined in the literature.

Project Owners	Study Content	Method	Truth
Aeshita Mathur et al. [17]	Containing 500 images for 26 English Alphabets	CNN VGG16	91.16%
Yao-Liang Chung [18]	Contains 4800 images for numbers 1–5 hand gesture recognition system	CNN VVGNET	95.61%
Müneer El Hammadi and team [19]	A sign language definition study was conducted with 5 words taken from 40 participants	3D CNN	96.69%
Molchanov and team [20]	20,000 images of 10 static digits	CNN	97.62%
Islam and team [21]	1075 image Bangla Sign Language 10 static digits	CNN	95%

Table 2. Sample images of numbers used in Turkish sign language.

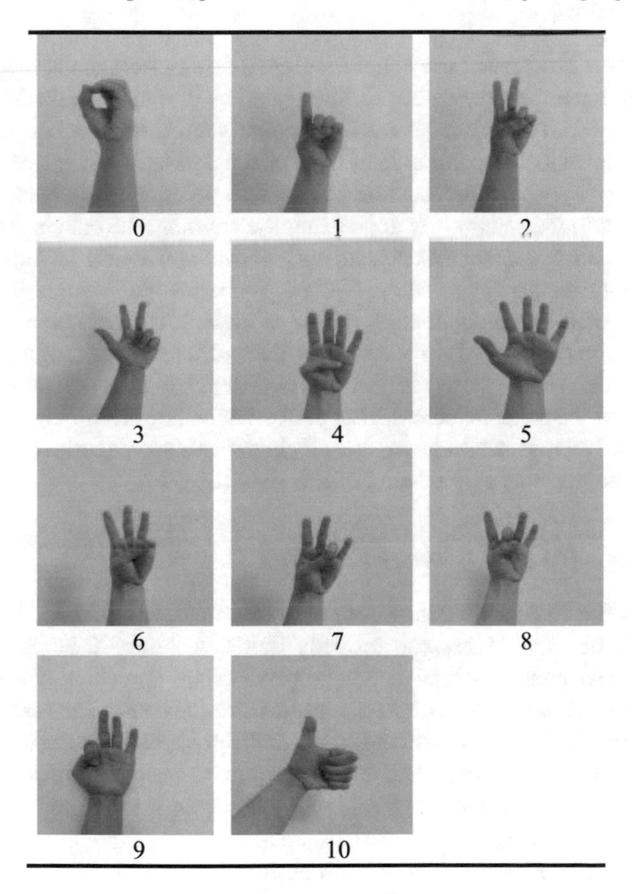

3 Material and Method

3.1 Dataset

The data set consists of 12981 images of Turkish sign language containing numbers between 0 and 10. Images were obtained with a 1080p resolution camera, at a frame rate of 30fps, at a focal length of 26 mm, in a plain background font, in RGB color. Images are in 3000x4000 pixel size, jpeg file format. In the study, 12981 images of 11 classes, taken from a person, from a background, and at distances ranging from 0 to 30 cm from the camera, were obtained on a plain white background and in-room conditions. The dataset is divided into 10695 images for train and 2286 images for test.

3.2 Deep Learning

Deep learning is a data modeling system developed as a sub-branch of machine learning in which different neuron layers are used, influenced by the human brain. Multiple neuron layers are designed with different methods and analyzed with different parameters.

In systems developed according to the size of the dataset, the depth of the network is optimized accordingly, and high accuracy is tried to be achieved in complex classifications. According to the weights created during the training, the data is processed in the hidden layers, their features are extracted, and predictions are made. Unlike traditional machine learning methods, the whole system evaluates the features it automatically obtains instead of manually extracting the data. Thus, it can also be applied in multiclass and complex structures. The class and complex structures. The algorithm used in all these processes can be supervised or unsupervised depending on the state of the data. As the similarities of the pictures increase, the performance of CNN architectures decreases [22]. In addition, representation learning, and classification categories are automatically discovered in the machine's processing of raw data [23].

3.3 Convolutional Neural Network (CNN)

CNN is one of the basic architectures of Deep Learning. Inspired by the way animals see. Its use has become widespread in many fields. Although it is preferred in fields such as audio processing, natural language processing, and biomedical, it gives the highest accuracy in image processing. It simplifies complex operations using convolution filters. CNN image classifications take input data, process it, and categorize it under certain categories [24]. A standard CNN architecture is a set of feedforward layers that implement convolutional filters and pool layers. After the last pooling layer, CNN uses several fully connected layers to classify the map features of the previous layers. Does not require a feature extraction before CNN architecture is implemented [25].

In a feedforward neural network, any middle layer is called hidden because the activation function and the final convolution layer mask the input and output. The layers that enable the formation of convolutions contain hidden layers [26].

In summary, ESAs consist of several trainable parts in a row. Then, classification is done with an educational classifier. In ESA, the training process starts with receiving the input data and continues by processing layer by layer. This is how the training process ends. A block diagram of the Turkish Sign Language recognition system is given in Fig. 1.

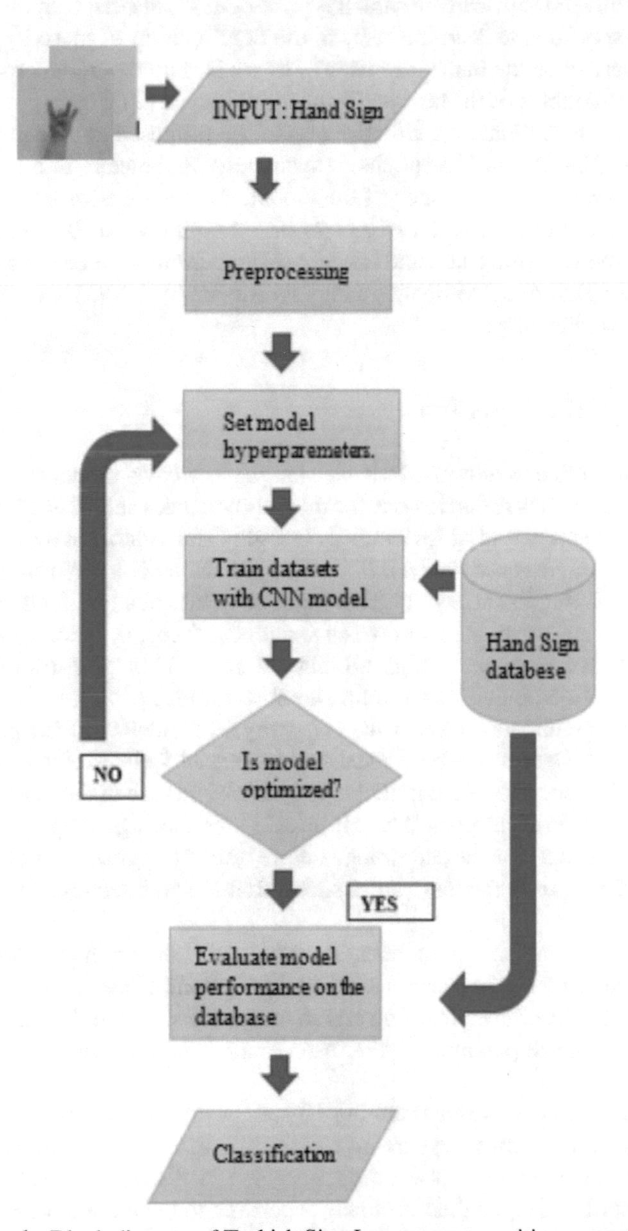

Fig. 1. Block diagram of Turkish Sign Language recognition system.

As the last operation, it gives the final output for comparison with the correct result. It is also very important how many pieces of data will be processed simultaneously in the training of the model. Here batch size values need to be defined in each architecture. Like the epoch number, the batch size is a hyperparameter with no rules. Conversely, if a batch size is too large, it may not fit in the memory of the computational sample used for training and will tend to overfit the data. It is important to note that batch size affects other hyper-parameters such as learning rate, so the combination of these hyper-parameters is just as important as the batch size itself. The main purpose of hyperparameters is to determine the variables of the best performance achieved [27].

To evaluate and validate the effectiveness of the proposed approaches, experiments were performed in 10 and 30 epochs, respectively. Parameters and hyperparameters were tried to raise the performance of the model, and the ones with the highest results were used in the project. Basic parameters like the number of layers, filters, and the optimization applied were taken into account in the performance analysis. Classification processes were performed on images using the Resnet-50, VGG16, Densenet-201, and InceptionV3 architectures.

4 Results and Discussion

For the detection of hand movements in sign language, a CNN model was created and the classification process was carried out for the numbers between "0–10". Google Colaboratory environment was used for training. Google Colaboratory is a system created for use in artificial intelligence studies. It is a Jupyter Notebook environment running in the cloud. By making the necessary distinction and classification in the data set Google Drive environment, for numbers between 0 and 10, "zero: 0, one: 1, two: 2, three: 3, four: 4, five: 5, six: 6, seven: 7, eight: 8, nine: 9, ten: 10" in the form of train and test. The dataset is divided into 80% for train and 20% for test.

In the study, architectures were run in two ways, 10 epochs and 30 epochs. An epoch ends when the dataset has been transmitted back and forth over the neural network exactly once. The success of the model is tested. Weights are updated accordingly. This process is repeated in each epoch. There is no rule for choosing the epoch number. This is a hyperparameter that must be determined before training begins. As the accuracy values increase, the decrease in the loss values shows that the architectures make a successful classification.

Train accuracy and validation accuracy results are shown in the same graph. As a result of the experiment, the success and loss-loss graphs of the test and training data at the end of 30 epochs of deep learning architectures are shown in Fig. 2.

Basic performance parameters were used to evaluate the validity and performance of this study.

The results of the algorithms running 10 epochs and 30 epochs for each evaluation metric and the architectures used are given in detail in Table 3. When the performances of the models were compared, it was seen that some models reached 100% accuracy and some architectures changed their accuracy according to the number of epochs. Looking at Table 3, it appears that the accuracy of VGG-16 and DenseNet-201 architectures increased after 30 epochs.

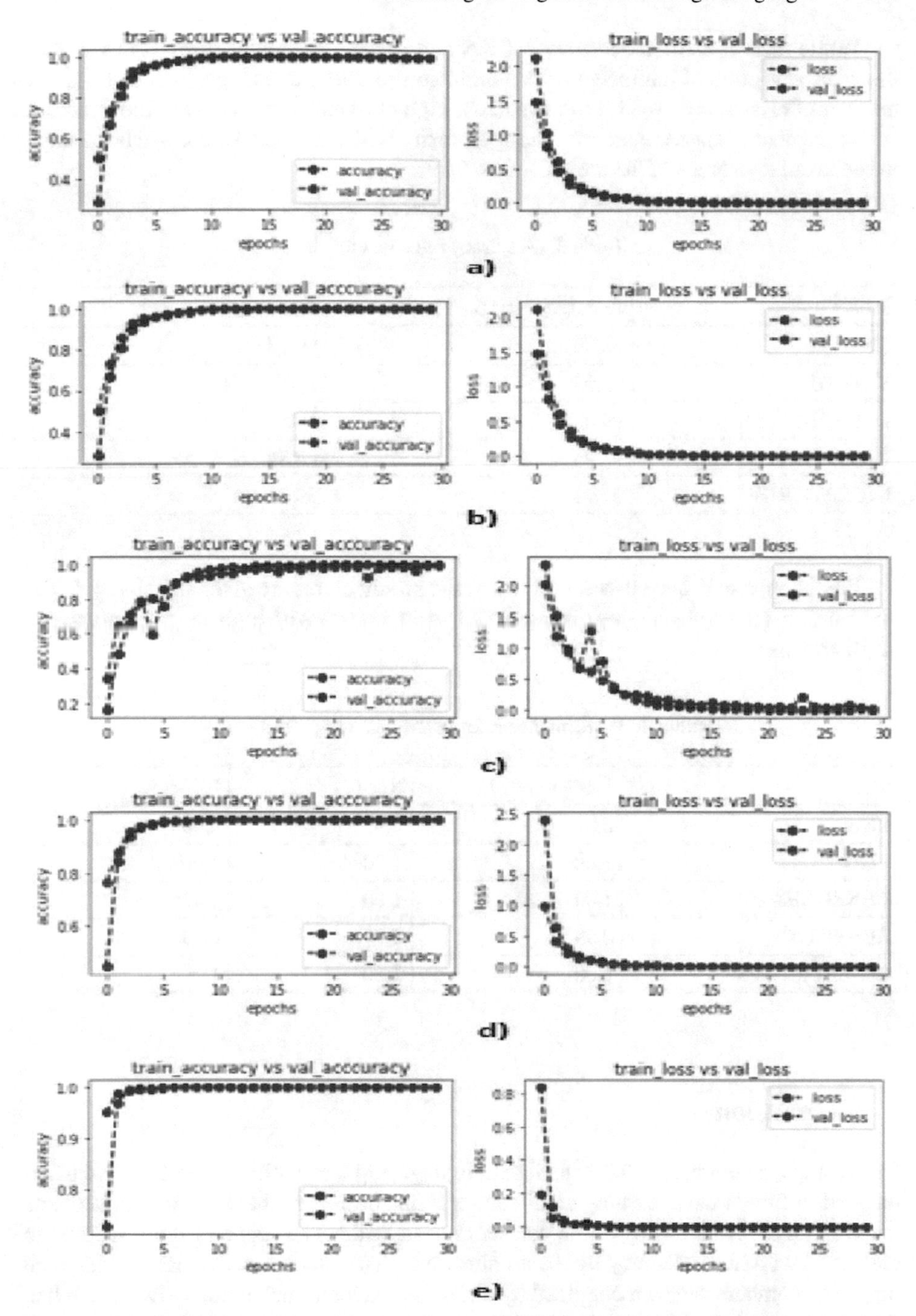

Fig. 2. Performance analysis graphs of models a) CNN b) VGG16 c) Resnet50 d) DenseNet201 e) InceptionV3

While there is a slight decrease in CNN and Resnet-50 architectures, it is seen that the accuracy value of the Inception-V3 architecture does not change. The fact that the number of classes is eleven is effective in the high classification success of the numbers. As the number of classes increases, the complexity will increase, so there will be changes in the accuracy values of the architectures.

Table 3. Accuracy rates of models.

Model	Accuracy (10 epoch)	Accuracy (30 epoch)
CNN	98,99	99,83
VGG-16	99,95	100
DENSENET-201	99,65	100
RESNET-50	98,95	98,86
INCEPTION-V3	99,91	99,91

Precision, recall, and F1-score values of the architectures are given in Table 4. It was observed that the architectures distinguished all 11 classes with high accuracy at the end of 30 epochs.

Table 4. Performance rates of models with 30 epochs.

Model	Precision	Recall	F1 Score
CNN	0,99	0,99	0,99
VGG-16	1,00	1,00	1,00
DENSENET-201	1,00	1,00	1,00
RESNET-50	0,99	0,99	0,99
INCEPTION-V3	1,00	1,00	1,00

5 Conclusion

In this study, recognition of Turkish Sign Language numbers with hand gestures is carried out with different deep learning architecture. It has been seen that the hand movements taken from different angles in the dataset created within the scope of the research are classified with high accuracy by the architectures. This study reveals that Turkish sign language numbers can be recognized by different convolutional neural networks. When the experimental results were examined, it was observed that the recognition performance of CNN architectures increased considering that the images were shot in a similar environment. Application of the proposed system with different hand movements and under different conditions may change the results.

It has been classified with higher accuracy than studies conducted in other different languages. The basic parameters that the architectures use while determining this affect the classification success. When the models were examined, VGG-16 and Densenet-201 were the architectures that gave the highest accuracy with 100% accuracy. Next comes the Inception-V3 architecture with 99.91%. In future studies, studies can be carried out with images including letters and numbers in different conditions. Turkish sign language numerals and the Turkish sign language alphabet can be compared with different datasets with new methods and models. In addition, different CNN architectures that can increase accurate prediction can be used in studies.

References

1. Sangeethalakshmi, K., Shanthi, K.G., Raj, A.M., Muthuselvan, S., Taha, P.M., Shoaib, S.M.: Hand gesture vocalizer for deaf and dumb people. Mater. Today Proc. **80**, 3589–3593 (2023)
2. Yildiz, Z., Yildiz, S., Bozyer, S.: İşitme Engelli Turizmi (Sessiz Turizm): Dünya ve Türkiye Potansiyeline Yönelik Bir Değerlendirme. Süleyman Demirel Üniversitesi Vizyoner Dergisi **9**(20), 103–117 (2018)
3. Theunissen, S.C., et al.: Self-esteem in hearing-impaired children: the influence of communication, education, and audiological characteristics. PLoS ONE **9**(4), e94521 (2014)
4. Morata, T.C., Themann, C.L., Randolph, R.F., Verbsky, B.L., Byrne, D.C., Reeves, E.R.: Working in noise with a hearing loss: perceptions from workers, supervisors, and hearing conservation program managers. Ear Hear. **26**(6), 529–545 (2005)
5. Gordon, R.G., Jr. ed.: Ethnologue: Languages of the World. Fifteenth Edition, SIL International, Dallas, TX (2005)
6. Tahir, T., Gence, C., et al.: Early software defects density prediction: training the international software benchmarking cross projects data using supervised learning. IEEE Access **11**, 141965–141986 (2023)
7. Cevik, T., et al.: Facial recognition in hexagonal domain – a frontier approach. IEEE Access **11**, 46577–46591 (2023)
8. Waziry, S., et al.: Performance comparison of machine learning driven approaches for classification of complex noises in quick response code images. Heliyon **9**(4) (2023)
9. Ahuja, R., et al.: Convolutional neural network based American sign language static hand gesture recognition. Int. J. Ambient Comput. Intell. **10**(3), 60–73 (2019)
10. Abdulhussein, A.A., Raheem, F.A.: Hand gesture recognition of static letters American sign language (ASL) using deep learning. Eng. Technol. J. **38**(6), 926–937 (2020)
11. Bhagat, N.K., Vishnusai, Y., Rathna, G.N.: Görüntü İşleme ve Derin Öğrenmeyi Kullanarak Hint İşaret Dili Hareket Tanıma. In: 2019 Digital Image Computing: Techniques and Applications (DICTA), pp. 1–8 (2019)
12. Sevli, O., Kemaloğlu, N.: Turkish sign language digits classification with CNN using different optimizers. Int. Adv. Res. Eng. J. **4**(3), 200–207 (2020)
13. Núñez Fernández, D., Kwolek, B.: Hand posture recognition using convolutional neural network. In: Mendoza, M., Velastín, S. (eds.) Progress in Pattern Recognition, Image Analysis, Computer Vision, and Applications (CIARP 2017). LNCS, vol. 10657, pp. 441–449. Springer, Cham (2018). https://doi.org/10.1007/978-3-319-75193-1_53
14. Rahmat, R.F., et al.: A study on dynamic hand gesture recognition for finger disability using multi-layer neural network. J. Theor. Appl. Inf. Technol. **96**(11), 3413–3425 (2018)
15. Wadhawan, A., Kumar, P.: Deep learning-based sign language recognition system for static signs. Neural Comput. Appl. **32**(12), 7957–7968 (2020)

16. Saleh, Y., Issa, G.F.: Arabic sign language recognition through deep neural networks fine-tuning.: Int. J. Online Biomed. Eng. **16**(5), 71–83 (2020)
17. Mathur, A., Singh, D., Chhikara, R.: Recognition of American sign language using deep learning. In: 2021 International Conference on Industrial Electronics Research and Applications (ICIERA), pp. 1–5. IEEE (2021)
18. Chung, H.Y., Chung, Y.L., Tsai, W.F.: An efficient hand gesture recognition system based on deep CNN. In: 2019 IEEE International Conference on Industrial Technology (ICIT). IEEE (2019)
19. Al-Hammadi, M., et al.: Hand gesture recognition for sign language using 3DCNN. IEEE Access **8**, 491–505 (2020)
20. Wangchuk, K., Riyamongkol, P., Waranusast, R.: Real-time Bhutanese sign language digits recognition system using convolutional neural network. ICT Express **7**(2), 215–220 (2021)
21. Islam, S., Mousumi, S.S.S., Rabby, A.S.A., Hossain, S.A., Abujar, S.: A potent model to recognize Bangla sign language digits using convolutional neural network. Procedia Comput. Sci. **143**, 611–618 (2018)
22. Barbhuiya, A.A., Karsh, R.K., Jain, R.: CNN based feature extraction and classification for sign language. Multimed. Tools Appl. **80**(2), 3051–3069 (2021)
23. Bengio, Y.: Learning deep architectures for AI. Found. trends® Mach. Learn. **2**(1), 1–127 (2021)
24. Liang, G., Hong, H., Xie, W., Zheng, L.: Combining convolutional neural network with recursive neural network for blood cell image classification. IEEE Access **6**, 36188–36197 (2018)
25. Zarándy, Á.: The art of CNN template design. Int. J. Circuit Theory Appl. **27**(1), 5–23 (1999)
26. Albawi, S., Mohammed, T.A., Al-Zawi, S.: Understanding of a convolutional neural network, 2017. In: 2017 International Conference on Engineering and Technology (ICET), vol. 10, pp. 1–6 (2017)
27. Andonie, R., Florea, A.C.: Weighted random search for CNN hyperparameter optimization (2020). arXiv preprint arXiv:2003.13300

Calculation of Bit-Error Probability for Direct-Sequence Spread-Spectrum Communications with Multiple-Access Interference of Rayleigh Distribution

Mohammed Albekairi(✉)

Department of Electrical Engineering, College of Engineering, Jouf University, Sakakah, Saudi Arabia
msalbekairi@ju.edu.sa

Abstract. Even with a small user base, recent cellular mobile systems showed capacity saturation in major urban areas. Further generations of wireless systems will offer higher data rates and flexibility. This demand required a large capacity increase. Multiple digital methods were used to solve the cellular mobile system capacity problem. Different users can share a fixed-spectrum resource using two digital strategies. One uses different frequencies (FDMA), and the other uses different time slots (TDMA). FDMA, TDMA, and hybrid capacities are well-defined. When RF channels or time slots are unavailable, no more customers can be served. Military applications of spread spectrum (SS) have been successful for decades. This spread spectrum uses Code Division Multiple Access, a new multiple access method. CDMA's higher capacity and multipath resistance make it an attractive scheme. The biggest factor limiting CDMA capacity is Multiple Access Interference (MAI). In this paper, we examine how MAI affects DS-CDMA system Bit Error Probability. Many methods have been reported for calculating the DS-CDMA bit error probability. They include three methods: Standard Gaussian Approximation (SGA), Improved Gaussian Approximation (IGA) and simplified IGA. These methods use the Central Limit Theorem (CLT), which approximates the MAI distribution as a Gaussian with a zero mean. We model MAI as a Rayleigh-distributed random variable. This model estimates the average BEP in an asynchronous DS-CDMA system well. SGA with a nonzero mean is used to compare our methods to previous work.

Keywords: Bit Error Probabilities · Code-Division Multiple-Access · Multiple Access Interference · Rayleigh Distribution

1 Introduction

Today, mobile communications are in high demand. Due to their use of traditional multiple access techniques (FDMA, TDMA, or a combination of them), existing mobile systems like AMPS and GSM have limited capacity. Urban business areas' severe mobile communication spectrum congestion highlights the need for a new cellular system that

J. Rasheed et al. (Eds.): FoNeS-AIoT 2024, LNNS 1035, pp. 11–21, 2024.
https://doi.org/10.1007/978-3-031-62871-9_2

uses the spectrum more efficiently. Spread spectrum (SS) communications began in the 1950s for military guidance and communication. The SS technique is named so because its transmitted bandwidth is much larger than the minimum needed to send information. SS was initially used for its noise and jamming resistance. As a multiple-access technique, the spread spectrum technique is very convenient. CDMA is a multiple-access method [1–5]. CDMA was created as a hybrid of time slots and frequency bands. The goal is to address mobile telecommunications capacity shortages. CDMA allows multiple users to share the same frequency band without interfering if mobile station (MS) transmitted power is carefully controlled. All neighboring cells can use the same frequency. BSs can serve an unlimited number of active users. When active users exceed the design value, service quality can be lowered to provide more traffic channels. Usually called soft capacity. Thus, CDMA systems may have unlimited capacity [6].

Digital communications systems like cell phones and wireless personal communications use CDMA technology. Commercial CDMA technology improves capacity, coverage, and voice quality, creating a new generation of wireless networks.

Multiple users can transmit over the same radio frequency (RF) bandwidth with SS multiple access techniques like CDMA. Different users' spread signals interfere unless their transmissions are perfectly synchronized and orthogonal spreading sequences (codes) are used. In most practical wireless systems, user synchronization is difficult in uplinks, and we may not want to use orthogonal codes. Thus, some users' interference with others may cause multiple access interference (MAI) [7]. We want to study how MAI affects communication system performance. MAI is the main factor limiting system performance and capacity. Thus, CDMA research has focused on MAI's impact on system performance. Bit error probability (BEP) measures this effect on system performance [8–12].

It is hard to figure out how well DS-CDMA systems with the matched filter receiver work for bit error even when there is additive white Gaussian noise (AWGN) in the channel. Usually, we use limits and approximations. The standard Gaussian approximation (SGA) is one of the most popular [13]. This approximation uses a central limit theorem (CLT) to approximate the sum of the MAI signals as an additive white-Gaussian process with zero mean additional to the background Gaussian noise process. This receiver uses a conventional single-user-matched filter (correlation receiver) to detect the desired user signal. The filter's output signal-to-noise ratio (SNR) is calculated using the MAI's average variance overall operating conditions.

Due to its simplicity, the SGA is widely used, but performance analyses based on it often overestimate system performance, especially when the number of users is small. Some derivatives of the SGA have been proposed to overcome these limitations, in particular the Improved Gaussian approximation (IGA) and the Simplified IGA (SIGA).

This paper looks at how MAI affects BEP performance in DS-CDMA systems with users that are spread out and interfere, and it comes up with a good estimate. A reanalysis of the SGA with a non-zero mean is also examined. Consider all relevant calculation methods and compare our results (MAI with Rayleigh distribution) to all others. We consider this distribution for MAI because the Rayleigh distribution is often used in mobile radio channels to describe the statistical time-varying received envelope of a flat fading signal or a multipath component. It is well known that the envelope of the sum of two quadrature Gaussian interfering signals follows a Rayleigh distribution. Additionally, this distribution describes radar target detection techniques' fluctuation. In this paper, the MAI is assumed to be the only source of bit errors, but additive white Gaussian noise (AWGN) can be included.

This paper is organized as follows: Sect. 2 introduces the mathematical foundations allowing the MAI to be considered as a Rayleigh distributed random variable. Section 3 shows experimental results and performance interpretation. Finally, Sect. 4 outlines the main contribution of this work and the future challenges.

2 MAI as a Rayleigh Distributed Random Variable

In mobile radio channels, the Rayleigh distribution is commonly used to describe the statistically time-varying nature of the received envelope of a flat-fading signal or the envelope of an individual component of a multipath. It is well known that the envelope of the sum of two quadrature Gaussian interfering signals obeys a Rayleigh distribution [14]. Moreover, this distribution describes the fluctuating nature of radar target detection techniques. Our object in this section is to calculate the BEP based on the assumption that the MAI is modeled as a random variable with a Rayleigh distribution that has a probability density function (pdf) of the form [15]:

$$f_{\psi_n}(\psi_n) = \frac{\psi_n}{\alpha^2} e^{-\psi_n^2/2\alpha^2} \qquad \psi_n \geq 0 \tag{1}$$

where ψ_n represents the interference introduced by the nth user. The intensity of that interference may vary from one user to another. To simplify our mathematical analysis, the characteristic function (CF) of this distribution must first be calculated because it plays an important part in handling our computation of BEP. The characteristic function (CF) corresponding to this pdf is defined as:

$$F_{\psi_n}(\omega) = \int_{-\infty}^{\infty} f_{\psi_n}(\psi_n)\, e^{-j\omega\psi_n} d\psi_n \tag{2}$$

$$F_{\psi_n}(\omega) = \int_{0}^{\infty} \frac{\psi_n}{\alpha^2} e^{-\psi_n^2/2\alpha^2} e^{-j\omega\psi_n} d\psi_n \tag{3}$$

$$F_{\psi_n}(\omega) = \int_{0}^{\infty} \frac{\psi_n}{\alpha^2} e^{-\psi_n^2/2\alpha^2} \cos\ \psi_n\omega\ d\psi_n - j\int_{0}^{\infty} \frac{\psi_n}{\alpha^2} e^{-\psi_n^2/2\alpha^2} \sin\ \psi_n\omega\ d\psi_n \tag{4}$$

The integration of the second term is given by:

$$\int_{0}^{\infty} \frac{\psi_n}{\alpha^2} e^{-\psi_n^2/2\alpha^2} \sin\ \psi_n\omega\ d\psi_n = \sqrt{\frac{\pi}{2}}\, \alpha\omega e^{-\omega^2\alpha^2/2} \tag{5}$$

while the first integration term is:

$$\int_0^{\infty} \frac{\psi_n}{\alpha^2} e^{-\psi_n^2/2\alpha^2} \cos \psi_n \omega \; d\psi_n = 1 - (\omega\alpha)^2 + \frac{(\omega\alpha)^4}{3} - \frac{(\omega\alpha)^6}{3.5} + \frac{(\omega\alpha)^8}{3.5.7} + \ldots \quad (6)$$

$$\approx e^{-(\omega\alpha)^2}$$

Then, Eq. (4) can be expressed as:

$$F_{\psi_n}(\omega) \approx e^{-(\omega\alpha)^2} - j\sqrt{\frac{\pi}{2}}\alpha\omega \; e^{-\omega^2\alpha^2/2} \quad (7)$$

The BEP can be calculated directly from:

$$P_e = \frac{1}{2} \int_{I_0}^{\infty} p_{\psi}(\psi) d\psi \quad (8)$$

where $p_{\psi}(\psi)$ is the joint pdf of the (k − 1) users. The ω-domain representation is given by:

$$P_{\psi}(\omega) = F_{\psi_1}(\omega) F_{\psi_2}(\omega) \cdots F_{\psi_{k-1}}(\omega) \quad (9)$$

If the interference produced by (k − 1) users is i.i.d. (Independent and Identically Distributed), the above equation can be simplified to

$$P_{\psi}(\omega) = \left[F_{\psi}(\omega)\right]^{k-1} \quad (10)$$

Substituting Eq. (7) into Eq. (10), gives

$$P_{\psi}(\omega) = \left[e^{-(\omega\alpha)^2} - j\sqrt{\frac{\pi}{2}}\alpha\omega \; e^{-\omega^2\alpha^2/2}\right]^{k-1}$$

$$= \sum_{\ell=0}^{k-1} \binom{k-1}{\ell} \left(\sqrt{\frac{\pi}{2}}\alpha\right)^{\ell} (-1)^{\ell} (j\omega)^{\ell} \left(e^{-\omega^2\alpha^2}\right)^{k-1-\frac{\ell}{2}} \quad (11)$$

Based on the properties of the Fourier transform [16]:

$$(j\omega)^{\ell} M(\omega) \Leftrightarrow \frac{d^{\ell}}{d\psi^{\ell}}(m(\psi)) \quad (12)$$

Let $M(\omega) = \left(e^{-\omega^2\alpha^2}\right)^{k-1-\frac{\ell}{2}}$ then:

$$m(\psi) = \frac{1}{\alpha\sqrt{4\pi\left(k-1-\frac{\ell}{2}\right)}} e^{-\psi^2/4\alpha^2\left(k-1-\frac{\ell}{2}\right)} \quad (13)$$

Substituting Eq. (13) into Eq. (12), the inverse Fourier transform of Eq. (11) can be evaluated as, ℓ th derivation of $m(\psi)$ w.r.t. ψ. In other words:

$$\frac{d^{\ell}}{d\psi^{\ell}}(m(\psi)) = \frac{d^{\ell}}{d\psi^{\ell}}\left(\frac{1}{\alpha\sqrt{4\pi\left(k-1-\frac{\ell}{2}\right)}}e^{-\psi^2/4\alpha^2\left(k-1-\frac{\ell}{2}\right)}\right) \tag{14}$$

The key to simplification of the resulting formula is the Hermite polynomial, which is defined as:

$$H_{\ell}(x) = (-1)^{\ell}\, e^{x^2}\, \frac{d^{\ell}}{dx^{\ell}}\left(e^{-x^2}\right) \tag{15}$$

$H_{\ell}(x)$ is called a Hermite polynomial. Rearranging the equation to become:

$$\frac{d^{\ell}}{dx^{\ell}}\left(e^{-x^2}\right) = (-1)^{-\ell} e^{-x^2} H_{\ell}(x) \tag{16}$$

Now, let $x = \frac{\psi}{2\alpha\sqrt{k-\frac{\ell}{2}}}$.
The using of the developed polynomial in Eq. (14) makes it as

$$\frac{d^{\ell}}{d\psi^{\ell}}\left(e^{\ \psi^2/4\alpha^2\left(k-1-\frac{\ell}{2}\right)}\right) = (-1)^{-\ell}\left(\frac{1}{2\alpha\sqrt{k-1-\frac{\ell}{2}}}\right)^{\ell} e^{-\psi^2/4\alpha^2\left(k-1-\frac{\ell}{2}\right)} H_{\ell}\left(\frac{\psi}{2\alpha\sqrt{k-1-\frac{\ell}{2}}}\right) \tag{17}$$

Using, the pdf of ψ can be easily evaluated as:

$$p_{\psi}(\psi) = \sum_{\ell=0}^{k-1}\binom{k-1}{\ell}\left(\sqrt{\frac{\pi}{2}}\alpha\right)^{\ell}(-1)^{\ell}\frac{1}{\alpha\sqrt{4\pi\left(k-1-\frac{\ell}{2}\right)}}(-1)^{-\ell}\left(\frac{1}{2\alpha\sqrt{k-1-\frac{\ell}{2}}}\right)^{\ell} e^{-\psi^2/4\alpha^2\left(k-1-\frac{\ell}{2}\right)} H_{\ell}\left(\frac{\psi}{2\alpha\sqrt{k-1-\frac{\ell}{2}}}\right) \tag{18}$$

Substituting Eq. (18) into Eq. (8), the average BEP is found to be

$$P_e = \sum_{\ell=0}^{k-1}\binom{k-1}{\ell}\frac{1}{\alpha\sqrt{16\pi\left(k-1-\frac{\ell}{2}\right)}}\left(\sqrt{\frac{\pi}{2}}\alpha\right)^{\ell}\left(\frac{1}{2\alpha\sqrt{k-1-\frac{\ell}{2}}}\right)^{\ell}\int_{I_0}^{\infty} e^{-\psi^2/4\alpha^2\left(k-1-\frac{\ell}{2}\right)} H_{\ell}\left(\frac{\psi}{2\alpha\sqrt{k-1-\frac{\ell}{2}}}\right)d\psi \tag{19}$$

Since the above integral equation has no closed form, we calculate it numerically.

3 Experimental Results

In this section, we are interested in assessing the obtained analytical results to show the validity of our proposed model. The numerical results given here are evaluated for the most important parameter α^2. This value is calculated as $\alpha^2 = \frac{\sigma_{\varsigma}^2}{2-\frac{\pi}{2}}$, where:

$$\sigma_{\varsigma}^2 = \frac{NT_c^2}{6}\sum_{n=1}^{k-1} P_n \tag{20}$$

$T_c = 1$ and $P_0 = P_1 = \ldots = P_{k-1} = 2$.

Figure 1 shows the variation of BEP with the number of users K. The curves in this figure are parametric in N (processing gain). As the processing gain increases, the BEP decreases. This behavior is logical since increasing the processing gain increases the immunity of the system to interference, and correspondingly, the BEP becomes lower than in the case of smaller processing gains. The results of Fig. 1 demonstrate this statement.

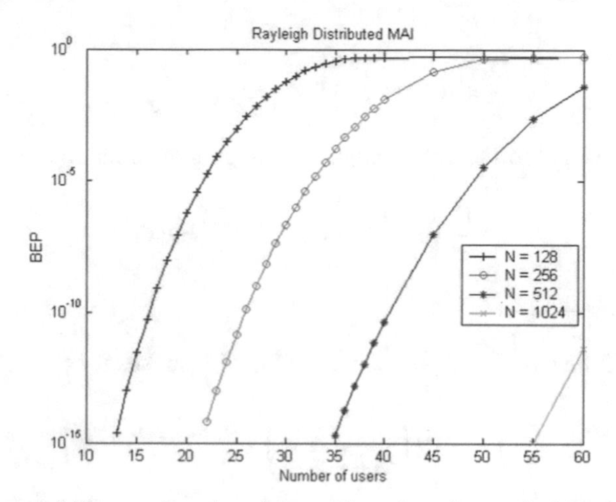

Fig. 1. Bit error probability as a function of the total number of users for N = 128, 256, 512, and 1024 when MAI obeys the Rayleigh distribution $\alpha^2 = \sigma_\varsigma^2 \big/ 10(2 - \frac{\pi}{2})$.

It is important to note that the family of curves of this figure is drawn for:

$\alpha^2 = \frac{\sigma_\varsigma^2}{2-\frac{\pi}{2}}$, $\alpha^2 = \frac{\sigma_\varsigma^2}{10\left(2-\frac{\pi}{2}\right)}$, and $\alpha^2 = \frac{\sigma_\varsigma^2}{20\left(2-\frac{\pi}{2}\right)}$.

To show this effect of α^2 on the behavior of BEP, we redraw the curves of Fig. 1 for smaller and greater values of α^2, as indicated in Fig. 2 and Fig. 3 respectively. As the results of these figures demonstrate, the parameter α^2 plays an important role in determining the BEP as a function of the number of users. When α^2 has smaller values, the BEP has correspondingly smaller values. This is predicated since smaller α^2 means smaller variance of MAI. In other words, increasing α^2 will give higher values for BEP. In any case, BEP increases as the number of users increases because increasing the number of users means increasing the effect of MAI on the desired user. The variance of MAI affects the values of BEP directly. To demonstrate this statement, we plot BEP as a function of K for different values of this variance, and the results are shown in Fig. 4. The family of curves in this figure has the same behavior, with the exception that as the variance increases, the start of its corresponding curves is shifted towards a larger number of users. The asymptotic value of BEP is approximately 50% in any case. To reduce the effect of this variance, one of the proposed solutions is to increase the processing gain. This conclusion is depicted in Fig. 5.

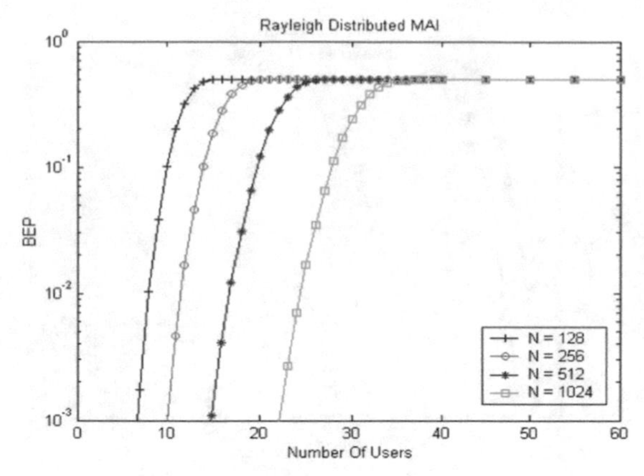

Fig. 2. Bit error probability as a function of the total number of users for N = 128, 256, 512, and 1024 when MAI obeys the Rayleigh distribution with $\alpha^2 = \sigma_\varsigma^2 \big/ \left(2 - \frac{\pi}{2}\right)$.

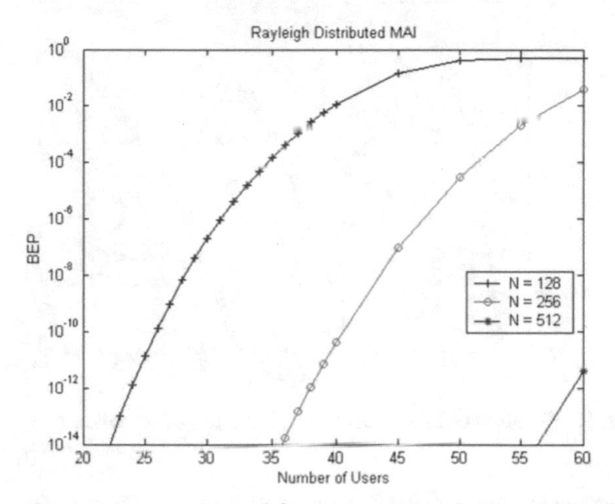

Fig. 3. Bit error probability as a function of the total number of users for N = 128, 256, 512, and 1024 when MAI obeys the Rayleigh distribution with $\alpha^2 = \sigma_\varsigma^2 \big/ 20\left(2 - \frac{\pi}{2}\right)$.

To compare our proposed model with the previously evaluated techniques, we calculate some values of BEP using our technique and preview techniques. The results are displayed in Fig. 6(a) for N = 128 and $\alpha^2 = \sigma_\varsigma^2 / \left(2 - \frac{\pi}{2}\right) 10$.

Based on this figure, we can show that the Rayleigh fading channel gives the highest BEP values when the number of users is small. As the number of users increases, the BEP also increases, with a rate that varies depending on the processing technique. The reference (practical) technique has the highest rate until the number of users reaches 17, beyond which our technique gives higher values than SGA with a non-zero mean. These higher values remain until the number of users reaches 28, beyond which the two techniques give the same value for BEP, which is higher than that obtained with

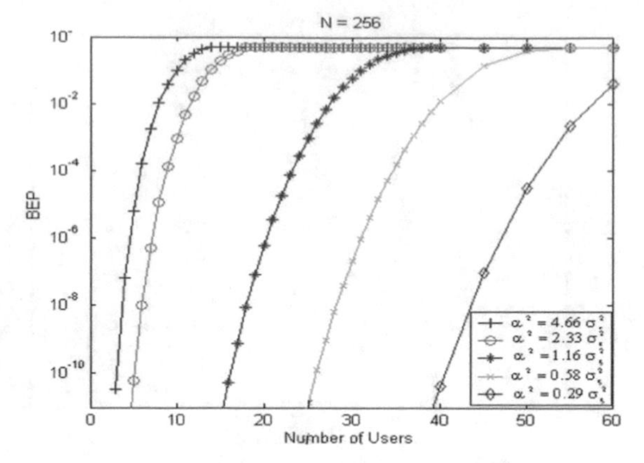

Fig. 4. Variation of BEP with K as a function of α^2 when N = 256.

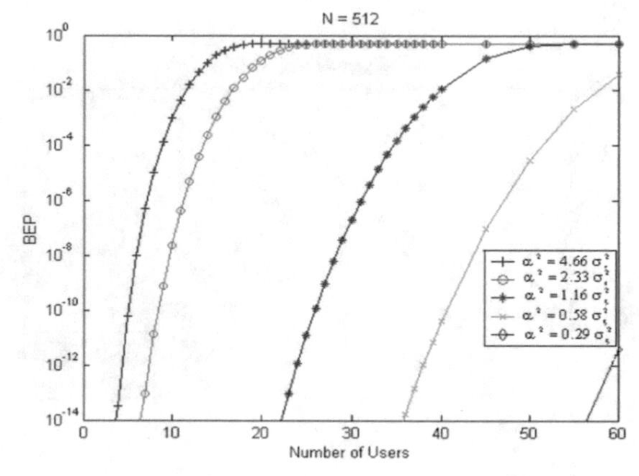

Fig. 5. Variation of BEP with K as a function of α^2 when N = 512.

the reference and the Rayleigh fading channel techniques. On the other hand, the SGA with a zero mean has smaller values for BEP. Since the proposed Rayleigh distribution has a nonzero mean, it is evident that the nearest approach to it is the SGA with a nonzero mean. For smaller values of K, the distribution gives smaller values for BEP than those obtained by SGA with a nonzero mean. As the number of users increases, the two approaches give the same values of BEP. This behavior is logical according to the central limit theorem, which states that the sum of a specified number of independent RVs in any distribution tends to be a Gaussian distribution. The curves in this figure are drawn for fixed values of processing gain and the variance of MAI. To show the effect of the variance of MAI on the behavior of these curves, we repeat them for another value of α^2 as shown in Fig. 6(b). In this case, we note that the reference (practical) technique has the highest rate until the number of users reaches 28, beyond which our technique gives higher values than SGA with a non-zero mean. These higher values remain until the number of users reaches 40, beyond which the two techniques give the same value for BEP, which is higher than that obtained with the reference and the Rayleigh fading channel techniques. To show the effect of the processing gain on the behavior of these curves, we repeat them for another value of N, as shown in Fig. 6(c) and Fig. 6(d). Based on the analysis of these figures, it can be inferred that the SGA with an average value of zero is not feasible. This is attributed to the fact that the mean of the Multiple Access Interference (MAI) does not equal zero, rendering the Signal-to-Interference-plus-Noise Ratio (SINR) with a zero mean as invalid.

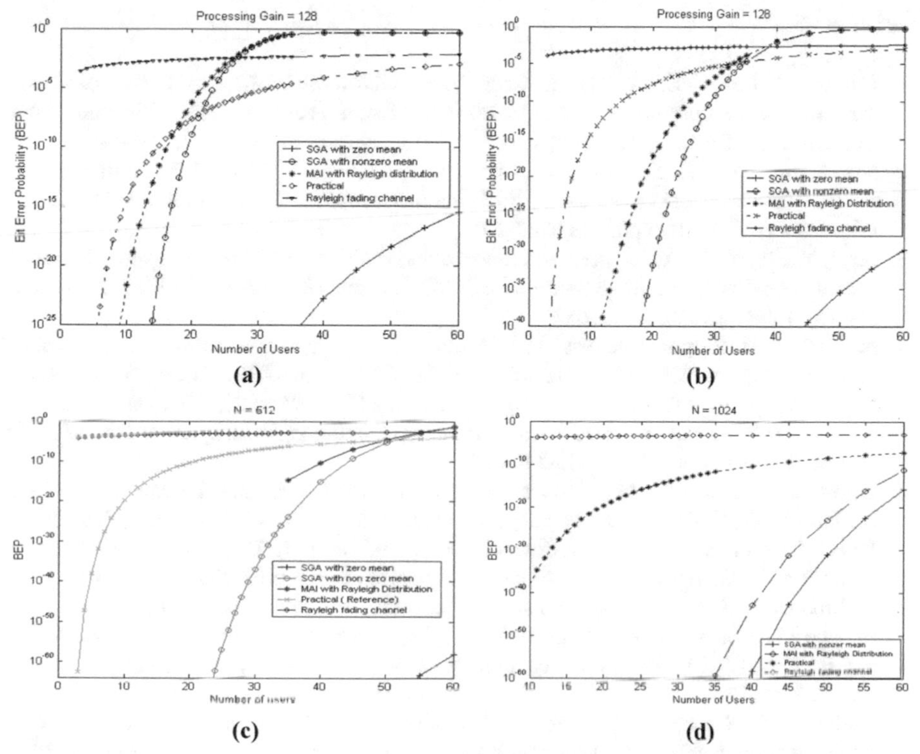

Fig. 6. Comparison of BEP results, evaluated for different processing techniques when: **(a)** $\alpha^2 = \sigma_\varsigma^2 \big/ (2 - \frac{\pi}{2})10$ and N = 128, **(b)** $\alpha^2 = \sigma_\varsigma^2 \big/ (2 - \frac{\pi}{2})20$ and N = 128, **(c)** $\alpha^2 = \sigma_\varsigma^2 \big/ (2 - \frac{\pi}{2})10$ and N = 512, **(d)** $\alpha^2 = \sigma_\varsigma^2 \big/ (2 - \frac{\pi}{2})10$ and N = 1024.

4 Conclusion

The Rayleigh distribution, a new MAI distribution, is proposed in this paper. For this distribution, we analyze errors caused by other users for the desired user. Calculating the CF of the proposed distribution, the basis of our analysis, yields a nearly exact BEP formula. Our numerical results prove the proposed distribution is valid for small user numbers. According to the CLT, the large number of interfering users tends to have a

Gaussian distribution. BEP behavior depends on some parameters. MAI variance and processing gain are examples. We examined how processing gain and MAI variance affect BEP. When processing gain rises, BEP falls. This is logical because increasing processing gain increases system immunity to interference and lowers BEP. Increasing MAI variance raises BEP. For smaller user numbers, our model's numerical values are closest to the standard curve. Instead of single-user detection, we can expand our research to include multi-user detection.

References

1. Lin, M.-C., Lee, S.-K., Fan, H.-H., Chen, Y.-H.: GDMA, LDS-CDMA, and HDS-CDMA for uplink communication systems. In: 2023 VTS Asia Pacific Wireless Communications Symposium (APWCS) (2023). https://doi.org/10.1109/apwcs60142.2023.10234063
2. Ma, T., Xiao, Y., Lei, X., Xiong, W., Xiao, M.: Spreading CDMA via RIS: multipath separation, estimation, and combination. IEEE Internet Things J. **10**(13), 11396–11413 (2023). https://doi.org/10.1109/jiot.2023.3242864
3. Shen, Y., Xu, Y.: Multiple-access interference and multipath influence mitigation for multicarrier code-division multiple-access signals. IEEE Access **8**, 3408–3415 (2020). https://doi.org/10.1109/access.2019.2962633
4. Moinuddin, M., Zerguine, A., Sheikh, A.: Multiple-access interference plus noise-constrained least mean square (MNCLMS) algorithm for CDMA systems. IEEE Trans. Circuits Syst. I Regul. Pap. **55**(9), 2870–2883 (2008). https://doi.org/10.1109/tcsi.2008.923166
5. Torrieri, D.: Principles of Spread-Spectrum Communication Systems. Springer, Cham (2022). https://doi.org/10.1007/978-3-319-14096-4
6. Evans, J., Everitt, D.: Infinite server traffic models for CDMA cellular mobile networks. In: Jabbari, B., Godlewski, P., Lagrange, X. (eds.) Multiaccess, Mobility and Teletraffic for Personal Communications. The Kluwer International Series in Engineering and Computer Science, vol. 366. Springer, Boston (1996). https://doi.org/10.1007/978-1-4613-1409-7_12
7. Mahmood, K., Asad, S.M., Moinuddin, M., Zerguine, A., Cheded, L.: Multiple access interference in MIMO-CDMA systems under Rayleigh fading: statistical characterization and applications. EURASIP J. Adv. Sig. Process. **2016**(1), (2016). https://doi.org/10.1186/s13634-016-0338-y
8. Kaddoum, G., Chargé, P., Roviras, D., Fournier-Prunaret, D.: A methodology for bit error rate prediction in chaos-based communication systems. Circuits Syst. Signal Process. **28**(6), 925–944 (2009). https://doi.org/10.1007/s00034-009-9124-5
9. Zhang, L., Lu, H., Wu, Z., Jiang, M.: Bit error rate analysis of chaotic cognitive radio system over slow fading channels. Ann. Telecommun. Annales des Télécommunications **70**(11–12), 513–521 (2015). https://doi.org/10.1007/s12243-015-0472-9
10. Fujisaki, H., Yamada, Y.: On bit error probabilities of SSMA communication systems using spreading sequences of Markov chains. In: Eighth IEEE International Symposium on Spread Spectrum Techniques and Applications - Program and Book of Abstracts (IEEE Cat No. 04TH8738) (2004). https://doi.org/10.1109/isssta.2004.1371842
11. Fujisaki, H., Keller, G.: Approximations for bit error probabilities in SSMA communication systems using spreading sequences of Markov chains. In: 2005 IEEE International Symposium on Circuits and Systems (2005)
12. Ding, X., Xiong, H., Gong, S., Peng, M., Tang, J.: M-ary orthogonal signal and BER performance in wireless communication systems. In: 2018 IEEE 4th International Conference on Computer and Communications (ICCC) (2018). https://doi.org/10.1109/compcomm.2018.8780668

13. Luque, J.R., Morón, M.J., Casilari, E.: Analytical and empirical evaluation of the impact of Gaussian noise on the modulations employed by bluetooth enhanced data rates. EURASIP J. Wirel. Commun. Netw. **2012**(1), (2012). https://doi.org/10.1186/1687-1499-2012-94
14. Rappaport, T.S.: Wireless Communications: Principles and Practice. Prentice Hall Communications Engineering and Emerging Technologies Series, 2nd edn. Prentice Hall, Hoboken (2002)
15. Vorobiyenko, P., Ilchenko, M., Strelkovska, I.: Current Trends in Communication and Information Technologies. Springer, Cham (2021). https://doi.org/10.1007/978-3-030-76343-5
16. Glover, I., Grant, P.M.: Digital Communications. Pearson Education, London (2010)

Designing and Simulation of Three Phase Grid-Connected Photovoltaic System

Mohamed Abdullahi Mohamed, Abdirahman Ali Elmi, Nour Abdi Ahmed, Yakub Hussein Mohamed, and Abdulaziz Ahmed Siyad(✉)

Department of Electrical Engineering, Faculty of Engineering, Jamhuriya University of Science and Technology, 28P2+H7J, Digfeer RD, Mogadishu, Somalia
{abdirahman.ali,yakub,abdulaziz}@just.edu.so

Abstract. PV power generation systems connected to the grid make the power they produce more useful. But both the utility grid installation and the photovoltaic system must meet the technical requirements to keep the PV installer safe and the utility grid responsible. Photovoltaic systems connect to the grid with the help of an electrical converter, which changes the DC power made by photovoltaic modules into the AC power that is used to power most electrical equipment. This study aims to design and simulate a three-phase grid-connected photovoltaic system that provides a reliable and stable source of electricity for loads connected to the grid. The primary areas of study include maximum power point tracking (MPPT), Boost converters, and bridge inverters. A boost converter, bridge inverter, and ultimately an inverter linked to the three-phase grid are used to interface the maximum power point tracking. This results in a load that introduces the photovoltaic module and provides a reliable and stable source of electricity for the grid.

Keywords: PV-array · DC-DC converter · grid-connect Photovoltaic system

1 Introduction

Electricity is a service that everyone needs, including businesses, homes, farms, etc. requires it to work perfectly; so, an electrical amount of energy is required for the development and growth of every nation. Increasing population, urbanization, and industrialization contribute to a daily rise in the global electricity demand [1].

Since the worldwide energy crisis, people, governments, and academics have become interested in renewable energy sources. The inclusion of power generated from renewable sources into the existing power grid to add clean energy without breaking the local grid rules and regulations is a significant technical issue. The system is set up so that the three inverters that are connected to the utility grid can send both active power and the amount of reactive power that is needed to the grid. There will be two different control loops in these kinds of systems: one to control active power and another to control reactive power. These two control loops should not be connected [2]. Renewable energy is gaining popularity worldwide because people are concerned about climate change and require a dependable source of electricity. Photovoltaic, or photovoltaic, panels are one

J. Rasheed et al. (Eds.): FoNeS-AIoT 2024, LNNS 1035, pp. 22–31, 2024.
https://doi.org/10.1007/978-3-031-62871-9_3

of the most significant methods to get energy from the sun, and more and more people are installing them because they are good for the environment. As the price of batteries continues to drop and the feed-in tariff increases, grid-connected photovoltaic systems become increasingly common. Because of this, intermittent photovoltaic generation changes depending on the weather [3]. The photovoltaic system has acquired tremendous opportunity as a new type of generating power to fulfill the increased need for electric energy as a result of the deregulation of electricity markets and attempts to limit emissions of greenhouse gases from existing electric power generating systems. Solar cells are becoming more efficient, manufacturing technology is improving, and economies of scale have contributed to this price drop. So, in the future, the photovoltaic system looks promising. In recent years, photovoltaic grid-connected systems have emerged as one of solar energy's most consequential uses [3–5].

2 Literature Review

Energy is the primary input for nearly all economic operations and has become essential for improved life quality. All of the equipment is powered by electricity. Today, the amount of energy a country uses is often seen as a measure of how developed it is, 24 percent of all the energy used in the world goes to transportation, 40 percent for industry, 30 percent for home and commercial reasons, and the remaining 6 percent for other uses, including agriculture [6].

The smart Grid technologies and making use of the benefits of PV's distributed nature can open up new avenues for value discovery. By improving PV contributions to grid support functions like frequency regulation, a modern PV system with energy storage and two-way communications can generate significant value. In this research, the authors modeled a PV system coupled to the grid and equipped with an enhanced frequency regulation scheme in MATLAB/Simulink [7]. The system was designed to supply auxiliary services to the grid, most notably frequency regulation. A photovoltaic power plant, battery storage, and a three-phase inverter are all part of this model's grid-connecting setup. A bidirectional DC-DC converter is needed to connect the battery system to the grid. Battery storage systems were found to be effective in simulations for regulating utility grid frequencies. The findings demonstrated that the battery system is capable of instantaneous response and participation in grid frequency regulation by the commanded signals and the prevailing power availability and load demands. The system's control methods were decoupled, straightforward, and simple to implement [8].

This research's main focus is on investigating variables such as voltage sag, voltage flicker, harmonics, voltage imbalance, and frequency shift. In addition, methods for meeting these demands and compliance controls are created. As a result, a sizable, three-phase PVPP that is linked to the grid is being created. The sage problem can be fixed and the low-voltage ride-through requirement is met with a modified inverter controller that doesn't need any extra parts. Grid-connected PV systems (GCPS) face new challenges due to PVPPs' differences from conventional power plants [9]. Furthermore, the stability, security, dependability, and quality of the power system started to change as a result of the substantial use of this renewable energy source. To guarantee the security of the power grid, issues with voltage fluctuations, voltage sag, harmonics, flicker, power factor, and voltage imbalance at the point of common coupling (PCC) should all be addressed. MATLAB/Simulink 2019a was used to create a large-scale GCPS simulation model. The conclusions show that the suggested controller complies with current standards for preventing power quality (PQ) issues. Through the optimization of design, operation, and control strategies for high PQ and green power, this research can aid in making the integration of PVPPs easier [10].

Say that MATLAB shows how to get the parameters out of a typical 60-W solar panel and how to test the model. This model is used to find out how temperature and light affect the range of maximum power points. The direct connection to a stable voltage inverter is also discussed, along with the buck, and helps most extreme power point tracker (MPPT) topologies. This research looks at the MPPT (most PowerPoint following) method, a support converter, and the "worry and watch" approach to the design and redesign of a photovoltaic system. In addition to examining the framework for solar matrices, this study also investigates the design and simulation of a three-phase inverter in MATLAB SIMULINK. It has been proven that these components can be modeled. Solar irradiance and temperature are two factors that have been studied about a PV system's output into the grid. To maintain a stable voltage at the inverter's output and a frequency that is in phase with that of the power grid, regulators and phase-locked loops have been developed and modeled. The modeling approach is straightforward, and it can be used to investigate the behavior of a system under varying degrees of heat and sunshine [11].

One of the main reasons for this fast growth is the German Renewable Energy Act, which encourages people to put PV systems in their own homes. This is one of the main reasons for this fast growth. As a consequence of this, there are now a significant number of producers using the low-voltage network. The purpose of this project is to investigate the influence that solar photovoltaic systems with a single phase have on low-voltage power grids. This bachelor's thesis investigates issues like harmonic distortion and grid asymmetry to better understand them. The output harmonic distortion of a single-phase PV system has been simulated using the program MATLAB. NEPLAN is used to analyze grid asymmetry through the lens of a worst-case scenario through calculations and simulations. Harmonic distortion from the inverter increases with a low voltage supply, as determined by both the simulations and the literature review. Harmonic distortion is not increased when PV and battery storage systems are combined. The results also highlight the role that single-phase systems play in escalating grid asymmetry. In the worst case, the current and voltage will affect the other phases when PV system production is high and all systems are connected to the same phase [6, 12].

This examines the current MMI topologies and how they function in PV settings. The goal of this article is to offer a thorough examination of the numerous novel submodule circuits employed in MMI topologies. There will be in-depth research on maximum power point tracking (MPPT) control strategies for PV inverters. To give a comprehensive view of the balancing methods, scopes, and capacities of the submodules in both balanced and unbalanced grid settings, we will describe and contrast the various control strategies of PV MMIs in the last section. This research shows that the best MPPT architecture for solar applications combines the benefits of solid stability, quick computation, and fine accuracy [13]. This review discusses the CHB and MMC families since they are suitable for this purpose because of their modularity, dependability, grid support, and high power density. To provide superior SM balancing, enable distributed MPPT capabilities, and guarantee the requisite galvanic isolation, more enhancements to the CHB- and MMC-based topologies are needed. Due to its ease of implementation, the PQ method for grid-side MMI control has become the industry standard [7]. The PQ approach will require some honing before it can be used as part of multi-layer MPPT algorithms on the input PV side. These controllers need to account for input and output imbalances in photovoltaics (PV). This paper examines the literature on CHB and MMC balancing strategies and finds that when PV power generation is unbalanced, ZSC and ZSV control techniques perform exceptionally well [14].

3 Methodology

This chapter describes the methodology used in this study; A MATLAB Simulink is used for designing and simulating a grid-connected photovoltaic system. It is discussed in detail in the following sections, which include the System Specification, Block diagram of grid-tied PV system, Methodology Flow Chart, maximum power point tracking, DC-DC converters, Boost Converters, and System Design which require the Components required to design a grid-tied PV, MATLAB & MATLAB Simulink.

3.1 System Specification

Figure 1 lists the elements of the typical grid-connected PV system. As seen in the image, DC-DC converters are regarded as the most important element in maximizing power and are required to do so. Boost converters are used to modify the PV array's terminal voltage in the context of this thesis. Thus, optimum tracking of the power point is achieved.

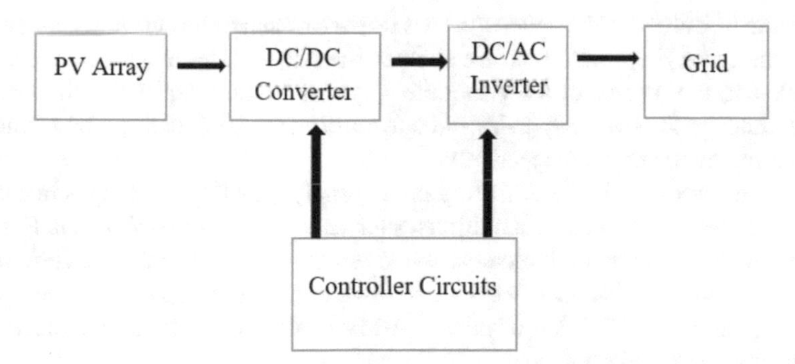

Fig. 1. Block diagram of grid-connection PV system.

A MATLAB-based grid-connected PV system is defined in this piece. To assess the grid-connected PV system, Simulink is employed. The model parts (Fig. 2):

- PV array of maximum **capacity 3000 kW** at 25 °C and 1000 W/m^2 & peak sunshine hour (6–6.5 h in Mogadishu Somalia), Depth of Discharge 75% and Temperature efficiency 80%.
- DC-DC boost converter.
- Power filter and Grid inverter.

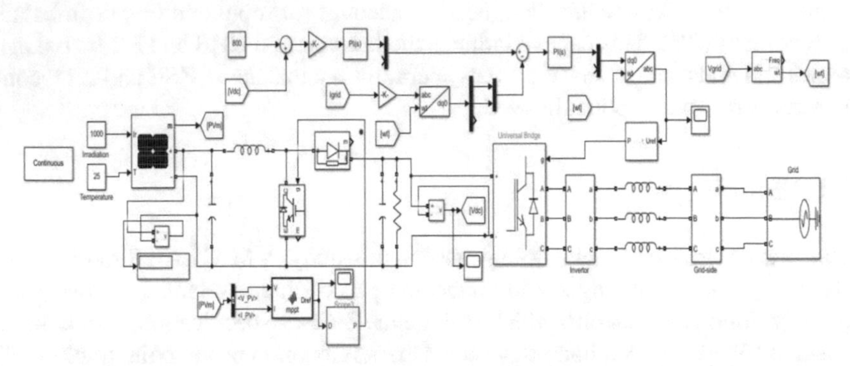

Fig. 2. A portion of the simulation diagram created using the MATLAB/Simulink system.

A 360-W PV panel is used. The array is set up with six parallel strings of modules, each of which has 13 modules in sequence (6*13*360 = 27.612 kW).

The DC-DC converter, inverter, and power filters transmit the electricity produced by the PV array to the grid. Depending on the needs of the system, the array voltage is raised to the desired amount using the DC-DC boost converter.

4 Simulation Results and Discussion

A PV system that is tied to the power grid has its performance and defining characteristics analyzed under varying conditions. We analyze the effects of temperature and irradiance on the I-V and P-V curves of a PV module. Figures 3 and 4 depict the changes in PV power modules' power-voltage (P-V) and current-voltage (I-V) features as a function of irradiance and temperature, respectively.

The power produced by the PV array is extremely sensitive to changes in solar irradiation and other environmental conditions, such as temperature, as shown in Fig. 4. The yielded voltage and power both increase when temperature decreases at a certain irradiance level. When the PV array is working in the current study at a constant temperature of 25 C, it generates 27.777 kw of power. Additionally, when the temperature lowers, the electricity generated rises. as seen in Fig. 4.

The three-phase 3000 kW PV system may interface with the broader power distribution system via the grid inverter and DC-DC boost converter. The DC-DC converter's MPPT tracker controls the reference current using the P&O technique. The waveforms of the current and voltage are shown in Fig. 5 for the grid and inverter. The voltage and current are in perfect phase with one another. Figure 5 serves as an example. The DC values were converted using a grid inverter and a three-phase RLC filter into pure sinusoidal grid current and voltage. The relationship between the AC voltage and current may be seen clearly in the illustration.

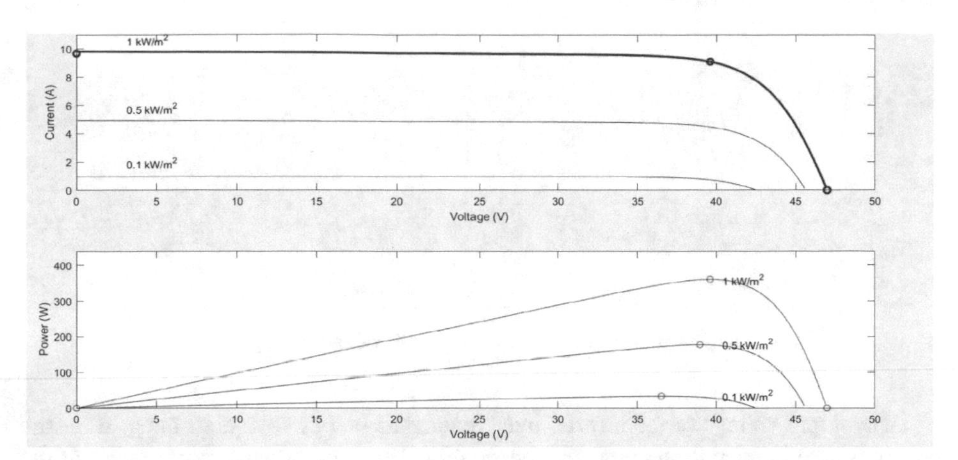

Fig. 3. Effect of radiation on the I-V and P-V properties of a PV power module.

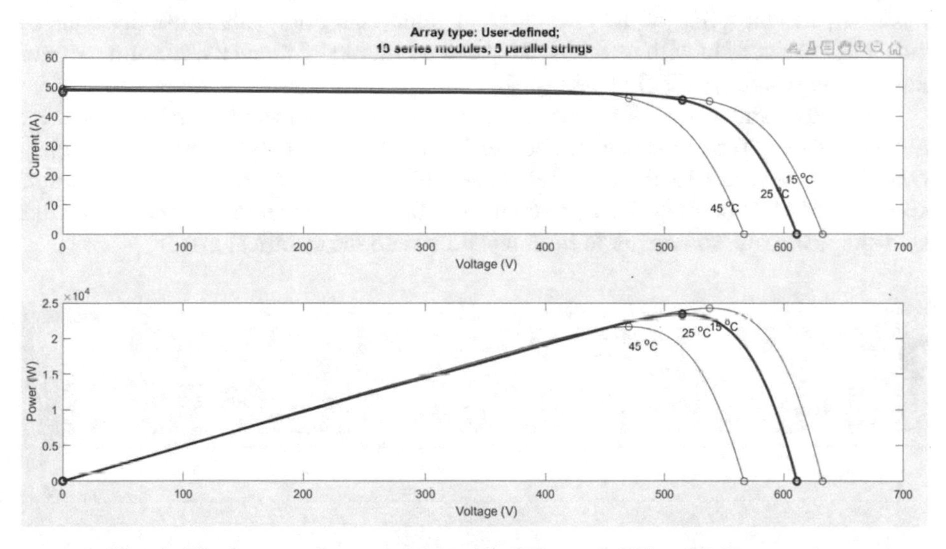

Fig. 4. The impact of temperature on the PV array's I-V and P-V properties

The grid's output voltage is usually a three-phase alternating current (AC) voltage with a frequency of 50 Hz. To ensure stable and reliable power system operation, the inverter's output voltage must match the grid's output voltage. The grid voltage level varies depending on location and time of day, but it is typically maintained within a 10% range of the nominal voltage. After the three-phase grid-connected PV system is connected, the grid output current is the alternating current that flows through the electrical grid. The grid's output current is usually within 10% of the nominal current, depending on location and time of day. The grid's output current must be handled by the inverter for system safety and efficiency. PWM or MPPT current control methods are usually used to achieve this.

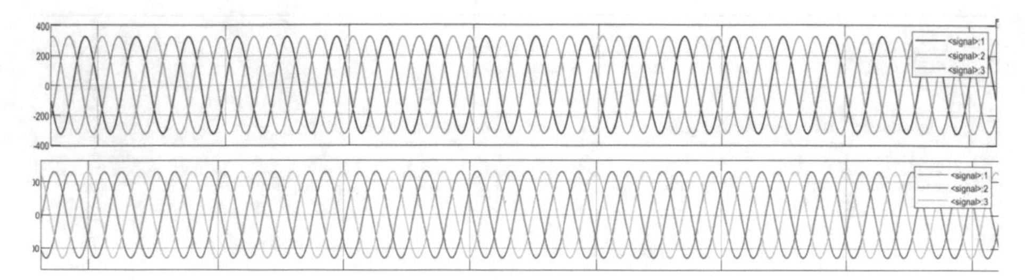

Fig. 5. The plot of the Gate pulse Output current and voltage of the grid.

The output voltage of the inverter must match the output voltage of the grid for the power system to operate steadily and dependably. The operational circumstances of the PV modules, such as solar radiation and temperature, may affect the inverter's output voltage. The inverter's output current must be regulated so that it doesn't go over the maximum current rating of the inverter or the electrical grid. This is usually done by controlling the current with methods like pulse width modulation (PWM) or maximum power point tracking (MPPT) (see Fig. 6).

The output voltage of a PV solar system is the voltage level generated by the PV modules when exposed to solar irradiance. In this case, the output voltage is Vamp = 39.6 V, and there are 13 PV modules connected in series, resulting in a total output voltage of 514.8 V (see Fig. 7). This voltage is a DC voltage that must be converted into AC voltage through an inverter to be compatible with the electrical grid.

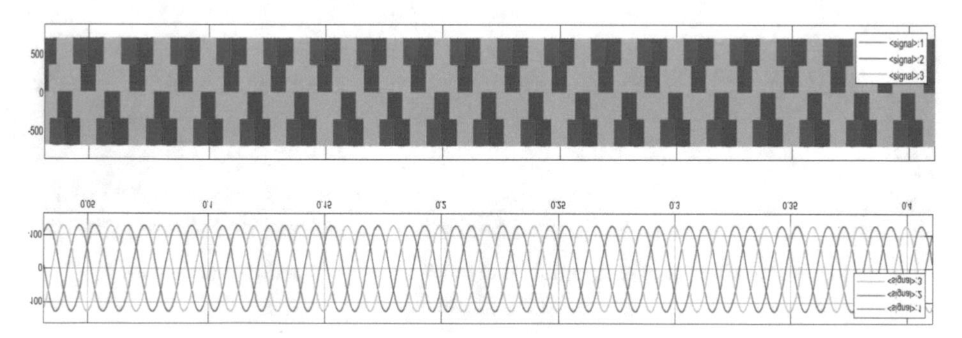

Fig. 6. Plot of the Gate pulse Output Current and voltage of the Inverter.

PV solar systems have an output current equal to the current generated by the PV modules when they are exposed to solar irradiation. With these 6 PV modules in series, the output current is Imp = 9.10 A (see Fig. 8). Because their currents are added together in series, the 6 PV modules can produce a higher combined output current of 54.6A. The

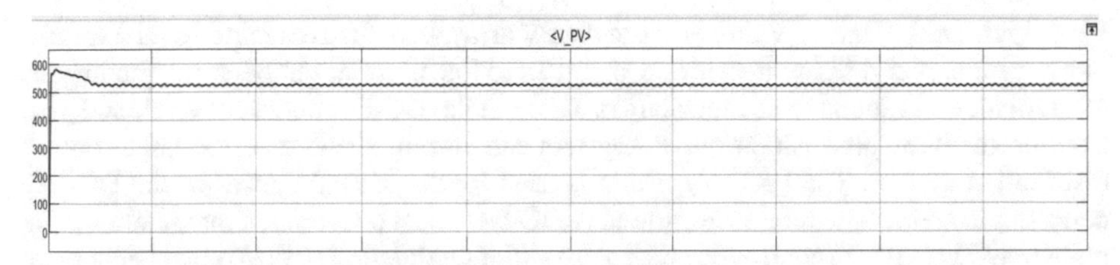

Fig. 7. Plot of the Gate pulse Output current and voltage of the grid.

safe and efficient operation of the power system depends on the output current being regulated.

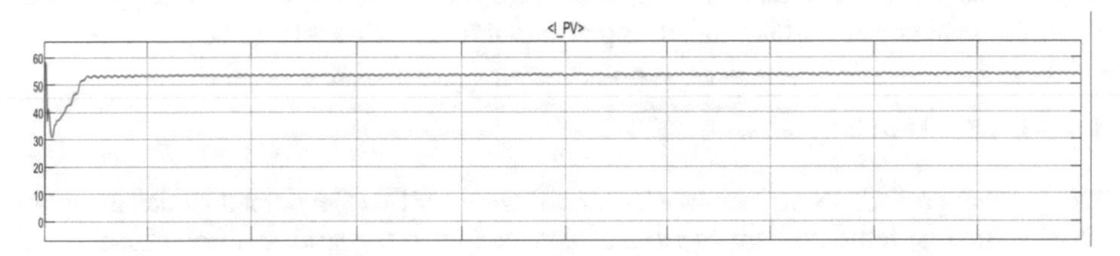

Fig. 8. Plot of the Gate pulse Output current and voltage of the grid

The PV modules generate a total of 28.080 kilowatts of electrical power that is fed into the electrical grid through a three-phase inverter (see Fig. 9). The output of the PV module is the required load that is introduced, and it is a reliable and stable source of electricity for the loads connected to the grid.

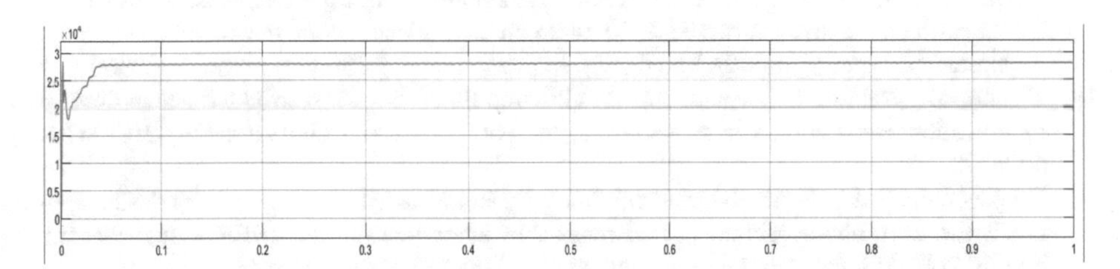

Fig. 9. Plot of the Gate pulse Output power of the PV module

5 Conclusion

This thesis presents the results of a three-phase grid-connected PV system simulation run in MATLAB/Simulink. Using a combination of interactive Simulink blocks, supplementary algorithms, and coding codes, the PV system's components were modeled. Plotting the Power against Voltage characteristics and Current versus Voltage characteristics of the solar PV array reveals that solar radiation and temperature have the most effects on

PV output. The I-V and P-V properties of the PV array were then examined over a variety of temperatures and light intensities. Compared to the effect of temperature, the impact of irradiance was found to be much larger. Through the DC-DC boost converter and grid inverter, the three-phase 3000 kW PV system can communicate with the larger power distribution system. The P&O algorithm is used by the MPPT tracker of the DC-DC converter to control the reference current. A DC-DC boost converter, a maximum power point tracking (MPPT) controller, and a pulse width modulation (PWM) signal generator are utilized to maximize the performance of the PV array. The DC values were converted using a grid inverter and a three-phase RLC filter into pure sinusoidal grid current and voltage. The figure makes it very clear that the AC voltage and current are in phase with one another. Pure and in-phase waveforms best describe them. MATLAB's SIMULINK was used for all simulations. This project will develop hardware suggestions for three-phase Solar PV systems that connect to the grid. Increasing the flow of energy to and from the local power grid is another step toward a more stable energy curve.

6 Future Work

During this project, recommendations for software will be developed to design solar photovoltaic systems that are capable of connecting to the grid in three phases, and analysis harmonics.

References

1. Mondal, A.H.: Daffodil international study on grid connected solar department of electrical and electronic engineering (2021)
2. Chandrakar, A.K.: Indian scenario an assignment on energy resources (2015)
3. Al-Shetwi, A.Q., Hannan, M.A., Jern, K.P., Alkahtani, A.A., Pg Abas, A.E.: Power quality assessment of grid-connected PV system in compliance with the recent integration requirements. Electronics **9**(2), 366 (2020). https://doi.org/10.3390/electronics9020366
4. Bishnoi, D., Prakash, O., Chaturvedi, H.: Utilizing flared gas for distributed generation: an optimization based approach. AIP Conf. Proc., vol. 2091 (2019) https://doi.org/10.1063/1.5096498
5. Khajehoddin, S.A., Karimi-Ghartemani, M., Jain, P.K., Bakhshai, A.: A control design approach for three-phase grid-connected renewable energy resources. IEEE Trans. Sustain. Energy **2**(4), 423–432 (2011). https://doi.org/10.1109/TSTE.2011.2158457
6. Mahmud, M.A., Hossain, M.J., Pota, H.R., Roy, N.K.: Robust nonlinear controller design for three-phase grid-connected photovoltaic systems under structured uncertainties. IEEE Trans. Power Deliv. **29**(3), 1221–1230 (2014)
7. Rey-Boué, A.B., Guerrero-Rodríguez, N.F., Stöckl, J., Strasser, T.I.: Modeling and design of the vector control for a three-phase single-stage grid-connected PV system with LVRT capability according to the Spanish grid code. Energies **12**(15), 2899 (2019)
8. Rexhepi, F., Schrewelius, K.: Grid integrated PV systems in Germany (2015)
9. Alotaibi, S.: Modular multilevel converters for large-scale grid-connected, pp. 1–30 (2021)
10. Zhang, H., Zhou, H., Ren, J., Liu, W., Ruan, S., Gao, Y.: Three-phase grid-connected photovoltaic system with SVPWM current controller. In: 2009 IEEE 6th International Power Electronics and Motion Control Conference, vol. 3, pp. 2161–2164. IEEE (2009). https://doi.org/10.1109/IPEMC.2009.5157759

11. Mahmud, M.A., Pota, H.R., Hossain, M.J.: Dynamic stability of three-phase grid-connected photovoltaic system using zero dynamic design approach. IEEE J. Photovolt. **2**(4), 564–571 (2012). https://doi.org/10.1109/JPHOTOV.2012.2195551
12. Rahimi, R., Farhangi, S., Farhangi, B., Moradi, G.R., Afshari, E., Blaabjerg, F.: H8 inverter to reduce leakage current in transformerless three-phase grid-connected photovoltaic systems. IEEE J. Emerg. Sel. Top. Power Electron. **6**(2), 910–918 (2017)
13. Showers, O.S.: Enhanced frequency regulation functionality of grid-connected PV system. Doctoral dissertation, Cape Peninsula University of Technology (2019)
14. Nandurkar, M.S.R., Rajeev, M.M.: Design and Simulation of three phase inverter for grid connected photovoltaic systems. Power **10**, 30KW (2012)

Security and Reliability Concerns of AI on Critical Embedded Systems

Ahmed Alkhoori, Abdulrahman Alkhoori, Abdulla Alkhoori, and Omar Ahmed(✉)

Abu Dhabi, UAE
omarkoori@hotmail.com

Abstract. Rapid progression in AI (Artificial Intelligence) and deep learning technologies has led to integration into IoT (Internet of Things), raising the concept of AIoT (Artificial Internet of Things). AIoT enhances capabilities by merging AI's decision-making with IoT's network of interconnected devices and systems. However, this integration also presents significant cybersecurity challenges. Understanding these challenges necessitates a focus on the core principles of cybersecurity: the protection of data Confidentiality, Integrity, and Availability. As IoT networks expand, securing a vast number of devices becomes crucial. Yet, AIoT introduces inherent complexities and new security concerns. Compromised deep learning algorithms could manipulate AIoT operations, thus threatening the overall security posture. Moreover, as devices continually communicate and share data, protecting this information against breaches, manipulations, or disruptions is critical. In an increasingly connected world, stakeholders must prioritize safeguarding these advanced systems. Emphasizing cybersecurity principles in the AIoT landscape is essential to realize its full potential without compromising security.

Keywords: AI · Machine Learning · Deep Learning · Neural Networks · IoT · AIoT · Sustainability · Cyber Security · RFID

1 Introduction

The emerging new era goals of simplicity and sustainability in systems, Machine Learning, and Neural Networks, that take part in AI, are the answers to those requirements [1]. With the power of AI, many possibilities and solutions have emerged as a perfect choice, yet it has shaped a whole new way of approaching problems and refining systems to perform in an effective method and algorithm.

AI initiated an automated approach to finding solutions, refining outputs, and inputs into a certain system or operation to be completed [2, 3]. With a whole new vector of operation in systems and devices as with the introduction of AI, certain issues do arise with it, as with AI also issues do arise, as in terms of security and the reliability of the AI in such systems and operations, which yields a critical outcome throughout operation [4], as with the infusion of AI in everyday systems and IT (Information systems), while also OT (Operation systems), a deliberate AI error or an outside vector of attack is much

J. Rasheed et al. (Eds.): FoNeS-AIoT 2024, LNNS 1035, pp. 32–45, 2024.
https://doi.org/10.1007/978-3-031-62871-9_4

of a security concern that need to be addressed and mitigated for the course of AI, in such risks are to be known and mitigated to acceptable level of operations, thus allowing AI to operate with high operational reliability and sustainability within systems.

Two new trends have been infused lately, the use of AI in IoT (Internet of Things) [5], which are both yet new subjects within the IT and OT world, that are implemented within old and new systems and operations, concerns, implementation, and adaptation are both combined as both are introduced within systems and operations throughout their cycle, therefore a whole new vector is initialized. AIoT (Artificial Internet of Things), is based on the implementation of AI technologies in the IoT sector, from industrial IoTs to home IoT systems and operations. That also enabled the deployment of wider solutions and appliances to be monitored and integrated, yet also with its advantages as an example of high performance, adaptation issues, and concerns do rise, that would be analyzed and categorized further down.

Many vulnerabilities and exploits are introduced with the implementation of both AI and IoT together [6], which are aligned against the main three aspects which are Confidentiality, Integrity, and Availability of those system's infrastructure and the information that is stored, processed, and accessed by those systems.

2 Background

2.1 AI

Artificial intelligence allows a certain system or machine to analyze its input and output, and to reason its parameters while also operating on that information gained or analyzed. AI in the world of information technology is used to analyze big data which are complex huge amounts of wide information [7], AI will perform on those big data through the use of multiple methods to find patterns and structured data to further analyze and adjust, also AI is used to automate operational tasks as automatic command execution when a certain variable is given or to adjust on existing commands and baseline operations, AI also proved remarkable in the prediction of threats and analyzation of anomalies [8], that is mostly used in firewalls systems to be able to combat unknown threats or day zero threats on the systems. AI is also been implemented widely on other platforms, such as within operational technologies in industrial areas, as in ICS (Industrial Control Systems), which allows the detection of malfunctioning equipment and the need and request for maintenance, as automatic requestion of replacement parts, another usage is the communication and controlling of industrial equipment as speeding up or reducing down the process streamline of an industry line [9]. Thanks to AI's fast calculation power, it can also read and differentiate data into a more approachable and understandable language that tweaks the industry while also that it can be understandable for humans.

To achieve those goals, certain methods are to be implemented and initiated to reach the designated outcome and working method of AI, each with its positive aspects and drawbacks, with many available methods, AI can be utilized to perform the tasks given, with higher efficiency and with the desired outcome. Those methods are different in how they approach their outcome, as each uses its method, equation, and algorithm.

2.2 AI and Machine Learning

AI is a multidisciplinary approach to the field of computer science that is implemented through the generation and implementation of algorithms [10], which allows systems and machines to approach and perform certain operations and tasks with ultimate accuracy and feasibility, regarding the method of learning as learning methods varies output widely, as each method do yield an output regarding its method. Machine learning is a sub-sector of AI, that is a method used to train or enhance the abilities of the AI throughout the field, as it processes data it has gathered, and keeps incrementing or adjusting as with the feasible information gathered, rather than strictly following an algorithm, therefore an algorithm is suspectable to change and the behavior of the AI would differ as the data has been processed and acted accordingly to the information. As in machine learning, categorization is based on two methods, supervised machine learning uses a labeled dataset that requires actual human interpretation throughout the phases of the process, and it is used to achieve multiple objectives as in classification and regression, in which classification is setting up data in datasets or categories throughout a classification algorithm that is implemented on the data given, as it produces models or datasets of input data and output data, it's used to create decision trees and predict outcomes for those decision trees, also supports vector machines, which are used to classify data regarding classes and create a vector that do support each of the vectors included [11]. Classification algorithms are also the keystone of the neural network training method of AI. Regression is used to predict an accurate continuous output within a given input from the data given, multiple types of outputs can be generated through regression methods as in linear regression which finds and captures linear relationships between variables throughout the data given, as in polynomial regression can be achieved which gives the non-linear outcome of data, which finds out nonlinear variables that do have an effect in regression [12]. As in unsupervised machine learning it is done without the use of labeling and dataset labels, therefore they tend to be less accurate than supervised machine learning, yet they are used without training. Due to their nature, they are used for clustering which is done by the use of algorithms to categorize unlabeled plain data into categories through multiple various cluster-based algorithms, as in "K-Means", which is a method that is focused on assigning similar data into clusters, through multiple calculated irritations of clustering, it creates clusters with a center of mass, other uses of clustering is Hierarchical clustering, that uses methods and algorithms to merge related data into clusters, until one large cluster is made, and path those clusters made merging them through a dendrogram, another hierarchical clustering is done through distance measurement between data, through the type of algorithms used, it will cluster data depending on the distance between data [13].

Another use of unsupervised machine learning is Dimensionality reduction, which is based upon reducing variables and input that do have a slight impact on the outcome, by either reducing small factors or removing random variables and input that exist in the data given, therefore it allows simplification of data, models and allow better performance for the AI, many techniques do exist for this type of machine learning, each with its own set of objectives to fulfill. PCA (Principal Component Analysis) is a linear-based method that simplifies complex graphs of data into cluster-based graphs, through the use of orthogonal axes, which allows simplification of data, through the reduction of dimensions. t-SNE (t-Distributed Stochastic Neighbor Embedding), which is chosen mostly due to its virtualization method and algorithm it can produce from the data received and processed, as it reduces multi-dimensional data into two-dimension or three-dimension data by finding related data and simplifying them, it uses algorithms and methods to calculate data nodes relation, on huge scales until it can be optimized down to gradient descent to allow visualization of such data [14].

Unsupervised Machine Learning allows finding associations between data and their datasets, as in which variable or data do have an impact on other data or variables throughout the sets, allowing it to find anomalies between data gathered. It uses variables to calculate the relation as Support, Confidence, and Lift. In Apriori Algorithm, one of the pioneer algorithms in association, as it's based on finding and mining frequent sets of data that are related and associating them on a larger dataset, through the larger data set, it tries to find and group smaller datasets that are related throughout the datasets [15].

Deep learning is a branch of machine learning that focuses and processes information closer to the human brain rather than machines, through the use of algorithms and functions such as neural networks, that keep automatically updating and revising their data to reach a higher level of accuracy and precision regarding its data. It relies heavily on artificial neural network nodes, those nodes would set data as inputs, and apply weights which allows the use of an activation function which would allow an output to be considered out of a network. Deep learning tends to digest a large amount of data to reach a high level of accuracy, due to its algorithm and methods used to find patterns from that information, it would create multiple patterns to achieve the accuracy of the objective required and would compare which pattern do satisfy, its deeper than machine learning due to its deep networks and nodes to achieve an output. Different types of neural networks do exist [16] (Table 1).

2.3 IoTs

IoTs are systems, devices, and hardware that use sensors, controllers, and software to be able to sense their environment while also communicating and exchanging data with other devices and systems throughout the network [18]. IoTs do exist in nearly every building, from home appliances connected to networks to HVAC (Heating, Ventilation, and Air Conditioning) systems and other industrial equipment, their popularity is gained thanks to their connectivity and ability to monitor and remote control, due to their nature of communication throughout the systems. IoT systems are mainly small in size and design and only limited to performing specific tasks, such as sensing the environment and sending data or actuating an action throughout the data received from a sensor or doing both and acting as a controller to a hardware or system. Multiple IoT networks do exist, each with its channel of communication and networking between the nodes (Table 2).

Many protocols do exist that support the use of IoTs in networks, each aims and supports a specific goal to be accomplished, and due to the many uses, types, and implementations of IoTs, each requires a tailored approach to satisfy the requirements and needs of those IoTs in those architectures. While most IoT topologies are alike and use the same bandwidth, they still work on different methods of operation between

Table 1. Neural Networks Types

Neural Networks	
Type	Description
FNN (Feedforward Neural Networks)	Basic type of neural network, each node passes its data to the next node, from the input node to the weighted nodes throughout the networks of weighted nodes, and lastly the output node in a straight path
Backpropagation Algorithm	A supervised method of neural network that includes error correction, it compares the output with the given objective output it should achieve, as it uses an algorithm to compare if it achieved the required accuracy or similarity, if satisfaction is not achieved, a backward pass would be used and update of its weighted nodes until it reaches the requirement
CNN (Convolutional Neural Networks)	A grid-based network type, which is based on three main classes, Convolution class which is the first class that scans and applies filters to the input, that would detect smaller patterns at the input, the number of filters applied scales with how deep the neural network goes, next class is the pooling layer, which is about sampling through dimension reduction, reducing redundant data while maintaining the more accurate, it does not weight any nodes or input, yet aggregate data gained from the earlier class, it uses methods and algorithms as max value of aggregation and average value of aggregation. The last class is the Fully connected layer, and the classification of those layers and filters are set in this last layer [17]

other protocols and operations, some favoring security and reliability while others favor bandwidth and speed of communication between the nodes of the IoT systems (Table 3).

The Cisco Reference Architecture is a blueprint that shows the different layers and structures that work together to form a robust and reliable IoT architectural solution [22], which divides the full architecture into several smaller layers to be able to create a baseline for engineers and architects to better understand and perform accordingly. The Physical Devices & Controllers layer is the first layer of the connection, which is based on the actual IoT devices and their controllers, It includes the sensors that sense the environment, the actuators that act upon the system, and the controller that control and exchange data through the sensor to actuator. It's the layer that translates from the physical world to the digital world. After that Connectivity layer takes place, which is the layer where most of the protocols take place as WI-FI Zigbee and so on, which is responsible for connecting the IoT together while also connecting the IoTs to further systems and components throughout the next layers of the architecture. Edge or fog computing is systems or devices that collect data from the IoTs and send it

Table 2. IoT Network Protocols

IoT Network Protocols	
Network Protocol	Description
Wi-Fi (IEEE 802.11)	Through the use of antennas, WNICs (Wireless Network Interface Card), and radio frequency a connection between IoTs can be achieved through the use of internet protocol, Wi-Fi allows for a high bandwidth speed, as with the support of 5G Wi-Fi. Wi-Fi IoTs are mostly used in cloud environments due to their design and structure that allow ease of setup regarding networks and the internet, which makes them common in-home networks and buildings [19]
Bluetooth and Bluetooth Low Energy (BLE) (IEEE 802.15)	A short-range radio frequency that requires low power to operate due to the usage of slot transmission, rather than a continuous transmission, which makes it a perfect choice for IoTs as it is lightweight by design, Bluetooth operates on 2.4GHz ISM (Industrial, Scientific, and Medical). Bluetooth uses piconets architecture, which uses Master-Slave Relation between the piconets [20]
Zigbee (IEEE 802.15.4)	A low-power wireless protocol that is favored due to its low-power mesh topology and networks, it uses 2.4 GHz and other frequencies in regions that allow it, such as the 868 MHz frequency band in Europe. Zigbee is mostly common in-home appliances and healthcare due to its mesh capability and confined space which allows it to easily create low-power mesh topologies throughout the IoTs[20]
LoRa (Long Range)	A CSS (Chirp Spread Spectrum) based Protocol, that allows long-range communication with low usage of power, due to its nature of shifting frequencies up and down on a spectrum, therefore allows it to combat noise and reduces the drawbacks that do occur with multipathing. Those advantages make it a perfect choice for IoTs, due to its low-power communication over long distances [20]
NB-IoT (Narrowband IoT)	Radio-based standard, that is focused on using cellular tele-comms bands that are made to allow IoTs to be able to use those bands with a low cost and low power, as it focuses on the use of LTE (Long-Term Evolution). NB-IoT serves the purpose of allowing IoT to connect to LTE communication, as it is used on IoTs that require little bandwidth and higher reliability and security
RFID (Radio-Frequency Identification)	The usage of radio waves to identify and track radio frequency tags, which those tags have electronic-based data. RFID's main component is the RFID Tag which responds when communicated with, tags exist in active and passive, active tags use power to transmit while passive only to be pinged or scanned by an RF reader, an RFID reader decodes the data that has received from the tag. In the world of IoT, it's used to identify objects and automate data capturing or receiving while also allowing integration between the physical world and the digital world

(continued)

Table 2. (*continued*)

IoT Network Protocols	
Network Protocol	Description
Thread	A low-power wireless mesh meant mostly for security and reliability in home automation, which allows IoT to send information back and forth even if part of the network is offline, thread uses IP-based protocol with Thread, which allows connection to a network or the internet easier than other similar Protocols, it requires no gateway or hub to connect to the internet, it can directly connect to the internet with their IP [21]
Sigfox	An LPWAN (low-power wide-area network), that uses UNB (Ultra-Narrow Band) through its communication, as with this method, has a reduced noise while also being able to send over longer distances, which allows it to be ideal in noisy environments. Its one-way communication method allows it to be easier and costs less to use and implement [20]

Table 3. Cisco Reference Architecture for IoT

7	Collaboration & Process
6	Application
5	Data Abstraction
4	Data Accumulation
3	Edge Computing
2	Connectivity
1	Physical Devices & Controllers

over to a system or device that is closer to the source, which makes the collection of data faster and more efficient, architectures that are meant to be on near real-time, or require large bandwidth tend to reinforce this layer as it increases the efficiency of those communication between the layers. Routers and gateways can be considered as fog computing in an IP Network of IoTs, also systems that filter out data or segregate data can be considered at the fog computing layer. At fourth layer, the Data Accumulation the layer that serves as the main connection between data processing layers and the layers of data collection, this layer is mainly about storing raw data and preserving it for further layers and analyses, which the layer could do minimal data filtering and aggregation which would reduce workload in the next phases of operation. The next layer Data Abstraction is the layer in which data is processed and stored in data sets rather than raw data, at this layer AI usage is highly intensive as the use of big data analysis that has been gathered earlier from the IoT devices that went through all the layers before. Data transformation occurs in this stage which data gets transformed into the required format or set for the next layers to adjust and perform upon. The application layer is the layer that contains the API and the front end to the actual users, it's the first interactable layer to the end user, which mostly would include dashboards and data presentation to the user mostly in graphs. The last layer is the Collaboration & Process, which includes the orchestration and workflow of the processes, and how the layers communicate and act together to form the whole architecture and systems, the layer also integrates other IoT systems and allows data sharing between other architectures of IoT systems, which allows collaboration between them as it support collaboration tools and logic handling of the collaboration.

2.4 Combination of AI and IoTs

The two fields are new fields when compared to other IT & OT fields, yet they complete each other perfectly, IoTs are capable of generating huge amounts of data from sensors by sensing around their environment, and they can migrate the physical world into the digital world, which this do generate a huge amount of data that require segregating, filtering and organizing [23]. Thanks to AI improvement over the last decade such huge data can be analyzed and handled at much higher efficiency, which increases the proficiency of IoTs as their data and information can be processed, while AI requires huge data to identify patterns and relations, they can perform well. AI can also be able to predict events, as with IoT sensing out the environment and surroundings AI can get a higher precision regarding those predictions, or even allow better decision making as resources can be allocated where it's needed due to the number of sensors deployed throughout a region. AI can use its knowledge to predict maintenance and hardware that is getting degraded over time, which would allow systems to maintain their efficiency over time period. AI and IoT are the next steps to smart city architecture as they allow this architecture to become true, as the city can be observed digitally with IoT, and AI can identify and understand the smart city component and act upon the received data and information.

2.5 AIoT Networks Sustainability

Creating a sustainable AIoT and its network would require in-depth planning to be in effect, due to its challenges, a sustainable AIoT would require prioritizing long-term strategy to allow a sustainable implementation, as the requirement to have a reduced overall energy consumption [24], and the usage of a high-grade material that would allow the IoTs to sustain over a long term, a scalable network, and IoT systems, it would allow easier to sustain such systems, even when upgrade is required or a regional upgrade to be done. When it comes to AI in the AIoT, prediction of future events is one of the most important aspects of a sustainable system, as events would be known before their occurrence it will allow an early warning concept, and allow maintaining the initiative of throughout an event, therefore allowing to maintain sustainability, as knowing when maintenance required due to performance or efficiency drops. Also, AI should be reliable and capable with wide machine learning algorithms and neural networks, to be able to continue learning even in the future without the constant requirement of updating and patching and to be able to resist future events. The usage of feedback loops or feedback algorithms for AI as in the Backpropagation Algorithm would allow AI to be much more sustainable and can be resilient against events, as it allows automated fixing and patching in certain parameters and operations.

Network sustainability of AIoT would be considered by using multiple protocols per network or architecture which would allow IoTs and systems to communicate even if a channel has been disrupted or outdated, therefore communication can still be achieved. As for networks, mesh topology is highly advised to be considered, as mesh topology is extremely resistant to downfalls and nodes falling, due to their interconnection between nodes, there are always multiple paths between nodes, and interoperability between IoTs and the systems within architecture is a requirement, which systems and protocols should be easy to integrate and established between nodes [25]. Edge and fog computing can allow networks to be sustainable over some time, as they allow IoTs to be less demanding in terms of computation power and bandwidth, therefore allowing less stressful events on IoTs and also allowing less capable IoTs to withstand future data processing and generation.

3 Cybersecurity Core Principles and AIoTs

Confidentiality, Integrity, and Availability are the three main terms that are the core principles in the world of cyber security, that should be addressed and safeguarded when practicing cyber security. Confidentiality stands for ensuring that the information and data are hidden from exposure and that it is accessed only by to correct personnel or systems. A hazard to confidentiality would be unauthorized access to sensitive data or information leakage to unauthorized personnel or the public. Integrity is about preserving the accuracy of the data and the systems, ensuring that the data is not to be altered, modified, or fabricated and that it is trustworthy and accurate. Integrity is focused on the system that holds such data that should not be tampered with or altered and that system files are safe. An action such as tampering with data or system, modification of data is a violation to integrity. Availability is the principle of having the request for information or access always satisfied when needed, which ensures authorized users can gain access whenever they require access. Violation of Availability is a DOS (Denial of Service) attack, which overloads buffers, sessions, and computation power to deny legitimate users from accessing the systems.

Applying those terms and core principles to AIoT networks [26], multiple concerns do arise, as IoT is known to be vulnerable by design due to size and power limitations while also remote location limitations, which makes IoT vulnerable by design to all three core principles. As for AI, algorithm collision or logical errors can yield an unexpected outcome and decision or may lead to a non-accurate prediction that may be acted on. As in Confidentiality, IoT transmission of data or between IoT layers can be exposed which would lead to a violation of Confidentiality, and with the multiple systems and devices that pass throughout an AIoT architecture while also protocols a higher chance of exposure to happen within IoTs. As for AI, a single logic error may lead to exposure of data to unauthorized nodes, which would display sensitive information to non-authorized systems. As for integrity, IoTs are designed to be average in computation power, therefore heavy usage of security software is hard to enforce, which makes them vulnerable to both physical or logical tampering, and with the multiple phases of IoT sending data throughout the network, those channels can be tampered with, which allow data to be modified throughout the process of operation of IoTs. Within AI, Adversarial Attacks that aim at tampering with learning algorithms or that poison data may lead to a violation of integrity as data inputs are fabricated and not real [27], which may lead to incorrect learning methods being implemented. As for Availability, IoTs would send their data over multiple devices as controllers, routers, gateways, or any edge computing system, a cutoff of data can happen because IoTs rely heavily on radio frequency waves such as Bluetooth, Wi-Fi, ZigBee, and other. Radio frequency can be impacted by the noise of nearby machines or even natural radio frequency occurrence; therefore, it may affect the availability of data to reach edge computing, edge computing is also a vector of attack in this architecture as if edge computers are down or got denied from service, IoTs would not be able to send data and may start to lose data that they cannot store anymore.

4 Literature Review

As with IoT networks and AI combined into a single architecture, they tend to combine their advantages which allows for an architecture that allows it to be used in multiple solutions, as it benefits from its high versatility and utility. Therefore, due to such a combination, the exploits and vulnerabilities of both IoT and AI also arise, which includes a range wide range of concerns. The medical domain is one of the hugest contributors and users of IoTs while also in cases where AI is to be used. In the medical field, IoTs are abbreviated into IoMT (Internet of Medical Things), which stands for medical-based IoTs that are implanted sensors and other wearable medical equipment. As IoMT focuses on wireless networks, they are always suspected of attacks due to the open communication method that it is applied, attacks mainly target the confidentiality of those IoMTs, as they do contain sensitive information about the patients through devices used for monitoring, or even the device medical functions, which can be extracted for other malicious uses. Those types of attacks can happen even from a passive attack, which is an attack that does not affect the transmission of the IoMTs and is harder to detect due to its passive nature, it would usually capture the traffic passing from the wireless signals and analyze it, while due to most IoMTs, they got basic to none security measures as in encryption, as they are mostly small in size and compact while also straight forward to their operations, therefore simple data transmission between two nodes can be sniffed by any traffic capture device or program. Other attacks can happen as through the IoMT, an attacker can access other IoMTs throughout the network of the devices, therefore gaining a wider surface of attack and being able to gain access on multiple devices at once, which mostly would contain patients' information and data that are sensitive to the patients. Using an IoMT as the attack vector can also lead the attacker to the network, which allows the attacker the ability to modify, fabricate, or even wipe data regarding the patients, which would lead to further and deeper problems in the system and the medical field [28]. Due to new methods of monitoring patients without the requirement of keeping the patient premises, the use of IoTs has been in demand, as it allows keeping patients under 24/7 monitoring without the requirement of constraining the patient, through the use of smartphones, cloud systems, and IoTs, all combined in use of IoTs in the medical domain. With that, the use of AIoT-H (Artificial

Internet of Things Healthcare) would be considered widely, due to its design, which both the data in IoTs and the data cloud should be considered in its protection. Security measures should be anticipated to enhance the confidentiality of the data, allow the availability of the information when required, and preserve the integrity of the data and information from unauthorized modification, while also monitoring and logging the data and information as they get processed in the system [29]. Other critical facilities are also under threat from attacks occurring on such IoTs, ICS (Industrial Control Systems) which are both hardware and software that are used to control and monitor industrial operations and processes either remotely or on-premises, that is common on power plants and other facilities that hold critical operations and maintain valuable resources. Due to their nature of acting on such critical infrastructure and industry, ICS is always the target, they do have multiple attack vectors from communication or network attacks as in transport channels attacks, which focus on OSI (open systems interconnection) transport layer level, at this level protocols as TCP and UDP operates, therefore attacks can happen on such networks which may allow an attacker to either infiltrate the network or even stop the network from operations as in performing a denial of surface attack upon the network between the IoT under the ICS systems and the main node of operations. Other attacks do target the ICS component itself as the software running or the hardware components In the ICS systems, as most ICS systems are customized per request, they tend to act in a different method than other ICS components, such as understanding the working method and documenting it is a of great issue to the engineers [30].

SCADA (Supervisory Control and Data Acquisition) systems, a subset of ICS, have been one of the attack vectors for a while, as SCADA are Industrial IoTs that allow control and monitoring over industrial processes and operations within the industrial sectors, as it allows control over industrial machinery and infrastructure while also process information back and forth between the central management and the remote locations of the SCADA Systems, as in real-time data of machinery, log files of operations and other devices that control flow of industry throughout the industry. Due to those factors, they are a high target, which also so sports a large surface of attack. Due to their scatter over large regions of geographic location, it tends to be hard to enforce security on such devices, which remote locations are hard to monitor and secure. SCADA also doesn't receive regular updates due to its rigid design of being lightweight and able to sustain very long hours of running without stopping, which patches would stop the running time of such systems that would affect the machinery or industry it controls. SCADA systems architecture is widely known throughout the world, and much of its public information is available to public access, therefore many well-defined architectures of how SCADA performs, and through such information attackers can gain better information about how SCADA works and its weak points [31]. IoT hardware origin is of a great requirement when pointing at security and securing such systems, which a level of trust should be acknowledged regarding the hardware in the IoTs, obtaining Hardware Root of Trust, includes TPM (Trusted Platform Modules) which is a hardware designed to produce and integrate cryptographic keys for further operations throughout the system, also HSM (Hardware Security Modules) is required, which is a hardware electronic that holds and manages digital keys for cryptographic operations throughout the system. With that, we can achieve a chain of trust, as we ensure that the IoTs are made and their architecture is made and designed to be secure rather than relying on other equipment and devices to secure such systems. Most IoT do lack hardware security in terms, of secure boot, which allows the boot loader to boot directly from the boot sector without anything interfering or adding extra codes to the boot sector of the firmware or IoT that is aligned to, another hardware security is a memory lock when dealing with cryptographic operations, therefore even memory dumping won't allow gaining access to what is happening throughout the memory operations. IoTs do usually lack hardware that generates TRNG (True Random Number Generator), which are number generators that use true random rather than pseudorandom when generating values for crypto keys, therefore it allows a higher level of brute force or cryptoanalysis resistance. A usually forgotten part is that IoTs when deployed may still have a debugging mode still active or a bypass security method is still intact within its deployment mode, which allows attackers to engage and use those ports to gain access over the IoTs [32]. IoT has a great positive influence over critical infrastructure and operations, which leads the agricultural field to implement and infuse its operation with IoTs, as they allow real-time monitoring of multiple aspects of agricultural operations such as crops monitoring as they grow, irrigation IoTs, which allows smarter monitoring over the irrigation systems and other automation. The use of AIoT in agriculture should be of certain specifications that are aligned with the agricultural community as systems should be reliable, secure, and easy to use and access. As agricultural infrastructure is mostly in the open, those IoTs could lack the reliability of having accurate data as many aspects may occur, as passing by animals, or even environmental factors, which may lead to incorrect data regarding the AIoT operations and yield incorrect data into the machine learning that happens on the agricultural AIoT that are within the agricultural infrastructure, which would lead to incorrect prediction and operations [33].

Other issues also arise with the implementation of IoTs in agricultural segments, as numerous IoT devices in the sector for different purposes may cause coupling issues, which are that multiple signals and electromagnetic force happen on the same spectrum, which leads to range loss and data loss of such IoTs that would lead in less efficiency of the IoT systems implemented [34]. With many types of IoT, some IoTs are meant to be simple small, and feasible, which are deployed in LLNs (Low-power and Lossy networks), IoTs in those networks are contaminated in computation power due to energy consumption, small memory, and basic processor, which do use lower energy than IoTs deployed in conventional places or typical IoT networks. Therefore, they tend to be harder in terms of securing, as adding extra security features is against their main design, which would increase power usage and also require components that require a higher computation power. Usually, such IoT contains RFIDs, which are expected to be attacked, by either gaining information through scanning those RFIDs or violating the availability of the RFID by jamming or burning out the circuits of the RFID by a radio frequency signal [35].

5 Conclusion

This study investigated the security challenges arising from the integration of AI into the IoT world to AIoT, with a focus on industrial automation, healthcare informatics, and precision agriculture. Our analysis revealed that as AI and IoT technologies converge, the attack surface significantly expands, introducing complex challenges across various industries. In industrial IoT, there is a heightened risk of system-wide breaches and operational downtimes. The healthcare sector faces critical issues in data privacy and the integrity of diagnostics, while precision agriculture could experience supply chain disruptions and yield inconsistencies. Our findings indicate that the AIoT infrastructure is vulnerable to multi-vector attacks, potential zero-day exploits, and advanced cyber threats. Addressing these vulnerabilities requires proactive threat modeling and the development of robust cybersecurity frameworks. The study emphasizes the need for continued innovation in AIoT security, aiming to enhance resilience against cyber threats and ensure a secure digital future.

References

1. Fan, Z., Yan, Z., Wen, S.: Deep learning and artificial intelligence in sustainability: a review of SDGs, renewable energy, and environmental health. Sustainability **15**, 13493 (2023). https://doi.org/10.3390/su151813493
2. Uraikul, V., Chan, C.W., Tontiwachwuthikul, P.: Artificial intelligence for monitoring and supervisory control of process systems. Eng. Appl. Artif. Intell. **20**(2), 115–131 (2007)
3. Farooq, M.S., et al.: A conceptual multi-layer framework for the detection of nighttime pedestrian in autonomous vehicles using deep reinforcement learning. Entropy **25**(1), 135 (2023)
4. Kreutzer, R.T., Sirrenberg, M.: What is artificial intelligence and how to exploit it? In: Understanding Artificial Intelligence: Fundamentals, Use Cases and Methods for a Corporate AI Journey, pp. 1–57 (2020)
5. Wallace, A.A.: When AI meets IoT: AIoT. In: The Emerald Handbook of Computer-Mediated Communication and Social Media, pp. 481–492. Emerald Publishing Limited (2022)
6. Kuzlu, M., Fair, C., Guler, O.: Role of artificial intelligence in the Internet of Things (IoT) cybersecurity. Discov. Internet Things **1**, 1–14 (2021)
7. Bhat, S.A., Huang, N.F.: Big data and AI revolution in precision agriculture: survey and challenges. IEEE Access **9**, 110209–110222 (2021)
8. Rousopoulou, V., et al.: Cognitive analytics platform with AI solutions for anomaly detection. Comput. Ind. **134**, 103555 (2022)
9. Çınar, Z.M., Nuhu, A.A., Zeeshan, Q., Korhan, O., Asmael, M., Safaei, B.: Machine learning in predictive maintenance towards sustainable smart manufacturing in industry 4.0. Sustainability **12**(19), 8211 (2020)
10. Beaudouin, V., et al.: Flexible and context-specific AI explainability: a multidisciplinary approach (2020). arXiv preprint arXiv:2003.07703
11. Alloghani, M., Al-Jumeily, D., Mustafina, J., Hussain, A., Aljaaf, A.J.: A systematic review on supervised and unsupervised machine learning algorithms for data science. In: Supervised and Unsupervised Learning for Data Science, pp. 3–2 (2020)
12. Nasteski, V.: An overview of the supervised machine learning methods. Horizons **4**, 51–62 (2017)

13. Grira, N., Crucianu, M., Boujemaa, N.: Unsupervised and semi-supervised clustering: a brief survey. Rev. Mach. Learn. Tech. Process. Multimedia Content **1**(2004), 9–16 (2004)
14. Lee, J.A., Verleysen, M.: Unsupervised dimensionality reduction: overview and recent advances. In: The 2010 International Joint Conference on Neural Networks (IJCNN), pp. 1–8. IEEE (2010)
15. Naeem, S., Ali, A., Anam, S., Ahmed, M.M.: An unsupervised machine learning algorithms: comprehensive review. Int. J. Comput. Digit. Syst. (2023)
16. Nielsen, M.A.: Neural Networks and Deep Learning. Determination Press, San Francisco, vol. 25, pp. 15–24 (2015)
17. O'Shea, K., Nash, R.: An introduction to convolutional neural networks (2015). arXiv preprint arXiv:1511.08458
18. Smys, S.: A survey on Internet of Things (IoT) based smart systems. J. ISMAC **2**(04), 181–189 (2020)
19. Elkhodr, M., Shahrestani, S., Cheung, H.: Emerging wireless technologies in the internet of things: a comparative study (2016). arXiv preprint arXiv:1611.00861
20. Morin, E., Maman, M., Guizzetti, R., Duda, A.: Comparison of the device lifetime in wireless networks for the internet of things. IEEE Access **5**, 7097–7114 (2017)
21. Unwala, I., Taqvi, Z., Lu, J.: Thread: an IoT protocol. In: 2018 IEEE Green Technologies Conference (GreenTech), pp. 161–167. IEEE (2018)
22. Modarresi, A., Gangadhar, S., Sterbenz, J.: A framework for improving network resilience using SDN and fog nodes. In: 2017 9th International Workshop on Resilient Networks Design and Modeling (RNDM), pp. 1–7. IEEE (2017)
23. Tzafestas, S.G.: Synergy of IoT and AI in modern society: the robotics and automation case. Robot. Autom. Eng. J. **31**, 1–15 (2018)
24. Bronner, W., Gebauer, H., Lamprecht, C., Wortmann, F.: Sustainable AIoT: how artificial intelligence and the internet of things affect profit, people, and planet. In: Connected Business: Create Value in a Networked Economy, pp. 137–154 (2021)
25. Ok, D., Ahmed, F., Agnihotri, M., Cavdar, C.: Self-organizing mesh topology formation in internet of things with heterogeneous devices. In: 2017 European Conference on Networks and Communications (EuCNC), pp. 1–5. IEEE (2017)
26. Babu, P.D., Pavani, C., Naidu, C.E.: Cyber security with IoT. In: 2019 Fifth International Conference on Science Technology Engineering and Mathematics (ICONSTEM), vol. 1, pp. 109–113. IEEE (2019)
27. Qiu, S., Liu, Q., Zhou, S., Wu, C.: Review of artificial intelligence adversarial attack and defense technologies. Appl. Sci. **9**(5), 909 (2019)
28. Fiaidhi, J., Mohammed, S.: Security and vulnerability of extreme automation systems: the IoMT and IoA case studies. IT Prof. **21**(4), 48–55 (2019)
29. Huang, X., Nazir, S.: Evaluating security of internet of medical things using the analytic network process method. Secur. Commun. Netw. **2020**, 1–14 (2020)
30. Emake, E.D., Adeyanju, I.A., Uzedhe, G.O.: Industrial control systems (ICS): cyber attacks & Security Optimization. Int. J. Comput. Eng. Inf. Technol. **12**(5), 31–41 (2020)
31. Yadav, G., Paul, K.: Architecture and security of SCADA systems: a review. Int. J. Crit. Infrastruct. Prot. **34**, 100433 (2021)
32. Chmiel, M., Korona, M., Kozioł, F., Szczypiorski, K., Rawski, M.: Discussion on IoT security recommendations against the state-of-the-art solutions. Electronics **10**(15), 1814 (2021)
33. Abraham, A., Dash, S., Rodrigues, J.J., Acharya, B., Pani, S.K. (eds.): AI, edge and IoT-based smart agriculture. Academic Press (2021)
34. Tomar, P., Kaur, G. (eds.): Artificial intelligence and IoT-based technologies for sustainable farming and smart agriculture. IGI Global (2021)
35. Jurcut, A.D., Ranaweera, P., Xu, L.: Introduction to IoT security. In: IoT Security: Advances in Authentication, pp. 27–64 (2020)

A Survey of Machine Learning Assistance in Seismic Interpretation

Mohammed Al Anbagi[1] and Zaid Kamoona[2](✉)

[1] Kufa University, Najaf, Iraq
mohammeda.yousif@uokufa.edu.iq
[2] Ministry of Oil, Baghdad, Iraq
zaid_kamoona@oec.oil.gov.iq

Abstract. The exploration and drilling of oil and gas are so expensive and takes a high responsibility for the geoscientists and interpreters to make their decision about the accurate place for drilling and this decision depends on a lot of underground factors. As researchers, we decided to conduct a survey about how to assist the geoscientists and seismic interpreters in making the right decision, with some of the underground characteristics that oil detection depends on. Our survey explains the use of machine learning algorithms to find geological characteristics such as Horizon, Fault, and Stratigraphy. The results show that interpretation by using machine learning algorithms has high accuracy and so close to the real seismic interpretations. Finally, it is noticed that there are a few experiments in this field due to the lack of available data set, although the result is acceptable, it needs time and a lot of experiments to apply on this important part.

Keywords: Seismic · Horizon · Facies · Fault · Deep Learning · Semantic Algorithms

1 Introduction

Oil and gas fields play an important role in the economy and industry of the world, Seismic interpretation is the only way that is used in the exploration of oil and gas in the subsurface. If geology can describe the subsurface accurately it means a great factor for the exploitation. Seismic interpretation is not an easy method, it needs a lot of time to record subsurface with depth up to 15 km and requires special software for processing the data acquired from seismic crews and analyzing software such as (Petrel) which needs an expert interpreter for interpretation. The recording of seismic depends on the reflection of the underground subsurface which only gives an indirect image of the subsurface geology [1, 2]. It is known that expert seismic interpreters take all the responsibility for the decision of subsurface exploration, similar to the machine learning role in the various domains (medical domain [3], anomaly detection [4], and game and software industry [5, 6], etc.) we try to enter machine learning to help interpreters to be sure about some specific important geological features such as Faults, Horizon, and Stratigraphy.

J. Rasheed et al. (Eds.): FoNeS-AIoT 2024, LNNS 1035, pp. 46–56, 2024.
https://doi.org/10.1007/978-3-031-62871-9_5

2 Seismic Exploration and Interpretations

Seismic is the way that we can investigate oil and hydrocarbons underground and this process can be done by some steps. We have two types of seismic exploration, the first one, is 2D seismic, which is a less costly and quick way that can be used for the first time for reconnoitering, this way gives less accuracy than the other type. 3D seismic contains a lot of lines and a lot of cross lines, and this way gives more accurate results used for the development of oil fields. These two types, use different types of power sources (vibrators or dynamite) to send a wave signal underground 5–11 km and they also spread geophones above the surface of the ground to receive the reflection signal, by this different reflection of different layers underground that geophones collected as illustrated in Fig. 1 and send it by a cable line to recorder station, which saves the acquired data as (SEGD) destination file as shown in Fig. 2.

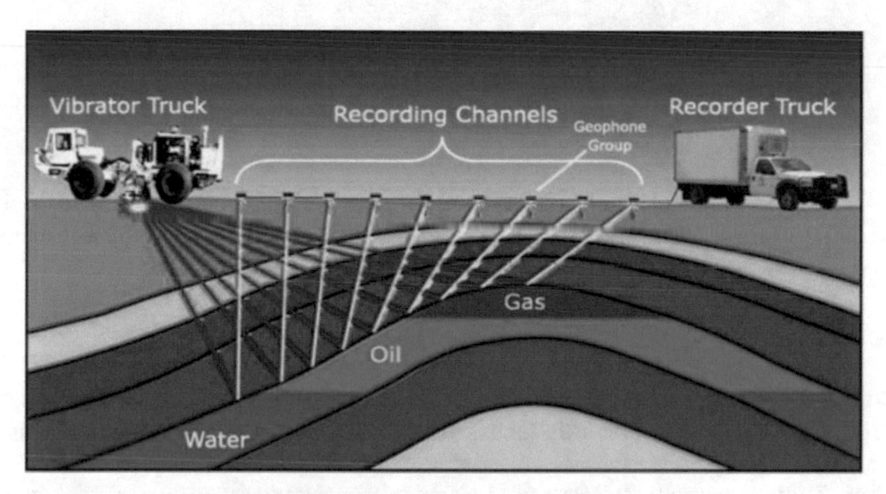

Fig. 1. Illustration of seismic acquisition process [7].

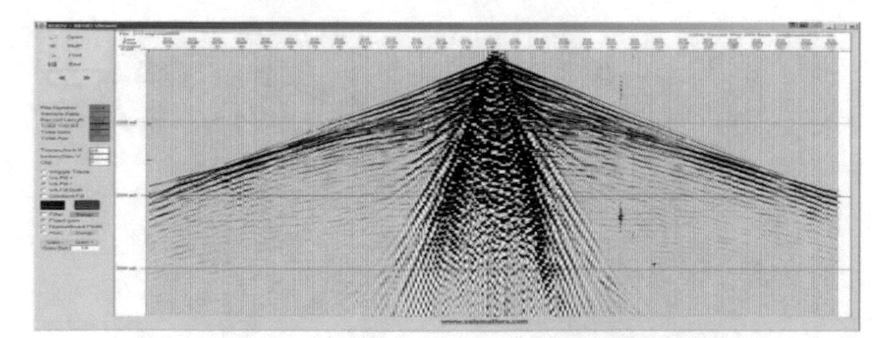

Fig. 2. Real data saved in a SEGD format [8].

After collecting all the data for different areas through the seismic acquisition process, all SEGDs files will filter from noises and other data that are not required by the processing machine to be used for seismic interpretation. Then the seismic interpreter tries to investigate and study the characteristics of the underground layers such as faults, horizons, reservoirs, etc. and all these characteristics give some valuable information to interpreters about underground structures and reservoirs.

Fig. 3. Perspective view of the 3D seismic volume [9].

Geologic surfaces like faults and horizons, as shown in Fig. 3, can be extracted from a three-dimensional seismic image. Because they offer structural maps of the subsurfaces, faults, and horizons are crucial for seismic structural interpretation [10].

2.1 Fault

As defined in many places, the fault is a fracture or zone of fractures between two blocks of rock (see Fig. 4). Faults allow the blocks to move relative to each other. This movement may occur rapidly, in the form of an earthquake - or may occur slowly, in the form of creep. Faults may range in length from a few millimeters to thousands of kilometers. Most faults produce repeated displacements over geologic time. During an earthquake,

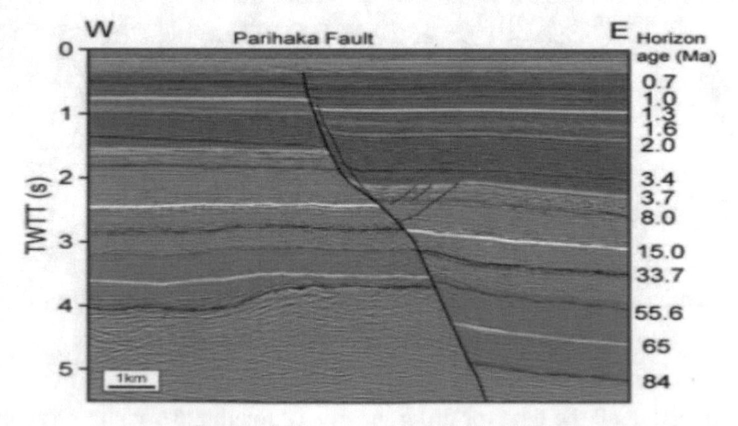

Fig. 4. Faults on seismic profiles [12].

the rock on one side of the fault suddenly slips concerning the other. The fault surface can be horizontal or vertical or some arbitrary angle in between [11].

2.2 Horizon

A horizon is an imaginary surface in the subsurface of the earth, usually thought of as representing a stratigraphic surface which is either lithostratigraphic or chronostratigraphic (see Fig. 5). Geoscientists often think of seismic horizons as geological horizons, or even as stratal surfaces, but this is an over-simplification. Seismic is prone to multiple reflections, interference effects, and distortion due to the velocity field [13].

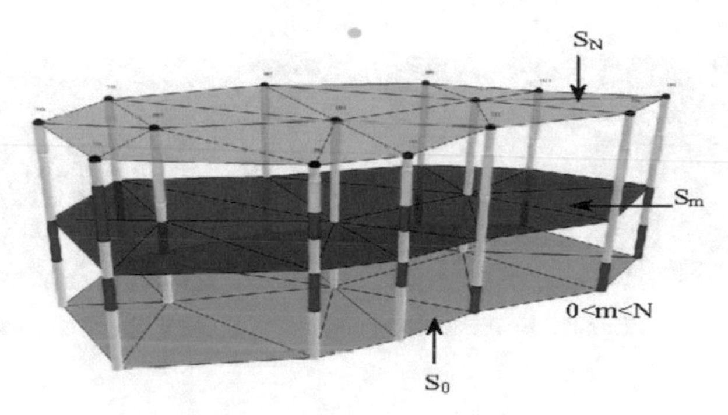

Fig. 5. Horizon surfaces in 3D environment [14].

3 Datasets and Machine Learning

In Seismic interpretation with machine learning, different machine algorithms such as deep neural networks, and CNNs were used for Faults, Horizon, and stratigraphy detection. The authors in [15–18] explain the benefits of using machine learning with seismic interpretation in both faults and horizons.

3.1 Faults and Stratigraphy

In the first data set (3D seismic dataset over the North Sea), Convolutional neural networks (CNNs) were used with seismic interpretation. The data is divided into two groups (Hard information and Soft information) depending on the result that is required, Hard information means the certainty information can be acquired from other sources or analyzing the dataset and Soft Information means the interpretational uncertainty and/or preference at specific zones [15].

In the first example, they trained the machine learning algorithm by fault information as hard information and the result helped to get a very good continuity of stratigraphy mapping near the fault information that they fed (see Fig. 6).

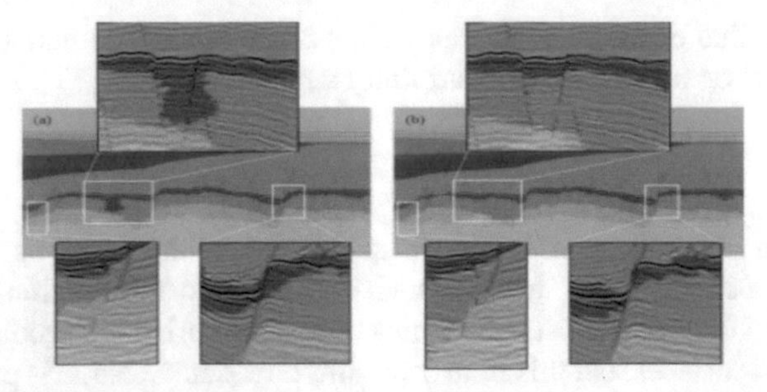

Fig. 6. Enhanced Stratigraphy Mapping: A Comparative Analysis of Interpretations with and without Fault Constraints in Testing Section

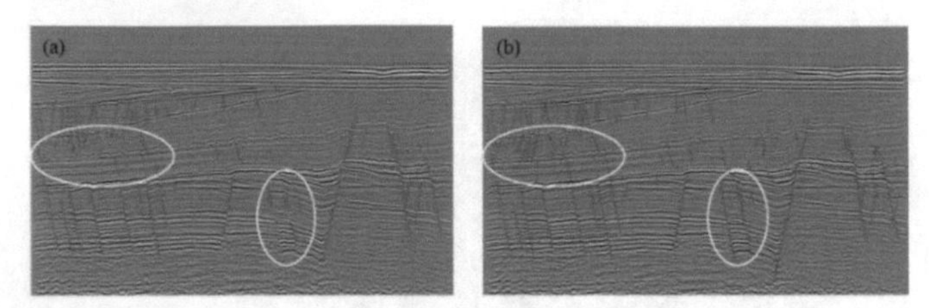

Fig. 7. Fault Detection Enhancement: A Comparative Analysis without and with Stratigraphy Information Constraints

In the second example, they fed the machine learning algorithm by stratigraphy information as hard information and tested the fault as a result. The result is so good with continuing fault and extended deeper as shown in Fig. 7.

Third example they want to create an algorithm for interpreting both fault and stratigraphy and the fault gets the higher priority.

The machine exhibits a stronger focus on stratigraphic sequences, primarily attributed to the imbalance in training data between stratigraphy and faults. Consequently, this concentration results in incomplete fault detection as shown in Fig. 8 [15].

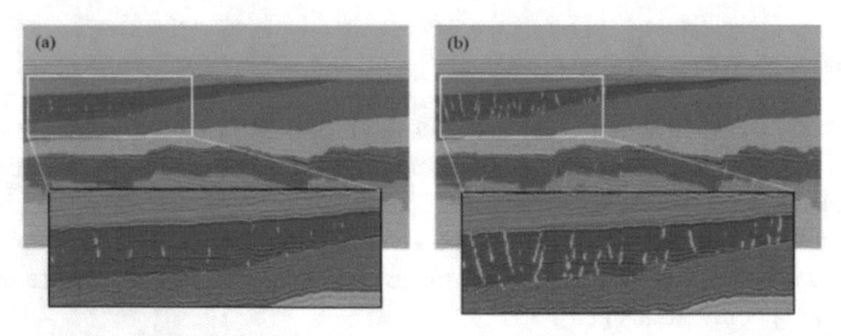

Fig. 8. Comparative Analysis of Fault Detection in a Testing Section with and without Fault Priority

In the fourth example, they want to get faults on specific target zones and stratigraphy information used as hard and soft, and the result as it shown in Fig. 9.

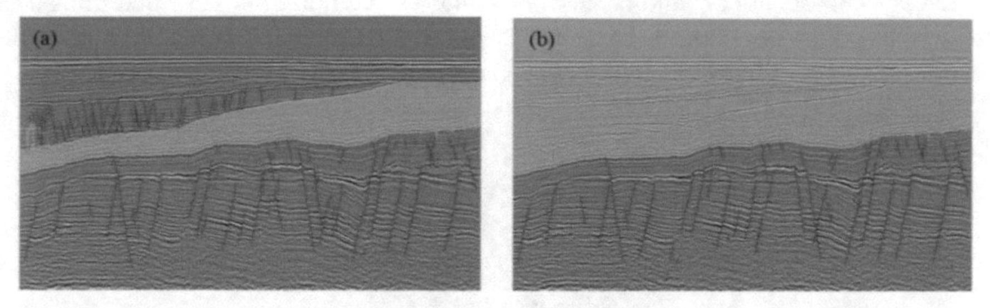

Fig. 9. Customized Fault Detection using Stratigraphy Information as Dual Constraints: Excluding Fault Look-alikes in the Chaotic Zone (a) and Focusing on the Bottom Zone (b)

The second data set, which is used in work, was collected from the interpretation data of students' Petrel projects at the University of Aberdeen, where 78 interpretations are classified. The data set of these interpretations was a 3D seismic dataset from the Gullfaks field located in the northern North Sea. The Top Ness horizon surfaces and fault were the main goals of the interpretation. Figure 10 shows an example of student interpretation containing the three major faults. In this study the authors discuss recent advancements in automating seismic interpretation through neural networks (NNs), noting their success with high-quality synthetic seismic data. However, NNs lack geological reasoning skills, relying on abstract features and struggling in low-quality seismic image areas. Analyzing uncertainties in the Gullfaks field 3-D seismic cube, fault interpretations in low-quality regions align with known geometries, suggesting influence from adjacent high-quality seismic data. This challenges the development of algorithms capturing human thought processes. While applying findings to automated interpretation is challenging, the study emphasizes the value of interdisciplinary approaches, integrating advanced algorithms with geological knowledge to automate the complex and uncertain task of 3-D seismic interpretation [16].

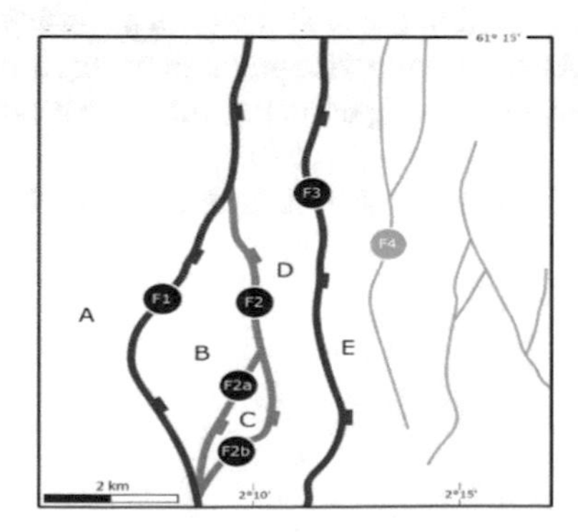

Fig. 10. Fault Mapping of the Statfjord Formation in the Gullfaks Field: Identification and Labeling of Major Faults

3.2 Horizon

First data set (Penobscot dataset), the author trained a deep neural network with a data set that is available at [17] and consists of 7 interpreted horizons They determined the intersection points between the horizon surfaces and individual seismic lines, assigning labels to pixels ranging from 0 to 7 based on each corresponding horizon interval. as shown in Fig. 11 [18].

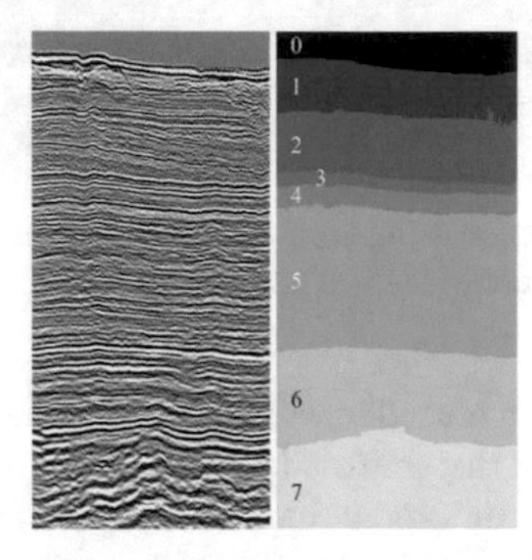

Fig. 11. Illustration of a trimmed inline section along with its corresponding labels.

Case I: Seismic Facies Classification
Every image was divided into tiles (40*40) pixels to generate the classification dataset and split these images into training and test sets and feed the algorithm. Every tile refers to class input in deep neural networks. They get a very good result with 25 slices and in 4 min they obtain 89% accuracy, and they notice the accuracy increase with the increase in slice and time.

Case II: Seismic Facies Segmentation
The same dataset was used with a semantic segmentation algorithm, they also used deep neural networks for this classification task, where images are also into tiles, but it is larger than the tiles that are used in case I and makes a comparison between more than one class. And then they generate pixel-wise predictions. Figure 12 indicates the model's generation of masks closely resembling the actual interpretation, demonstrating minimal discontinuity.

By using these algorithms, they get 97% accuracy, and the result is very similar to the real interpretation.

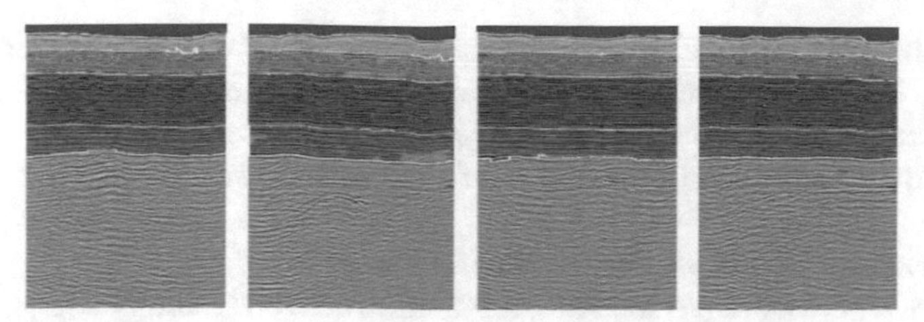

Fig. 12. Pixel-Level Semantic Segmentation of Penobscot Dataset with Seismic Horizons Highlighted

For the second data set (Netherlands), The author in this work trained deep neural networks algorithm with seismic data which is a public 3D seismic survey called Netherlands Offshore F3 Block and available at the Open Seismic Repository [19]. Nine horizons and one fault were interpreted as shown in Fig. 13. In this work, two applications were presented: classification and semantic segmentation of seismic images. This work made geoscientists and machine learning specialists work in the same field and make a comparison to their results [19].

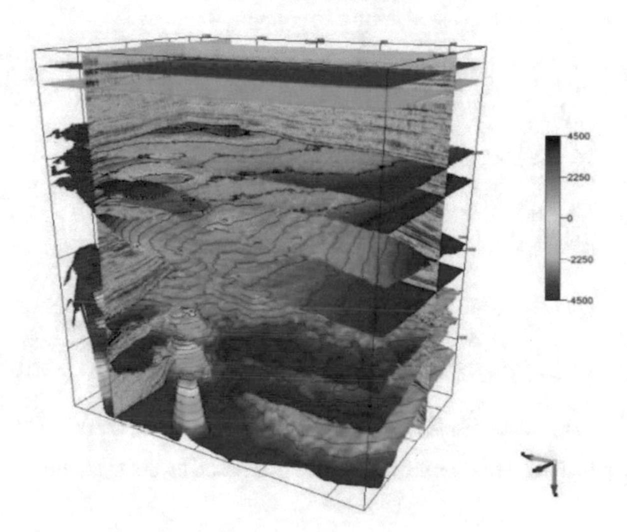

Fig. 13. Nine interpreted horizons are shown along with two seismic sections.

Case I: Classification of Rock Layers

The primary assumption is that various layers can be differentiated by their textural features, as elaborated in [20, 21]. Therefore, a model capable of classifying images based on their textural attributes could be employed for the classification of distinct rock types. In this study, the researchers trained deep neural networks that effectively distinguished between strata in the Netherlands dataset. They employed 9 seismic sections randomly chosen from the odd lines of the cube for training and validation. Within these sections,

80% was allocated for training purposes, while the remaining 20% served as the validation set. Testing was conducted on all even lines, resulting in 4,784 tiles per class after balancing. The ultimate accuracy achieved on the test dataset was 81.6%. Refer to Fig. 14 for the presentation of the final confusion matrix [19].

Case II: Semantic Segmentation of Seismic Images

The authors harnessed a pre-trained deep neural network initially designed for rock layer classification. Following this, they strategically pruned the network by removing its classifier tail and introduced an upscale module to enable pixel-wise predictions based on the key extracted features. Subsequently, the refined model underwent training using the Netherlands interpretation dataset. Notably, their approach involved dividing the input seismic section into smaller tiles, and by merging specific layers to prevent imbalance caused by thin layers, they applied the network across the entire image to generate the final prediction. Impressively, this methodology yielded a mean Intersection over Union (IoU) metric exceeding 98%, as showcased in Fig. 15, where the model's segmentations closely mirrored the actual interpretation with minimal discontinuity [19].

True classes \ Predicted classes	0	1	2	3	4	5	6	7	8	9
0	2776	18	134	101	182	268	125	287	371	522
1	0	4455	62	10	0	146	0	3	84	24
2	1	27	4421	61	73	144	5	43	9	0
3	0	0	51	4455	79	114	59	0	1	25
4	0	0	14	433	3916	254	113	22	2	30
5	0	0	58	69	282	3777	423	60	88	27
6	5	0	52	63	383	874	3061	48	148	150
7	0	1	0	15	6	200	33	4263	27	239
8	5	4	17	0	5	238	76	73	4073	293
9	0	1	1	0	75	235	191	152	279	3850

Fig. 14. The confusion matrix for the classification of rock layers using the proposed dataset.

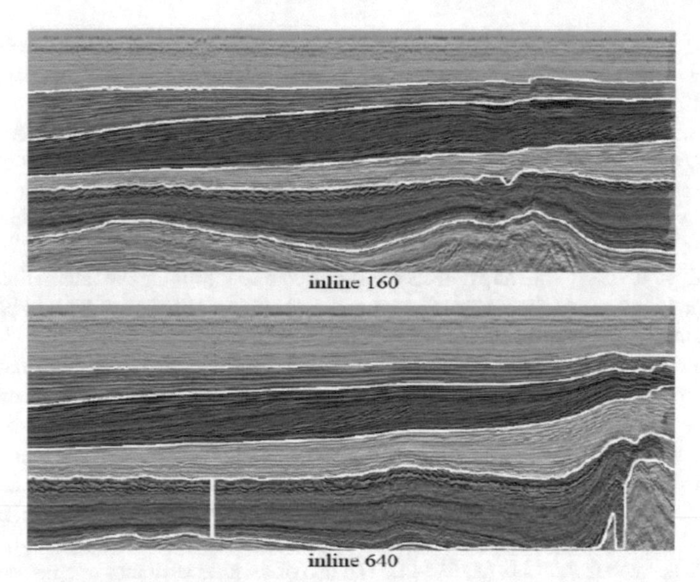

Fig. 15. Semantic Segmentation of Seismic Inline 160 (Top) and 640 (Bottom) from the Netherlands Dataset, with Overlaid Pixel Class Colors and Reference Seismic Horizons (Ground Truth) Highlighted by White Lines.

4 Conclusion

The application of machine learning, particularly deep neural networks, and CNNs, in seismic interpretation has demonstrated considerable success in detecting faults, horizons, and stratigraphy. The presented examples illustrate the effectiveness of training algorithms with specific geological information, emphasizing the importance of a balanced approach when interpreting fault and stratigraphy data. The challenges and nuances in automating seismic interpretation are acknowledged, highlighting the interdisciplinary nature of combining advanced algorithms with geological knowledge. Overall, these findings underscore the potential of machine learning to enhance and automate complex tasks in seismic interpretation, paving the way for further advancements in the field. The authors noted the intriguing potential of machine learning to support geological interpreters in their decision-making processes. However, a significant challenge in this endeavor is the limited availability of geological data. Currently, machine learning has not fully replicated the intricacies of human-brain activities in handling this aspect of the interpretation process.

References

1. Herron, D.A.: First steps in seismic interpretation. In: Society of Exploration Geophysicists (2011)
2. Clerk Maxwell, J.: A Treatise on Electricity and Magnetism, 3rd edn. Clarendon, Oxford (1892)

3. Rasheed, J.: Analyzing the effect of filtering and feature-extraction techniques in a machine learning model for identification of infectious disease using radiography imaging. Symmetry **14**(7), 1398 (2022). https://doi.org/10.3390/sym14071398
4. Kaya, S.M., Isler, B., Abu-Mahfouz, A.M., Rasheed, J., AlShammari, A.: An intelligent anomaly detection approach for accurate and reliable weather forecasting at IoT edges: a case study. Sensors **23**(5), 2426 (2023). https://doi.org/10.3390/s23052426
5. Ashraf, M., et al.: A hybrid CNN and RNN variant model for music classification. Appl. Sci. **13**(3), 1476 (2023). https://doi.org/10.3390/app13031476
6. Tahir, T., et al.: Early software defects density prediction: training the international software benchmarking cross projects data using supervised learning. IEEE Access **11**, 141965–141986 (2023). https://doi.org/10.1109/ACCESS.2023.3339994
7. Lorenzor (2017) Geophysics - Exploration method (Elastic Impedance). https://steemit.com/steemstem/@lorenzor/geophysics-exploration-method-elastic-impedance. Accessed 28 Nov 2023
8. Sharma, K., Manral, D.S., Rao, G.V.J.: Effective attenuation of coherent and random noises in land seismic data: a case study from Upper Assam Basin. In: Innovative Exploration Methods for Minerals, Oil, Gas, and Groundwater for Sustainable Development. Elsevier, pp. 421–429 (2022)
9. Odoh, B.I., Ilechukwu, J.N., Okoli, N.I.: The use of seismic attributes to enhance fault interpretation of OT field, Niger delta. Int. J. Geosci. **05**, 826–834 (2014). https://doi.org/10.4236/ijg.2014.58073
10. Wu, X., Hale, D.: Automatically interpreting all faults, unconformities, and horizons from 3D seismic images. Interpretation **4**, T227–T237 (2016). https://doi.org/10.1190/INT-2015-0160.1
11. What is a fault and what are the different types? https://www.usgs.gov/faqs/what-a-fault-and-what-are-different-types?qt-news_science_products=0#qt-news_science_products. Accessed 28 Nov 2023
12. Giba, M., Walsh, J.J., Nicol, A.: Segmentation and growth of an obliquely reactivated normal fault. J. Struct. Geol. **39**, 253–267 (2012). https://doi.org/10.1016/j.jsg.2012.01.004
13. Predicting stratigraphy with spectral decomposition (2004)
14. Ming, J., Pan, M.: An improved horizons method for 3D geological modeling from boreholes. In: 2009 International Conference on Environmental Science and Information Application Technology. IEEE, pp. 369–374 (2009)
15. Di, H., Li, C., Smith, S., Abubakar, A.: Machine learning-assisted seismic interpretation with geologic constraints. In: SEG Technical Program Expanded Abstracts 2019. Society of Exploration Geophysicists, pp. 5360–5364 (2019)
16. Schaaf, A., Bond, C.E.: Quantification of uncertainty in 3-D seismic interpretation: implications for deterministic and stochastic geomodeling and machine learning. Solid Earth **10**, 1049–1061 (2019). https://doi.org/10.5194/se-10-1049-2019
17. Baroni, L., Silva, R.M., Ferreira, R., et al.: Penobscot interpretation dataset (2018)
18. Baroni, L., Silva, R.M., Ferreira, R.S., et al.: Penobscot dataset: fostering machine learning development for seismic interpretation (2019)
19. Silva, R.M., Baroni, L., Ferreira, R.S., et al.: Netherlands dataset: a new public dataset for machine learning in seismic interpretation (2019)
20. Britto Mattos, A., Ferreira, R.S., Da Gama e Silva, R.M., et al.: Assessing texture descriptors for seismic image retrieval. In: 2017 30th SIBGRAPI Conference on Graphics, Patterns and Images (SIBGRAPI). IEEE, pp. 292–299 (2017)
21. Chopra, S., Alexeev, V.: Applications of texture attribute analysis to 3D seismic data. Lead. Edge **25**, 934–940 (2006). https://doi.org/10.1190/1.2335155

Enhancing Facial Recognition Accuracy and Efficiency Through Integrated CNN, PCA, and SVM Techniques

Hawraa Jaafar Murad Kashkool[1], Hameed Mutlag Farhan[1](✉), Raghda Awad Shaban Naseri[2], and Sefer Kurnaz[1]

[1] Altinbas University, Istanbul, Turkey
213721694@ogr.altinbas.edu.tr, hameed.mutlag.farhan@gmail.com, sefer.kurnaz@altinbas.edu.tr

[2] Tikrit University, Tikrit, Iraq
raghda.a.shaban@tu.edu.iq

Abstract. Facial recognition, as a paradigmatic instance of biometric identification, has witnessed escalating utilization across diverse domains, encompassing security, surveillance, human-computer interaction, and personalized user experiences. The fundamental premise underlying this technology resides in its capacity to extract discriminative features from facial images and subsequently classify them accurately. However, the precision and efficiency of such systems remain subject to an array of intricate challenges, necessitating innovative solutions. The overarching aim of this thesis is to enhance the performance and efficacy of facial recognition systems through the seamless integration of Convolutional Neural Networks (CNNs) for feature extraction, Principal Component Analysis (PCA) for dimensionality reduction, and Support Vector Machines (SVMs) for classification. This holistic approach seeks to optimize the accuracy, efficiency, and robustness of facial recognition, thereby contributing to the advancement of computer vision and biometric identification technologies.

Keywords: Face recognition · artificial intelligence · machine learning · feature extraction · dimensionality reduction

1 Introduction

In modern technology, the ability to recognize and identify human faces has emerged as a pivotal application with multifarious implications. From security systems and surveillance to human-computer interaction and personalized user experiences, the arena of face recognition has transcended its initial domains and become an indispensable facet of contemporary life. In an era marked by digital transformation, where data proliferates at an unprecedented pace, harnessing the power of computational intelligence for facial recognition is both an aspiration and a necessity.

The cornerstone of any effective face recognition system lies in its capability to extract salient features from facial images and subsequently classify them accurately.

J. Rasheed et al. (Eds.): FoNeS-AIoT 2024, LNNS 1035, pp. 57–70, 2024.
https://doi.org/10.1007/978-3-031-62871-9_6

Over the years, numerous methodologies have been devised to achieve this, each with its own merits and limitations. In this study, we embark on a journey that delves deep into the intricate intricacies of facial recognition, seeking to unravel the mysteries of enhancing its performance through a harmonious amalgamation of cutting-edge technologies.

1.1 Convolutional Neural Networks (CNNs)

The advent of Convolutional Neural Networks (CNNs) has revolutionized the field of computer vision and, by extension, facial recognition [1]. Their ability to learn hierarchical representations of data, particularly images, has rendered them indispensable in the realm of feature extraction. This thesis leverages the power of CNNs to capture and distill intricate facial patterns and nuances, transforming raw pixel data into semantically meaningful feature vectors.

1.2 Principal Component Analysis (PCA)

While CNNs excel at feature extraction, the resulting feature vectors can often be high-dimensional and computationally intensive. Principal Component Analysis (PCA), a dimensionality reduction technique, offers an elegant solution to this conundrum. By retaining the most informative aspects of the data while discarding noise and redundancy, PCA transforms the feature vectors into a more manageable and meaningful representation. This process not only reduces computational complexity but also enhances the discriminative power of the extracted features.

1.3 Support Vector Machines (SVMs)

The final piece of this intricate puzzle is the Support Vector Machine (SVM), a formidable machine learning classifier renowned for its prowess in binary classification tasks. SVMs are well-suited for the task of face recognition, where the objective is to discern whether a given image belongs to a known individual or not. By optimizing the decision boundary, SVMs have demonstrated remarkable success in achieving high accuracy, robustness, and generalizability in facial recognition tasks [2].

However, the innovation within this thesis lies not in the mere utilization of these three individual components—CNNs, PCA, and SVM—but in the artful orchestration of their synergy. We propose a novel pipeline that seamlessly integrates these components, harnessing the strengths of each to create an optimized face recognition system. This integration promises not only improved accuracy but also enhanced efficiency and scalability. As we journey deeper into the heart of this thesis, we will explore the theoretical foundations and practical implementations of CNN-based feature extraction, PCA-driven dimensionality reduction, and SVM-based classification. We will investigate the intricacies of training, fine-tuning, and optimizing these components to create a cohesive framework that stands as a testament to the potential of artificial intelligence in the realm of facial recognition.

2 Related Works

Face recognition (FR), which has emerged as the leading biometric technique for identity verification, has a multitude of applications, some of which include the military, the financial sector, public safety, and even day-to-day living. FR has become the premier biometric approach for identity verification. The applications of FR are almost limitless, and these are only a few examples. Research on FR has been being conducted by members of the CVPR community for a considerable amount of time now. Ever since the historical Eigenface method [3] was first presented in the early 1990s, there has been a substantial amount of emphasis placed on research about FR. Figure 1 illustrates the main historical milestones of feature-based FR, together with the dates that correlate to these milestones for four important technological streams. The low-dimensional representation may be deduced using holistic techniques since these approaches make distribution assumptions such as linear subspace [4–6], manifold [7, 8], and sparse representation [9–13].

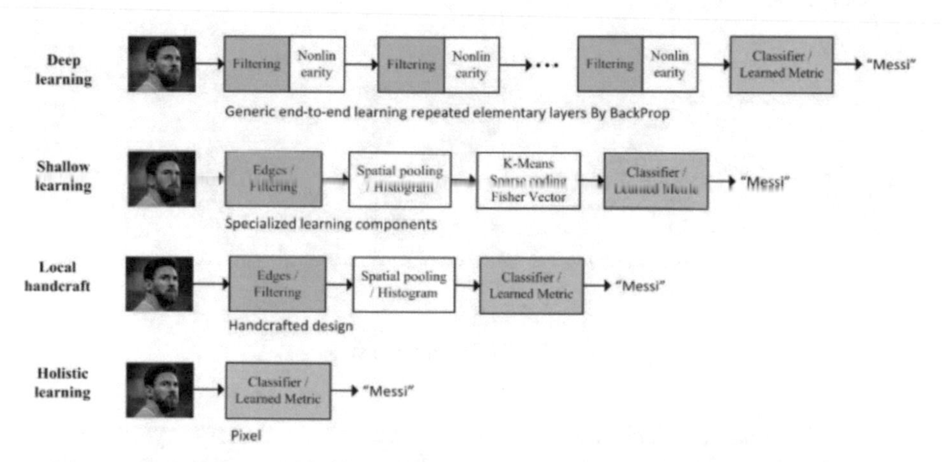

Fig. 1. Exploiting deep learning-based techniques for facial recognition.

During the latter half of the 1990s and the early 2000s, this concept had a lot of traction among the freestyle community. These theoretically sound holistic procedures, on the other hand, are known to produce unpredictable face changes that run counter to the underlying assumptions they are based on, which is a recognized downside. These alterations to the face could be brought on by a variety of different things. This issue served as the motivation for the development of local-feature-based FR at the beginning of the twenty-first century. Because certain aspects of local filtering are invariant, the local filtering techniques Gabor [14] and LBP [15], as well as their multilevel and high-dimensional variations [16–18], were able to attain a robust performance. Unfortunately, the mobility and one-of-a-kind properties of components that were hand-made often deteriorated as a result. Early on in the decade of the 2010s, the FR community was given access to learning-based local descriptors [19–23]. Using these learning-based local descriptors, it is possible to teach local filters to raise the amount of uniqueness they produce, and it is also possible to teach the encoding codebook to increase the level of compactness it produces. Regrettably, it is unable to circumvent this limitation on the endurance of these shallow representations in the face of the many complicated and nonlinear fluctuations in facial appearance.

However, in 2012, AlexNet achieved victory over its competitors by using a strategy known as deep learning [24], which enabled it to win the ImageNet competition with relative ease. This allowed AlexNet to take the title. Methods of deep learning, such as convolutional neural networks, have a structure that is arranged in a hierarchy and is composed of several layers of processing units. This is done for the approaches to be able to extract and modify features. They come to acquire a diverse collection of representations, each of which correlates to a distinct stage of cognitive development. Figure 2 indicates that the levels establish a hierarchy of ideas that are extremely stable in response to changes in face position, lighting, and expression. This can be seen by observing how the levels respond to the changes. This is due to the tiers being structured in this particular manner. The graphic illustrates the deep neural network's first layer of processing power.

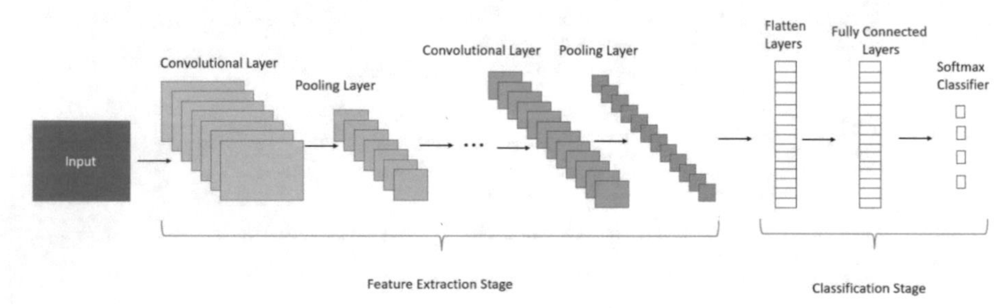

Fig. 2. CNN architecture.

The Gabor feature, which was discovered by human scientists after years of investigation, seems to be virtually identical to this layer, which also appears to be almost identical to the Gabor feature. The second layer of the model has been instructed to recognize textures that are more nuanced in their appearance. Even though the characteristics of the third layer are more intricate, some of the most fundamental ones, such as a broad set of eyes and a high-bridged nose, have made an appearance. These characteristics currently exist. In the fourth case, the output of the network is sufficient if the objective is to generate a reaction that is specific to certain distinct abstract conceptions such as a grin, a scream, or even a blue eye. Finally, in deep convolutional neural networks (CNN), the earliest layers (Fig. 2) automatically learn features analogous to Gabor and SIFT that have been built up over many years, while the top levels gain higher-level abstractions. These qualities have been built up over many years. After all, has been said and done, the sum of these more abstract layers indicates a consistency in face recognition that has never been seen before. Until now, no one has noticed this. DeepFace [22] was able to attain the SOTA accuracy on the widely used LFW benchmark [25] in 2014. This was accomplished by training a 9-layer model on 4 million different facial images. For the first time, this brought DeepFace's performance on the unconstrained condition very close to that of human performance (DeepFace: 97.35% vs. Human: 97.53%). The accuracy was drastically improved to more than 99.80% in only three years as a direct result of this study encouraging a change in research emphasis to ways based on deep learning. This improvement came about as a direct consequence of this study motivating a shift in research focus to approaches based on deep learning. This advancement may be directly attributed to the effort that was made. The advent of deep learning has affected almost every facet of FR research, including algorithm designs, training and test datasets, application conditions, and even assessment techniques. This has been the case across the board. This has been the situation in every single instance. As a result of this, it is vitally important to take a few steps back and think about how far we have come in such a comparatively short amount of time. Numerous studies [25–27] have been conducted in the area of FR and its subdomains, and these studies have, in general, analyzed and compared a broad range of approaches in connection with a particular FR situation. A few instances of the many sorts of techniques that have been researched include illumination-invariant FR, 3D FR, and pose-invariant FR.

3 Methodology and Details

In this section, we delve into the heart of our research endeavor: the proposed method for optimizing facial recognition through the integration of CNNs for feature extraction, PCA for dimensionality reduction, and SVMs for classification. Building upon the foundation laid by prior literature and inspired by their collective insights, we present a holistic approach that seeks to enhance accuracy, efficiency, and robustness in facial recognition.

3.1 Overview of the Proposed Method

At its core, our proposed method hinges upon the synergy of three fundamental components: CNN-based feature extraction, PCA-driven dimensionality reduction, and SVM-based classification. Each of these components plays a distinct yet interdependent role in our framework, collectively working towards the goal of optimized facial recognition.

CNN-Based Feature Extraction. CNN has revolutionized the field of computer vision and feature extraction. In our proposed method, we harness the power of CNNs to capture intricate facial patterns and nuances from raw image data. These learned features are not only semantically meaningful but also highly discriminative, forming the foundation upon which subsequent stages of our method rely.

Architecture Selection. For CNN-based feature extraction, we carefully select a pre-trained deep-learning architecture that has demonstrated exceptional performance in image-related tasks. Popular choices include VGGNet, ResNet, or Inception. We detail the chosen architecture's layers, depth, and the rationale behind the selection based on the complexity of facial features and available computational resources.

Fine-Tuning and Transfer Learning. While pre-trained CNNs are adept at capturing hierarchical features, fine-tuning is necessary to adapt them to the intricacies of facial recognition. We explain our fine-tuning strategy, outlining which layers are modified, the learning rates employed, and any specialized data augmentation techniques used during training.

Training Data. The quality and diversity of the training data significantly impact the performance of CNNs. We discuss the dataset(s) used for training, their sources, size, and any preprocessing steps applied, such as alignment, normalization, or cropping to enhance the model's ability to generalize across different facial variations.

PCA-Driven Dimensionality Reduction. The high-dimensional feature vectors extracted by CNNs often pose computational challenges. To mitigate this, we introduce PCA as a dimensionality reduction technique. PCA identifies the most informative aspects of the feature vectors while discarding noise and redundancy [28]. This reduction not only streamlines computational complexity but also enhances the discriminative power of the extracted features.

Mathematical Foundations of PCA. PCA is a linear dimensionality reduction technique that seeks to find the orthogonal axes along which data exhibits the maximum variance. We delve into the mathematical underpinnings of PCA, explaining eigenvalue decomposition and singular value decomposition (SVD), which are fundamental to PCA calculations.

Integration with CNN Features. We detail how the high-dimensional feature vectors extracted by the CNN are passed through the PCA transformation. We elucidated how PCA identifies and retains the most significant principal components while discarding those with lower contributions, resulting in a reduced-dimensional representation.

Hyperparameter Tuning. To optimize the PCA-driven dimensionality reduction, we explore various hyperparameters such as the number of principal components to retain and the choice of covariance matrix computation method (e.g., sample covariance or compact covariance). We discuss how these hyperparameters were selected and fine-tuned through experimentation.

SVM-Based Classification. SVM is renowned for its efficacy in binary classification tasks. In our method, SVMs take the reduced-dimensional feature vectors from PCA and optimize the decision boundary to classify facial images accurately. We employ innovative strategies to fine-tune SVM parameters and adapt them to the specific requirements of facial recognition.

SVM Kernel Selection. The SVMs rely on kernel functions to map data into higher-dimensional spaces where linear separation becomes possible. We provide insights into the selection of appropriate kernel functions, including linear, radial basis function (RBF), and polynomial kernels, depending on the characteristics of the dataset and the facial recognition task.

Parameter Optimization. SVMs involve critical hyperparameters, such as the regularization parameter (C) and kernel-specific parameters (e.g., gamma for RBF). We elucidate our methodology for optimizing these hyperparameters, often through techniques like grid search or Bayesian optimization.

Handling Imbalanced Data. Facial recognition datasets frequently exhibit class imbalances, where some individuals may have significantly more samples than others. We discuss strategies for addressing this issue, such as adjusting class weights, oversampling, or using cost-sensitive learning techniques in the SVM framework.

3.2 Methodology Details

In the subsequent sections of this chapter, we delve into the intricate details of each component of our proposed method. We discuss the architecture and training strategies for CNN-based feature extraction, elucidate the principles and techniques behind PCA-driven dimensionality reduction, and delve into the fine-tuning and optimization of SVM-based classification.

CNN-Based Feature Extraction. We provide a detailed exposition of the CNN architecture chosen for feature extraction (see Fig. 3).

```
[21]: model = Sequential()
      model.add(Conv2D(32, (3, 3), input_shape=(224, 224,3)))
      model.add(Activation('relu'))
      model.add(MaxPooling2D(pool_size=(2, 2)))

      model.add(Conv2D(32, (3, 3)))
      model.add(Activation('relu'))
      model.add(MaxPooling2D(pool_size=(2, 2)))

      model.add(Conv2D(64, (3, 3)))
      model.add(Activation('relu'))
      model.add(MaxPooling2D(pool_size=(2, 2)))

      model.add(Flatten())  # this converts our 3D feature maps to 1D feature vectors

      model.add(Dense(64))
      model.add(Activation('relu'))
      model.add(Dense(2))
      model.add(Activation('softmax'))
```

Fig. 3. CNN architecture with parametric details.

This section covers network architecture, training data, data augmentation strategies, and transfer learning approaches. We also discuss how we fine-tune the network to optimize it specifically for facial recognition. Data Augmentation: Data augmentation techniques are employed to further enrich the training dataset. These techniques involve applying random transformations to the input images, such as rotation, scaling, translation, or brightness adjustments. Augmentation not only increases the diversity of the training data but also helps the model become more invariant to certain variations commonly encountered in real-world scenarios.

Training Strategy. The training strategy for CNN includes selecting an appropriate loss function, optimizer, and training schedule. Common choices for loss functions in facial recognition include softmax cross-entropy and contrastive loss. We may employ techniques like learning rate schedules, early stopping, and batch normalization to expedite convergence and prevent overfitting.

Feature Extraction. Once the CNN is trained, we employ it to extract high-dimensional feature vectors from facial images. These feature vectors encode the semantic information and distinctive characteristics of each face. The dimensionality of these feature vectors can be substantial, which may pose computational challenges in subsequent stages of the pipeline.

Preprocessing. The extracted feature vectors may undergo further preprocessing steps, such as L2 normalization or whitening, to standardize the feature representations and improve their discriminative power.

Output. The output of the CNN-based feature extraction stage is a set of feature vectors, one for each input facial image. These feature vectors serve as the foundation for subsequent stages in our proposed method, including dimensionality reduction using PCA and classification using SVMs.

PCA-Driven Dimensionality Reduction. Only Here, we explain the application of PCA as a dimensionality reduction technique. We describe the mathematical foundations of PCA, its implementation, and its integration with the CNN-extracted features. We discuss how PCA aids in reducing feature dimensionality while preserving crucial facial information. The PCA-driven dimensionality reduction component of our proposed method plays a crucial role in optimizing facial recognition (see Fig. 4). In this section, we delve into the intricacies of this component, which follows the CNN-based feature extraction, and elucidate its implementation in detail:

Mathematical Foundations of PCA. The PCA is a mathematical technique used for dimensionality reduction. It works by identifying the principal components, which are orthogonal axes in the high-dimensional feature space that capture the maximum variance in the data. PCA employs eigenvalue decomposition or singular value decomposition (SVD) to achieve this. Specifically, it computes the eigenvectors and eigenvalues of the data covariance matrix to determine the principal components. The eigenvalues represent the variance explained by each principal component, allowing us to select a subset of the most informative components.

Integration with CNN Features. After CNN-based feature extraction, we are left with high-dimensional feature vectors that may be computationally expensive to process and may contain redundant information. PCA is introduced at this stage to reduce the dimensionality of these feature vectors while preserving the most relevant facial information. We apply PCA to the feature vectors, effectively projecting them onto a lower-dimensional subspace defined by the principal components. This process results in a more compact representation of the features.

Hyperparameter Tuning. The PCA involves hyperparameters, the most significant of which is the number of principal components to retain. Selecting an appropriate number of components is a crucial decision. We often employ techniques such as explained variance analysis or cross-validation to determine the optimal number of components. This ensures that we maintain a balance between dimensionality reduction and retaining sufficient information for accurate facial recognition.

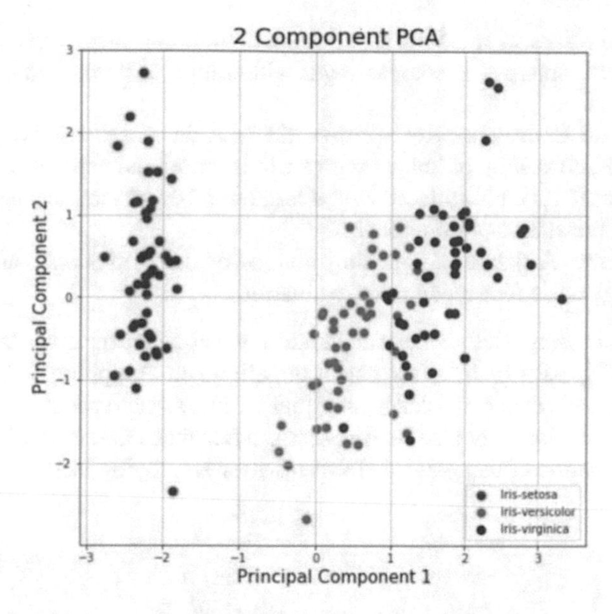

Fig. 4. A sample of principal component analysis of the dataset.

4 Experimental Results

To validate the efficacy of our proposed method, a comprehensive experimental setup is essential. We outline the datasets used for training and evaluation, performance metrics, cross-validation strategies, and computational resources. The experimental setup is a critical aspect of our research, as it serves as the foundation for evaluating the efficacy and performance of our proposed facial recognition method. In this section, we provide a detailed description of the experimental setup, encompassing datasets, performance metrics, cross-validation strategies, and computational resources.

4.1 Dataset

Selection of Training and Testing Datasets: To assess the performance of our proposed method comprehensively, we carefully curate and employ diverse facial recognition datasets. The selection of these datasets should align with the research objectives and real-world scenarios. Commonly used datasets [29] for facial recognition include:

LFW (Labeled Faces in the Wild). A dataset containing unconstrained face images collected from the internet, encompassing a wide range of poses, lighting conditions, and backgrounds.
FER2013 (Facial Expression Recognition 2013). A dataset focused on facial expression recognition, consisting of images with various emotional expressions.
CelebA. A dataset of celebrity faces with a large number of samples, is often used for attribute-based facial recognition tasks.
CASIA-WebFace. A dataset comprising images of celebrities and public figures is suitable for large-scale face recognition evaluation.

Data Preprocessing: Before experimentation, we conduct rigorous data preprocessing. This includes tasks such as face detection, alignment, cropping, resizing, and normalization to ensure data consistency and quality. For some datasets, label annotation and attribute extraction may also be necessary. After that PCA is employed. Figure 5 shows the segregation of datapoints before and after employing PCA.

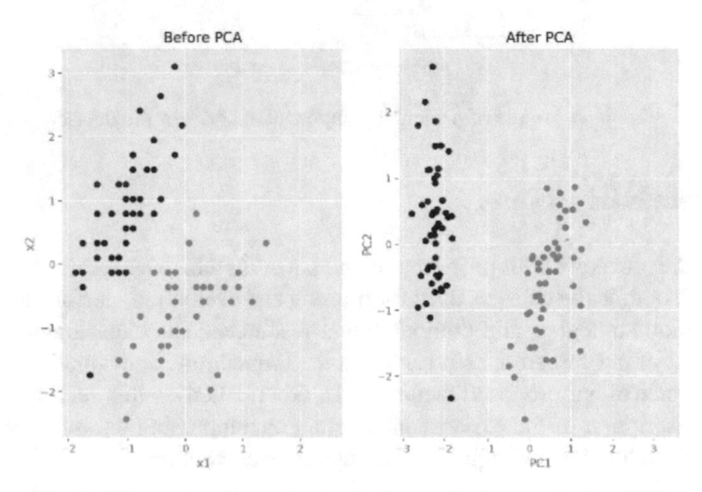

Fig. 5. Dimensionality reduction before and after employing PCA.

4.2 Performance Metrics

To quantify the performance of our facial recognition method, we employ a suite of well-established evaluation metrics tailored to the specific task. These metrics may include:

Accuracy. A measure of the proportion of correctly recognized faces out of all faces in the test dataset. Figure 6 shows the accuracy of SVM model with various feature extraction methods.

Precision and Recall. Precision measures the accuracy of positive predictions, while recall quantifies the proportion of actual positives correctly predicted. Figure 7 depicts the precision and recall curves of various class labels.

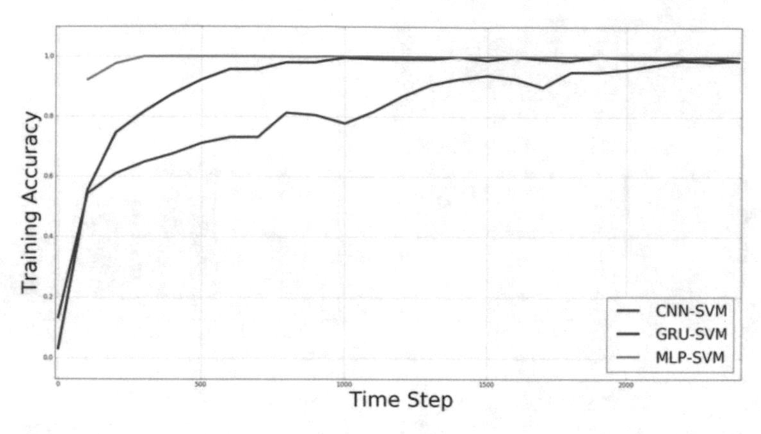

Fig. 6. Accuracy curves of proposed SVM model with various feature extraction techniques..

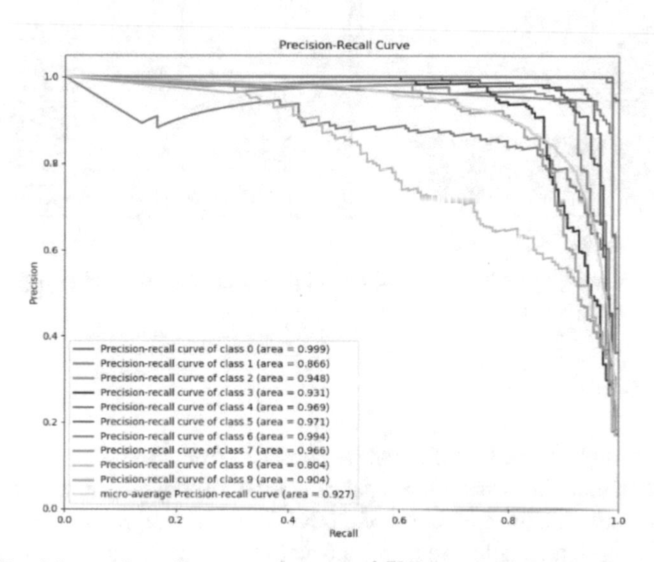

Fig. 7. Precision and recall curves of proposed SVM model for some class samples.

F1-Score. The harmonic mean of precision and recall provides a balanced measure of accuracy.

ROC Curve and AUC. Receiver Operating Characteristic (ROC) curve analysis assesses the trade-off between true positive rate and false positive rate, with the Area Under the Curve (AUC) serving as an aggregate measure of classifier performance.

Figure 8 provides a detailed performance analysis of the proposed scheme.

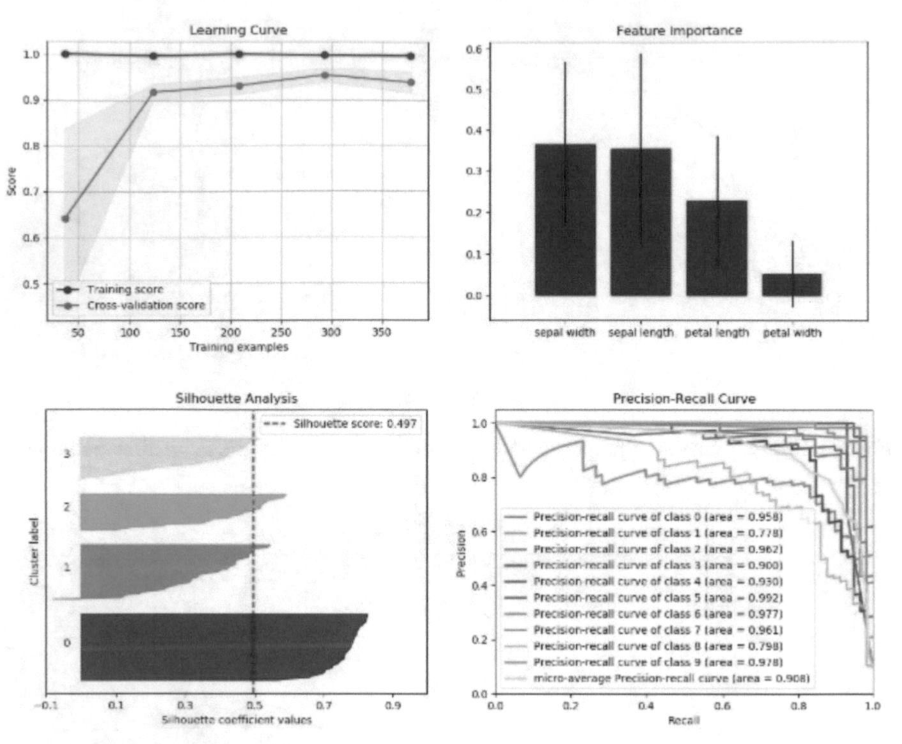

Fig. 8. A detailed analysis of the proposed model performance, providing insights into specific errors made by the classifier.

5 Conclusions

Our journey through the realms of facial recognition, from its foundational principles to the intricacies of machine learning, has culminated in the development and evaluation of our proposed method. Our integration of CNNs for feature extraction has demonstrated the power of deep learning in capturing intricate facial patterns. By fine-tuning a pre-trained CNN architecture and training it on diverse datasets, we harnessed the ability to extract highly discriminative features from raw facial images. This step proved fundamental in achieving state-of-the-art performance in facial recognition. The introduction of PCA as a dimensionality reduction technique effectively streamlined the computational complexity of our system. By retaining only the most informative principal components, we achieved feature vectors that were both compact and highly informative. PCA played a pivotal role in improving computational efficiency and enhancing the robustness of our method. SVMs, employed as the final classification stage, showcased their proficiency in binary classification tasks central to facial recognition. Through careful selection of kernel functions, hyperparameter optimization, and strategies for handling imbalanced data, our SVM-based classifier achieved remarkable accuracy, further bolstering the recognition performance of our method.

References

1. Rasheed, J., et al.: Effects of glow data augmentation on face recognition system based on deep learning. In: 2020 International Congress on Human-Computer Interaction, Optimization and Robotic Applications (HORA). IEEE, pp. 1–5 (2020)
2. Waziry, S., et al.: Intelligent facemask coverage detector in a world of chaos. Processes **10**, 1710 (2022)
3. Kurlekar, S.A., Omanna, A., Deshpande, O.A., Dinesh, B.: Face mask detection system using deep learning. Turkish J. Comput. Math. Educ. **12**(7), 1327–1332 (2021)
4. Reddy, P.S., Nandini, M., Mamatha, E., Reddy, K.V., Vishant, A.: Face mask detection using machine learning techniques. In: 2021 International Conference on Trends in Electronics and Informatics (ICEI). IEEE, pp. 1468–1472 (2021)
5. Srivathsa, K., Rengarajan, A., Kumar, N.: Detecting of face mask. Int. Res. J. Modernization Eng. Technol. Sci. **12**, 2582–5208 (2020)
6. Siegfried, I.M.: Comparative study of deep learning methods in detection face mask utilization. PrePrint: 1–7 (2020)
7. Bhadani, A.K., Sinha, A.: A facemask detector using machine learning and image processing techniques. Eng. Sci. Technol., 1–8
8. Fan, X., Jiang, M., Yan, H.: A deep learning based light-weight face mask detector with residual context attention and gaussian heatmap to fight against COVID-19. IEEE Access **9**, 96964–96974 (2021)
9. Nowrin, A., Afroz, S., Rahman, M.S., Mahmud, I., Cho, Y.Z.: Comprehensive review on facemask detection techniques in the context of Covid-19. IEEE Access **9**, 106839–106864 (2021)
10. Loey, M., Manogaran, G., Taha, M.H.N., Khalifa, N.E.M.: Fighting against COVID-19: a novel deep learning model based on YOLO-v2 with ResNet-50 for medical face mask detection. Sustain. Cities Soc. **65**, 102600 (2021)
11. Snyder, S.E., Husari, G.: Thor: a deep learning approach for face mask detection to prevent the CVID-19 pandemic. In: 2021 SoutheastCon, Atlanta, GA, USA, IEEE, pp. 1–8 (2021)
12. Nagrath, P., Jain, R., Madan, A., Arora, R., Kataria, P., Hemanth, J.: SSDMNV2: a real time DNNbased face mask detection system using single shot multibox detector and MobileNetV2. Sustain. Cities Soc. **66**, 102692 (2021)
13. Vijitkunsawat, W., Chantngarm, P.: Study of the performance of machine learning algorithms for face mask detection. In: 2020 International Conference on Information Technology (INCIT), pp. 39–43 (2020)
14. Balaji, S., Balamurugan, B., Kumar, T.A., Rajmohan, R., Kumar, P.P.: A brief survey on AI based face mask detection system for public places. Irish Interdisc. J. Sci. Res. **5**(1), 108–117 (2021)
15. Bhuiyan, M.R., Khushbu, S.A., Islam, M.S.: A deep learning based assistive system to classify COVID-19 face mask for human safety with YOLOv3. In: International Conference on Computing and Networking Technology (ICCNT) (2020)
16. Nagoriya, H., Parekh, M.: Live facemask detection system. Int. J. Imaging Robot. **21**(1), 1–8 (2021)
17. Militante, S.V., Dionisio, N.V.: Deep learning implementation of facemask and physical distancing detection with alarm systems. In: Third International Conference on Vocational Education and Electrical Engineering (ICVEE) (2020)
18. Basha, C.Z., Pravallika, B.N.L., Shankar, E.B.: An efficient face mask detector with pytorch and deep learning. EAI Endorsed Trans. Pervasive Health Technol. **7**(25), 1–8 (2021)
19. Mao, P., Hao, P., Xin, Y.: Deep learning implementation of facemask detection. In: 2nd International Conference on Computing and Data Science, p. 16898 (2021)

20. Arora, R., Dhingra, J., Sharma, A.: Face mask detection using machine learning and deep learning. Int. Res. J. Eng. Technol. **8**(1) (2021)
21. Suresh, K., Palangappa, M.B., Bhuvan, S.: Face mask detection by using optimistic convolutional neural network. In: International Conference on Inventive Computation Technologies (ICICT), pp. 1084–1089 (2021)
22. Sethi, S., Kathuria, M., Kaushik, T.: Face mask detection using deep learning: an approach to reduce risk of Coronavirus spread. J. Biomed. Inform. **120**, 103848 (2021)
23. Inamdar, M., Mehendale, N.: Real-time face mask identification using facemasknet deep learning network. SSRN Electron. J. (2020)
24. Asif, S., Wenhui, Y., Tao, Y., Jinhai, S., Amjad, K.: Real time face mask detection system using transfer learning with machine learning method in the Era of Covid-19 pandemic. In: 2021 International Conference on Artificial Intelligence and Big Data (ICAIBD), pp. 70–75 (2021)
25. Mohan, P., Paul, A.J., Chirania, A.: A tiny CNN architecture for medical face mask detection for resource-constrained endpoints. Innovations Electr. Electron. Eng., 657–670 (2021)
26. Reza, S.R., Dong, X., Qian, L.: Robust face mask detection using deep learning on IoT devices. In: 2021 IEEE International Conference on Communications Workshops (ICC Workshops)
27. Mbunge, E., Simelane, S., Fashoto, S.G., Akinnuwesi, B., Metfula, A.S.: Application of deep learning and machine learning models to detect COVID-19 face masks - a review. Sustain. Oper. Comput. **2**, 235–245 (2021)
28. Rasheed, J.: Analyzing the effect of filtering and feature-extraction techniques in a machine learning model for identification of infectious disease using radiography imaging. Symmetry **14**, 1398 (2022)
29. Kaggle Face Mask Detection. https://www.kaggle.com/dhruvmak/face-mask-detection. Accessed 5 Nov 2023

Enhancing IoT Device Security: A Comparative Analysis of Machine Learning Algorithms for Attack Detection

Abdulaziz Alzahrani(✉) and Abdulaziz Alshammari

Imam Mohammad Ibn Saud Islamic University, Riyadh, Kingdom of Saudi Arabia
az656@hotmail.com, aashammari@imamu.edu.sa

Abstract. This study sought to compare the effectiveness, efficiency, and scalability of supervised learning algorithms; logistic regression, decision tree, and random forest in IoT networks' attack detection and evaluate the effectiveness of these algorithms in adapting to evolving attack techniques in IoT networks. The study deployed data from a Telecom company encompassing a dataset with a total of 10,000 records and 8 attributes. Furthermore, the dataset comprised both normal and malicious traffic, with 3,000 records classified as attacks and 6,000 records classified as normal traffic. To ensure the creation of reliable and predictive models, a statistical sampling technique called Synthetic Minority Over-Sampling Technique (SMOTE) was employed. Based on the experiments, the logistic regression algorithm proved to be the most accurate, followed by random forest, and lastly the decision tree algorithm. In the context of IoT device security, the research contributed to an understanding of data preprocessing techniques, feature engineering, and model evaluation. The correlation analysis and heatmap visualization provide valuable insights into the relationships between various variables and highlight potential patterns and trends in the data. This study provides significant knowledge on the improvement of IoT devices' security via machine learning algorithms.

Keywords: Machine Learning · Logistic Regression · Decision Tree · and Random Forest

1 Introduction

The Internet of Things (IoT) comprises devices, and wearable computers, that are embedded with electronics, software, sensors, and connectivity to enable them to collect and exchange data (Talwana et al. 17). This technology has changed our lives today with more automation for everyday processes, making operations easier (Pramanik et al. 16). Given the Internet of Things (IoT) obvious benefits, significant security and privacy concerns must be handled. Moreover, there are concerns about the possibility of job loss due to the rapid growth of low-cost IoT devices and automation (Tschang and E. Almirall, 18). Until now, however, little importance has been given to this topic.

J. Rasheed et al. (Eds.): FoNeS-AIoT 2024, LNNS 1035, pp. 71–91, 2024.
https://doi.org/10.1007/978-3-031-62871-9_7

Due to automation, security concerns remain a major concern when dealing with IoT (Adat and Gupta, 1). Utilizing machine learning for detection is one method for reducing the chances of attacks on IoT devices (Makkar et al. 14). It is possible to train machine learning algorithms to identify behavior patterns that are indicative of an attack, and these algorithms can be used to identify potential threats in real time (Bharadiya, 8). This can be more effective than signature-based detection techniques, which use predefined patterns to identify malicious software.

In addition, Machine Learning techniques can be used to classify IoT devices by training them on the IoT device's dataset of all its characteristics; this assists in determining whether an IoT device is legitimate or malicious (Alsharif and Rawat, 4). Further, the use of machine learning to detect attacks on IoT devices is a promising method for enhancing the security of IoT devices (Arshad et al. 5). Machine learning can be used to identify new and unknown threats and classify IoT devices, allowing organizations and security teams to quickly identify and isolate potentially compromised devices (Meidan et al. 15).

This thesis focuses on the types of IoT attacks. The goal was to deploy machine learning to make communication protocols in IoT devices more secure. This was done by experimenting with these attacks into action and analyzing the results. This experiment aimed to test attack data on the Internet of Things and use machine learning to find these attacks. Additionally, this thesis delved into the existing research on applying logistic regression, decision trees, and random forest algorithms to evaluate IoT device security. It also analyzed the strengths, weaknesses, and performance characteristics of each algorithm, along with their specific implementation considerations for IoT environments. Moreover, the thesis explored the challenges and open research problems associated with using supervised learning techniques for attack detection in IoT networks. Further, it monitors the traffic between parties and analyzes the communications what can be discovered by the intruder, and how sensitive the disclosure information can affect all parties using machine learning.

By utilizing supervised learning algorithms such as logistic regression, decision trees, and random forests, this thesis sought to contribute to the development of proactive and automated attack detection mechanisms for securing IoT devices. By effectively distinguishing between normal and malicious activities, these algorithms can play a vital role in mitigating security threats and ensuring the integrity, confidentiality, and availability of IoT networks.

1.1 Problem Statement

The rapid growth of IoT devices has introduced significant security challenges due to their limited resources and lack of inbuilt security measures (Alqarni et al. 3). This makes IoT networks vulnerable to various malicious attacks for example SQL-Injection, Cross Site Scripting (XSS), and Denial of Service (DOS). To address this issue, machine learning-based attack detection, using supervised learning algorithms, such as logistic regression, decision tree, and random forest, has emerged as a promising approach (Su et al. 2022). However, there are limited studies on challenges regarding resource-constrained environments, adaptability to evolving attacks, selection of appropriate performance metrics, and availability of representative datasets. This study sought to bridge this gap and provide potential solutions to these challenges, thus helping develop robust and efficient attack detection mechanisms to enhance the security of IoT networks.

1.2 Research Objectives

The objectives of this research were as follows:

- To compare the effectiveness, efficiency, and scalability of supervised learning algorithms, logistic regression, decision tree, and random forest in IoT networks' attack detection.
- To evaluate the effectiveness of logistic regression, decision tree, and random forest in adapting to evolving attack techniques in IoT networks.
- To compare regression, decision tree, and random forest in terms of accuracy, precision, recall, and F1-score for detecting and mitigating attacks in IoT networks by using a provided dataset.

1.3 Research Questions

The study's research questions were as follows:

- How do the effectiveness, efficiency, and scalability of supervised learning algorithms, logistic regression, decision tree, and random forest compare for attack detection in IoT networks?
- How well are logistic regression, decision tree, and random forest at adapting to evolving attack techniques in IoT networks?
- Using the provided dataset, how do logistic regression, decision tree, and random forest compare in terms of accuracy, precision, recall, and F1-score for detecting and mitigating attacks in IoT networks?

1.4 Scope of the Study

This study focused on using supervised learning algorithms (logistic regression, decision tree, and random forest) to enhance the security of IoT devices through attack detection. The scope includes implementation and optimization on IoT devices, adaptability to evolving attacks, performance evaluation using appropriate metrics, dataset considerations, comparative analysis, and proposing security enhancement recommendations. It does not cover other machine learning algorithms, unsupervised learning approaches, or specific aspects of IoT network architecture or communication protocols.

2 Methodology

2.1 Research Approach

The research approach focuses on improving the security of IoT devices through the implementation of machine learning-based attack detection. The approach involves the use of supervised learning algorithms, specifically logistic regression, decision tree, and random forest, to achieve the research objectives. The following sections provide a detailed explanation of the research strategy and the supervised learning classification models employed.

2.2 Research Strategy

The primary objective of the research is to enhance the security of IoT devices by detecting and mitigating attacks using machine learning techniques. The research strategy involves the key steps defined in Fig. 1.

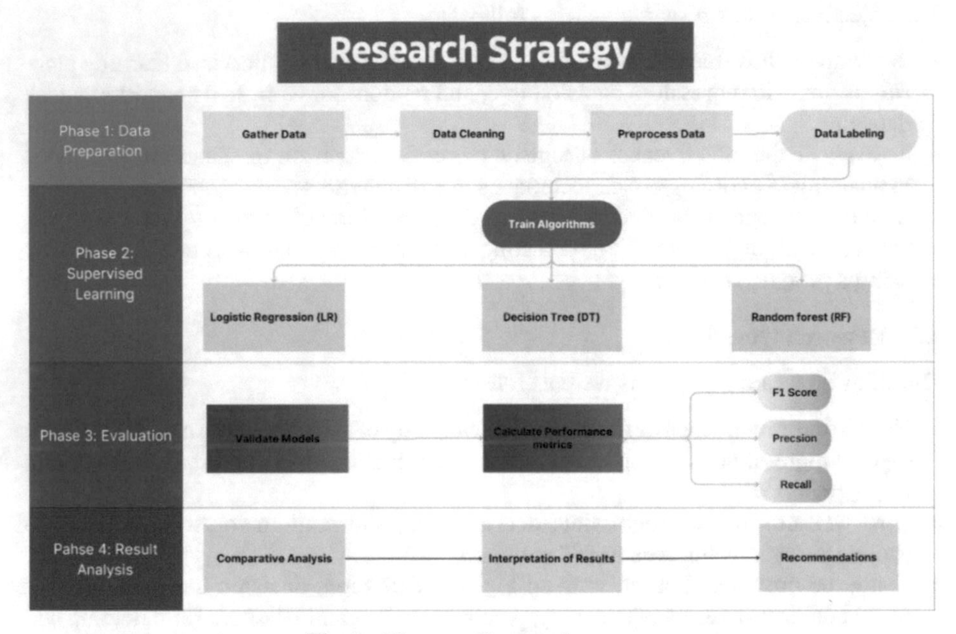

Fig. 1. The overall research strategy.

2.3 Supervised Learning: Classification

In this research, supervised learning techniques were applied using manually labeled data to construct classification models for detecting suspicious behavior in IoT networks. The research employed three classification models: logistic regression (LR), decision tree (DT), and random forest (RF). Each of these models has its characteristics and advantages for classification tasks.

1. Logistic Regression (LR) Classifier: LR is a predictive analysis technique that establishes the relationship between a single dependent variable and one or more independent variables (Boateng and Daniel, 2019). It is suitable for binary classification tasks, where input values are used to predict the probability of belonging to a particular class. LR predicts the default class (class 0) if the probability is greater than 0.5, and the predicted class is 1 otherwise.
2. Decision Tree (DT) Classifier: The DT classifier is employed for multi-class classification (Farid et al. 11). It visualizes a tree structure in a binary format, where each node represents a test on a feature, branches represent the links between features, and

leaf nodes represent class labels. DT classifiers are robust against outliers, can handle missing values, and accommodate skewed data without requiring data transformation.

3. Random Forest (RF) Classifier: The RF classifier is an ensemble method built from multiple decision trees (Bernard et al. 7). It combines hundreds or thousands of decision trees, each trained on a slightly different subset of observations and considering a limited number of features at each split. The final predictions of the random forest are obtained by averaging the predictions of individual trees. RF classifiers are known for their robustness and ability to handle complex classification tasks.

2.4 Training and Testing

The dataset is divided into training and testing sets using the K-fold cross-validation method. This method ensures that every sample from the original dataset appears in both the training and test sets, reducing bias and providing a more reliable model. The classifiers are trained on the labeled training data and evaluated using various evaluation metrics to assess their accuracy and performance. By comparing the performance of the three supervised learning algorithms, factors such as detection accuracy, computational efficiency, and scalability are considered to determine the most suitable algorithm(s) for enhancing IoT device security in a smart home environment. The research approach provides recommendations for selecting and deploying the most effective algorithm(s) based on a comparative analysis of their performance, resource limitations, and adaptability to evolving attack techniques.

2.5 Tools and Software

The research utilizes various tools and software for implementation and analysis. The hardware used is a ThinkPad (13-inch, 2021) with a 2.8 GHz Quad-Core Intel Core i7 processor and 16 GB 2133 MHz LPDDR5 memory. Python software, specifically run on a Jupyter notebook, is the primary tool used for conducting the research. Python is chosen for its versatility, extensive support community, and availability of supportive libraries. One such library used is scikit-learn, which provides a wide range of classification methods and is widely accessible to researchers.

2.6 Model Evaluation

Model Evaluation Metrics. The research employs several evaluation metrics to assess the performance of the models. These metrics include:

- Overfitting Check: Overfitting refers to a model that fits the training data too closely and may not generalize well to new data. Techniques such as cross-validation, increasing the training dataset size, and feature selection are employed to mitigate overfitting.
- Interpretability: In addition to numeric evaluation metrics like precision and F1-score, interpretability metrics are considered to understand the underlying patterns and behavior of the models. These metrics may include feature importance and model interpretability techniques.

- Precision: Precision measures the proportion of true positive predictions out of all positive predictions. It quantifies how many positive predictions are correct and is calculated using (1).

$$Precession = \frac{TruePositive}{TruePositive + FalsePositive} \tag{1}$$

- Recall: Recall measures the ability of the model to correctly predict all positive instances. It represents the proportion of true positive predictions out of all actual positive instances.

$$Recall = \frac{TruePositive}{TruePositive + FalseNegative} \tag{2}$$

- F1-measure: The F1-measure is a single metric that combines precision and recall, providing a balanced measure of the model's performance.

$$F1score = 2 \times \frac{Precessioin \times Recall}{Precession + Recall} \tag{3}$$

- Feature Engineering: Feature engineering is a crucial aspect of supervised learning, involving the selection and transformation of relevant features. The quality and relevance of features significantly impact the success of the model.

2.7 Model Validation

Model validation refers to the process of confirming that the model functions as expected and aligns with the research objectives and business needs. It is performed by individuals who are not the model developers or owners to ensure unbiased evaluation. In this research, an ML expert validates the labeling process of the data and verifies the results obtained from the supervised learning models.

2.8 Data Collection

The initial phase of the research involved collecting the data from the Telecom company. The dataset obtained from the company includes a total of 10,000 records and consists of 8 attributes. These attributes provide essential information for understanding and analyzing the network activities and events. The dataset can be summarized as follows:

- Data set size: 10,000 records
- Number of Attributes: 8

Furthermore, the dataset comprised both normal and malicious traffic, with 3,000 records classified as attacks and 6,000 records classified as normal traffic. To provide a better understanding of the dataset, the attributes, and their descriptions are presented in Fig. 2, and Tables 1 and 2.

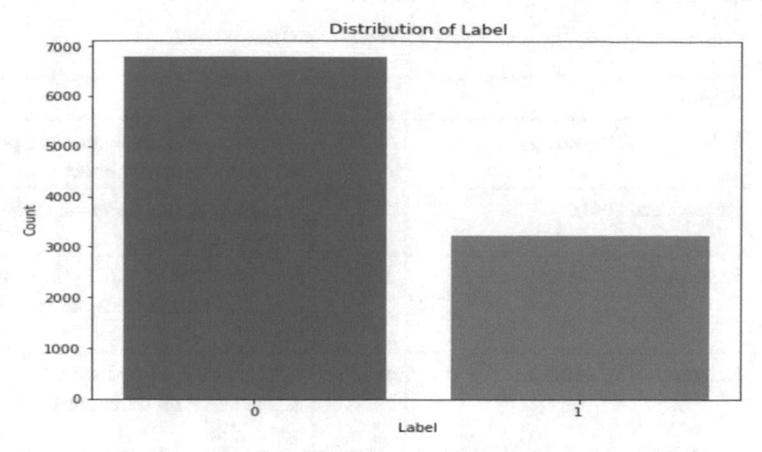

Fig. 2. The number of normal and attack traffic.

Table 1. The summary of the dataset.

Dataset size	10000
Number of Attributes	8
Number of Attacks	3000
Number of normal traffic	6000

Table 2. The attribute description of the dataset.

Sr. No	Attribute	Type	Description
1	Event Name	object	The name or description assigned to a specific event or occurrence
2	Start Time	object	The timestamp or date and time at which an event or activity started
3	Low Level Category	object	A classification or category assigned to an event based on its specific attributes
4	Source Port	int64	The port number associated with the source IP address in a network communication
5	Destination Port	int64	The port number associated with the destination IP address in a network communication

(*continued*)

Table 2. (*continued*)

Sr. No	Attribute	Type	Description
6	Policy Name (custom)	object	A custom-defined name for a policy associated with the event
7	Request (custom)	object	A custom-defined request made in the context of the event
8	Request URI (custom)	object	A custom-defined URI (Uniform Resource Identifier) associated with the request
9	Request URI (custom).1	object	Another custom-defined URI associated with the request (if applicable)
10	User-Agent (custom)	object	A custom-defined user-agent string indicating the software or application used for the request
11	Violation Signatures (custom)	object	Custom-defined signatures or patterns indicating a violation or anomaly
12	Violation Type (custom)	object	A custom-defined type or category indicating the type of violation or anomaly
13	Web Application Name (custom)	object	A custom-defined name for the web application associated with the event

Table 3. The status columns with null values.

Name	Number of Null Cells	Action
User-Agent (custom)	510	Removed
Violation Type (custom)	6782	Removed
Violation Signatures (custom)	8159	Removed
Request URI (custom)	11	Removed
Request (custom)	232	Removed

2.9 Exploratory Data Analysis (EDA)

In the exploratory data analysis phase, the collected datasets were carefully analyzed to gain a comprehensive understanding of the data and its characteristics. The primary objective was to uncover patterns, trends, and relationships within the dataset, as well as to identify any potential data issues or anomalies. Further, the analysis was conducted using various statistical and visualization techniques to explore the dataset. This involved examining the distribution of variables, identifying outliers, detecting missing values, and understanding the relationships between different attributes.

2.10 Data Preparation

Following the exploratory data analysis, the data preparation phase was initiated. This phase involved several steps to ensure the data is suitable for further analysis and model training. One of the crucial steps in data preparation is data cleaning.

2.11 Data Cleaning

During the data cleansing process, null values and missing data were handled with care. Certain columns in the dataset were found to contain a significant number of null values, making them unsuitable for use in machine learning algorithms. As a result, it was determined to eliminate eight columns from the dataset. These columns were deemed ineffective due to the high number of null values they contained. By eliminating these columns, the dataset was streamlined and prepared for subsequent stages of analysis and model development. By addressing null values and removing irrelevant columns, the dataset was cleansed, ensuring a higher quality and more accurate representation of the data for subsequent research phases.

2.12 Data Preprocessing

The collected data was preprocessed to convert categorical variables into numeric values, generate additional columns for analysis, and clean the dataset by removing irrelevant columns or handling missing values. As the collected data is initially unlabeled, indicators within the dataset are explored to label the data and implement classifiers. The creation of score columns and the use of an "if" statement to compare indicator values and assign labels based on predefined thresholds or conditions.

Since different machine learning algorithms are used in the research, it is recommended to scale the data to normalize the range of input features. This method enhances the performance of the models and prevents results from being biased as a result of varying feature scales. Before building predictive models, it is useful to reduce the number of input features. Feature selection techniques are used to refine accuracy, reduce overfitting, and enhance the models' efficacy.

2.13 Dataset Transformation

The dataset involved categorical columns, including Event Name, Low-Level Category, and Policy Name. To utilize these variables in machine learning models, they were converted into numeric values. This conversion allows the algorithms to process and learn from these categorical attributes effectively. By transforming the categorical columns into numerical representations, the dataset became more compatible with the chosen machine-learning techniques. Additionally, 13 additional columns were generated using Python to enhance the analysis and modeling process. These columns were created to provide further insights and features that could contribute to the detection and classification of attacks on IoT devices.

2.14 Data Labeling

Initially, the data was unlabeled, meaning it did not have predefined labels indicating whether an event or network activity was malicious or normal. To address this, the identified indicators within the dataset were explored and used to label the data and apply classifiers. A score column was created, and an "if" statement was applied to compare indicator values, enabling the assignment of labels based on predefined thresholds or conditions. This labeling process allowed for the supervised learning algorithms to learn from labeled data and classify new instances accurately.

2.15 Data Scaling

Given the utilization of different machine learning algorithms in the research, data scaling was recommended. Data scaling is a technique used to normalize the range of input features. By scaling the data, the features are brought to a similar scale, ensuring that no single feature dominates the learning process. This technique can improve the model's performance and prevent biased results due to feature scales.

2.16 Feature Selection

Feature selection is an important step to reduce the number of input features or variables in the dataset. By reducing the feature space, training time can be decreased, model performance can be improved, and over-fitting can be mitigated. In this research, the feature selection process aimed to refine accuracy and reduce overfitting.

2.17 Univariate Feature Selection

Univariate feature selection is a statistical method employed to assess the relationship strength between each feature and the target variable. This method evaluates each feature independently, considering its impact on the target variable. By selecting features that exhibit strong relationships with the target variable, the dataset can be optimized for model training and prediction.

2.18 Statistical Sampling

To ensure the creation of reliable and predictive models, a statistical sampling technique called Synthetic Minority Over-Sampling Technique (SMOTE) was employed. SMOTE is an over-sampling technique that balances the dataset by generating synthetic samples of the minority class (Xiaolong et al. 19). This technique takes the difference between a feature vector and its nearest neighbor, multiplies it by a random number between 0 and 1, and adds the result to the original feature vector. By applying SMOTE, a more balanced dataset was created, enabling the machine learning models to better handle imbalanced class distribution and improve the quality of predictions.

2.19 Data Limitations and Issues

Due to sensitivity and privacy concerns, it was challenging to obtain information from the agency. The relatively small size of the collected datasets, which consisted of only 10,000 records, was another limitation. This limited data set may not adequately represent the diversity and complexity of actual IoT network traffic and attack scenarios. However, several strategies were employed to address the study's constraints. For instance, the implementation of data quality assurance procedures ensures the accuracy and completeness of data through rigorous cleaning and validation. In addition, feature engineering and selection techniques were applied, utilizing domain knowledge to select appropriate characteristics and dimensionality reduction techniques to effectively manage the limited dataset. By acknowledging these limitations and implementing appropriate measures, the research aimed to reduce potential biases and improve the validity and reliability of the findings.

3 Results and Discussion

3.1 Introduction

This section presents and analyzes the study's findings after supervised machine learning methods have been conducted and applied to the dataset. The outcomes of the techniques and their evaluation methods are discussed. The analysis was conducted using the following procedures:

- Different preprocessing methods were applied to the data and the dataset. These methods included feature selection to select the most important features before modeling, SMOTE for class im-balance, overfitting, and bias-variance tradeoff.
- Additional machine learning classifiers were applied to multiple classes of data using DT, RF, and RL techniques to classify and identify suspicious traffic. Python was the programming language used. Each classifier's results were evaluated.
- Machine learning classifiers were applied to a multi-class of data using LR, DT, and RF, and techniques and Python programming language. The results of the classifiers were measured.

Before modeling, a simple analysis was conducted to gain a general understanding of the datasets, beginning with the Data understanding section. In Data Prepressing, the prepared datasets were inputs to the machine learning algorithms. In Modeling and Evaluation, the various experiments conducted during the research were presented.

3.2 Exploratory Data Analysis (EDA)

The collected datasets were analyzed to obtain the full picture concerning the existing data and to understand each column. The results are presented in the following sections.

2. Check for missing values in a DataFrame

```
missing_values = data.isnull().sum()
print(missing_values)
```

```
Event Name                          0
Start Time                          0
Low Level Category                  0
Source Port                         0
Destination Port                    0
Policy Name (custom)                0
Request (custom)                  232
Request URI (custom)               11
Request URI (custom).1             11
User-Agent (custom)               510
Violation Signatures (custom)    8159
Violation Type (custom)          6782
Web Application Name (custom)       0
```

Fig. 3. The missing value table.

3.3 Dataset

The dataset analysis considered the presence of null values in the records. Specifically, there were 13 columns in the dataset that contained null values. These columns and their corresponding records are outlined in Fig. 3.

Data Cleaning.

Dealing with null values. After analyzing the dataset, it was observed that several columns contained a significant number of null values. To ensure the quality and reliability of the data for machine learning algorithms, a decision was made to remove columns with a large number of null values. The following columns were eliminated from the dataset:

- User-Agent (custom): This column had 510 null cells. User-agent typically refers to the software or application used to request in the context of an event. Since this column had a significant number of missing values and may not directly contribute to the detection and classification of attacks, it was deemed not useful for deploying machine learning algorithms.
- Violation Type (custom): This column had 6782 null cells. Violation Type represents a custom-defined type or category indicating the type of violation or anomaly. Similar to the User-Agent column, the high number of missing values and the limited relevance of this attribute for attack detection led to its removal from the dataset.
- Violation Signatures (custom): This column had 8159 null cells. Violation Signatures refer to cus-tom-defined signatures or patterns indicating a violation or anomaly. Due to the large proportion of missing values and the potential lack of significance for attack detection, this column was eliminated.

- Request URI (custom): This column had 11 null cells. Request URI represents a custom-defined Uniform Resource Identifier associated with the request. Although the number of missing values was relatively small, it was decided to remove this column to simplify the dataset and focus on more relevant features.
- Request (custom): This column had 232 null cells. The request represents a custom-defined request made in the context of an event. Similar to the other columns, the high number of missing values and the potential limited impact on attack detection led to the decision to eliminate this column.

By removing these columns, the dataset was streamlined and the noise was reduced. This process aimed to enhance the performance of the machine learning model by focusing on more informative and impactful features. It also facilitated the handling of unstructured text data, as removing these columns eliminated the complexities associated with analyzing and incorporating such data into the model.

3.4 Enriching Data

In the data preprocessing stage, additional columns were added to the dataset to provide more detailed information for analysis and modeling purposes. These additional columns were generated using Python programming language and represent different low-level categories of attacks. Each column corresponds to a specific attack category, allowing for a more fine-grained analysis of the data. The added columns are as follows:

- Low-Level Category Buffer Overflow: This column indicates the occurrence of buffer overflow attacks, which happen when a program or system tries to store more data in a buffer than it can handle. Buffer overflow attacks can lead to system vulnerabilities and potential exploitation by attackers.
- Low-Level Category Code Injection: This column represents instances of code injection attacks, where malicious code is injected into a system or application to execute unauthorized commands or actions. Code injection attacks can allow attackers to gain unauthorized access or control over a system.
- Low-Level Category Command Execution: This column signifies the presence of command execution attacks, where unauthorized commands or scripts are executed on a target system. Command execution attacks can lead to unauthorized access, data manipulation, or system compromise.
- Low-Level Category Compliance Violation: This column indicates violations of compliance regulations or policies. Compliance violations may involve unauthorized access to sensitive data, breaches of privacy regulations, or non-compliance with security protocols.
- Low-Level Category Cross-Site Scripting: This column represents occurrences of cross-site scripting (XSS) attacks, where malicious scripts are injected into web pages viewed by other users. Cross-site scripting attacks can allow attackers to steal sensitive information or perform unauthorized actions on websites.
- Low-Level Category HTTP in Progress: This column indicates ongoing HTTP attacks, where the Hypertext Transfer Protocol (HTTP) is exploited to carry out different types of attacks. This can include HTTP parameter pollution, request smuggling, or other malicious activities.

- Low Level Category Information Leak: This column denotes instances of information leakage, where sensitive or confidential data is unintentionally or maliciously disclosed. Information leaks can lead to privacy breaches, data exposure, or unauthorized access to sensitive information.
- Low-Level Category Invalid Command or Data: This column represents the presence of invalid commands or data in network communication. It can indicate attempted attacks or abnormal behavior where commands or data do not conform to expected standards or protocols.
- Low-Level Category_Misc DoS: This column indicates miscellaneous Denial of Service (DoS) attacks, which involve overwhelming a target system or network with a high volume of requests or malicious traffic. DoS attacks aim to disrupt or deny access to services for legitimate users.
- Low-Level Category_Misc Exploit: This column signifies miscellaneous exploitation attempts or attacks that do not fall into specific categories but still pose a threat to the security of the system or network. These could include novel or unknown attack techniques.
- Low-Level Category_SQL Injection: This column represents instances of SQL injection attacks, where malicious SQL code is inserted into a web application's database query. SQL injection attacks can allow attackers to manipulate or retrieve sensitive data from the database.
- Low-Level Category Trojan Detected: This column indicates the detection of Trojan horse malware. Trojans are malicious software that disguises themselves as legitimate programs and can enable unauthorized access, data theft, or system control.
- Low Level Category Unknown Evasion Event: This column represents unknown or unidentified evasion events, where attackers attempt to bypass security measures or conceal their malicious activities. Unknown evasion events can indicate sophisticated or novel attack techniques.
- Low-Level Category Web Exploit: This column denotes the presence of web-based exploits, including vulnerabilities or weaknesses in web applications that can be exploited to gain unauthorized access, manipulate data, or compromise the system.

By adding these additional columns, the dataset was enriched with more specific information about different attack categories. This can help in the analysis and modeling process by capturing the nuances and patterns associated with various types of attacks, ultimately improving the accuracy and effectiveness of the models. The resulting heatmap shows the correlation between different columns, with higher values indicating stronger correlations. This can help identify relationships and patterns in the data.

The results from the correlational matrix analysis are important because they help us understand how different variables are related to each other. This gives us clues about what we should focus on and learn more about. To visually represent these relationships, we use a heatmap. A heatmap is like a colorful map depicted in Fig. 4 shows how strongly different columns in our dataset are related. we can see different colors that indicate how strongly the columns are related. Higher values and darker colors mean stronger correlations. This helps us identify any patterns or relationships in the data, which can be very useful for further analysis and understanding.

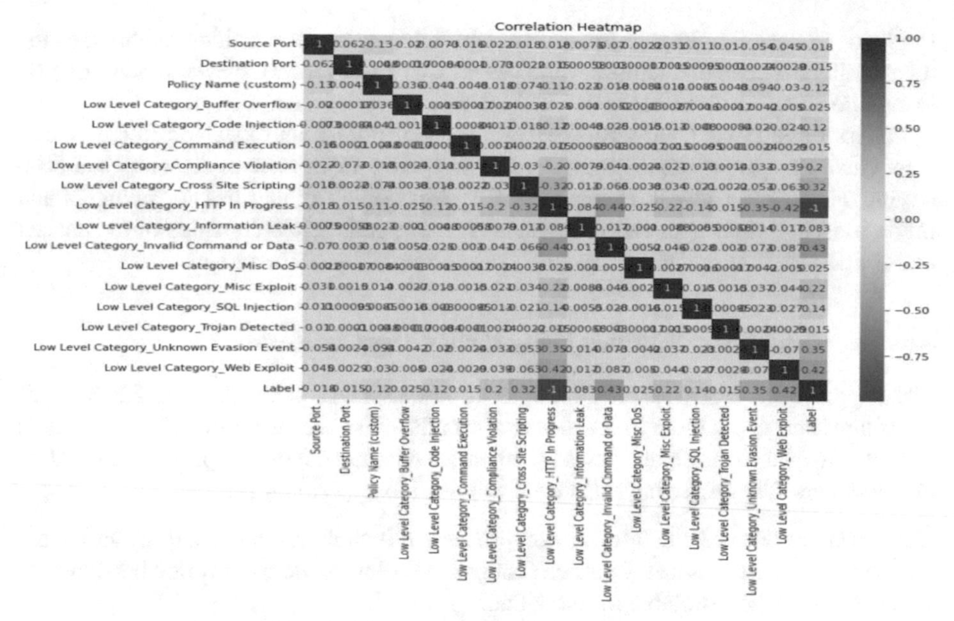

Fig. 4. The heatmap correlational matrix.

3.5 Brief Description of Models

Decision Trees (DTs). Decision trees are versatile machine-learning algorithms used for classification and regression tasks (Bari et al. 6). They recursively split the data based on different features, creating a tree-like structure where each internal node represents a feature-based decision and each leaf node represents the final prediction. Decision trees are capable of handling both categorical and numerical features and can capture complex relationships within the data. However, they are prone to overfitting, which can be mitigated using pruning techniques or by combining multiple decision trees into an ensemble model.

Random Forest (RF). Random Forest is an ensemble machine learning model that combines multiple decision trees to make predictions (Dai et al. 10). It uses the bagging technique, where each decision tree is trained on a random subset of the data. When making predictions, the final output is determined by aggregating the predictions of all individual decision trees (e.g., majority voting for classification). Random Forest helps reduce overfitting and variance, improves generalization, and provides robust predictions. It is particularly effective when dealing with high-dimensional datasets or when there are complex relationships between features.

Logistic Regression (LR). Logistic regression is a statistical modeling technique commonly used for binary classification tasks (Kirasich et al. 12). It estimates the probability of an event occurring based on predictor variables. Logistic regression applies the sigmoid function to the linear combination of predictors, mapping the result to a value between 0 and 1, representing the probability of the event (Alam et al., 2019). It assumes a linear relationship between predictors and the logarithm of the odds of the event.

Logistic regression is known for its interpretability and can provide insights into the relationship between predictors and the likelihood of an event. However, it may struggle to capture complex non-linear relationships present in the data.

These three models—Decision Trees, Random Forest, and Logistic Regression—were chosen as part of the research to evaluate their performance in detecting and classifying suspicious network events in IoT security. Each model has its strengths and limitations, and their application depends on the specific characteristics of the dataset and the desired interpretability, accuracy, and complexity of the model.

3.6 Experiment: Classification Using Multi-Class Classification

The objective of this experiment was to build classifiers using the DT, RF, and LR algorithms and apply them to the dataset for multi-class classification. The goal was to allocate the data to multiple classes based on the characteristics and patterns learned by the classifiers. The following steps were followed to experiment.

Data Preparation. The dataset was preprocessed, including data cleaning and feature engineering steps as discussed earlier. Categorical columns were converted into numeric values to make them suitable for modeling.

Feature Selection.

- Before building the classifiers, feature selection techniques were applied to reduce the dimensionality of the dataset and focus on the most informative features.
- Univariate feature selection, which tests the relationship strength between each feature and the target variable, was performed to select the most relevant features.

Model Building. The DT, RF, and LR algorithms were implemented to build the classifiers.

- Decision Trees: Decision tree classifiers were constructed to partition the data based on features and make predictions.
- Random Forest: Random forest classifiers were created by combining multiple decision trees and aggregating their predictions.
- Logistic Regression: Logistic regression models were trained to estimate the probabilities of different classes based on the input features.

Model Training. Figure 5 shows the performance accuracies of the exploited models.

- The dataset was divided into training and testing sets using the K-fold cross-validation technique, which ensures that every sample appears in both training and test sets.
- The classifiers were trained on the training set and evaluated on the testing set to assess their performance. The accuracy results demonstrate the performance of each model in correctly classifying instances.
- Performance evaluation metrics such as precision, recall, and F1-score were calculated to measure the effectiveness of the classifiers in predicting the correct classes.

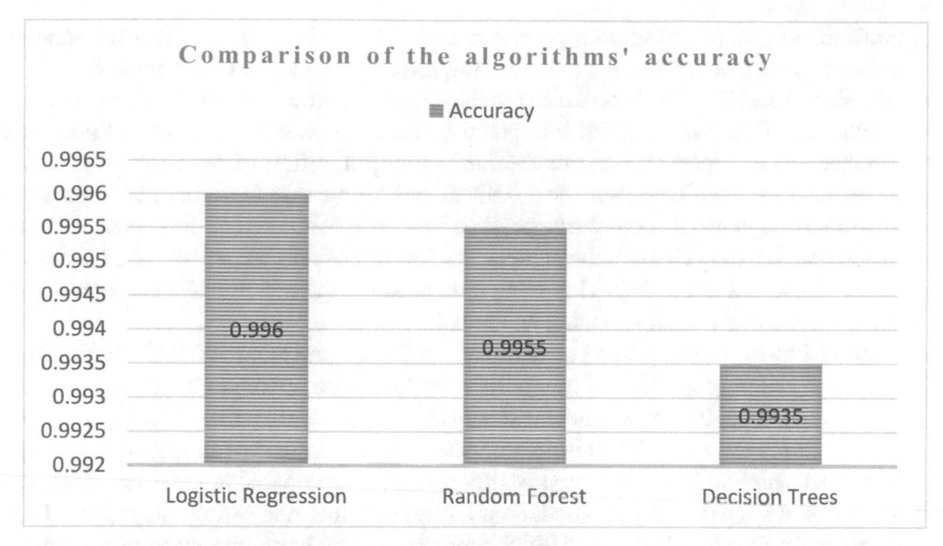

Fig. 5. The performance accuracies of the exploited models.

- The Logistic Regression model achieved a high accuracy of 0.996. This indicates that the model was able to correctly classify the majority of instances in the test data with a high degree of accuracy. The Logistic Regression model shows promise in accurately predicting the target variable based on the given features.
- The Decision Trees model achieved an accuracy of 0.9935. This implies that the model performed well in accurately classifying the test instances, although slightly lower than the Logistic Regression model. Decision Trees demonstrate their ability to capture complex decision boundaries and classify instances effectively.
- The accuracy of the Random Forest model was 0.9955. In terms of accuracy, this result suggests that the Random Forest model performed similarly to the Logistic Regression model and outperformed the Decision Trees model. Random Forest combines multiple decision trees to make predictions, resulting in enhanced generalization and robustness.

Model Evaluation. Based on these accuracy results, each of the three models performed well in correctly classifying instances. For a comprehensive analysis of model performance, it is essential to consider additional evaluation metrics and factors, such as precision, recall, F1-score, computational efficiency, and scalability, as described below.

- Logistic Regression: The Logistic Regression model achieved a precision of 0.996, indicating that it correctly identified a high proportion of true positive instances compared to the total instances predicted as positive. This suggests that the model has a low rate of false positives. The recall for Logistic Regression is also 0.996, indicating that the model successfully identified a high proportion of true positive instances compared to the actual positive instances. It suggests that the model has a low rate of false negatives. The F1 score, which combines precision and recalls into a single metric, is 0.996 for Logistic Regression. This indicates a high overall performance, considering both precision and recall, which implies that the model strikes a good balance between minimizing false positives and false negatives.

- Decision Trees: The Decision Trees model achieved a precision of 0.9935, indicating a relatively high proportion of true positive instances relative to predicted positive instances. It indicates that the model has a low proportion of false positives. The recall for Decision Trees was also 0.9935, indicating that the model successfully identified a high proportion of true positive instances compared to the actual positive instances. This indicates that the model had a low rate of false negatives. The Decision Trees F1 score was 0. 9935. This score indicates that the model had a strong overall performance in terms of precision and recall.
- Random Forest: The Random Forest model achieved a precision of 0.9955, indicating a relatively high proportion of true positive instances compared to the instances predicted as positive. It suggests that the model has a relatively low rate of false positives. The recall for Random Forest is also 0.9955, indicating that the model successfully identified a high proportion of true positive instances compared to the actual positive instances. It suggests that the model has a relatively low rate of false negatives. The F1 score for Random Forest is 0.9955, representing the harmonic mean of precision and recall. This score indicates a strong overall performance of the model in terms of precision and recall.

Based on these evaluation metrics, all three models (Logistic Regression, Decision Trees, and Random Forest) demonstrate strong performance. They achieve high precision, recall, and F1 scores, indicating their ability to accurately classify instances and strike a balance between minimizing false positives and false negatives. These findings highlight the effectiveness of the models in capturing the underlying patterns and distinguishing between different classes or categories.

However, it is essential to consider other factors such as computational efficiency, scalability, and domain-specific requirements when interpreting the results and selecting the most appropriate model for the specific problem and dataset. To determine the best model among logistic regression, decision trees, and random forest classifiers based on their accuracy scores, we compare the accuracy scores and identify the highest one. The highest accuracy score and the corresponding model name were extracted using tuple unpacking. Finally, the best model name and accuracy are printed using 'print()' statements.

In Fig. 6, we were able to identify the best model among the three based on the highest accuracy score achieved.

Comparative Analysis. The performance of the DT, RF, and LR classifiers was compared based on the evaluation metrics. Factors such as accuracy, precision, recall, and F1-score were considered to determine the strengths and weaknesses of each algorithm in multi-class classification.

```
[57]: best_model = max([(logreg_accuracy, "Logistic Regression"), (dt_accuracy, "Decision Trees"), (rf_accuracy, "Random Forest")])
      best_accuracy, best_model_name = best_model
      print("Best Model:", best_model_name)
      print("Best Accuracy:", best_accuracy)

      Best Model: Logistic Regression
      Best Accuracy: 0.996
```

Fig. 6. The overall accuracy of best performing models among exploited models.

4 Discussion and Conclusion

The results of the experiment provided insights into the performance of the DT, RF, and LR classifiers in allocating the data to multiple classes. The evaluation metrics assisted in determining the accuracy of each algorithm's classification predictions and its ability to recognize data patterns. Accuracy, precision, recall, and F1-score were taken into account in the comparative analysis of the classifiers. Each algorithm's performance was evaluated based on its ability to solve the multi-class classification problem and make accurate predictions across classes. The results analysis provided valuable insight into the strengths and weaknesses of each algorithm. It also assisted in the identification of the most efficient algorithm (s) for multi-class classification in the context of IoT security and attack detection. Overall, the purpose of this experiment was to investigate the capabilities of the DT, RF, and LR algorithms for multi-class classification and to gain insight into their performance when assigning data to multiple classes. The results of this experiment contribute to the comprehension of these algorithms' suitability for IoT security.

4.1 The Contribution of the Research

In the context of IoT device security, the research contributed to an understanding of data preprocessing techniques, feature engineering, and model evaluation. The correlation analysis and heatmap visualization provide valuable insights into the relationships between various variables and highlight potential patterns and trends in the data. Overall, the research contributes to the field by providing actionable knowledge and recommendations for enhancing the security of IoT devices through the application of machine learning techniques.

4.2 Challenges and Limitations in the Research

This study suffered several limitations. One of these challenges is data limitations. The availability and quality of the data used for this research may have limitations, such as the size of the dataset and its ability to accurately represent real-world IoT network traffic and attack scenarios. This may impact the generalizability of the results. Similarly, the research was conducted with the limited resources of IoT devices, including computational power and memory, in mind. Although this is a possible limitation, it might limit the performance and scalability of machine learning algorithms. Due to privacy and sensitivity concerns, obtaining data from agencies or organizations for research purposes, particularly in the domain of IoT security, can be difficult. This challenge restricted access to exhaustive and diverse datasets. Similarly, the complexity of Attacks limited this study. Briefly, the sophistication and complexity of attacks in IoT environments pose detection and classification challenges. It is possible that the research was unable to account for all potential attack scenarios, and future work could focus on overcoming this limitation.

4.3 Recommendations for Future Work

This study paves the way for various future studies. Among the possible areas for future research is the exploration of advanced machine learning techniques, such as deep learning, to enhance the precision and resiliency of attack detection in IoT environments. Deep learning models, such as convolutional neural networks (CNNs) and recurrent neural networks (RNNs), may provide improved performance in identifying complex patterns and behaviors. Similarly, there is a need to study Real-time Monitoring and Response. The development of real-time monitoring systems that can detect and respond to attacks in real-time, thereby enabling proactive security measures. This may involve integrating streaming data processing techniques and developing efficient algorithms for real-time analysis.

Further, future scholars may focus on Adversarial Attack Defense. Investigating techniques to detect and mitigate adversarial attacks specifically directed at machine learning-based IoT security systems. The development of robust defenses against adversarial attacks is crucial. These attacks aim to deceive or manipulate models. Additionally, future research should explore the integration of multimodal data sources, such as sensor data, network logs, and device logs, to improve detection capabilities and provide a holistic view of IoT device security. Moreover, studies in the future may concentrate on Real-World Deployments. Conducting evaluations and experiments in real-world IoT deployments to validate the efficacy and applicability of the developed models. This would require collaboration with industry partners and field trials in diverse IoT environments. Lastly, future research can build upon the findings of this thesis and contribute to the advancement of IoT device security by addressing these recommendations. There are plenty of possibilities for further exploration and innovation because the field is dynamic and evolving.

References

Adat, V., Gupta, B.B.: Security in internet of things: issues, challenges, taxonomy, and architecture. Telecommun. Syst. **67**, 423–441 (2018)

Allam, A., Nagy, M., Thoma, G., Krauthammer, M.: Neural networks versus Logistic regression for 30 days all-cause readmission prediction. Sci. Rep. **9**(1), 9277 (2019)

Alqarni, H., Alnahari, W., Quasim, M.T.: Internet of things (IoT) security requirements: Issues related to sensors. In: 2021 National Computing Colleges Conference (NCCC), pp. 1–6. IEEE (2021)

Alsharif, M., Rawat, D.B.: Study of machine learning for cloud assisted iot security as a service. Sensors **21**(4), 1034 (2021)

Arshad, A., et al.: A novel ensemble method for enhancing internet of things device security against botnet attacks. Decis. Anal. J. **8**, 100307 (2023)

Bari Antor, M., et al.: A comparative analysis of machine learning algorithms to predict Alzheimer's disease. J. Healthc. Eng. **2021** (2021)

Bernard, S., Heutte, L., Adam, S.: On the selection of decision trees in random forests. In: 2009 International Joint Conference on Neural Networks, pp. 302–307. IEEE (2009)

Bharadiya, J.: Machine learning in cybersecurity: techniques and challenges. Eur. J. Technol. **7**(2), 1–14 (2023)

Boateng, E.Y., Abaye, D.A.: A review of the logistic regression model with emphasis on medical research. J. Data Anal. Inf. Proc. **7**(4), 190–207 (2019)

Dai, B., Chen, R.C., Zhu, S.Z., Zhang, W.W.: Using random forest algorithm for breast cancer diagnosis. In: 2018 International Symposium on Computer, Consumer and Control (IS3C), pp. 449–452. IEEE (2018)

Farid, D.M., Rahman, M.M., Al-Mamuny, M.A.: Efficient and scalable multi-class classification using naïve Bayes tree. In: 2014 International Conference on Informatics, Electronics & Vision (ICIEV), pp. 1–4. IEEE (2014)

Kirasich, K., Smith, T., Sadler, B.: Random forest vs logistic regression: binary classification for heterogeneous datasets. SMU Data Sci. Rev. **1**(3), 9 (2018)

Mahmud, S.H., Hossin, M.A., Jahan, H., Noori, S.R.H., Bhuiyan, T.: CSV ANNOTATE: generate annotated tables from CSV file. In: 2018 International Conference on Artificial Intelligence and Big Data (ICAIBD), pp. 71–75. IEEE (2018)

Makkar, A., Garg, S., Kumar, N., Hossain, M. S., Ghoneim, A., Alrashoud, M.: An efficient spam detection technique for IoT devices using machine learning. IEEE Trans. Ind. Inform. **17**(2), 903–912 (2020)

Meidan, Y., et al.: Detection of unauthorized IoT devices using machine learning techniques. arXiv preprintarXiv:1709.04647 (2017)

Pramanik, P.K.D., Pal, S., Choudhury, P. (2018). Beyond automation: the cognitive IoT. artificial intelligence brings sense to the internet of things. In: Sangaiah, A., Thangavelu, A., Meenakshi Sundaram, V. (eds.) Cognitive Computing for Big Data Systems Over IoT. Lecture Notes on Data Engineering and Communications Technologies, vol. 14, pp. 1–37. Springer, Cham (2018). 10.1007/978-3-319-70688-7_1

Talwana, J.C., Hua, H.J.: Smart world of internet of things (IoT) and its security concerns. In: 2016 IEEE International Conference on Internet of Things (iThings) and IEEE Green Computing and Communications (GreenCom) and IEEE Cyber, Physical and Social Computing (CPSCom) and IEEE Smart Data (SmartData), pp. 240–245. IEEE (2016)

Tschang, F.T., Almirall, E.: Artificial intelligence as augmenting automation: implications for employment. Acad. Manage. Perspect. **35**(4), 642–659 (2021)

Xiaolong, X.U., Wen, C.H.E.N., Yanfei, S.U.N.: Over-sampling algorithm for imbalanced data classification. J. Syst. Eng. Electron. **30**(6), 1182–1191 (2019)

Crime Prediction Using Machine Learning

Hneah Guey Ling[1], Teng Wei Jian[1], Vasuky Mohanan[1], Sook Fern Yeo[2,3](✉), and Neesha Jothi[4]

[1] INTI International College Penang, Bayan Lepas, Malaysia
[2] Faculty of Business, Multimedia University, Jalan Ayer Keroh Lama, 75450 Melaka, Malaysia
yeo.sook.fern@mmu.edu.my
[3] Department of Business Administration, Daffodil International University, Dhaka 1207, Bangladesh
[4] Department of Computing, College of Computing and Informatics, Universiti Tenaga Nasional (UNITEN), Putrajaya Campus, Kajang, Malaysia

Abstract. The widespread occurrence of criminal activities poses a substantial threat to public safety and property. Hence, the proactive prediction of crimes is vital as it empowers law enforcement agencies to make decisions on resource allocation and targeted interventions based on the data, ultimately leading to a more secure and protected community. Additionally, such initiatives raise public awareness, encouraging vigilance during periods of heightened criminal activity. In this project, machine learning techniques are leveraged to forecast the crime rate in the city of Chicago. This research introduces a more efficient data preparation method, optimizing data representation to enable machine learning models to capture patterns and learn from the information provided effectively. After training the models using LightGBM, XGBoost, CatBoost, and Gradient Boosting, the models achieved R^2 scores of 0.8086, 0.8088, 0.8094, and 0.8084, respectively. An ensemble method combining these individual models was implemented to improve the prediction performance. Through the voting ensemble method, the final R^2 score for crime rate prediction was enhanced to 0.8104.

Keywords: Machine Learning · Time-Series Forecasting · Crime Prediction

1 Introduction

Crime is a prevalent social issue that has garnered substantial global attention due to its widespread impact and serious consequences. In the year 2022, the Federal Bureau of Investigation (FBI) received approximately 7.78 million reported crime cases in the United States [1]. A high crime rate may cause detrimental effects in the economic, social, and psychological aspects such as strained resources, decreased economic productivity, and heightened fear and insecurity of individuals [2].

Similar to classification or prediction in areas like music [3] and software defect [4], there are many research papers related to predicting crime using traditional methods, such as the research conducted by Masron [5], Pathak and Hossain Tasin [6]. However, employing traditional methods such as statistical methods, manual data analysis, and

J. Rasheed et al. (Eds.): FoNeS-AIoT 2024, LNNS 1035, pp. 92–103, 2024.
https://doi.org/10.1007/978-3-031-62871-9_8

rule-based systems for crime prediction proves challenging in effectively leveraging large datasets. These methods can be time-consuming and may not fully capture the complex and dynamic nature of crime patterns [7]. The traditional methods are unable to provide reliable insights for law enforcement agencies to take action accordingly. Therefore, there is a need to develop advanced machine-learning models to enhance the utilization of historical crime data to yield improved results [8].

This project aims to predict the crime rate at a specific spatial-temporal point using machine learning. Hence, law enforcement agencies can allocate resources proactively and implement targeted interventions based on the models. Law enforcement agencies can strategically determine the timing and locations that require heightened patrolling efforts and conduct campaigns at high-risk areas to raise the awareness of the public. This is vital to enhance public safety and create a peaceful society.

2 Related Works

This study analyzes current research on crime prediction using machine learning. It involves a comprehensive comparison of techniques utilized by various authors, evaluating their performance. The aim is to gain a deeper understanding of this domain while identifying potential solutions to address existing gaps in the literature.

Al Amin Biswas and Basak [9] with the research titled "Forecasting the Trends and Patterns of Crime in Bangladesh using Machine Learning Model" focuses on the prediction of crime rate using regression. The dataset was acquired from the Bangladesh Police's website. Linear regression (LR), polynomial regression (PR), and RF are implemented to predict the crime count. The R^2 scores are 0.82, 0.95, and 0.89, respectively.

Aziz et al. [10] conducted research with the title "Machine Learning-based Soft Computing Regression Analysis Approach for Crime Data Prediction". The dataset is obtained from the National Crime Records Bureau. The authors implemented regression models using LR, DT, and RF. The R^2 scores are 0.89, 0.57, and 0.96, respectively.

The research titled "Big Data Analytics and Mining for Effective Visualization and Trends Forecasting of Crime Data" conducted by Feng et al. [11] discussed crime prediction using the Prophet, NN, and long short-term memory (LSTM) models. The root-mean-square deviation (RMSE) for Prophet, LSTM, and NN is 48.73, 48.15, and 63.68 respectively for the Philadelphia crime dataset.

Alsayadi et al. [12] improved the performance of crime prediction using ensemble methods in the research titled "Improving the Regression of Communities and Crime Using Ensemble of Machine Learning Models". The authors used five algorithms, which are DT, NN, support vector regression (SVR), RF, and KNN. The R^2 scores are 0.92, 0.959, 0.96, 0.81 and 0.993, respectively. Then, the authors combined these five regression models into an ensemble which yielded a R^2 score of 0.969.

3 Methodology

OSEMN framework is used as the guide for the project. There are five key stages in this framework, including obtain, scrub, explore, model, and interpret. This framework provides a systematic approach to progress through different project stages to ensure that the project stays organized and logical.

3.1 Obtain

The main dataset for this project is acquired from the Chicago Data Portal, an open portal that provides access to a wide range of datasets about various aspects of the Chicago city. This crime dataset is extracted from the Chicago Police Department's Citizen Law Enforcement Analysis and Reporting system. The victims' full addresses in the dataset have been replaced with block-level information due to privacy concerns. There are 7777231 instances and 22 features contained in the dataset. It includes the reported incidents of crime that happened in Chicago from 2001 until 12 April 2023.

The weather data is acquired from the Chicago Data Portal. The data is collected from sensors which are located at beaches along Chicago's Lake Michigan lakefront. There exist three sensor stations, namely the 63rd Street Weather Station, Foster Weather Station, and Oak Street Weather Station. The sensors typically record the specified measurements on an hourly basis. There are 159034 instances and 18 instances in the dataset [13]. The required features for this project are air temperature in degrees Celsius and humidity in percentage.

The socioeconomic data is provided by the U.S. Bureau of Labor Statistics. It records the monthly unemployment rate in Chicago city from 1 January 1990 to 1 June 2023. The dataset contains a total of 403 rows of instances [14].

3.2 Scrub

A total of 991 crime records in the dataset share the same case number, indicating duplicates and have been removed. There are 22 features in the dataset however many of them are irrelevant for crime rate predictions. Hence, only the 'Date' and 'Community Area' features are retained. The 'Community Area' is selected to represent the location of the crime because the 'District' is too broad, making it challenging for the models to identify patterns as there are many variations in the frequency of cases across different parts of the district.

The 'Community Area' in the dataset is observed to have null values. The rows with missing values are deleted directly as 'Community Area' is an important feature and it should not be imputed by other values for data reliability purposes. Rows with primary types 'NON-CRIMINAL', 'NON – CRIMINAL', and 'NON-CRIMINAL (SUBJECT SPECIFIED)' have been removed from the dataset because they do not represent criminal incidents. The removal is necessary as the machine learning models aim to predict crime rates only.

The 'Date' column is converted into datetime using the pd.to_datetime() function to ensure the dates are handled properly. New features, which are 'Day', and 'Month', are then extracted from the 'Date' feature. Only crimes that occurred between 2018 and

2022 have been considered for the machine learning model, as these data points are more recent and relevant for the predictions.

The aggregation function is used to create a new feature, 'Crime Count', which contains the count of crimes for each combination of 'Community Area', 'Day', 'Month', and 'Year'. This feature is the target variable for the machine learning models. Then, the dataset is augmented with missing date entries for each community area and date combination. It ensures that the dataset contains a complete set of dates for each community area, filling in missing dates with a 'Crime_Count' of 0. This means that there is no crime happening at that spatial-temporal point.

Feature engineering is conducted to enhance the model's ability to discover underlying patterns and prevent overfitting. Several novel features are introduced, including the 'Day_of_week' derived from the date, a binary 'is_weekend' and 'Is_Holiday' indicator, and a 'season' feature to capture variations in crime occurrence across different timeframes. These features provide valuable insights into the temporal dynamics of crime, such as the influence of weekends, holidays, and seasonal changes on crime rates.

Furthermore, the feature engineering process includes the creation of 'Crime_Count_Lag364' and 'Crime_Count_Lag5' variables, representing the crime counts from 364 and 5 days ago respectively. The choice of 364 days is particularly relevant as it captures historical data from a year ago on the same day of the week, facilitating the identification of yearly patterns and trends. 'Crime_Count_Lag5' provides the recent history of crime counts. Additionally, a 'Rolling_Crime_Count' feature has been introduced to calculate the average crime count over a month.

The integration of socioeconomic data, 'Unemployment_Rate', adds a valuable contextual dimension to the models, enabling the consideration of economic conditions. Higher unemployment often correlates with elevated crime rates due to increased financial stress and desperation.

'Humidity' and 'Air Temperature' have been incorporated as features in the dataset to account for weather conditions, as these factors can influence crime rates. Research by Habibullah [15] shows that higher temperatures might affect human behavioral decisions and lead to more violent crimes. Missing values of weather data for a given date are imputed using data from the previous day.

The 'Day' and 'Month' features have been combined into a single feature known as 'DayOfYear', where each day is represented with a value ranging from 1 to 366. Therefore, the models are more likely to treat the entire year as a continuous spectrum and detect complex seasonal patterns that might not be apparent when 'Day' and 'Month' are considered as separate features.

'Community Area', 'Is_Holiday', 'is_weekend', and 'season' are categorical variables and need to be encoded. As one-hot encoding represents each category as a separate binary feature, this technique will yield a large data dimension. Besides, the label encoding method assumes the higher the categorical value, the better the category is, which is not suitable in this case. This causes the models to learn incorrect patterns and affect the performance. Therefore, the binary encoding technique is selected. Each category is represented by a unique binary value.

Sine and cosine transformation is utilized for encoding 'DayOfYear' and 'Day_of_week', which are cyclical time features. This method tells the machine time-related information. For example, the connection between January and February is stronger than between January and May.

A standardization technique is applied to scale the features in the dataset. This prevents the model from assigning more importance to features with larger values. All the features are treated equally.

3.3 Explore

A detailed analysis of the dataset is conducted to gain insights into its characteristics and identify the trends.

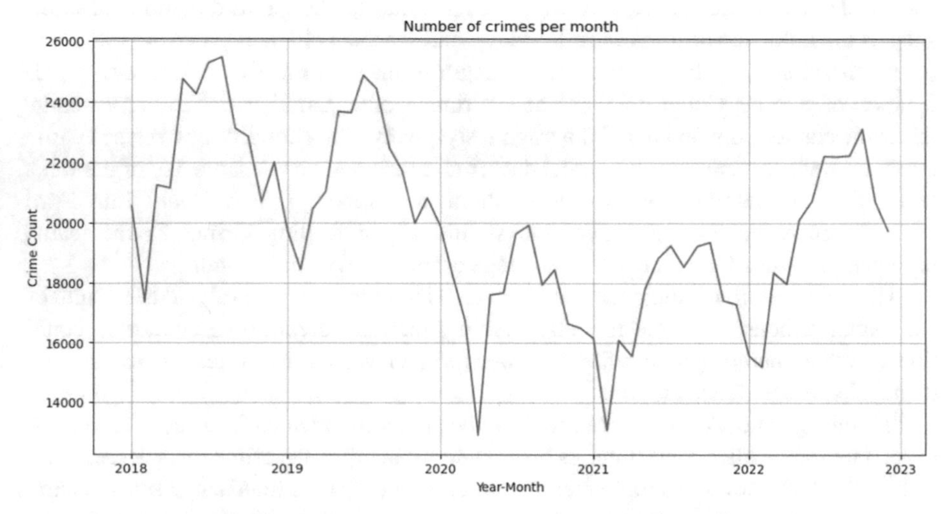

Fig. 1. Line chart for the number of crimes per month

The line chart in Fig. 1 shows the monthly crime rates in Chicago between 2018 and 2022. There is a trend in the data where the crime rates consistently rise monthly each year and are followed by a subsequent decrease.

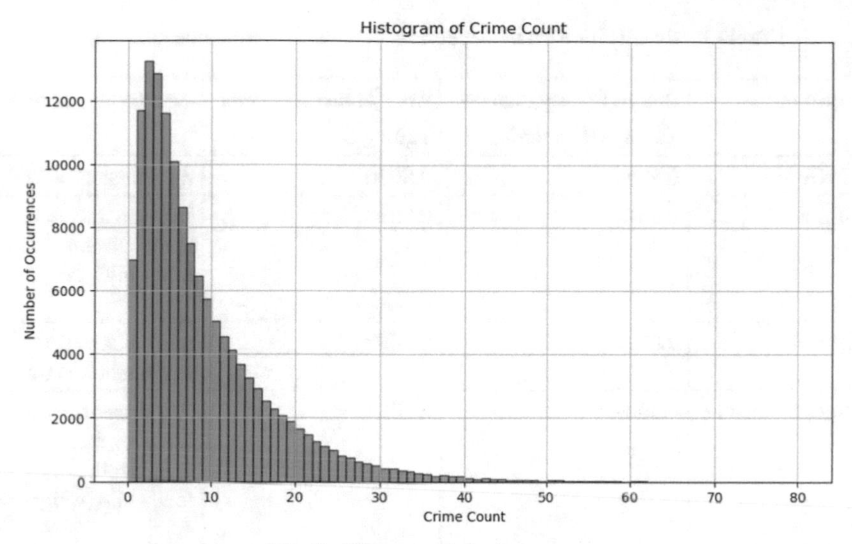

Fig. 2. Histogram of crime count

The histogram shown in Fig. 2 is used to visualize the distribution of the crime count. There are mostly less than 10 crimes for each spatial-temporal point.

3.4 Model

The data is sorted based on time 90% of the previous data is used for training and 10% is for testing. The traditional splitting method which splits data into training and testing sets randomly is not suitable for time-series forecasting tasks. The models will learn information from future data to predict past data, which can be considered unfair and produce overly optimistic results.

CatBoost, XGBoost, LightGBM, and Gradient Boosting are used to build the models for crime rate prediction. The hyperparameters for each model are tuned. Different configurations of settings, such as the learning rate, maximum tree depth, and the number of trees are explored to identify the optimal choice [16]. The predictions from each model are combined to get the final output using the voting ensemble method.

3.5 Interpret

The metrics used to evaluate the crime rate prediction models are R^2 score, mean absolute error (MAE), mean squared error (MSE), and root mean squared error (RMSE).

4 Result

Table 1 above shows the R^2 score of each model before and after hyperparameter tuning. The minimum number of data samples required in a leaf of LightGBM is reduced from 20 to 10, allowing it to capture more complex patterns. The strength of L1 and

Table 1. Results for each model before and after hyperparameter tuning

Algorithm	Before Hyperparameter Tuning (R^2 Score)	After Hyperparameter Tuning (R^2 Score)	Best Hyperparameters
LightGBM	0.8068	0.8086	n_estimators = 200,
XGBoost	0.8012	0.8088	learning_rate = 0.1, num_leaves = 30, min_child_samples = 10,
CatBoost	0.8025	0.8094	reg_alpha = 0.1, reg_lambda = 0.5
Gradient Boosting	0.8009	0.8084	n_estimators = 300, learning_rate = 0.1, max_depth = 5, min_child_weight = 4, subsample = 0.9, colsample_bytree = 0.9, gamma = 3

L2 regularization is increased to prevent the model from becoming too complex and overfitting. These hyperparameters lead to a result of 0.8086. For XGBoost, the number of estimators is increased while the learning rate is decreased to provide the model with more iterations to learn and ensure that it takes smaller steps toward optimizing the loss function. Gamma is applied to regularize the tree and the final result is 0.8088.

The border_count of the CatBoost algorithm has been increased from its default value of 254 to 512. This means that the feature will be divided into more bins while training and able to capture more detailed relationships. The R^2 score for the CatBoost algorithm is 0.8094. The learning rate, max_depth, and min_samples_split are increased and the final result for gradient boosting is 0.8084.

Table 2. Results for voting ensemble model

	Training	Testing
R^2 Score	0.8340	0.8104
MAE	2.2459	2.5712
MSE	10.2175	12.3263
RMSE	3.1965	3.5109

The four models are combined to form an ensemble. Table 2 above shows the performance metrics for the voting ensemble model. The training R^2 score is 0.8340 and the testing R^2 score is 0.8104. This model is the final model for the crime rate prediction.

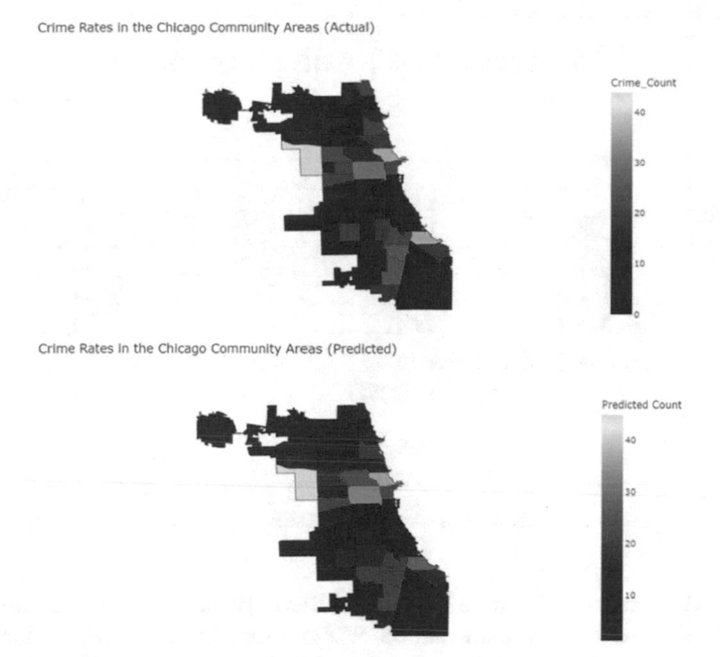

Fig. 3. Comparison between actual and predicted crime rate

Figure 3 above shows separate heatmaps for the actual and predicted crime rates in Chicago on 1 October 2022. The crime hotspots are predicted successfully using the voting ensemble model.

5 Discussion

In this section, the performance of the proposed model will be compared with the previous studies. The first part of this comparison will focus on the R^2 score, where the R^2 score of the proposed model will be compared with those reported in the works by Al Amin Biswas and Basak [9], Aziz et al. [10], and Alsayadi et al. [12]. The second part will center on the RMSE, and the RMSE of the proposed model will be compared with the value reported in the model developed by Feng et al. [11].

Figure 4 displays a comparison of testing R^2 score. The model proposed in this study is evaluated against the best model from three other works. The R^2 score of the proposed model is 0.81, while the models created by Al Amin Biswas and Basak [9], Aziz et al. [10], and Alsayadi et al. [12] are 0.95, 0.96, and 0.99, respectively.

The existing crime prediction models lack temporal granularity, as the predictions are based on a year-wise basis. Yearly crime rate prediction is unable to capture certain seasonal patterns. For example, an increase in crimes during the summer months when more people are outdoors and socializing. This limitation also hinders law enforcement agencies from obtaining precise insights into crime rates at specific times and identifying when additional patrols are required as the predicted crime rate reflects the cumulative rate throughout the year. This affects the effectiveness of resource planning.

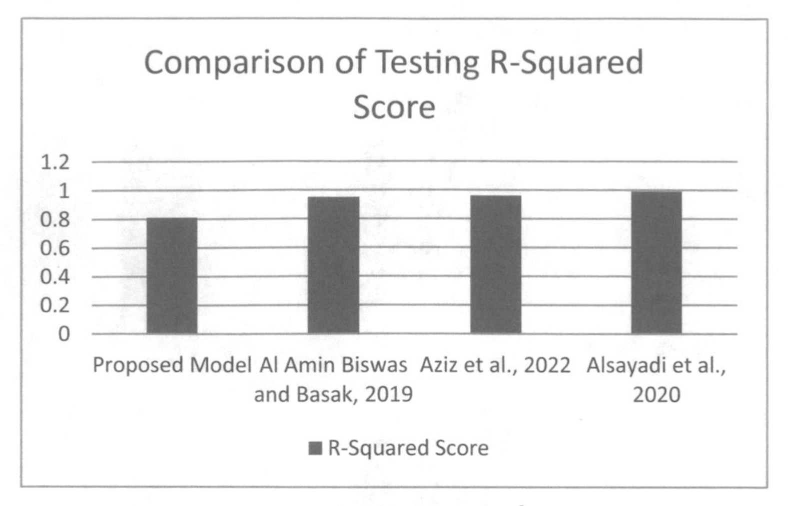

Fig. 4. Comparison of testing R^2 score

The model created by Al Amin Biswas and Basak [9] has a low spatial granularity. It predicts the crime rate for the entire nation. Without specific details about which areas of the country are experiencing higher crime rates at certain times, it is difficult to allocate resources to the areas that need the resources the most.

Therefore, the proposed model provides a higher spatial and temporal granularity, ensuring that the insight is meaningful and capable of capturing crime patterns more efficiently. The crime rate for a specific date and community area can be predicted using the proposed model.

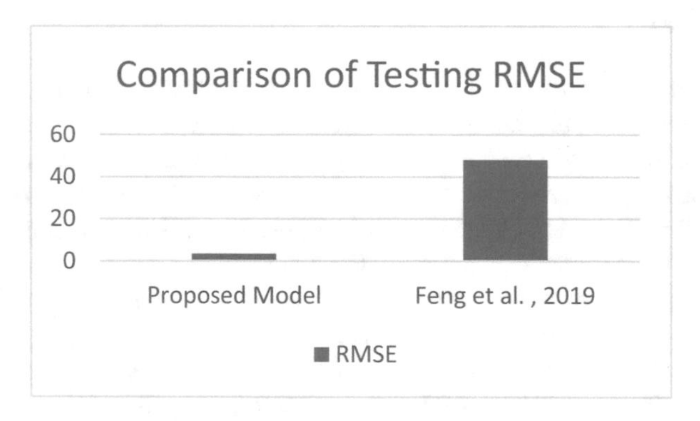

Fig. 5. Comparison of testing RMSE

Figure 5 shows the comparison between the testing RMSE of the proposed model and the model developed by Feng et al. [11]. The proposed model provides a lower RMSE, which is 3.51 compared to 48.15. This suggests that the predictions made by the proposed model are closer to the actual values, reflecting a better prediction.

Table 3. Comparison of testing accuracy with and without feature augmentation

	Original	With feature augmentation
LightGBM	0.7708	0.8086
XGBoost	0.7703	0.8088
CatBoost	0.7707	0.8094
Gradient Boosting	0.7446	0.8084
Ensemble Model	0.7701	0.8104

Table 3 shows the comparison of testing accuracy with and without feature augmentation. New features are incorporated into the dataset through feature engineering and data integration to provide the models with more patterns to learn and avoid overfitting.

The following features are added to the dataset:

1. Weekend: This is a binary indicator of whether the date is a weekend or not. Crimes often exhibit different patterns on weekdays as opposed to weekends. For example, crime rates are lower when people are at work or school during weekdays.
2. Season: The models can capture seasonal patterns and variations in crime rates.
3. Lag Values: The historical crime counts from 5 days and 364 days ago are incorporated into the dataset. A lag feature for 5 days ago provides the models with insights into short-term patterns. The lag value from 364 days ago helps the models in capturing long-term seasonal patterns. The yearly trend can be identified, allowing the models to determine whether crime rates are typically high or not during specific times based on trends from previous years.
4. Rolling Crime Count: The rolling average of crime counts over a month provides a more stable representation of the recent crime trend to the models. The noises from daily fluctuations are smoothed out.
5. Unemployment Rate: Economic factors can influence crime rates, with crime often increasing during periods of economic hardship.
6. Holiday: This feature indicates whether the specific date is a holiday. Holidays might lead to increased crime due to social activities or people being away from home.
7. Humidity and Air Temperature: Weather conditions affect human behavioral decisions and can affect crime rates.
8. Day of Week: Certain crimes occur more frequently on specific days of the week.

6 Conclusion

This research found that tree-based algorithms, which are XGBoost, LightGBM, CatBoost, and Gradient Boosting work well in crime rate prediction. These models are capable of capturing the non-linear and complex relationship in the dataset. Besides, the ensemble characteristic of these algorithms helps to reduce errors. Tree-based algorithms can determine and select the most important features to make decisions. The models are combined to form an ensemble using the voting ensemble method and the final R^2 score is 0.8104. The findings show the effectiveness and practical implications of using machine learning for crime prediction.

7 Future Work

The availability of the dataset for public use is notably limited currently. It is a challenging task to acquire additional data for integration into the machine learning models. In the future, collaboration with organizations can be made to access the required data.

The following are additional data sources that could be considered for incorporation into the dataset:

1. Weather Data: The accuracy and reliability of weather data can be significantly improved by collecting it from multiple sensor locations across the 77 community areas, rather than relying on just the three sensors near the sea to represent the weather for the whole of Chicago city. Chicago's large geographical area and diverse microclimates may cause variations in weather conditions across different community areas. For example, rainfall in one part of the city may not necessarily mean rain in another.
2. Demographic Data: The income, and education level of the population provide information that helps identify high-risk areas. Populations with lower education or income may experience higher crime rates.
3. Spatial Data: The number of police stations, the intensity of street lights, and surveillance coverage in each community area are factors that may affect crime by influencing the deterrence of criminal activities. The potential criminals will reconsider before committing a crime due to the environment. The urbanization of the community area can be incorporated as different social dynamics and opportunities may influence criminal behavior. Moreover, the proximity between community areas can be computed. This is because the crime patterns for neighboring community areas might be similar due to shared socioeconomic characteristics or spillover effects.

Experiments with more advanced algorithms, such as deep learning can be conducted. Convolutional neural networks (CNNs) and recurrent neural networks (RNNs) are examples of deep learning architectures that are useful for time series data and able to extract features or patterns from crime data efficiently.

References

1. Federal Bureau of Investigation https://cde.ucr.cjis.gov. Accessed 22 Oct 2023
2. Universal Class https://www.universalclass.com/articles/business/the-impact-of-crime-on-community-development.htm. Accessed 02 Oct 2023
3. Ashraf, M., et al.: A hybrid CNN and RNN variant model for music classification. Appl. Sci. **13**, 1476 (2023). https://doi.org/10.3390/app13031476
4. Tahir, T., et al.: Early software defects density prediction: training the international software benchmarking cross projects data using supervised learning. IEEE Access **11**, 141965–141986 (2023). https://doi.org/10.1109/ACCESS.2023.3339994
5. Masron, T., Nordin, M.N., Yaakub, N.F., Jubit, N.: Spatial analysis of crime hot-spot in the Northeast Penang Island District and Kuching District, Malaysia. Planning Malaysia (2021).https://doi.org/10.21837/pm.v19i19.1057
6. Pathak, A., et al.: Belief rule-based expert system to identify the crime zones. Adv. Intell. Syst. Comput. 237–249 (2021).https://doi.org/10.1007/978-3-030-68154-8_24
7. Genpact https://www.genpact.com/insight/the-evolution-of-forecasting-techniques-traditionalversus-machine-learning-methods

8. Craig Wisneski https://www.akkio.com/post/5-reasons-why-machine-learning-forecasting-is-better-than-traditional-methods#:~:text=Machine%20learning%20can%20adapt%20to%20changes%20quickly&text=As%20the%20data%20set%20changes. Accessed 02 Oct 2023
9. Biswas, A.A., Basak, S.: Forecasting the trends and patterns of crime in Bangladesh using machine learning model. In: 2019 2nd International Conference on Intelligent Communication and Computational Techniques (ICCT) (2019). https://doi.org/10.1109/icct46177.2019.8969031
10. Aziz, R.M., Hussain, A., Sharma, P., Kumar, P.: Machine learning-based soft computing regression analysis approach for crime data prediction. Karbala Int. J. Mod. Sci. **8**, 1–19 (2022). https://doi.org/10.33640/2405-609x.3197
11. Feng, M., et al.: Big data analytics and mining for effective visualization and trends forecasting of crime data. IEEE Access **7**, 106111–106123 (2019). https://doi.org/10.1109/access.2019.2930410
12. Alsayadi, H.A., Khodadadi, N., Kumar, S.: Improving the regression of communities and crime using ensemble of machine learning models. J. Artif. Intell. Metaheuristics **1**, 27–34 (2022). https://doi.org/10.54216/jaim.010103
13. Chicago Data Portal https://data.cityofchicago.org/Parks-Recreation/Beach-Weather-Stations-Automated-Sensors/k7hf-8y75/about_data
14. U.S. Bureau of Labor Statistics https://fred.stlouisfed.org/series/CHIC917URN
15. Habibullah, M.S.: The effects of weather on crime rates in Malaysia. Int. J. Bus. Soc. (2017).https://doi.org/10.33736/ijbs.482.2017
16. Hsu, H.-Y., et al.: Personalized federated learning algorithm with adaptive clustering for Non-IID IOT data incorporating multi-task learning and neural network model characteristics. Sensors **23**, 9016 (2023). https://doi.org/10.3390/s23229016

A Web-Based Disease Prediction System Using Machine Learning Algorithms and PCA

Anushey Khan[1(✉)] and Ilham Huseyinov[2]

[1] Artificial Intelligence and Data Science, Graduate Institute of Science, Istanbul Aydin University, Istanbul 34295, Turkey
anusheykhan17@gmail.com

[2] Software Engineering Department, Istanbul Aydin University, Istanbul 34295, Turkey
ilhamhuseyinov@aydin.edu.tr

Abstract. This work presents a web-based disease prediction system to diagnose diseases using machine learning algorithms and principal component analysis (PCA). Using web technology made the system accessible from any place, at any time. An ensemble of classifiers such as Gaussian Naive Bayes, Support Vector Machines, and Random Forest, have been applied to predict the diseases. Also, PCA is utilized to find influential symptoms for a certain type of disease. The performance of the system is evaluated before and after applying PCA to the dataset in terms of accuracy, precision, recall, and F1-score. As a result, the impact of PCA on the system performance is illustrated.

Keywords: Web Technology · Gaussian Naive Bayes · Support Vector Machines · Random Forest · PCA · Mode Prediction

1 Introduction

In recent years, the development of machine learning algorithms has opened new avenues for disease diagnosis and prediction [1], enabling automated systems to accurately identify diseases based on symptoms. These systems can greatly assist healthcare professionals in improving treatment efficiency [2]. The applied web technology has made this assistance accessible from any place, at any time. One type of medical service is that users enter symptoms, and the system predicts the likelihood of diseases. For patients, the system serves as an intelligent self-diagnosis service, while for medical professionals, it serves as assistance in the diagnosis and treatment process, enhancing the efficiency of the treatment.

There are many research studies devoted to the problem of medical disease prediction [3], environmental issues [4], safety in autonomous vehicles [5], and the media industry [6]. While a few studies have produced encouraging results, it is frequently the case that these studies lack methodologies for feature dimensionality and thorough evaluation of classifiers used. Because of over-fitting, researchers in [7] employed the Naive Bayesian (NB) method primarily to forecast an illness, making it challenging to assess accuracy. Therefore, to decrease feature dimensionality and evaluate the effect of

J. Rasheed et al. (Eds.): FoNeS-AIoT 2024, LNNS 1035, pp. 104–112, 2024.
https://doi.org/10.1007/978-3-031-62871-9_9

principal component analysis (PCA) on the performance of the system, this study uses an ensemble of classifiers along with the PCA technique.

This study has developed a web-based disease Prediction system to diagnose diseases using machine learning algorithms and a PCA. The system can be accessed at any time and from any location. To forecast diseases, the study combines classifiers such as Gaussian Naive Bayes (GNB), Support Vector Machines (SVM), and Random Forest (RF). Additionally, PCA is applied to find influential symptoms for certain types of diseases. The performance of classifiers is evaluated in terms of accuracy, precision, recall, and F1-score before and after applying PCA to the dataset. As a result, the impact of PCA on the system performance is illustrated.

2 Related Work

Researchers in [8] have emphasized the importance of machine learning techniques for medical database analysis such as deriving symbolic rules, using background knowledge, and sensitivity and specificity of induced descriptions. The study presented in [9] has employed the Naïve Bayesian algorithm for recommendation systems. Predictions for the vetted items, for E-Learning, are produced through the NB classifier algorithm. The author in [10] talked about a global medicine recommender system. They have concluded that SVM is the best medicine recommendation model because of its accuracy, and precision with the dataset they used.

Reference [11] is a review of twenty-three studies and over five thousand patients, which indicates NBNs have the best performance in a majority of the disease prediction model. It works much better than other methods and can help healthcare providers make decisions regarding diagnosis. Another research [12] discusses the need for recommender systems in healthcare and provides an overview of prevalent research. The existing research classifies the heath recommender system according to four concepts: method; technique; recommendation area and knowledge representation.

Another study [13] explores the use of supervised machine-learning algorithms for disease prediction using health data. It compares different algorithms and finds that SVM is the most frequently used, followed by NB. However, Random Forest (RF) demonstrates superior accuracy, topping the performance in 53% of the cases where it was applied. The study offers valuable insights for researchers in choosing appropriate algorithms for disease prediction studies.

The author in [14] exploited PCA along with many filtering techniques to extract the best features for diagnosing infectious diseases. The author extracted useful features and compared certain image processing filters combined with advanced machine learning algorithms, and concluded that the combination of PCA with machine learning drastically improved the performance. The research [15] explains how to predict cardiac disease using supervised machine learning and PCA. It emphasizes the value of data mining methods, especially for predicting the development of diabetes and cancer. This article made a comparison between machine learning algorithms and concluded that the SVM was successfully applied the most, followed by the Naïve Bayes algorithm and then the RF algorithm in the disease prediction software. Out of which RF gave the highest accuracy. Following the same concept from this study we mapped our project. (Comparing different supervised machine learning algorithms for disease prediction).

3 Workflow Diagram of the Disease Prediction System

The workflow diagram of the disease prediction and medicine recommendation system developed in this paper is presented in Fig. 1. The workflow diagram for Disease prediction using machine learning can be summarized as follows:

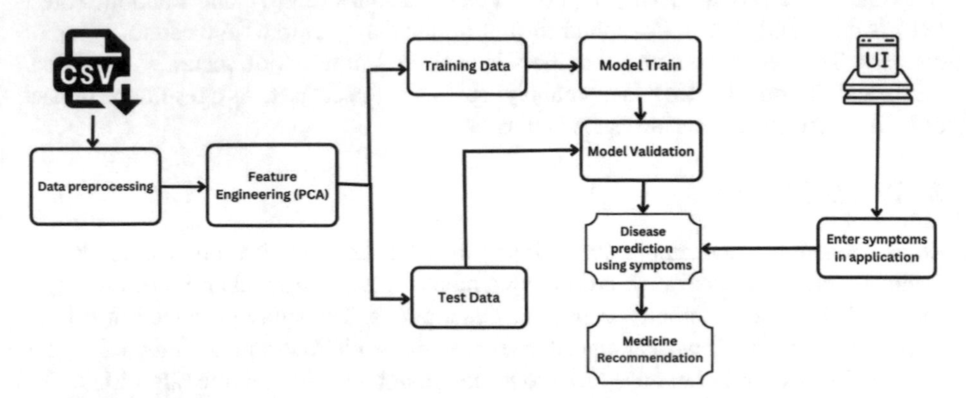

Fig. 1. The proposed architecture.

3.1 Data Preprocessing

This step cleans and handles missing values or any inconsistencies in the dataset.

3.2 Preprocessing

This ensures that the data is in a suitable format for training the machine learning models.

3.3 Feature Engineering

Feature engineering using the PCA step reduces the number of features while preserving most of the relevant information. It transforms the original features into a new set of uncorrelated variables (principal components), which are used as input features for the models.

3.4 Train-Test Data Splitting

This step splits the dataset into training and testing datasets. The training dataset is used to train the machine learning models, while the testing dataset is used to evaluate their performance.

3.5 Model Training and Validation

This step trains classifiers Gaussian Naive Bayes (GNB), SVM, and RF on the transformed feature set using PCA and then validates them using the testing dataset.

3.6 Disease Prediction

This step predicts diseases based on symptoms entered by users through the User Interface (UI). Each model makes predictions based on the same input symptoms.

3.7 Ensemble Voting

This step uses an ensemble voting method to combine the predictions from the three classifiers. The mode of the three individual classifier predictions is calculated to determine the final prediction for each input.

4 Data Analysis, and Preprocessing

The dataset is a CSV file containing symptom information and corresponding disease labels. In the pre-processing stage, data encoding was applied to convert categorical disease labels into numerical values. The encoding was performed using a dictionary-based approach, where each disease label was mapped to a corresponding numeric value. For example, disease names like 'Fungal infection,' 'Allergy,' 'GERD,' etc., are encoded into the integers 0, 1, and 2 respectively.

Encoding the disease labels in this manner ensures that the classifiers can interpret and process them during training and prediction. It allows the algorithms to establish relationships between the symptom features and the numerical disease labels, facilitating disease prediction based on input symptoms. Data is summarized in Fig. 2.

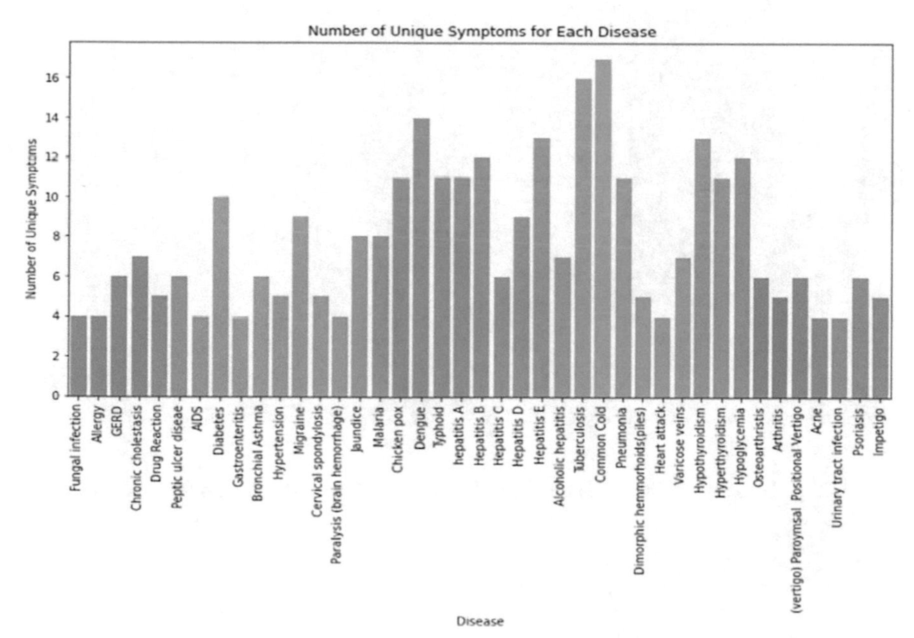

Fig. 2. Dataset composed of samples from various diseases along with counts of unique symptoms.

4.1 Why PCA?

When dealing with many symptoms in disease classification, PCA can help avoid the "curse of dimensionality" and improve the classifiers' capacity to generalize well to new, unseen data. Figure 2 shows that the dataset shows unique diseases have symptoms ranging from 4–16, which means the same symptoms are occurring in multiple diseases, hence it is making the models over-fit. To combat this problem a few experiments were conducted where the number of components was reduced according to their PCA value, and the accuracy was measured through k-cross validation after each update [16–18].

5 Experimentation and Evaluation of Results

5.1 Experimental Analysis with PCA

Figure 3 shows the data distribution of the number of symptoms across the disease before applying PCA, while Fig. 4 depicts the distribution of the number of symptoms across the disease after applying PCA. The absence of a discernible pattern of normal distribution is evident in the data.

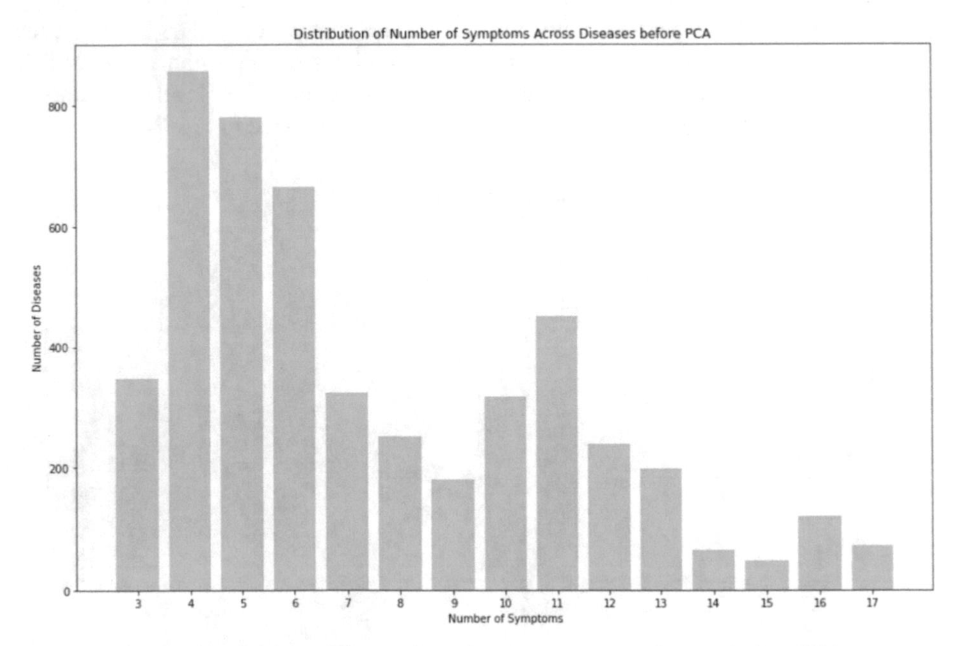

Fig. 3. Distribution of the number of symptoms across diseases before PCA.

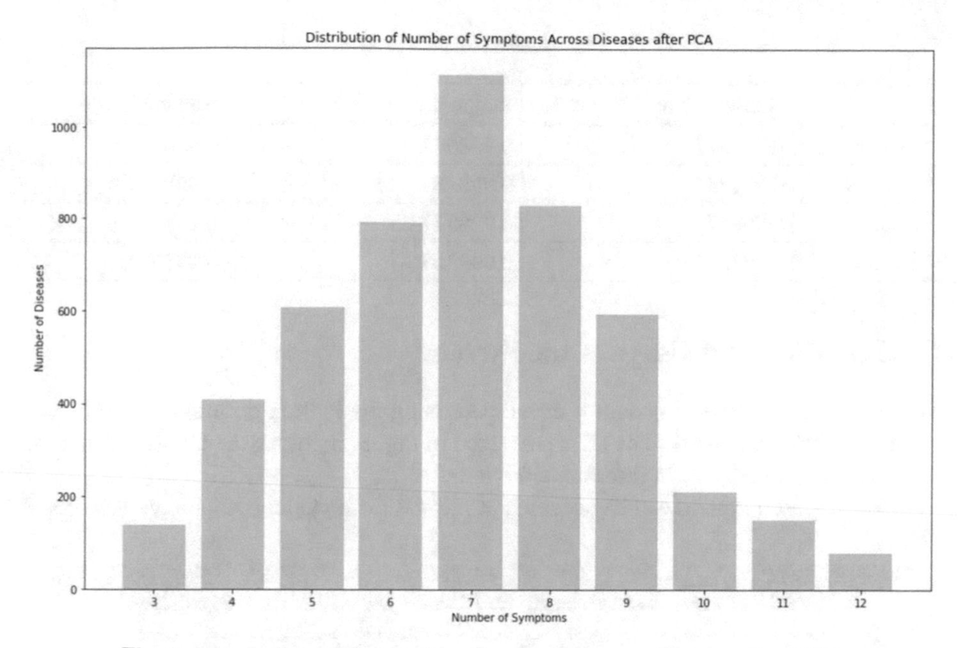

Fig. 4. Distribution of the number of symptoms across diseases after PCA.

Table 1 lists the performance of exploited algorithms when all the features are fed to the model. Evidently, a discernible pattern of normal distribution is observable in this data. This characteristic holds substantial promise for machine learning applications, as it provides a solid foundation for accurate model training and predictive analysis.

Table 1. Performance for non-PCA-reduced data.

	Gaussian Naïve Bayes	Support Vector Machine	Random Forest
Accuracy	1.0	1.0	1.0
Precision	1.0	1.0	1.0
Recall	1.0	1.0	1.0
F1-score	1.0	1.0	1.0

Table 2 shows the performance of exploited algorithms when the number of components was reduced to less than 45. Accuracy is affected because overfitting is overcome. After this experimentation and a few trials, it was deduced that the model best responded to components being set between 45–50 as the accuracy remained 100% but the disease predicted were the most relevant.

Table 2. Performance for PCA-reduced data.

	Gaussian Naïve Bayes	Support Vector Machine	Random Forest
Accuracy	0.998577	0.998577	0.998577
Precision	0.998734	0.998734	0.998734
Recall	0.998577	0.998577	0.998577
F1-score	0.998563	0.998563	0.998563

6 Operation and Usage of the System

The proposed web-based system is developed using the Python programming environment the Flask framework. The UI is presented in Fig. 5. By using the UI users can enter input symptoms and obtain disease predictions.

The proposed system can be accessed at http://anusheykhan.pythonanywhere.com/

Fig. 5. Distribution of the number of symptoms across diseases after PCA.

7 Conclusion

Through a meticulous process of data analysis and exploration, we successfully addressed the challenge of over-fitting in our disease prediction model. Our investigation revealed that certain symptoms exhibited overlapping patterns across multiple diseases, potentially leading to inaccurate predictions. To mitigate this issue, we leveraged the power of Principal Component Analysis (PCA), a dimensionality reduction technique. By carefully selecting the optimal number of components— guided by iterative trials—we achieved remarkable results. Notably, the highest accuracy and improved performance emerged when retaining between 45 to 50 components. This strategic implementation of PCA not only refined our model's predictive capabilities but also underscored the importance of thoughtful feature selection in enhancing the model's robustness and precision.

In the future, we would like to overcome these issues, work with a bigger dataset, and explore other machine-learning techniques.

References

1. Yahyaoui, A., et al.: Performance comparison of deep and machine learning approaches toward COVID-19 detection. Intell. Autom. Soft Comput. **37**, 2247–2261 (2023). https://doi.org/10.32604/iasc.2023.036840
2. Rasheed, J., Alsubai, S.: A hybrid deep fused learning approach to segregate infectious diseases. Comput. Mater. Continua **74**, 4239–4259 (2023). https://doi.org/10.32604/cmc.2023.031969
3. Rasheed, J., Shubair, R.M.: Screening lung diseases using cascaded feature generation and selection strategies. Healthcare **10**, 1313 (2022). https://doi.org/10.3390/healthcare10071313
4. Waseem, K.H., et al.: Forecasting of air quality using an optimized recurrent neural network. Processes **10**, 2117 (2022). https://doi.org/10.3390/pr10102117
5. Farooq, M.S., et al.: A conceptual multi-layer framework for the detection of nighttime pedestrian in autonomous vehicles using deep reinforcement learning. Entropy **25**, 135 (2023). https://doi.org/10.3390/e25010135
6. Ashraf, M., et al.: A hybrid CNN and RNN variant model for music classification. Appl. Sci. **13**, 1476 (2023). https://doi.org/10.3390/app13031476
7. Uddin, S., Khan, A., Hossain, M.E., Moni, M.A.: Comparing different supervised machine learning algorithms for disease prediction. BMC Med. Inform. Decis. Mak. **19**, 281 (2019). https://doi.org/10.1186/s12911-019-1004-8
8. Bao, Y., Jiang, X.: An intelligent medicine recommender system framework. In: 2016 IEEE 11th Conference on Industrial Electronics and Applications (ICIEA). IEEE, pp 1383–1388 (2016)
9. Assegie, T.A., Nair, P.S.: The performance of different machine learning models on diabetes prediction. Int. J. Sci. Technol. Res. **9**, 2491–2494 (2020)
10. Lavrač, N.: Selected techniques for data mining in medicine. Artif. Intell. Med. **16**, 3–23 (1999). https://doi.org/10.1016/S0933-3657(98)00062-1
11. Özcan, M., Temel, T.: New recommender system using naive Bayes for E-Learning. In: The Eurasia Proceedings of Educational and Social Sciences, pp 309–312 (2016)
12. Kanchan, B.D., Kishor, M.M.: Study of machine learning algorithms for special disease prediction using principal of component analysis. In: 2016 International Conference on Global Trends in Signal Processing, Information Computing and Communication (ICGTSPICC). IEEE, pp 5–10 (2016)
13. Tran, T.N.T., Felfernig, A., Trattner, C., Holzinger, A.: Recommender systems in the healthcare domain: state-of-the-art and research issues. J. Intell. Inf. Syst. **57**, 171–201 (2021). https://doi.org/10.1007/s10844-020-00633-6
14. Rasheed, J.: Analyzing the effect of filtering and feature-extraction techniques in a machine learning model for identification of infectious disease using radiography imaging. Symmetry **14**, 1398 (2022). https://doi.org/10.3390/sym14071398
15. Pincay, J., Teran, L., Portmann, E.: Health recommender systems: a state-of-the-art review. In: 2019 Sixth International Conference on eDemocracy & eGovernment (ICEDEG). IEEE, pp 47–55 (2019)
16. Jaadi, Z.: Principal Component Analysis (PCA) Explained | Built In (2022). https://builtin.com/data-science/step-step-explanation-principal-component-analysis

17. Dey, A., Singh, J., Singh, N.: Analysis of supervised machine learning algorithms for heart disease prediction with reduced number of attributes using principal component analysis. Int. J. Comput. Appl. **140**, 27–31 (2016). https://doi.org/10.5120/ijca2016909231
18. Moghbeli, M.: Applying naive bayesian networks to disease prediction: a systematic review. Acta Inform. Medica **24**, 284 (2016). https://doi.org/10.5455/aim.2016.24.284-289

Smart Cities, Sustainable Paths: Energy Harvesting and Mobility Solutions for Tomorrow's Urban Landscapes

Md Ashraful Islam[1](✉), Shahela Akter[1], Abdulla Al Mamun[1], Balakumar Muniandi[2], Mashfiq Ahasan Hridoy[3], Emdad Ullah Khaled[1], Md. Sabbir Alam[1], and M. M. Naushad Ali[1]

[1] Green University of Bangladesh, Dhaka, Bangladesh
mdashrafulislam3210@gmail.com
[2] Lawrence Technological University, Southfield, MI 48075, USA
[3] Centre for Policy Dialogue, Dhaka, Bangladesh

Abstract. This study presents a transformative guide for Purbachal City, strategically integrating smart energy, infrastructure, and mobility solutions. The study selects PV panels and converters through meticulous assessments and simulations, ensuring the efficacy of a solar energy system that boasts an annual production of 7,317.34 MWh. To foster resilience within cities, the smart infrastructure plan suggests installing solar panels on streetlight poles and foot-over structures. Notably, the Type D Electric Bus is a key contributor to sustainable urban mobility as an alternative to conventional fossil fuel-powered buses. Economic viability, as demonstrated by the HOMER simulation with a low LCOE of 0.02 USD, an Internal Rate of Return (IRR) of 11%, a Return on Investment (ROI) of 7.7%, and a simple payback period of 8.4 years. 84,479 tons of CO2 will be saved by implementing this system shows a significant environmental benefit. PVsyst simulation has been conducted to verify the result by HOMER. This meticulous strategy and comprehensive approach will position Purbachal City as a beacon of smart and sustainable urban development, aligning with global trends and fostering environmental responsibility, economic growth, and an enhanced quality of life.

Keywords: Renewable Energy · Smart city · Sustainable city · HOMER · PVsyst

1 Introduction

Cities worldwide are undergoing a transformative shift towards smart urban environments, integrating technology to enhance living standards. Driven by escalating urbanization, smart cities employ digital technologies for optimized transportation, energy consumption, and infrastructure. Sustainability takes center stage in global urban development, addressing the strain on resources and the environment. Within smart cities, a critical focus lies on responsible resource use, environmental impact reduction, and creating adaptable urban spaces. Energy harvesting and sustainable mobility emerge as

J. Rasheed et al. (Eds.): FoNeS-AIoT 2024, LNNS 1035, pp. 113–124, 2024.
https://doi.org/10.1007/978-3-031-62871-9_10

key priorities, responding to global reconsideration of conventional energy sources. In Bangladesh, heavily reliant on traditional energy, integrating energy harvesting technologies becomes a pivotal step towards sustainability. Sustainable urban mobility, exemplified by electric vehicles, addresses air quality and greenhouse gas concerns [1]. Initiatives like escalators on foot-over bridges further enhance responsible urban behavior, shaping a comprehensive approach to smart city development. Globally, major urban centers are investing heavily in smart city initiatives. Cities like Singapore, Copenhagen, Amsterdam, Copenhagen, and Barcelona serve as benchmarks for sustainable urban development. The integration of technology into urban environments, often referred to as the development of smart cities, has garnered significant attention in recent literature. Scholars highlight the transformative potential of technology in enhancing the quality of life, addressing environmental concerns, and improving overall urban efficiency. Niloofar et al. [2]. Discusses the role of smart transport in urban planning and the transition from traditional to smart cities in developing countries with sustainability requirements. It mentions that smart transport solutions can address challenges such as traffic congestion and air pollution. Dastan Bamersigye and Petra Hlavackova [3] provide a comprehensive analysis of sustainable transport for smart cities, emphasizing the interconnectedness of transportation with environmental, social, and economic aspects. It highlights the urgent need for collaboration among stakeholders at various levels to address the unsustainability of current transportation systems. Mahesh Babu *et al.* [4] discuss by including energy management, smart transportation systems can improve the quality of life for locals, reduce traffic congestion and pollution, and increase the overall efficiency of urban settings. But *et al.* [5] study shows the importance of smart infrastructure in the development of smart cities, where digital technologies are used in various areas such as energy, security, transport, and resource management. The paper discusses the importance and benefits of smart infrastructure in cities. Hadi Amini et al. [6] analyze the importance of information and communication technologies (ICTs) as the backbone of smart cities, enabling efficient control and optimization of resources in various domains such as transportation and energy systems. In the context of Bangladesh, a country experiencing rapid urbanization, the imperative for smart city solutions is a must. Purbachal City, situated in Dhaka, emerges as a microcosm of this global shift, where the integration of energy harvesting, and mobility solutions becomes a strategic response to the challenges of urbanization. In response to the global trend towards smart cities, this study aims to align Purbachal with the best global practices while addressing local challenges. The integration of energy harvesting technologies, such as harnessing renewable sources like solar and wind energy, stands out as a pivotal step towards sustainability in a country heavily reliant on conventional energy. This strategic move reduces the carbon footprint and enhances the city's resilience by diversifying its energy sources. Furthermore, the focus extends to smart mobility solutions, including adopting electric buses and installing escalators on foot-over bridges. These initiatives not only address air quality concerns but also contribute to the reduction of greenhouse gas emissions, creating a more responsible and efficient urban transportation system. The focus on smart energy, infrastructure, and mobility aligns Purbachal City with international best practices, fostering sustainability and resilience. As we delve into the subsequent sections, the study unfolds the specific initiatives undertaken, positioning Purbachal City as a noteworthy contender in the global landscape of smart and sustainable urban development.

2 Methodology

Embarking on the journey to transform Purbachal City into a smart and sustainable urban hub begins with a thorough site assessment. Figure 1 illustrates the structured methodology for this study.

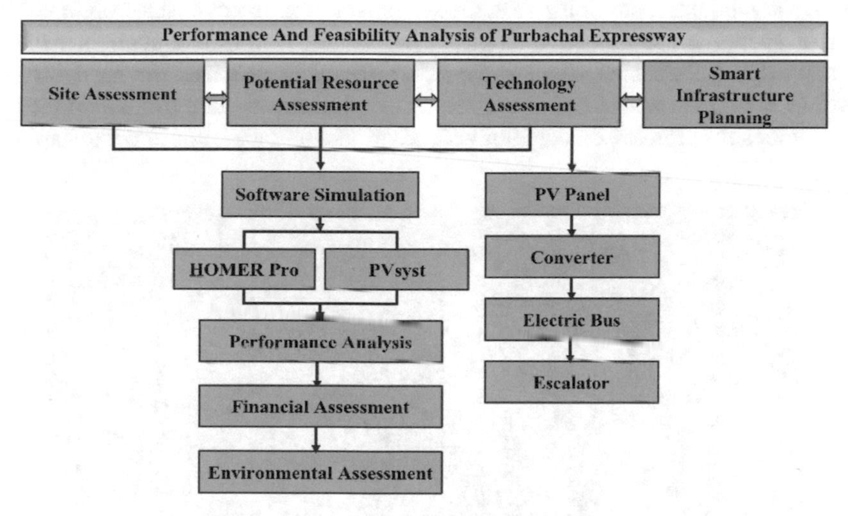

Fig. 1. Methodology flowchart for this study

The initial step involves a thorough examination of Purbachal City's geographical, environmental, and infrastructural aspects. This site assessment forms the basis for identifying suitable areas for implementing smart energy, infrastructure, and mobility solutions. Simultaneously, potential resources, such as sunlight exposure, are evaluated to determine the feasibility of energy harvesting technologies, laying the groundwork for the integration of solar panels. The study assesses key technologies essential for smart infrastructure, ensuring their alignment with Purbachal City's context and energy needs. The goal is to create an environmentally responsible infrastructure that enhances urban efficiency. The team then formulates a smart infrastructure plan, outlining the strategic integration of technologies identified in the assessment. This plan includes smart lighting, EV charging station design, and solar panel integration on foot-over bridges, prioritizing scalability and interoperability. The focus is on creating adaptive infrastructure that promotes sustainability and resilience to meet the city's evolving needs. In the second step, the study employs software simulations with HOMER Pro and PVsyst. HOMER Pro optimizes the energy system, considering resource availability and demand, while PVsyst validates outcomes, especially the Levelized Cost of Electricity (LCOE). This dual-step simulation ensures a robust representation of the energy system's performance. Moving to the third step, the study conducts a comprehensive analysis of implemented smart city initiatives. Performance analysis assesses integrated technologies, economic analysis evaluates financial viability, and environmental analysis measures sustainability impact. Through assessments, simulations, and analyses, the study aims to lay the groundwork for Purbachal City's emergence as a beacon of smart and sustainable urban development.

2.1 Site Assessment

The selection criteria for Purbachal City and its Expressway focus on maximizing the effectiveness of smart city initiatives. Geographical assessment ensures optimal placement of energy harvesting technologies, considering factors like sunlight exposure. Acknowledging Purbachal City's rapid urbanization emphasizes the need for smart solutions to address associated challenges. Current infrastructure evaluation identifies opportunities for seamless integration with smart technologies, minimizing disruption.

Fig. 2. Purbachal Expressway on the map

2.2 Potential Resource Assessment

The solar energy irradiation data depicted in Fig. 2 for the Purbachal Expressway provides valuable insights into the solar potential of the region throughout the year. The daily irradiation values, ranging from a minimum of 4.02 kWh/m2/day in September to a maximum of 5.76 kWh/m2/day in April, illustrate the varying intensity of solar radiation [7]. The clearness index values, reflecting the atmospheric transparency, show a fluctuation from 0.382 in July to 0.643 in December. This data serves as a crucial foundation for the feasibility assessment of implementing solar panels along the expressway, contributing to the overall smart energy strategy for Purbachal City (Fig. 3).

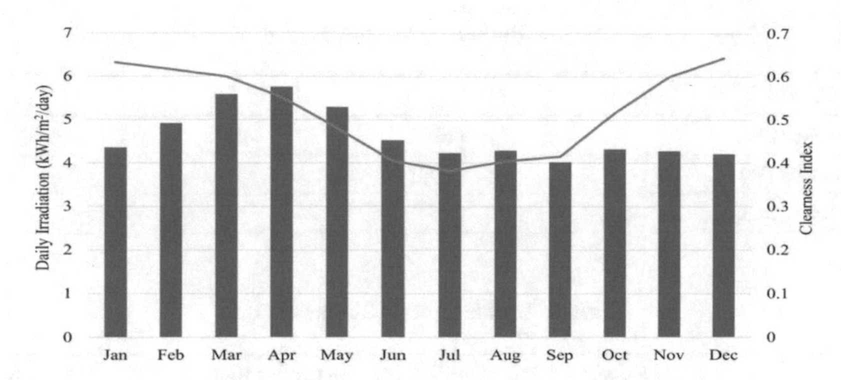

Fig. 3. Available solar resources in the Purbachal City and expressway.

2.3 Technology Assessment

Technology assessment holds paramount importance in guiding strategic decision-making and ensuring the successful implementation of smart city initiatives.

PV Panel and Converter

The assessment of PV panels and converters is crucial for determining the efficacy of solar energy harvesting on the Purbachal Expressway. Table 1. details the specifications for the selected PV panel (EVO 6 Pro SE6-66HBD) and converter (Canadian Solar CSI-1KTL1P-GI-FL). The PV panel, with a capacity of seven hundred Wp, dimensions of 2384 x 1303 x 35 mm, and a 25-year lifespan, emphasizes extended operational life. In contrast, the chosen converter boasts a 60kW capacity, an impressive 97.2% efficiency, and a 15-year lifespan [8]. This careful selection aligns with the smart energy strategy for Purbachal City, prioritizing longevity and efficiency.

Table 1. Specification of the selected PV panel and Converter

PV Panel	
Model	EVO 6 Pro SE6-66HBD
Capacity (Wp)	700
Dimensions of Module (mm)	2384 x 1303 x 35
Weight (kg)	34.8
Efficiency (%)	22.5
Lifetime (year)	25 years
Converter	
Model	Canadian Solar CSI-1KTL1P-GI-FL

(*continued*)

Table 1. (*continued*)

PV Panel	
Capacity (kW)	60
Efficiency (%)	97.2
Lifetime (year)	15

Table 2. Deciding factor for Electric Bus

Model	Power Consumption (kWh/km)	Range (km)	Battery Capacity (kWh)	Passenger Capacity
Olectra C9	1.3	250	324	49
Olectra k9	-	300	-	40
BYD	1.29	232	324	58
PMI URBAN	1.13	180	204	35
K11M	1.86	311	578	55
Type D Electric Bus	1.02	250	230	84
Solaris Urbino 15 LE	-	270	470	65
K11M	1.86	311	578	55

Suitable Electric Bus Selection

Among the electric bus models evaluated in Table 2., the Type D Electric Bus emerges as the most suitable option for Purbachal City's smart mobility strategy. With a power consumption of 1.02 kWh/km, a range of 250 km, a battery capacity of 230 kWh, and an impressive passenger capacity of 84, the Type D Electric Bus offers a balanced combination of energy efficiency, range, and passenger accommodation. This makes it a robust choice for addressing the transportation needs of Purbachal City, ensuring a sustainable and efficient solution for the urban mobility landscape. Type D Electric Bus specification is given in Table 3..

Escalator

The infrastructure planning section focuses on the strategic installation of solar panels on streetlight poles to generate electricity for lighting, escalators, and electric vehicles. This approach aims to reduce dependence on conventional energy sources by directing solar-generated power to escalators, promoting an environmentally friendly strategy. The inclusion of integrated charging stations for electric vehicles aligns with the global trend towards sustainable urban development. Utilizing solar energy not only enhances productivity but also contributes to a cleaner urban environment. This integrated approach reflects a progressive and environmentally sustainable direction in line with modern global sustainability initiatives [9].

Table 3. Type D Electric Bus specification

Model	Type D Electric Bus
Manufacture	BYD, China
Length (M)	Iron Phosphate
Battery Type	12.34
Top Speed (Km/h)	105
Motor Capacity (kW)	150*2
Charging Capacity (kW)	110
Charging Time (Hours)	2.1 -2.6

Smart Infrastructure Planning

The infrastructure planning section focuses on the strategic installation of solar panels on streetlight poles to generate electricity for lighting, escalators, and electric vehicles. This approach aims to reduce dependence on conventional energy sources by directing solar-generated power to escalators, promoting an environmentally friendly strategy. The inclusion of integrated charging stations for electric vehicles aligns with the global trend towards sustainable urban development. Utilizing solar energy not only enhances productivity but also contributes to a cleaner urban environment. This integrated approach reflects a progressive and environmentally sustainable direction in line with modern global sustainability initiatives. Figure 4 illustrates the 3D view of the proposed smart infrastructure.

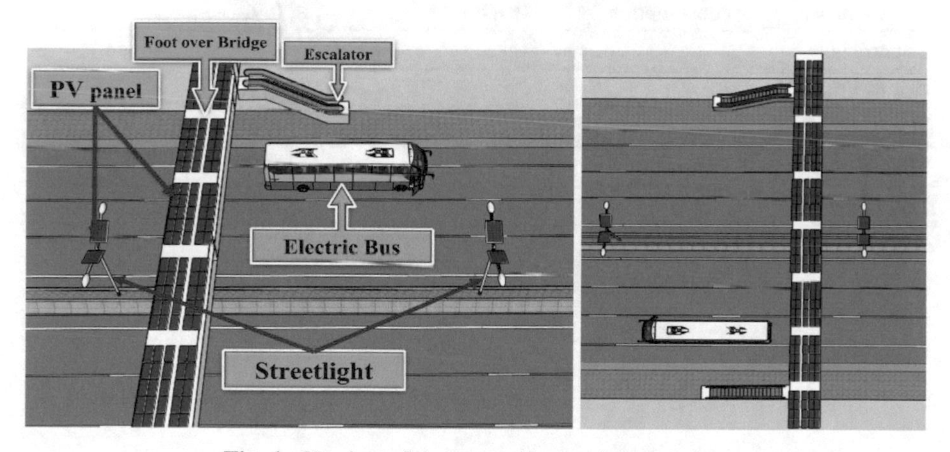

Fig. 4. 3D view of the proposed smart infrastructure

2.4 Load Profile and Economic Input Parameters

The load calculation in Table 4. provides a comprehensive analysis of power requirements for various components in the urban infrastructure. Six-foot-over bridges with 24 escalators collectively demand 1632 kW/day. With 136 trips covering a daily distance of 3155.2 km, the total power requirement for electric buses amounts to 2902.784 kW/day. Operating on 1170 light poles with 2340 lights for 12 h daily, the streetlights necessitate 280.8 kW/day. This analysis illuminates the electricity demands of key components in the urban infrastructure, aiding in efficient energy management and planning for the proposed smart city initiatives in Purbachal City.

Table 4. Load calculation for all types of loads.

Entry	Unit
Escalator	
Number of foot-over bridge	6
Number of Escalators	24
Average hourly electricity consumption (kWh)	4 [9]
Escalator operating time per day (hour)	17
Electric Bus	
Operating time per day (hour)	17
Number of trips per day	136
Route distance (km)	23.2
Per kilometer power consumption by Bus (kWh/km)	0.92
Streetlights	
Number of light poles	1170
Number of lights	2340
Operating time per day (hour)	12
Light (W)	120*2 = 240

Table 5. Cost information for used components.

Entry	Amount	Unit
PV panel price	187	$/piece
PV O&M cost [10]	7	$/kW
Converter price	2200	$
Battery	550	$/kW
Discount rate [10]	10	%
Inflation rate [10]	5.59	%

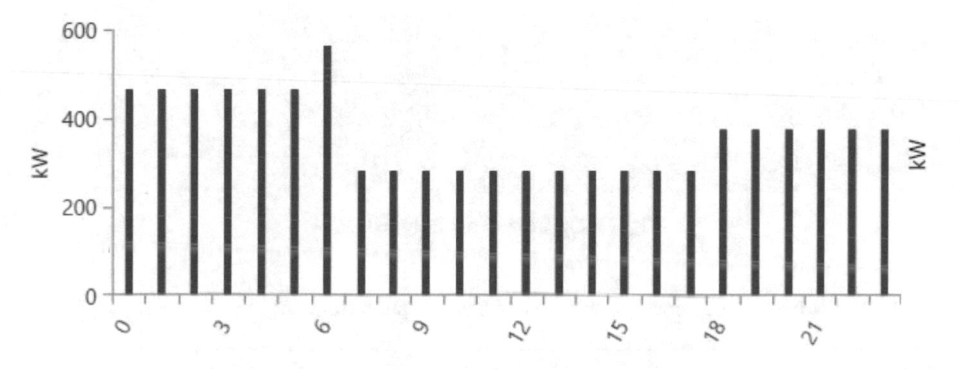

Fig. 5. Daily load profile for HOMER

3 Result Analysis

3.1 HOMER Simulation

Figure 6 presents the system architecture for the proposed smart cities. This system consists of a PV panel, converter, and Battery with a grid connection. Load requirements for the entire system are 8722 kWh/day with a peak load of 563 kW. Figure 7 depicts the optimized result by the HOMER. The result shows that the on-grid PV system without the storage system is the most optimal. To meet the daily load requirements 6,687 kWp PV system, and a 2,532 kW converter are required. The total Net Present cost(NPC) for the system is 2.16M with an impressive Cost of Energy(COE) is USD 0.02 only.

Table 6. provides a technical summary of production and consumption by HOMER for the smart energy system. The EVO 6 Pro SE6-66HBD contributes 84% of the total production, generating 10,287,690 kWh/year. Grid purchases account for 16% of the production. In terms of consumption, the AC Primary Load uses 34.3%, while grid sales constitute 65.7%. Excess electricity amounts to 2,771,008 kWh, with a renewable fraction of 78.8%. Table 7. outlines economic metrics by HOMER, featuring a Levelized Cost of Electricity (LCOE) at 0.02 USD/kWh, an Internal Rate of Return (IRR) of 11%, a Return on Investment (ROI) at 7.7%, and a Simple Payback period of 8.4 years.

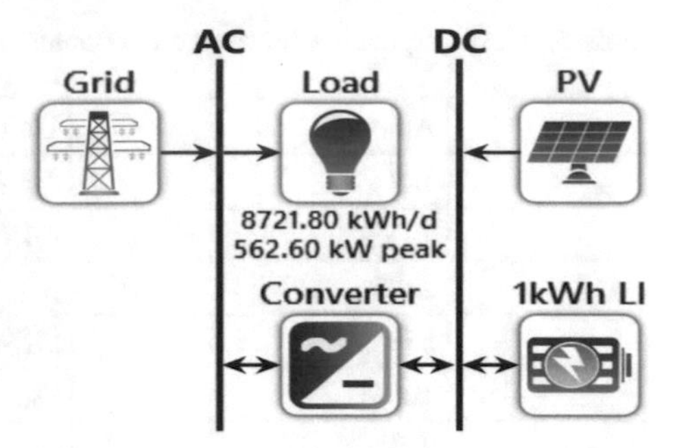

Fig. 6. System architecture in HOMER.

PV (kW)	1kWh LI	Grid (kW)	Converter (kW)	Dispatch	NPC ($)	COE ($)	Operating cost ($/yr)	Initial capital ($)	Ren Frac (%)
6,587		999,999	2,532	CC	$2.16M	$0.0152	-$88,869	$3.52M	78.8
6,394	18	999,999	2,532	CC	$2.18M	$0.0155	-$84,949	$3.48M	78.6
		999,999		CC	$4.88M	$0.100	$318,346	$0.00	0
	14	999,999	1.10	CC	$4.90M	$0.100	$318,706	$8,469	0

Fig. 7. Optimized result by HOMER

Table 6. Technical Production and Consumption Summary by HOMER.

Production	kWh/year	Percentage (%)
EVO 6 Pro SE6-66HBD	10,287,690	84.0
Grid Purchases	1,964,740	16
Total	12,252,430	**100**
Consumption		
AC Primary Load	3,183,457	34.3
Grid Sales	6,087,497	65.7
Total	9,270,954	**100**
Quantity		
Excess Electricity	2,771,008	22.6
Renewable Fraction		78.8

Table 7. Economic metric by HOMER.

Entry	Value
LCOE	0.02 USD/kWh
Internal Rate of Return (IRR)	11%
Return On Investment (ROI)	7.7%
Simple Payback	year

Table 8. PVsyst simulated system configuration.

System Summary	
PV Array	
No. of modules	6,282 Units
P_{nom} total	4,397 kWp
Inverters	
No. of units	73 Units
P_{nom} total	4380 kWac
P_{nom} ratio	1
Result Summary	
Produced Energy	7317.34 MWh/year
Performance Ratio	87.67%
Module area	19514 m^2
LCOE	0.02 USD/kWh
GHG reduction	84,479 tons

3.2 PVsyst Simulation

The PVsyst simulation presents a comprehensive overview of the solar energy system's configuration and performance. With 6,282 modules and a total Pnom of 4,397 kWp, the PV array is efficiently complemented by 73 inverters with a total Pnom of 4380 kWac, maintaining a Pnom ratio of 1. The simulation yields a promising annual energy production of 7,317.34 MWh, accompanied by an impressive Performance Ratio of 87.67%. The module area spans 19,514 m2, and the Levelized Cost of Electricity (LCOE) is calculated at 0.02 USD/kWh. Also, this system is capable of saving around 85k tons CO2 over its lifetime which surely will create a positive impact on the environment. This PVsyst result aligns with the HOMER simulation, indicating consistency in the system's performance and reinforcing the reliability of the overall smart energy strategy for Purbachal City.

4 Discussion and Conclusion

The study outlines a comprehensive strategy for transforming Purbachal City into a smart and sustainable urban center. Through rigorous assessments, technology choices, and simulations using HOMER and PVsyst, we integrated smart energy, infrastructure, and mobility solutions. Notable selections include PV panels and converters, ensuring the efficiency of the solar energy system. The smart infrastructure plan includes deploying solar panels on streetlight poles, facilitating a resilient urban environment. Load calculations guide efficient energy use for escalators, electric buses, and streetlights, with the Type D Electric Bus, identified as a key contributor to sustainable urban mobility. Economic viability, demonstrated by the HOMER simulation with a low LCOE of 0.02 USD/kWh, an IRR of 11%, an ROI of 7.7%, and a simple payback period of 8.4 years, was corroborated by PVsyst, affirming the robustness of the proposed smart energy strategy. In conclusion, the study establishes a solid foundation for Purbachal City's evolution into a smart and sustainable urban hub. Aligned with global sustainability trends, integrating renewable energy, and emphasizing efficient urban mobility, the proposed initiatives showcase a holistic approach. Purbachal City is poised to embody smart city principles, promoting environmental responsibility, economic growth, and an enhanced quality of life for its residents.

Acknowledgment. This work was supported in part by the Center for Research, Innovation and Transformation (CRIT) of the Green University of Bangladesh (GUB).

References

1. Islam, M.A., Ali, M.M.N., Nahian, A.J., Maruf, M.H., Lipu, M.S.H., Shihavuddin, A.S.M.: Prospect of electric boat in Dhaka: an alternate way of transportation. In: 2022 4th International Conference on Sustainable Technologies for Industry 40, STI 2022 (2022). https://doi.org/10.1109/STI56238.2022.10103245
2. Shojarazavi, N., et al.: The Role of Smart Transport in Urban Planning and the Transition from Traditional to Smart Cities in Developing Countries with Sustainability Requirements (2023). https://doi.org/10.20944/PREPRINTS202306.0637.V1
3. Bamwesigye, D., Hlavackova, P.: Analysis of sustainable transport for smart cities. Sustainability 2019 **11**, 2140 (2019). https://doi.org/10.3390/SU11072140
4. Babu, A.M., Akhil, B., Pochampally, N.K.: Smart cities and intelligent transport systems. Int. J. Appl. Struct. Mech., 22–25 (2023). https://doi.org/10.55529/ijasm.31.22.25
5. But, T., Mamotenko, D., Hres-Yevreinova, S.: Smart-infrastructure in the sustainable development of the city: world experience and prospects of Ukraine. European Scientific Platform (Publications) (2023)
6. Amini, M.H., Arasteh, H., Siano, P.: Sustainable smart cities through the lens of complex interdependent infrastructures: panorama and state-of-the-art. Stud. Syst. Decis. Control **186**, 45–68 (2019). https://doi.org/10.1007/978-3-319-98923-5_3/COVER
7. HOMER - Hybrid Renewable and Distributed Generation System Design Software. https://www.homerenergy.com/. Accessed 29 Aug 2023
8. Canadian Solar – Global. https://www.canadiansolar.com/. Accessed 14 Sep 2023
9. Al-Sharif, L., Eng C Lift and Escalator Energy Consumption
10. Islam, M.A., et al.: Integrating PV-based energy production utilizing the existing infrastructure of MRT-6 at Dhaka, Bangladesh. Heliyon, e24078 (2024). https://doi.org/10.1016/J.HELIYON.2024.E24078

A Mobile Robot with an Autonomous and Custom-Designed Control System

Brwa Abdulrahman Abubaker[1,2](✉), Jafar Razmara[2], and Jaber Karimpour[2]

[1] Department of Computer Science, Bayan University, Erbil, Iraq
brwa.abubaker@bnu.edu.iq

[2] Department of Computer Science, University of Tabriz, Tabriz, Iran
{razmara,karimpour}@tabrizu.ac.ir

Abstract. Teaching autonomous mobile robots (AMRs) to acquire knowledge independently has been a formidable challenge, characterized by protracted convergence times and computational intensity within traditional methods. This research introduces an innovative paradigm employing a customized spiking neural network (SNN) to address these challenges, fostering autonomous learning and control of AMRs within unfamiliar environments. The proposed model amalgamates spike-timing-dependent plasticity (STDP) with dopamine modulation to augment the learning process. Incorporating the biologically inspired Izhikevich neuron model imparts adaptability and computational efficiency to the control systems, particularly in response to dynamic environmental alterations. Evaluation through simulations elucidates initial challenges during the training phase, where the infusion of brain-inspired learning, dopamine modulation, and the Izhikevich neuron model introduces intricacies, notably manifesting in difficulties adapting to diverse obstacle scenarios. Initial performance metrics reveal a 73% accuracy rate in reaching the target with a 27% collision rate in single obstacle scenarios. However, progressing to the testing phase demonstrates substantial enhancement, culminating in a remarkable 98% accuracy in reaching the target and a marked reduction in collisions to 2% in single obstacle scenarios. These outcomes underscore the model's adaptive prowess and proficiency in navigating complex environments with varied obstacles. The innovative application of the customized SNN, integrating STDP and dopamine modulation, showcases promising potential in surmounting the challenges associated with reinforcement learning in AMRs.

Keywords: SNN · AMRs · Reinforcement Learning · STDP

1 Introduction

This research strengthens autonomous navigation by merging machine-learning techniques for object recognition [1, 2] and mechanical control. It builds on the foundational work of Goyal and Benjamin [3] and incorporates reinforcement learning as described by Sutton and Barto [4], with practical applications similar to those by Yang et al. [5] and Farooq et al. [6]. This methodology enables the system to continually enhance its driving strategy for more efficient and safe navigation. The combined machine-learning

J. Rasheed et al. (Eds.): FoNeS-AIoT 2024, LNNS 1035, pp. 125–133, 2024.
https://doi.org/10.1007/978-3-031-62871-9_11

approaches provide a robust framework for autonomous vehicles, significantly improving environmental perception and adaptability. The system not only identifies objects with high accuracy but also navigates smoothly, mirroring natural driving behavior and prioritizing passenger comfort and safety. This dual emphasis on visual perception and control dynamics establishes a new standard in self-driving technology development. Despite its strengths, the system faces challenges, particularly when obstructions affect sensor accuracy. The work of Li, Pan, and Chu [7] on applying reinforcement and deep learning for lateral control in autonomous vehicles offers insights that could be applied to mobile robotics to refine sensor data interpretation and enhance robotic system efficacy.

This article proposes an innovative SNN-based approach to address the challenges of applying Q-learning and DQNs in robotic control, particularly in new and dynamic environments. Traditional methods face issues with long learning processes, complex control structures, slow convergence, high computational demands, and difficulty adapting to changes, often requiring extensive training data. The proposed Autonomous Mobile Robot (AMR) learning and control system, grounded in the biologically accurate Izhikevich neuron model [8], aims to improve navigation and functionality in unfamiliar settings. Previous research has shown the effectiveness of neural networks in robotic mobility and navigation, with studies by Arena et al. [9], Shamsfakhr and Bigham [10], and Pandey et al. [11] demonstrating their application in sensory-motor integration and obstacle detection. The SNN in this work utilizes STDP and dopamine modulation to enhance learning and adaptability, closely resembling biological learning processes. This advancement marks a significant contribution to the field of autonomous robotics, emphasizing performance improvement through hyperparameter tuning and a new algorithm (Abubaker et al., 2023) [12].

This article is divided into four sections to thoroughly examine robot behavioral learning. The first section surveys existing research and sets the stage for our work. The second section outlines our approach, covering both theory and application. The third section offers extensive testing to confirm our solution's effectiveness in actual settings. The concluding section recaps our results, considers their impact, and suggests directions for further inquiry and innovation. Visual aids are included to help readers easily follow our research's development.

2 Methodology

2.1 Navigation Enhancement via Sensor Integration

The Mobile Robot Control (MRC) system's navigation is significantly enhanced by integrating advanced sensors, including precise color detection and reliable ultrasonic sensors for detailed navigation. Ultrasonic sensors detect obstacles and measure distances using sound waves, while color sensors identify objects based on reflected light. The Sensor Neural Network (SNN) processes this data to guide the robot's movements accurately. The effectiveness of these sensors in obstacle recognition and avoidance is established (Roberts et al., 2021) [13], with their performance attributes—frequency, scope, precision, and adaptability—being crucial. The SNN integrates sensory data to enable informed and precise decision-making for accurate navigation. Additionally, the system's mapping capabilities allow the robot to generate detailed environmental maps, aiding in optimal pathfinding and hazard avoidance. The SNN-based navigation system runs on a robust platform capable of handling the computational demands of neural networks and sensor data processing (Kumar et al., 2022) [14], supporting robot control, actuation, and smooth communication with external devices.

2.2 Background in Biology

Recent advances in neuroscience have deepened our understanding of the complex human brain. Neurons, the basic units of the nervous system, transform electrical impulses into signals within their networks, as illustrated in Fig. 1. The neuron's structure, including its membrane voltage changes and action potential creation, is vital for its operation. Neurons communicate through axons and synapses, which are key in signal processing and crucial for learning and adaptation. Factors influencing synaptic activity, such as neurotransmitter release and plasticity, are fundamental for neural information flow. These mechanisms support a spectrum of neural functions across species, showcasing the diversity and complexity of neural systems (Li, Pan, & Chu, 2020) [7].

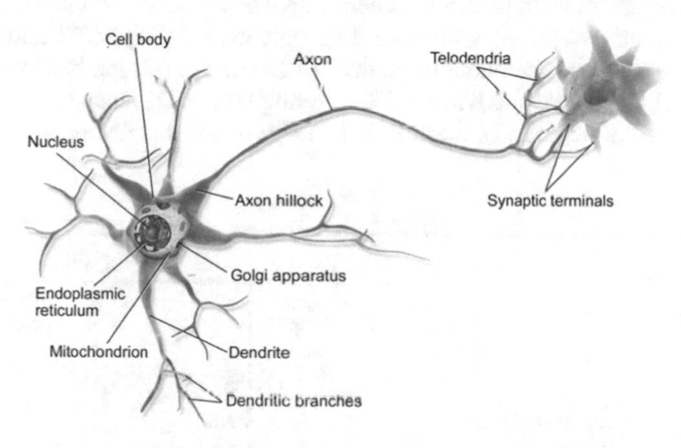

Fig. 1. A biological neuron's structure [13].

2.3 Izhikevich Neuron Model

The Izhikevich Neuronal Model (INM), created by Eugene M. Izhikevich [16] in 2003, is a key mathematical model for simulating spiking neurons in computational neuroscience. It offers a biologically plausible and computationally efficient framework for mimicking various natural neuronal firing behaviors. At its core are differential equations that describe the membrane potential 'v', recovery variable 'u', and input current 'I', with 'DV/dt' and 'du/dt' representing their time derivatives. Parameters 'a', 'b', and 'c' dictate the firing dynamics, allowing the model to capture the spiking activity of neurons. When 'v' hits a threshold, a spike occurs, followed by a reset of 'v' and 'u'. The INM's flexibility enables the simulation of multiple neuronal behaviors, such as regular spiking and bursting, by adjusting 'a' and 'b', making it a versatile tool for studying diverse neuronal types and their firing patterns.

$$\frac{dv}{dt} = 0.04 * \mathrm{v}^2 + 6 * \mathrm{v} + 140 - \mathrm{u} + \mathrm{I} \tag{3-1}$$

$$\frac{du}{dt} = \mathrm{a(bv - u)} \tag{3-2}$$

2.4 STDP with Dopamine Modulation

STDP is a neuroplasticity mechanism that modifies synaptic strengths based on the timing of neuronal firing, strengthening connections when a presynaptic neuron fires before a postsynaptic one, and weakening them when the order is reversed. This timing-dependent adjustment helps neurons associate related events within a short time frame, playing a crucial role in learning and memory. However, STDP alone doesn't fully account for learning involving rewards and punishments. Dopamine, a neuromodulator, enhances STDP by providing a reinforcement signal based on prediction errors, reinforcing synaptic inputs associated with rewarding outcomes through LTP, and weakening those linked to negative predictions through LTD. Dopamine release, tied to reward presence and anticipation, interacts with STDP, targeting D1 and D2 receptors to respectively amplify LTP and promote LTD, creating an STDP window that favors inputs predicting rewards (see Fig. 2).

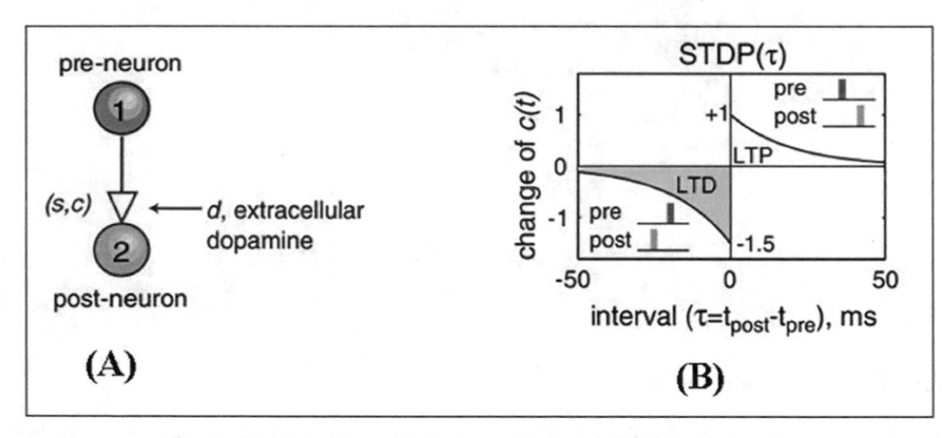

Fig. 2. (A) A postsynaptic Neuron, (B) A the STDP function [17].

When a presynaptic neuron fires before a postsynaptic neuron (t-pre and t-post, respectively), the synaptic variable 'c' changes according to the STDP function STDP(τ), with 'τ' being the time difference between their spikes. This change in 'c' is depicted in Fig. 2. Synchronous firing sets τ to 1 and resets 'c' to zero. The rate of decline of 'c' is critical for adjusting synaptic weight sensitivity to reward timing, thus enhancing learning. 'c' influences the 'd' factor, which adjusts synaptic weights. Notably, 'c' decreases rapidly, diminishing dopamine's impact within five seconds post-STDP, aligning with experimental findings.

$$\dot{d} = -d/\tau_d + DA(t) \tag{3-3}$$

Dopamine, indicated by DA(t), is essential in neural signaling and stems from midbrain dopaminergic neuron activity (Cass & Gerhardt, 1995) [17]. In a stable environment, dopamine's absorption time is constant, around 0.1. In SNN simulations, Izhikevich used a slightly higher constant of 0.2, below that of the Prefrontal Cortex. The exact value of this constant can differ among neuron types (Wightman & Zimmerman, 1990) [18].

2.5 Target Tracking and Obstacle Avoidance

The combination of sensor technology with SNNs has resulted in an effective mobile robot navigator. Roberts et al. (2021) showcased the use of color detection sensors for accurate target identification and ultrasonic sensors for obstacle detection [13]. Through continuous learning, the robot improves its navigation capabilities, achieving precise target location and effective obstacle circumvention. The integration of SNNs with sensor inputs is key in mobile robot navigation, enabling advanced target tracking and obstacle avoidance. The system's neural network controller demonstrates emergent behaviors in these tasks without relying on predefined algorithms (see Fig. 3).

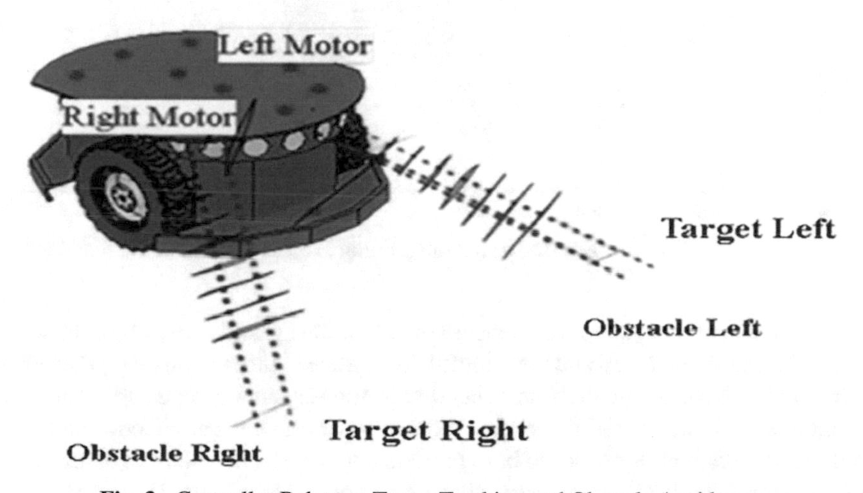

Fig. 3. Controller Balances Target Tracking and Obstacle Avoidance

In Fig. 3, the controller consists of sensory neurons that respond to the target's proximity (e.g., Target Left, Target Right) and activate as the target gets closer. Neurons designed to sense obstacles (e.g., Obstacle Left, Obstacle Right) activate to prevent collisions. The robot's motor actions, combined with STDP, strengthen the synaptic links between target-detecting neurons and motor outputs, directing the robot toward the target. Simultaneously, connections between obstacle-sensing neurons and motors are reinforced, improving the robot's obstacle-avoidance abilities.

3 Results and Discussion

In this first stage, our study focuses on a two-dimensional (2D) robot design. This concept is shown in Fig. 4, which shows a mobile robot with two wheels and independent velocities for its left (VL) and correct (VR) wheels. (θ) is the orientation of the robot, and 'b' is the separation between its wheels. This graphic representation effectively conveys the kinematic behaviors of the robot. The static obstacle is sporadically inserted into the 2D environment, making the robot's pathfinding task more difficult. The robot's primary goal is to find its way autonomously to a designated area, also referred to as the "Target" or "Goal" location. The robot starts its mission at a designated "Start" point. A crucial performance indicator is the navigation time, which measures the amount of time needed for the robot to make the journey from the starting "Start" position to the "Goal" destination.

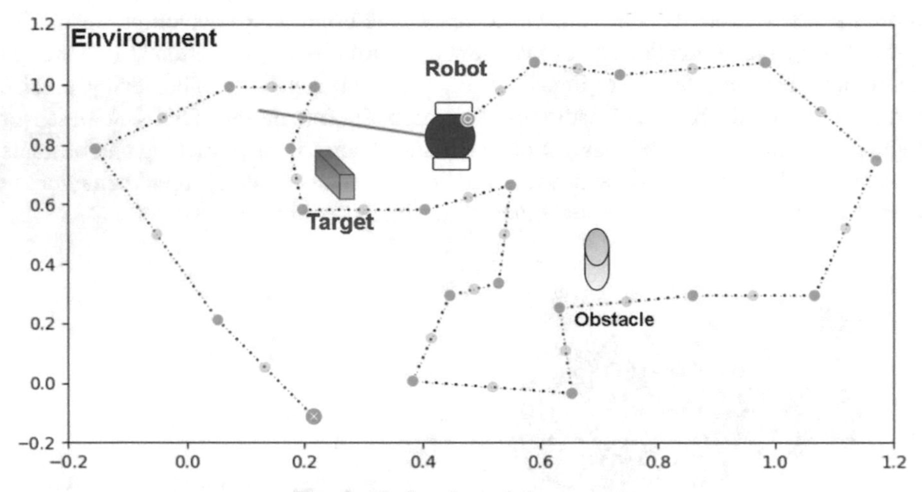

Fig. 4. During the training process

The training of the system commences with all initial synaptic weights set to one (see Fig. 5). Over time, a marked and meaningful divergence in the synaptic weights between the left and right motor neurons associated with the sensors becomes apparent. These adaptations are indicative of the system's operational effectiveness. Consequently, the robot exhibits distinct responses to both barriers and targets. A pattern of sensor activations captures the robot's initial, somewhat erratic movements toward the target, as well as its responses to encountered obstacles. These target and obstacle detections occur at regular intervals. Figure 4 provides a graphical representation of the robot's trajectory throughout this phase of training, visually chronicling its progress and behavioral adjustments.

3.1 Training Time and Synaptic Weight

The robot employs a Hebbian learning paradigm known as STDP to refine its navigational abilities. Through this process, it adjusts the synaptic connections between its sensors and motors based on the timing of their respective signals. A sensor signal preceding a motor signal results in a strengthened connection, while the reverse sequence leads to a weakening of the link. This dynamic learning approach significantly improves the robot's ability to adeptly maneuver through complex and ever-changing environments by allowing for real-time adjustments based on sensory feedback. Figure 5 illustrates the synaptic modifications that take place during the initial training phase of the robot's SNN, triggered by the initial dopamine release upon the robot's successful target acquisition. At this early stage of experimentation, we observe that a substantial number of synaptic weights undergo modification. However, it is essential to note that these changes are nascent and generally subtle, with values remaining close to their starting point of 1.0. While these initial shifts are encouraging, further exploration and experimentation are necessary to achieve more pronounced changes and to optimize the robot's learning system fully.

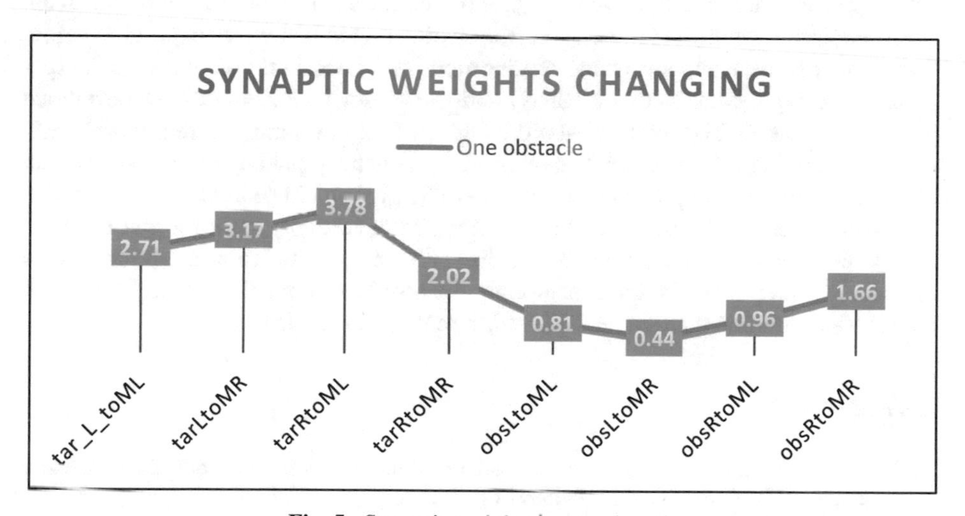

Fig. 5. Synaptic weights in one obstacle

The synaptic weights between sensors and motors in a robot's navigation system reveal how it has learned to respond to stimuli. A weight of 2.71 for 'tarLtoML' shows a strong connection that activates the left motor when a target is detected on the left, facilitating straightforward movement towards the target. 'tarLtoMR' has an even stronger weight of 3.17, indicating that the robot is likely to turn towards a left-detected target. Conversely, 'tarRtoML' at 3.78 signals a strong activation of the left motor in response to a target on the right, aiding in the robot's turning motion. 'tarRtoMR' at 2.02 is a weaker connection that still positively influences the right motor for refined target orientation. For obstacle avoidance, 'obsLtoML' at 0.81 suggests a decrease in left motor activation to steer away from obstacles on the left, while 'obsLtoMR' at 0.44 indicates a slight right motor activation for avoidance. 'obsRtoML' at 0.96 shows a moderate left motor response for dodging right-side obstacles, and 'obsRtoMR' at 1.66 points to a strong right motor activation to evade obstacles detected on the right. These weights collectively enable the robot to approach targets and navigate around obstacles effectively.

4 Conclusion

This research introduces a cutting-edge method for the training and guidance of AMRs in unfamiliar settings. Our strategy harnesses a specialized SNN that melds biologically inspired mechanisms, including STDP with dopamine signaling and the Izhikevich neuron model, to facilitate autonomous learning and navigation. Extensive simulations were conducted within the robotic simulation environment to evaluate the proposed system rigorously. The empirical data showcases the strategy's proficiency, achieving a remarkable 98% success rate in target acquisition with a minimal 2% incidence of collisions during encounters with an obstacle over 1000 simulated seconds. These results underscore the method's advanced navigational prowess for AMRs. A distinctive feature of the proposed model is its integration of ultrasonic and color detection sensors, forming a comprehensive sensor fusion framework. This fusion significantly bolsters the algorithm's performance in navigating through intricate scenarios. Future work based on the given conclusion would be to transition from simulation to real-world experimentation. This would involve equipping actual Autonomous Mobile Robots (AMRs) with the specialized Spiking Neural Network (SNN) and the sensor fusion framework developed during the research. These robots would then be tested in a variety of real-world environments, ranging from industrial settings with predictable patterns to dynamic urban landscapes with more unpredictable elements. The goal would be to validate and refine the model based on its performance in real-world conditions, ensuring that the high success rate and low collision incidence observed in the simulation are replicated in practice. This step is crucial for transitioning the research from a theoretical framework to a practical solution that can be deployed in real-world applications.

References

1. Cevik, T., Cevik, N., Rasheed, J., et al.: Facial recognition in hexagonal domain—a frontier approach. IEEE Access **11**, 46577–46591 (2023)
2. Rasheed, J., Waziry, S., Alsubai, S., Abu-Mahfouz, A.M.: An intelligent gender classification system in the era of pandemic chaos with veiled faces. Processes **10**, 1427 (2022)
3. Goyal, S., Benjamin, P.: Object recognition using deep neural networks: a survey. CoRR abs/1412.3684 (2014)
4. Sutton, R.S., Barto, A.G.: Introduction to Reinforcement Learning, 1st edn. MIT Press, Cambridge (1998)
5. Yang, G.S., Chen, E.K., An, C.W.: Mobile robot navigation using neural Q-learning. In: Proceedings of 2004 International Conference on Machine Learning and Cybernetics (IEEE Cat. No.04EX826), pp. 48–52 (2004)
6. Farooq, M.S., Khalid, H., Arooj, A., et al.: A conceptual multi-layer framework for the detection of nighttime pedestrian in autonomous vehicles using deep reinforcement learning. Entropy **25**, 135 (2023)
7. Li, Y., Pan, Q., Chu, F.: Reinforcement learning and deep learning based lateral control for autonomous driving. Neurocomputing **416**, 316–328 (2020)
8. Izhikevich, E.M.: Solving the distal reward problem through linkage of STDP and dopamine signaling. Cereb. Cortex **17**(10), 2443–2452 (2007)
9. Arena, P., Fortuna, L., Frasca, M., Patané, L.: Learning anticipation via spiking networks: application to navigation control. IEEE Trans. Neural Netw. **20**(2), 202–216 (2009)

10. Shamsfakhr, F., Bigham, B.S.: A neural network approach to navigation of a mobile robot and obstacle avoidance in dynamic and unknown environments. Turk. J. Electr. Eng. Comput. Sci. **25**(3), 1629–1642 (2017)
11. Pandey, A., Parhi, D.R.: Optimum path planning of mobile robot in unknown static and dynamic environments using Fuzzy-Wind Driven Optimization algorithm. Defence Technol. **13**(1), 47–58 (2017)
12. Abubaker, B.A., Razmara, J., Karimpour, J.: A novel approach for target attraction and obstacle avoidance of a mobile robot in unknown environments using a customized spiking neural network. Appl. Sci. **13**, 13145 (2023)
13. Roberts, J., et al.: Enhancing navigation capabilities through sensor integration in spiking neural networks for MRs. IEEE Trans. Rob. **38**(2), 1–14 (2021)
14. Kumar, A., et al.: Sensor integration for mobile robot navigation using spiking neural networks. Robot. Auton. Syst. **141**, 1–12 (2022)
15. Jiang, Z., Gao, S.: An intelligent recommendation approach for online advertising based on hybrid deep neural network and parallel computing. Clust. Comput. **23**, 1987–2000 (2020)
16. Izhikevich, E.M., Desai, N.S.: Relating STDP to BCM. Neural Comput. **15**(8), 1511–1523 (2003)
17. Cass, W.A., Gerhardt, G.A.: In vivo assessment of dopamine uptake in rat medial prefrontal cortex: comparison with dorsal striatum and nucleus accumbens. J. Neurochem. **65**(1), 201–207 (1995)
18. Wightman, R.M., Zimmerman, J.B.: Control of dopamine extracellular concentration in rat striatum by impulse flow and uptake. Brain Res. Rev. **15**(2), 135–144 (1990)

Blockchain-Based Secured Estate Property Registration System: Baghdad Estate Registration as a Case Study

Tiba W. Al-dulaimy(✉) and Saad N. Alsaad

Department of Computer Science, Mustansiriyah University, Baghdad, Iraq
{tiba_waleed,dr.alsaadcs}@uomustansiriyah.edu.iq

Abstract. Estate property is a very important document to guarantee the rights of the owners. The issue of maintaining these documents is an arduous issue. Because the owner of a property may change frequently, it is challenging to maintain exhaustive and lengthy records of ownership transfers. The problems are exacerbated by falsified or incomplete records that are difficult to track over time. In this research, a system based on blockchain technology is proposed to create and implement a smart contract to facilitate the work of the system. It is implemented on the Binance Smart Chain platform to keep estate property records and to prevent manipulation and fraud. The system uses the keccak256 algorithm to create a unique hash for each bond based on its information. This hash helps to verify the validity and integrity of the transaction and to ensure that there are no two identical bonds. The proposed system is designed for Baghdad estate registration as a case study that uses formal information as practical implementation and as the initial stage for developing the whole system. The system increases reliability and makes the circulation of real estate transactions transparent, efficient, and secure.

Keywords: Binance Smart Chain · Blockchain · Estate Property · IPFS · Keccak256

1 Introduction

Estate property registration systems capture and preserve property information to present ownership proof, speed up transactions, and prevent illegal activities. But as the world around us is changing more quickly, maintaining the integrity of properties has become a major concern [1]. The property registration system currently in use may have significant issues concerning sensitive data security, fraud, and data falsification [2, 3]. As is the case in the real estate registration system in the city of Baghdad, where real estate data relied on paper preservation, which made it suffer from many problems such as the loss of documents proving ownership or false persons claiming ownership. Although there are attempts to move towards digitization and rely on the process of keeping real estate records in central databases, fraud. Data security and data protection still pose challenges. Ultimately, whether records are kept in traditional offices or a centralized system, losing them would be devastating [4]. Generally, the Estate Property Registration

J. Rasheed et al. (Eds.): FoNeS-AIoT 2024, LNNS 1035, pp. 134–144, 2024.
https://doi.org/10.1007/978-3-031-62871-9_12

system is crucial to protecting property rights and it is crucial to keep its data accurate and unchangeable [5]. An organization's success is measured by its use of modern technology systems for data management. These systems simplify data processes and provide efficient solutions for accessing and retrieving reports [6, 7]. Distributed ledgers are groups of digital data that are geographically scattered, duplicated, synchronized, and shared across several places. Through dispersed network nodes, this technology permits the generation and distribution of transactional data [8]. In distributed ledgers, there is no central authority, but each node has access to and a copy of the ledger [5]. Blockchains are a technology used to implement these ledgers that allow a block of data to be created and linked with other blocks, each block contains transactional information that is stored in a structure designed [8]. For estate property registration, blockchain technology is very dependable and trustworthy due to its transparency and immutability [9].

In this research, blockchain technology was proposed for the estate property registration system. It is used to protect property information and to maintain the complete historical record of the property. The features of the Binance SmartChain platform were exploited to deploy the smart contract and store the data on the blockchain. The interplanetary file system IPFS was also used to keep the original copies of the real estate bond images.

2 Related Work

Generally, the property registration system (PRS) is highlighted in several literary works. Below we review some important articles that dealt with this topic: Ramya et al. implemented PRS using a private blockchain. It includes the implementation of the document state registration in the blockchain. The system verifies using the data stored in the digital locker and thus reduces the percentage of document forgery [10]. The problem concerns the hashing technique that is greatly affected even by minor changes in the original document (for example Character fading, ink marks, etc.). Nizamuddin et al. proposed a completely decentralized system based on the blockchain, eliminating dependence on a trusted third party. The smart contract controls all interactions and transactions between members [11]. Krishnapriya et al. designed a system that uses a blockchain platform for land registration using a SHA256 algorithm that produces a unique hash for each block. They also use the Proof of Work (PoW) method to verify the transaction before distributing blocks to nodes in the network, which can make the system costly as well [12]. The authors in [13] designed a smart contract for land and real estate registration in the blockchain network. They devised a method for adding all property owners to the blockchain network as members. Similar to this, a record of the property that contains information on the real estate is kept in the NoSQL DB. Nandi et al. proposed designing a record-keeping mechanism using the Ethereum platform based on blockchain technology to convert physical assets into liquid and secure blockchain-based token assets [14]. In [15], the authors used the Interplanetary File Storage System to create a scheme for storing farmers' information and land records. To prevent tampering and forgery, the blockchain technology stores the address of the data hash generated by IPFS [15]. Yadav et al. proposed a system to digitize real estate transactions using the blockchain to reduce the possibility of document falsification and fraud. They also proposed a consensus algorithm to reduce the upstream transmission. The authors compared it with five prominent consensus algorithms and concluded their speed according to the results of the comparison [16]. Vashist et al., based on Blockchain, designed a land registry system using modern technologies to speed up the registration process. They also used the "Will" feature as an additional feature for registered land information [17]. Gaikwad et al. present a blockchain-enabled decentralized data warehousing system and its interactions with an Ethereum Virtual Machine (EVM) to create a reasonable smart contract for land management on the blockchain [4]. Finally, the authors in the paper [18], created a system for recording property rights and land data certificates. They also used a QR code scan to verify the authenticity of the document, so that the system removed the encryption and matched it with the document data coming from the network to complete the registration process.

3 Background

3.1 Blockchain

A blockchain is a decentralized, open-source, public digital ledger that is unchangeable once it has been recorded [19]. It is distributed and shared across a network of nodes where each node maintains a record of all transactions [20]. Blockchain uses methods of time signature, digital encryption, and distributed consensus to carry out trusted transactions between untrusted parties [21]. Blockchain is a list of blocks connected, the first block to be created in the chain without a parent is called the "genesis block" [22]. Each block contains transaction information, and through cryptographic hashing, these blocks are linked to each other [19].

Recent researches show that blockchain is an effective solution to deal with issues such as high cost, insecure data storage, and low efficiency [23]. Blockchain technology has many characteristics that make it suitable for building many applications that require maintaining the security and integrity of data. These characteristics can be summarized as follows [24]:

- Transparency: Blockchain's contents are accessible to all subscribers, and they are automatically updated when new blocks are added.
- Immutability: Once data is recorded within the blockchain it becomes very difficult to modify. Essentially a blockchain is a chain of data that can be distinguished by a unique code for each block called a "hash". By referring to the hash of the preceding block, each resultant block is linked to earlier blocks. The avalanche effect refers to the fact that any slight modification to the plaintext might result in a significant change to the ciphertext. Meaning that any change to the block's content will result in an avalanche, which will modify the block's hash.
- No Single Point of Failure: A blockchain database, in contrast to centralized databases, is not maintained and stored in a single location, making the records and data it holds truly public and simple to verify. The database is replicated across all network nodes, so a single blockchain can include millions of copies.
- Decentralization: Due to the peer-to-peer network architecture, the blockchain network is distributed and decentralized. Since information or copies of the blockchain

are spread over millions of nodes in the network, the hacker cannot modify the information of a particular block, he must make the same modifications to all copies on every node in the network. Due to the need for large amounts of computing resources, the possibility of this attack is limited.

3.2 Binance Smart Chain (BSC)

Ethereum provides solutions to many of the difficulties with the Bitcoin-First Blockchain, however, it has concerns with cost, scalability, and block time. The BSC, an upgraded version of the original Binance Chain, was released in 2020. [25]. It was created as an additional platform to the original one. Similar to Ethereum, BSC provides choices for decentralized applications (app) developers, allowing them to implement smart contracts and deploy on other Blockchains that support EVM [26], The disruptive architecture known as BSC has features including smart contract support, cheaper costs, extremely high throughput, higher scalability, and shorter block times. A robust Proof-of-Authority stack consensus technique that combines Proof-of-Authority and Proof-of-Stake is used by BSC [25]. Both Binance Chain and Binance Smart Chain were created with cross-chain compatibility in mind. Due to the quick transaction capabilities of the original version and the smart contracts of the expanded version, assets can be transferred between blocks via BSC [26]. Table 1 represents the comparison between Binance Smart Chain and Ethereum.

Table 1. Ethereum and Binance smart comparison

Features	Binance Smart Chain	Ethereum
Launch	2020	2015
Consensus mechanism	Proof of stacked authority	Proof of work
Chain architecture	Dual chain	Single chain
Block time	~3 s	~7–15 s
Number of validators	21	Thousands
Validators selection	Taking turns	Solving puzzles
Throughput	High	Low
Transaction cost	Low	High
Scalability	high	low
Reward token	BNB	ERC20
Validators type	Private	Public

3.3 Interplanetary File System (IPFS)

IPFS (InterPlanetary File System) is a decentralized and distributed file system that connects computers and enables them to share a common file system. It operates on a peer-to-peer network similar to BitTorrent, allowing users to exchange files directly with each other [19, 23]. Unlike the traditional HTTP protocol, IPFS provides a decentralized alternative that is resistant to censorship and offers a multitude of advantages. One of the key features of IPFS is its reliance on a Distributed Hash Table (DHT) to retrieve node connection data and locate files. This decentralized approach ensures that there is no single point of failure in the system. Even if a device is unplugged or disconnected, the files stored on IPFS can still be accessed, making it highly resilient [19]. IPFS also utilizes content addressing, which means it provides a storage model based on blocks of content that are efficiently linked together. This content-rich bridging enables highly efficient content retrieval and distribution [23, 27]. By ad-dressing files based on their content, rather than their location, IPFS enhances data integrity and availability. The benefits of IPFS extend beyond resilience and efficient content addressing. It offers enhanced security, as files on IPFS are identified by their cryptographic hashes, ensuring the integrity and authenticity of the data. Moreover, IPFS enables bandwidth optimization by utilizing a distributed network of nodes for file distribution, reducing the load on individual servers and improving overall network performance [28].

4 Scheme Methodology

For secure estate property registration, a blockchain-based system is proposed. The main objective of the system is to secure estate property information from manipulation and forgery and to easily maintain estate records that facilitate the process of tracking ownership change transactions. Figure 1 illustrates the system architecture that describes the main components and the connections between them. It starts from the upper corner and ends at the right lower corner.

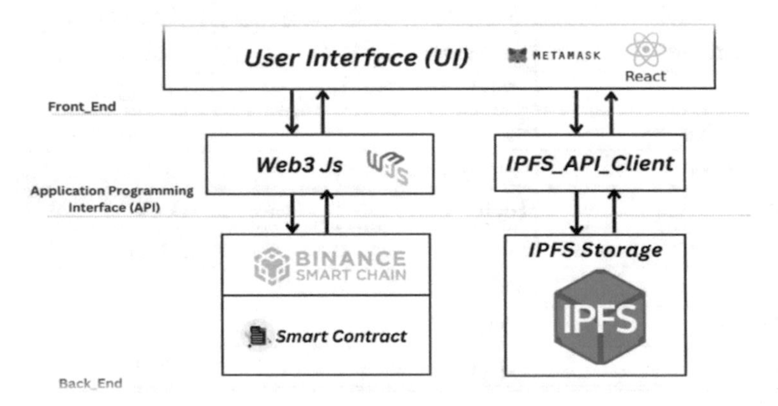

Fig. 1. General system architecture.

The main components used in the system can be summarized as follows:

- Smart Contract: A smart contract has been implemented in the system to facilitate secure and efficient storage and to manage estate property records. It responsible

party that has the authority to regulate and control access to the system, including adding documents and searching the registry.
- Binance Smart Chain: The Binance Smart Chain Testnet (BSC) network was used to deploy the smart contract. One of the reasons for choosing this platform over others was the fact that BSC had chosen Proof of Staked Authority or PoSA which reduces transaction cost and shortens block time.
- Meta Mask Wallet: It is a widely used cryptocurrency wallet that allows a person to access multiple wallets such as his bsc wallet. It has been used to securely connect to this decentralized application, and manage and store account keys.
- Keccak256: Keccak-256 is a crucial tool for ensuring security in smart contracts, as it ensures the generation of unique keys and addresses for each user and contract, verification of digital signatures, and protection of data from hacking and tampering. Since our smart contract relies heavily on security and reliability, employing Keccak-256 assist us in meeting these goals.
- IPFS: To enhance the accuracy and reliability of real estate records, the Interplanetary File System(IPFS) was employed to hold an image of the original copy of estate property title deeds. The image is stored in IPFS, and its hash value is required to add the title deeds to the blockchain. This measure ensures that only valid and authorized title deeds are added to the blockchain.

The main steps of the system with highly abstracted details are described in Algorithm 1.

Algorithm 1: high abstracted steps of the proposed system.

Input: property bond.

Output: a record of Encrypted data and secure form to be stored in the blockchain system.

Step 1: Connection: Connect to the authorized Meta Mask wallet.

Step 2: Submit Upload Folder: upload the image and store it on IPFS, then compute the hash value.

Step 3: Hash Generation: Compute the hash of keccak256 for the estate property bond data.

Step 4: Saving: save the estate property bond data file with the IPFS hash value and keccak256 hash to the blockchain.

5 Simulation of the Scheme

The implemented system has undergone testing using real-life documents from the estate registration in Baghdad. Figure 2 provides an illustrative example of one such document that was utilized during the testing phase. Below, we outline the tracking steps involved in the process:

- **Generate IPFS hash:** the uploaded image of the original estate property document i is converted to hash value during IPFS. The hash value is considered as the identifier of the document image on the distributed ipfs web. Below is the hash value to the document: (TfU1WUPyM513fZwHm2odKsPWeNZQDN7de5Jzgj34dL2ywP79)

- **keccak256 hash Generation:** After the estate property information and IPFS hash value are completed, the keccak256 hash is generated automatically in smart contract using keccak256(abi.encodePacked()) (keccak256 hash value of the bond is "0x7f2e54845aa0ec7028720358d674aad6845ff2daba308611").

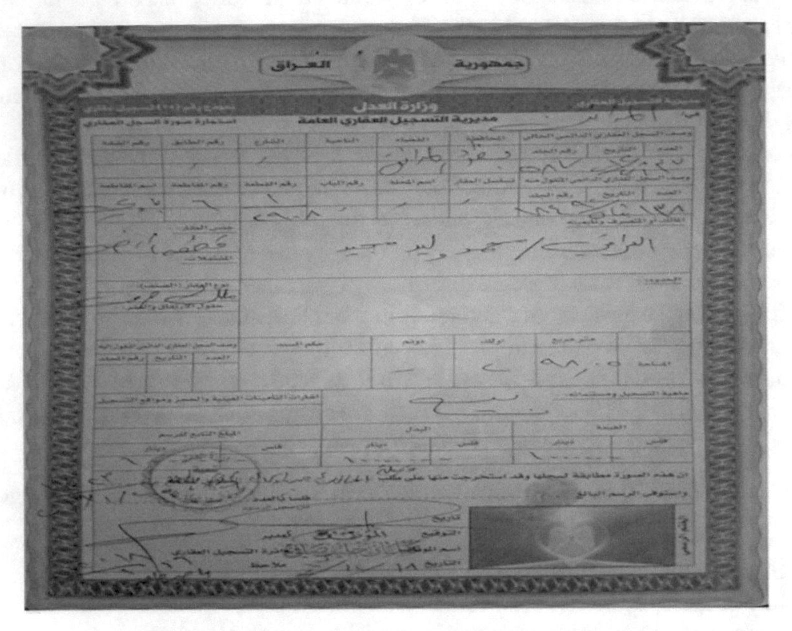

جمهورية العراق

وزارة العدل

مديرية التسجيل العقاري العامة

Fig. 2. Practical bond of Baghdad estate registration.

- **Create a new block:** Ownership data, ipfs hash, and keccak256 hash are saved in the blockchain. A direct request to obtain permission to add will be sent through the MetaMask wallet as shown below in Fig. 3. After approval of the process, a message will appear as in Fig. 4 showing the status of acceptance of the request and completion of the process of adding bonds to the Blockchain.

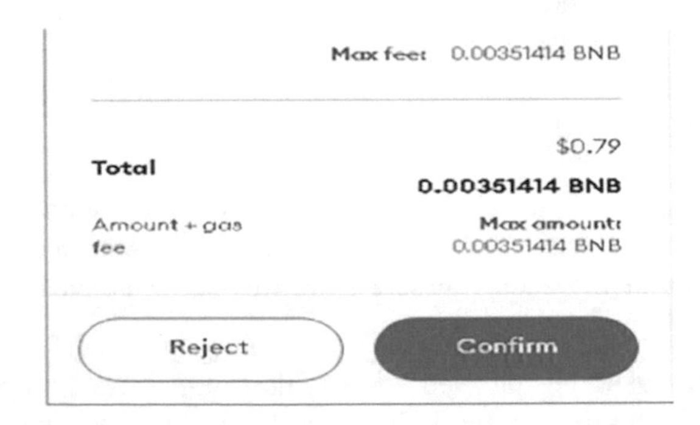

Fig. 3. Request permission to add a block.

Fig. 4. Message of the request status.

- **Search process:** Once the storage process of real estate documents on the blockchain is completed, authorized parties gain the ability to search and retrieve information from the registry. This search process allows users to inquire about specific documents by entering relevant search criteria, such as the name of the owner or the property number. By doing so, the information stored on the blockchain regarding the requested owner or property becomes accessible. Figure 5 illustrates the user interface for the Search by Owner Name feature. This screen provides a designated space where users can input the name of the owner they wish to inquire about. Alternatively, users may also perform a search using the property number. Upon initiating the search, the system retrieves the corresponding information associated with the owner or property from the blockchain.

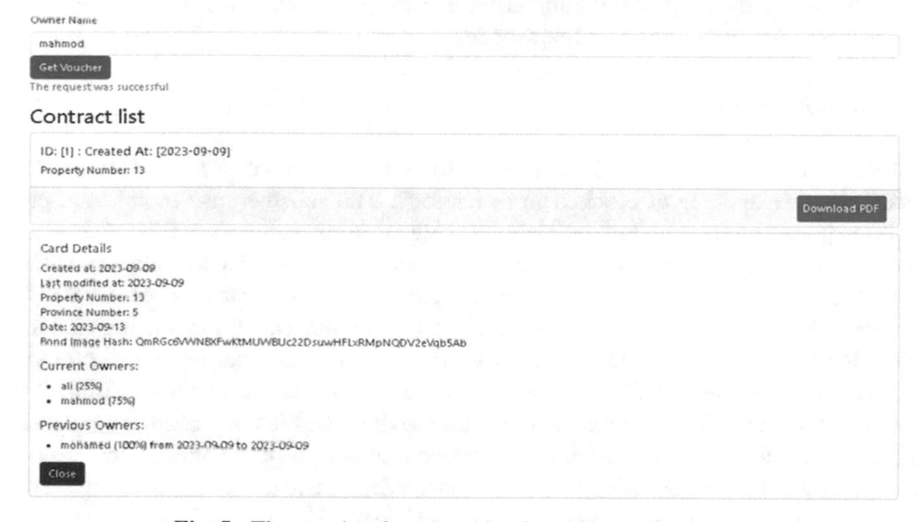

Fig. 5. The voucher form resulting from the search process.

- **Create a PDF file:** A downloadable and printable PDF file is generated using the React-PDF library that includes the bond information from the blockchain and the corresponding bond image stored on IPFS. The PDF file can be downloaded and printed by clicking the download button shown in Figure 5, providing a comprehensive document containing details of the bond, owner's name register, dates of modifications, and an associated image.

6 Results and Discussion

The proposed system has been developed as a web application for real estate registration in Baghdad. It utilizes the Binance SmartChain Blockchain as a secure database and incorporates a smart contract with designated rates and functions. The user interface is built using ReactJS, including libraries like React-Bootstrap and React-PDF for a flexible and user-friendly experience. The Ethereum JS API libraries, such as web3.js, connect the back end to the front end. Additionally, the interplanetary file storage system (IPFS) is used to store original title deed images via the ipfs-http-client library.

After testing and simulating the system, the following results were obtained:

- Efficient Database: Blockchain efficiently stores and manages large amounts of information, making it suitable for storing extensive records like real estate bonds.
- Fraud Prevention: The immutable and decentralized nature of blockchain provides a secure environment for property registration, minimizing the risk of fraud.
- Easy Access: The system allows easy access to previously stored real estate bonds, providing a detailed historical record of updates with corresponding dates.
- Speed and Cost Reduction: Compared to traditional systems, the proposed solution reduces resource consumption (e.g., paper, ink, storage servers) and time and effort spent on storage, search, and query operations. It leverages the Binance Smart Chain network as a database, facilitating faster and more cost-effective transactions while ensuring ease of use within a few seconds.

7 Conclusions

We have proposed a decentralized scheme to safeguard estate property information, leveraging the features of blockchain technology. The scheme ensures security, protection against tampering, and efficient tracking of information over time. Our smart contract promotes transparency, protection, access control, and identification of authorized parties for utilizing services like data addition and real estate record searches. To ensure data integrity, we employed the keccak256 function, a deterministic function that consistently produces the same output given the same inputs. Traditional real property record-keeping systems lack a comprehensive historical record, making it difficult to obtain a complete picture of real property transactions and leaving room for potential fraud and the inclusion of fraudulent transactions. In our proposed system, this issue is resolved by leveraging the advantages of Binance Smart Chain (BSC), enabling secure storage and efficient querying of estate property history and associated bond information at a lower cost. BSC has demonstrated its ability to process transactions swiftly through the utilization of a Proof of Staked Authority (PoSA) consensus mechanism, ensuring high throughput and low latency. Moreover, BSC offers lower transaction fees compared to Ethereum, making it a more cost-effective solution in terms of storage. This is particularly important considering the frequent updates of property title bonds. By addressing these challenges, we expect that our proposed system for protecting real estate property will contribute to the development of decentralized storage and safeguarding of estate property bonds.

References

1. Sharma, R., Galphat, Y., Kithani, E., Tanwani, J., Mangnani, B., Achhra, N.: Digital land registry system using blockchain. In: Proceedings of the 4th International Conference on Advances in Science & Technology (ICAST2021) (2021). https://doi.org/10.2139/ssrn.3866088
2. Wardak, A.B., Rasheed, J.: Bitcoin cryptocurrency price prediction using long short-term memory recurrent neural network. Eur. J. Sci. Technol. **38**, 47–53 (2022). https://doi.org/10.31590/ejosat.1079622
3. Albahadily, H.K., Jabbar, I.A., Altaay, A.A., Ren, X.: Issuing digital signatures for integrity and authentication of digital documents. Al-Mustansiriyah J. Sci. **34**(3), 50–55 (2023). https://doi.org/10.23851/mjs.v34i3.1278
4. Gaikwad, D., Hambir, A., Chavan, H., Khedkar, G., Athawale, D.: Real estate land transaction system using blockchain. Int. J. Res. Appl. Sci. Eng. Technol. **10**(3), 307–311 (2022)
5. Ladić, G., Milosavljević, B., Nikolić, S., Sladić, D., Radulović, A.: A blockchain solution for securing real property transactions: a case study for Serbia. ISPRS Int. J. Geo Inf. **10**(1), 35 (2021)
6. Al-Kendi, W.B., Al-Nayyef, H.H.: Data management via QR code using android smart devices. Al-Mustansiriyah J. Sci. **31**(3), 95–100 (2020). https://doi.org/10.23851/mjs.v31i3.853
7. Hashim, H.B.: Challenges and security vulnerabilities to impact on database systems. Al-Mustansiriyah J. Sci. **29**(2), 117–125 (2018)
8. Natarajan, H., Krause, S., Gradstein, H.: Distributed ledger technology and blockchain. World Bank, Washington, DC (2017)
9. Yadav, A.S., Agrawal, S., Kushwaha, D.S.: Distributed ledger technology-based land transaction system with trusted nodes consensus mechanism. J. King Saud Univ.-Comput. Inf. Sci. **34**(8), 6414–6424 (2022)
10. Ramya, U.M., Sindhuja, P., Atsaya, R.A., Bavya Dharani, B., Manikanta Varshith Golla, S.S.: Reducing forgery in land registry system using blockchain technology. In: Luhach, A.K., Singh, D., Hsiung, P.-A., Hawari, K.B.G., Lingras, P., Singh, P.K. (eds.) ICAICR 2018. CCIS, vol. 955, pp. 725–734. Springer, Singapore (2019). https://doi.org/10.1007/978-981-13-3140-4_65
11. Nizamuddin, N., Salah, K., Azad, M.A., Arshad, J., Rehman, M.H.: Decentralized document version control using Ethereum blockchain and IPFS. Comput. Electr. Eng. **76**, 183–197 (2019)
12. Krishnapriya, S., Sarath, G.: Securing land registration using blockchain. Procedia Comput. Sci. **171**, 1708–1715 (2020)
13. Ali, T., Nadeem, A., Alzahrani, A., Jan, S.: A transparent and trusted property registration system on permissioned blockchain. In: 2019 International Conference on Advances in the Emerging Computing Technologies (AECT), pp. 1–6. IEEE (2020)
14. Nandi, M., Bhattacharjee, R.K., Jha, A., Barbhuiya, F.A.: A secured land registration framework on blockchain. In: 2020 third ISEA conference on security and privacy (ISEA-ISAP), pp. 130–138. IEEE (2020)
15. Devi, D., Rohith, G.S., Hari, S.S., Ramachandar, K.S.: Blockchain-based mechanism to eliminate frauds and tampering of land records. In: ITM Web of Conferences, vol. 37, p. 01011 (2021)
16. Yadav, A.S., Kushwaha, D.S.: Blockchain-based digitization of land record through trust value-based consensus algorithm. Peer-to-Peer Netw. Appl. **14**(6), 3540–3558 (2021)
17. Vashist, R., Swarup, S., Sood, S., Chauhan, A.: My land my will - a novel blockchain based land registry system (2020)

18. Kusuma, M.A., Sukarno, P., Wardana, A.A.: Security system for digital land certificate based on blockchain and QR code validation in Indonesia. In: 2022 International Conference on Advanced Creative Networks and Intelligent Systems (ICACNIS), pp. 1–6. IEEE (2022)
19. Alah, K., Alfalasi, A., Alfalasi, M.: A blockchain-based system for online consumer reviews. In: IEEE INFOCOM 2019 - IEEE Conference on Computer Communications Workshops (INFOCOM WKSHPS), pp. 853–858. IEEE (2019)
20. Thakur, V., Doja, M.N., Dwivedi, Y.K., Ahmad, T., Khadanga, G.: Land records on blockchain for implementation of land titling in India. Int. J. Inf. Manage. **52**, 101940 (2020)
21. Xu, R., Zhang, L., Zhao, H., Peng, Y.: Design of network media's digital rights management scheme based on blockchain technology. In: 2017 IEEE 13th International Symposium on Autonomous Decentralized System (ISADS), pp. 128–133. IEEE (2017)
22. Benisi, N.Z., Aminian, M., Javadi, B.: Blockchain-based decentralized storage networks: a survey. J. Netw. Comput. Appl. **162**, 102656 (2020)
23. Muwafaq, A., Alsaad, S.N.: Design scheme for copyright management system using blockchain and IPFS. Int. J. Comput. Digit. Syst. **10**, 613–618 (2021)
24. Savelyev, A.: Copyright in the blockchain era: promises and challenges. Comput. Law Secur. Rev. **34**(3), 550–561 (2018)
25. Monga, S., Singh, D.: MRBSChain a novel scalable medical records binance smart chain framework enabling a paradigm shift in medical records management. Sci. Rep. **12**(1), 17660 (2022)
26. Le Quoc, K., Trong, P.N., et al.: Letter-of-credit chain: cross-border exchange based on blockchain and smart contracts. Int. J. Adv. Comput. Sci. Appl. **13**(8), 890–898 (2022)
27. Khan, S.N., Loukil, F., Ghedira-Guegan, C., Benkhelifa, E., Bani-Hani, A.: Blockchain smart contracts: Applications, challenges, and future trends. Peer-to-Peer Netw. Appl. **14**, 2901–2925 (2021)
28. Sarıtekin, R.A., Karabacak, E., Durgay, Z., Karaarslan, E.: Blockchain based secure communication application proposal: cryptouch. In: 2018 6th International Symposium on Digital Forensic and Security (ISDFS), pp. 1–4. IEEE (2018)

Detecting BGP Routing Anomalies Using Machine Learning: A Review

Ali Hassan Muosa[1(✉)] and A. H. Ali[2]

[1] Department of Computer Science, Faculty Science Computers and Mathematics, University of Thi-Qar, Nasiriyah, Iraq
ali.h.m@utq.edu.iq

[2] Department of Electronic and Communication, University of Kufa, Najaf, Iraq

Abstract. The Border Gateway Protocol (BGP) is a protocol for exchanging IP prefixes online. It allows for incremental path-vector routing protocol and reachability between Autonomous Systems (ASes), enabling efficient global internet activity. BGP anomalies have been known to cause ASes to malfunction in various ways; therefore, detecting them is critical. Machine Learning (ML) solutions currently guarantee improved BGP irregularity recognition based on BGP update messages volume and manner features, which are frequently boisterous and bursty. This work organizes these anomalies and provides the most recent approaches for detecting discrepancies. We also look into a few crucial requirements for the up-and-coming detection of internet routing abnormalities methods.

Keywords: BGP · Anomaly Detection · Machine Learning · Inter-Domain Routing · Autonomous Systems

1 Introduction

BGP Cornerstone of Internet Routing and Contributions to Anomaly Detection The Internet, the lifeblood of this digital age, relies on a precise and complex routing system at its core, known as the Border Gateway Protocol (BGP). This protocol connects thousands of administrative areas, called autonomous systems (ASes), where access to the Internet Protocol (IP) address space is exchanged [1].

Given the widespread use of BGP and the sensitivity of the data being replicated in independent systems, any misconfiguration or protocol failure puts the stability of the Internet at risk. Therefore, detecting anomalies in the BGP system is crucial to maintaining the stability and integrity of the Internet [2]. This field of research has witnessed great activity, and many methods have been developed to detect anomalies. These approaches differ significantly, especially in the type of data processed and the detection algorithms used. Detection algorithms are based on fields as diverse as statistics, probability, information theories, and computer science [3].

In this context, our work presents a classification of anomaly detection methods in BGP based on the nature of the data used in the initial detection. We also review some machine learning (ML) concepts to provide established detection methods in this

J. Rasheed et al. (Eds.): FoNeS-AIoT 2024, LNNS 1035, pp. 145–164, 2024.
https://doi.org/10.1007/978-3-031-62871-9_13

field. Machine learning is the area directly related to our most important contributions. Detecting anomalies in the BGP protocol ensures the stability of the Internet we all depend on. This area of research represents an enormous technical challenge due to the complex nature of BGP. Contributions to the field of BGP anomaly detection can significantly improve the reliability and security of the Internet [4].

A classification of BGP anomaly detection methods based on the nature of the data used, a review of machine learning concepts relevant to BGP anomaly detection, and the development of new anomaly detection methods using machine learning are important contributions of the research.

Future research focuses on developing more accurate anomaly detection methods using deep machine learning. Research new ways to analyze multi-source data (such as BGP logs and other network data) to detect anomalies more effectively. Improve collaboration between researchers and network operators to improve the effectiveness and widespread dissemination of anomaly detection methods. Understanding the BGP protocol and the importance of anomaly detection is essential to ensuring the sustainability of the Internet on which we all depend. Research in this field contributes significantly to improving the security and reliability of the Internet's basic infrastructure.

2 BGP Background

Despite the expectation behind these anomalies, regardless of whether they are noxious, for example, misconfiguration, power outage, and BGP tend to be anomalies that prevent having achieved successful conversion of attainability messages and allow great in size or quantity capacity of anomalous update messages [2], worms or focused on assaults, misconfiguration, or connection disappointments, assess several infringements and anomaly detection mechanisms [3]. There is growing interest in detecting and resolving BGP discrepancies by observing BGP traffic [4] rather than depending on the massive scope and sending arrangements, such as Resource Public Key Infrastructure (RPKI), to limit the danger of hijacking [5, 6]. Therefore, various adjustments are recommended to enhance BGP security [7] (Moscow Power Blackouts) link failures [8], Internet Protocol (IP) prefix hijacks [9], and misconfiguration directing [10]. Most of the research devoted to outlier detecting [11–13] is working to diagnose or identify anomalous behavior in BGP messages and to alert this. The features are extracted using and updating the raw BGP data, and various anomalies are classified [14, 15] using a tool to convert raw BGP messages into time series. A dataset was created that helps detect anomalies [16].

2.1 Border Gateway Protocol (BGP)

BGP is working on Transmission Control Protocol (TCP), one AS routing protocol. BGP is the only one to handle the internet protocol network size and the only one to handle routing domain protocol among multiple connections properly. BGP is based on the EGP (Exterior Gateway Protocol) [1] experience. The BGP system's primary function is communicating network reachability information with other BGP systems. The network reachability information includes the given AS information [2]. This information effectively builds the AS interconnection's topological diagram, thus eliminating routing loops. At the same time, policy decisions can be implemented at the AS level [11] (see Fig. 1).

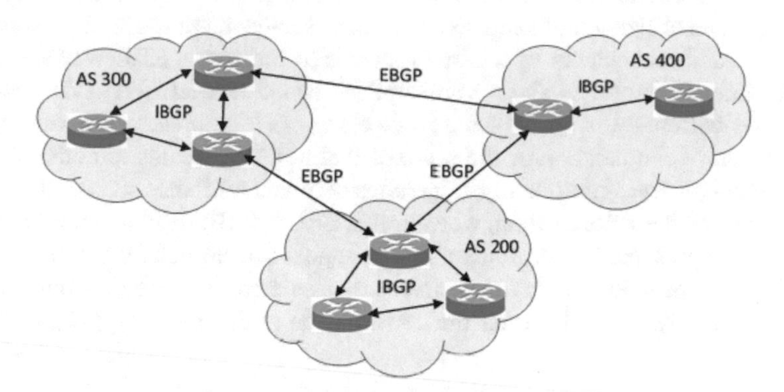

Fig. 1. Overview of a sample border gateway protocol topology.

Since it is possibly connected to various ASes, numerous border routers are probably working BGP inside an AS [13]. IBGP (Internal/Internal BGP) is a type of BGP that works between two or more peer entities on the same AS. BGP working among peer entities among various ASes is named EBGP (External/Exterior BGP) [13]. The BGP neighbor relation (or peer entity) is realized from manual configuration [15], and the peer entities' data were exchanged over TCP (via port number 179) sessions [11]. BGP routers at any AS periodically send keep-alive messages that have a size of 19-byte to preserve the connection relationship (the default periodic be 30 s). Among routing protocols [16], Only uses BGP TCP as the Transport Layer Protocol [17].

BGP is an inter-domain routing or external protocol. Path Vector is very useful or important information to looking for AS number to BGP route update effectively and efficiently avoiding loops. There are no restrictions on network topology performed by BGP [18]. BGP does not need to transfer full routing database information in every routing update message, but it just necessarily transfers full information once in startup running [19]. Announces updated routing messages only when network information changes later [20]. BGP supports CIDR and VLSM techniques [12]. The network prefix and subnet mask describe and announce networks. BGP allows the sender to aggregate routing information and single-use entry to mention multiple destination networks and economize network bandwidth [21]. BGP also allows the receiver to investigate the sender's identity at authentication and message authentication [22].

2.2 BGP Message Type

The Border Gateway Protocol (BGP) has four primary message types [2], each comprising a BGP header message, (see Fig. 2) [22]. The primary four messages can be summarised as follows: The following is an open message transmitted by a router in an attempt to establish communication with its peer. Upon receiving this communication, the recipient will subsequently transmit an OPEN message to verify and authorize the establishment of the Border Gateway Protocol (BGP) session. This communication enables a router to publicly declare its Autonomous System (AS) number, identifier, and available functionalities, including but not limited to IPv6 capability, as depicted in (see Fig. 3). The utilization of keepalive messages serves the purpose of preserving the session. The default behavior of a Cisco router is to transmit a KEEPALIVE message at regular intervals of 60 s. The session will be ended if a KEEPALIVE message is not received within 180 s, as depicted in (see Fig. 4). The present message provides a comprehensive enumeration of the prefixes that a router intends to communicate or withhold from its peer. Additionally, it includes pertinent attributes such as the AS path, the next hop, and the route's origin, whether it is EBGP or IBGP. Please refer to the (see Fig. 5) visual representation. A notification message is transmitted by a Border Gateway Protocol (BGP) peer to inform the recipient that an error has been encountered. The session is promptly concluded after the transmission of this message [23], as depicted in (see Fig. 6).

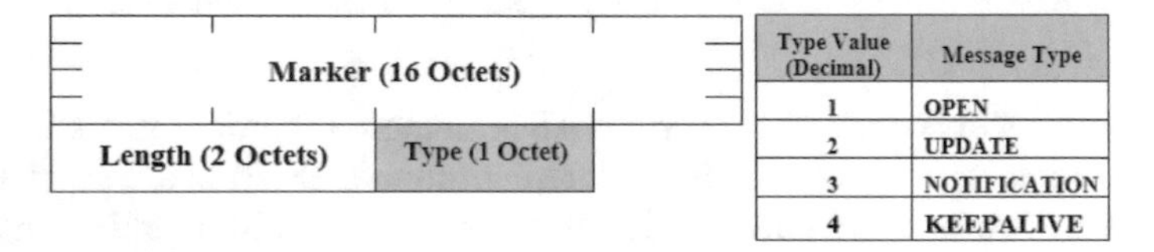

Type Value (Decimal)	Message Type
1	OPEN
2	UPDATE
3	NOTIFICATION
4	KEEPALIVE

Fig. 2. Header message format in the BGP.

Version (1 Octet)
My Autonomous System (2 Octets)
Hold Time (2 Octets)
BGP Identifier (4 Octets)
Opt Param Length (1 Octet)
Optional Parameters ...

Fig. 3. Open message format in the BGP.

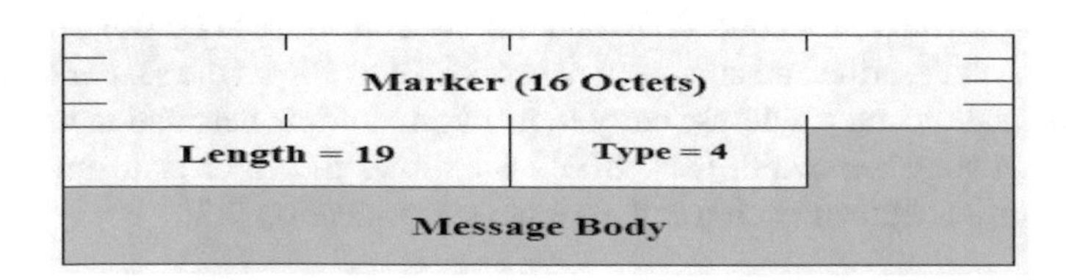

Fig. 4. Keepalive message format in the BGP.

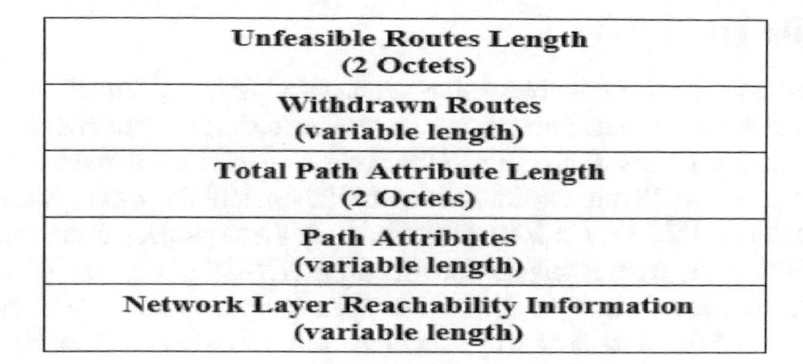

Fig. 5. Update message format in the BGP.

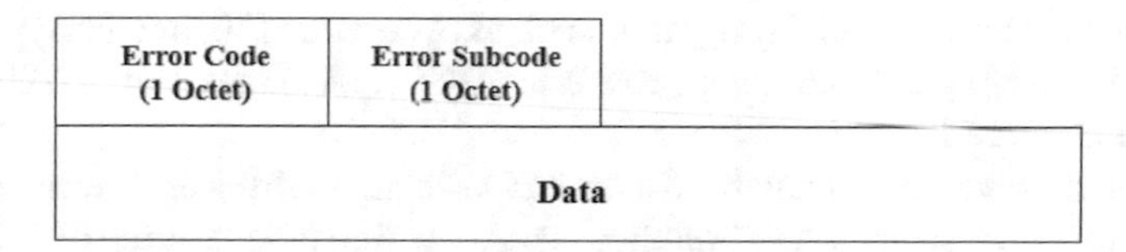

Fig. 6. Notification message format in the BGP.

2.3 BGP Anomalous Behavior

In [2] The classification of BGP anomalies can be organized into four primary categories:

- Direct Intended: The widely recognized planned anomaly is a hijacking attack of BGP. That happens while the assailant professes to claim a prefix or sub-prefix, which is placed with another AS to divert the path to the attacker. The attacker might objectively catch a trafficking victim, blackhole, or traffic disposal [24].
- Direct Unintended: BGP routers can identify and report anomalous prefixes and sub-prefixes that arise due to misconfigurations. These anomalies tend to be relatively short-lived as all parties involved, including the initiating entity and affected peers, are eager to promptly intervene and disrupt the anomaly [25].
- Indirect Anomalies: Some occasions aren't immediately identified with internet routing or BGP but may incredibly influence BGP elements and AS reachability. The most well-known cause of a backhanded anomaly is assaults that generate large traffic, clogging AS routers and forcing them to hibernate. Flutter routes are caused by a failure to reply to KEEPALIVE messages for a brief length of time. To lessen the impact of this situation, most suppliers assign a greater priority to BGP traffic [26].
- Network Failures: Differently structured materials comprise the network (e.g., ASes, Routers). There exist potential failures that could occur and subsequently have a direct impact on the functioning of the Internet [27].

3 BGP Datasets and Feature Selection

Many anomaly detection techniques in BGP use different data sources. This is datasets preprocessed and valuable information extracted from which we obtain several features to be used as inputs into the techniques to detect BGP anomalies.

3.1 BGP Datasets

The datasets' sources are accessible to all researchers without any limitations. Acquiring the most dependable internet routing data is a prevalent undertaking in this domain, offering highly valuable information. Noteworthy projects in this regard include the Routing Information Service (RIS) project, initiated by the Réseaux IP Européens Network Coordination Centre (RIPE NCC) in 2001 [28], the RouteViews project at the University of Oregon [29], and the dataset location provided by BCNET in Canada, specifically in British Columbia and Vancouver [15]. Three open-source initiatives have been developed to gather and retain route data by using the visibility of internet topology. Routing reflector clients (RRCs) collect Border Gateway Protocol (BGP) update messages in a predetermined format, which are then stored. The Multi-Threaded Routing Toolkit (MRT) utilizes a binary format. The user has provided a numerical reference without any accompanying text. Special programs, such as pybgpdum [30] and libbgpdump [31], are utilized for reading it. A sample of converted BGP update messages to ASCII format is presented in Table 1.

The datasets were made available for access without requiring an internet connection by utilizing the three primary data warehouses. According to the source [32], Rout View periodically refreshes its BGP data every 15 min and generates a routing table every two hours. In contrast to RIPE, the primary repository, which formerly issued a BGP update message every 15 min and a routing table every eight hours until 2003, subsequently transitioned to a BGP update message frequency of every 5 min [33]. Furthermore, a more advanced iteration of RouteViews was introduced, namely BGPmon developed by Colorado State University. This system is seamlessly integrated with the internet and operates on up-to-the-minute data. XML can effortlessly interpret this data without necessitating any conversion to an alternative format [34].

Table 1. BGP Update Message Example.

Field	Values
Time-Stamp	2021 07 17 19:00:55
Length	89
From AS	1313
To AS	31722
From IP	192.64.184.13
To IP	194.18.4.29
BGP packet type	Update
Origin	0(IGP)
AS-Path	313 4420 7186 16670 7255 7255
Next-hop	192.64.184.13
Announced NLRI prefix	197.166.188.0/24
Announced NLRI prefix	197.166..0/24

The updated iteration does not offer dataset processing capabilities via additional utilities such as Cyclops [35]. Utilizing many sources of BGP update messages represents an optimal approach for the identification of BGP abnormalities. Various features are derived from the aforementioned data, which are further analyzed to extract significant information utilized in the identification of Border Gateway Protocol (BGP) abnormalities, as shown in (see Fig. 7).

3.2 BGP Update Message Features Selection

The BGP update message comprises a set of slightly complex structures, so detecting anomalies in these message threads is complex and challenging [36].

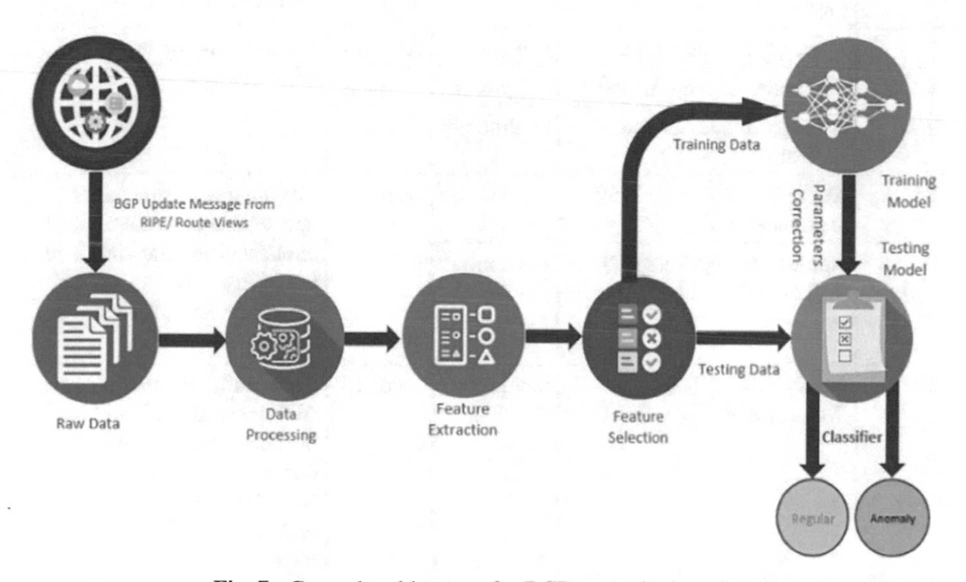

Fig. 7. General architecture for BGP anomaly detection.

It is also impossible to discover the cause of the occurrence and location of the anomalies if these messages are individual because analyzing a series of messages may be more useful in giving sufficient information about that [37]. Thus, extracting several features is the first step in classifying BGP anomalies. Among the most important and popular features are the volume of BGP update messages and the length of the AS-Path [38]. There are often complex, repetitive, and full of data changes for further classification of anomalies. It is also necessary to find useful features from the original data to greatly use them. The AS-Path property of a BGP update message provides the best path that packets take to reach their destination [39]. If a feature is extracted from this AS-Path, it can be called a path feature; conversely, it is classified as a volume feature. The features [40] can be divided into three main parts: binary, continuous, and categorical, extracted AS-path and volume features, and the definition of the extracted features [41], shown in Table 2.

Table 2. Features Extraction from BGP Update Messages.

Feature	Name	Type	Category	Definition
1	Number of Announcements	continuous	volume	Number of routes available for data delivery from source to destination
2	Number of Withdrawals	continuous	volume	The number of routes no longer reached
3	Number of Announced Prefixes	continuous	volume	Messages with the Type field set to announcement/withdrawal in BGP update messages
4	Number of Withdrawn Prefixes	continuous	volume	
5	Average AS-Path Length	categorical	AS-path	Various AS-path lengths of a BGP message
6	Maximum AS-Path Length	categorical	AS-path	
7	Average unique AS-Path Length	continuous	AS-path	
8	Number of Duplicate BGP announcements	continuous	volume	The number of times BGP update messages have been duplicated and the kind field has been set to announcement/implicit withdrawal
9	Number of Duplicate BGP Withdrawals	continuous	volume	
10	Number of Implicit Withdrawals	continuous	volume	For already declared prefixes, BGP looks at the kind field in the announcement as well as different AS-path parameters. Both the current and prior updates are associated with the same BGP session
11	Average AS-Path Edit Distance	categorical	AS-path	Number of edit distances between two attributes of the AS-path of all messages that were performed to match the attributes
12	Maximum AS-Path Edit Distance	categorical	AS-path	Within a one-minute interval, the greatest edit distance between the AS-path features of all messages

(*continued*)

Table 2. (*continued*)

Feature	Name	Type	Category	Definition
13	Inter-Arrival Time	continuous	volume	The average time between arrivals for all messages awaiting a one-minute time limit
14–24	Maximum Edit Distance = n = (7, …, 17)	binary	AS-path	Maximum edit distance
25–33	Maximum AS-path Length = n = (7, …, 15)	binary	AS-path	Maximum AS-path length
34	Number of (IGP) Packets	continuous	volume	IGP, EGP, or unknown sources generate BGP update messages
35	Number of (EGP) Packets	continuous	volume	
36	Number of Incomplete Packets	continuous	volume	
37	Average Packet Size	continuous	volume	In bytes, the average of all BGP update messages

4 BGP Anomalies Detection

BGP inter-domain routing abnormal events include attacks, configuration errors, and large-scale power failures. These anomalies lead to erroneous or uncontrollable routine behaviors, affecting the global routing infrastructure [20]. The two most common BGP routing anomalies are route hijacking and route leakage. Statistics show that about 25% of routing hijacking and routing leaks last less than 10 min, but they can affect 85% of the internet within 2.5 min [12]. Some scholars have researched the abnormal characteristics of BGP, applied them to pattern recognition and ML methods, and achieved good results [21].

However, it is important to acknowledge that these studies do have certain limitations. Specifically, the detection of small-scale anomalies can be challenging despite their higher occurrence rate. Furthermore, it is important to note that these small-scale abnormal events have a limited duration and exclusively impact a certain IP address prefix. Additionally, the number of affected IP addresses is quite modest [22]. The observed connection among the samples exhibits a high degree of strength. If there is a high correlation among samples collected during the occurrence of an irregular event, the incidental factors influencing the features of the samples may likely be magnified [22]. The aetiologies of various abnormalities vary and necessitate distinct treatment approaches based on the specific type of abnormality within the context of the research process [42]. Route hijacking is a malicious technique wherein an Autonomous System (AS) engages in the unauthorized announcement of an IP address prefix. Route leakage commonly occurs as a result of setup problems, which in turn lead to a route-forwarding strategy that contravenes established commercial ties. Various categories of anomalies may exhibit distinct characteristics. The act of categorizing diverse types of exceptions together is deemed undesirable [43]. There exist numerous methodologies for anomaly identification, with the most significant ones being depicted (refer to Fig. 8).

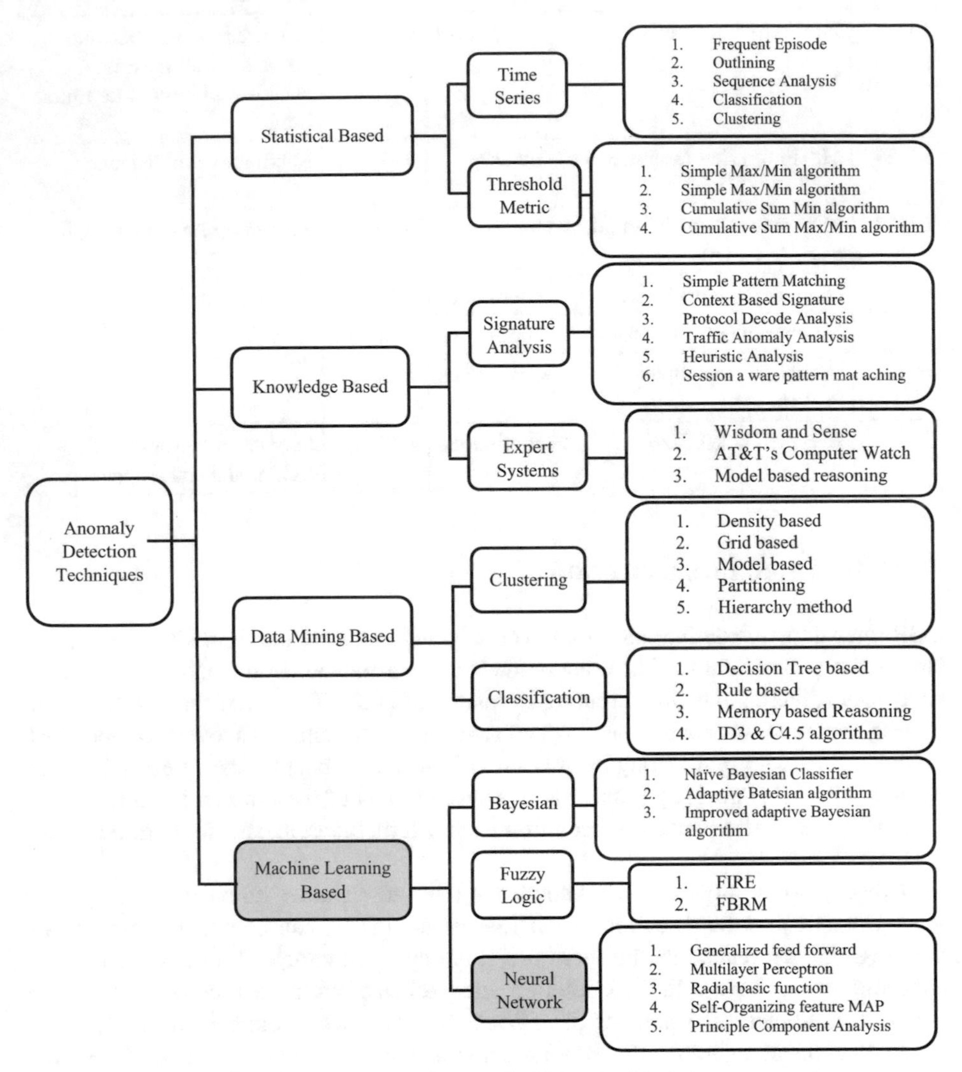

Fig. 8. Taxonomy of Anomaly Detection Techniques.

5 The Applications and Methods Based on ML in Networking

Machine learning-based analysis tools are excellent for learning what normal behavior looks like on the network and highlighting anomalies and deviations from that basis [44]. This detection capability makes ML useful and safe [45]. ML analysis can help spot the first signs of an impending failure on network equipment or even those on which such signs should appear soon [46]. The concept being discussed is known as predictive maintenance. There is an increasing trend among equipment makers to incorporate this form of analysis into their administrative tools, particularly those that are centered around a Software-as-a-Service (SaaS) offering [47]. The identification of anomalies in network behavior plays a crucial role in enabling cybersecurity teams to detect numerous compromised physical nodes exhibiting malicious conduct within the network [48]. ML techniques have significantly advanced the field of behavioral analysis and the detection and remediation of Distributed Denial of Service (DDoS) [49, 50].

Various algorithms have been developed, and depending on the properties of the learning data, they can be divided into three categories [51–54]:

- Supervised Learning Approaches: The training dataset contains regular and anomalous samples for the detection model. The sample classification training model determines whether samples are regular or anomalous.
- Semi-Supervised Learning Approaches: Only regular examples are accessible in the learning set. The user is unable to obtain data about abnormalities. When the behavior of unknown samples differs from that of regular samples, they are labeled as outliers.
- Unsupervised Learning Approaches: The researchers have no descriptive information for any of the samples in the training data. Regular and anomalous samples can be found in the training set, but the categorization of each sample is uncertain.

Much named training data is required for supervised learning [55–57]. Also, aggregate regular and anomaly samples are complex and time-consuming [58]. Further, it isn't easy to detect new outliers using a training model in known outliers [59]. Unsupervised learning does not require named information to be added to the training data, but it has a greater false alarm rate and a lower detection rate [60].

The present most unusual aspect pertains to the design intended for the limited data set, to address the difficulties associated with expanding dimensions [61]. The straight application of these methodologies to big data sets may yield suboptimal outcomes [62]. The methods commonly employed to tackle this difficulty include the utilization of high-dimensional data, reduction of dimensionality through mapping in lower-dimensional subspaces, and the application of traditional detection algorithms to process new data [63]. Several techniques have been proposed for dimensionality reduction, including feature packaging [64], genetic algorithm [65], linear discriminant analysis [66, 67], principal component analysis [68], and machine learning [63, 69].

6 A Review of BGP Anomaly Detection ML-Based Approaches

Several studies have used ML approaches to improve the accuracy of detecting BGP anomalies. We'll go over the most significant ones.

Li et al. [19] presented a working mechanism for detecting anomalies in BGP by using ML algorithms of Recurrent Neural Network (RNN) and extracting 37 features classified as AS-Path and Volume features. Despite our efforts, we possess the necessary qualifications to identify anomalies in the Border Gateway Protocol (BGP) through the detection of two distinct categories of irregularities. These include indirect anomalies as well as a third category of link failures resulting from the activities of malicious entities such as Slammer, WannaCrypt, and Moscow Blackout. However, it did not address the issue of anomaly locating, and the work could not detect direct anomalies in the BGP. Karimi et al. [70] It was proposed to develop a Neural Network Classifier to identify anomalous Border Gateway Protocol (BGP) occurrences within a network that are attributable to Worm Attacks. Use Single Classifier to discover three types of worm attacks: Nimda, Slammer, and Code-Red Worms; by extracting eight features to build BGP update messages, it did not solve the problem of locating the attacker that was anomaly causing, nor the detection of direct anomalies. Cosovic et al. [71] To categorize instances of Routing Table Leak, it is recommended to employ two-tiered Artificial Neural Network (ANN) models that are founded on the Backpropagation Algorithm, generally misconfiguring a router like Routing Leak AS9121, Indesit Routing Table Leak and Malaysian Telecom AS 4788. AWS Route Leak 15 AS-Path and Volume features were extracted over five days. This job cannot detect all other anomalies except for Link Failures.

Sanchez et al. [72] Define various graph features for BGP anomaly detection using different ML algorithms, Support Vector Machines (SVM), Multi-Layer Perceptron (MLP), Decision Trees (DT), Random Forests (RF), and Naive Bayes classifiers are some examples of classifiers (NB), to specifically BGP path leaks detection. The study relied on extracting 14 features every five minutes. It cannot identify the source of the anomaly.

Manna and Alkasassbeh [73] SNMP-MIB Big Data Set uses Random Forest, J48 (Decision Tree), and REP Tree classifiers ML algorithms. Although it consumes time and resources, they extract eight features to classify the data and detect BGP Denial of Service (DOS) anomalies. This work does not identify the source of the anomaly and cannot detect direct anomalies. Li et al. [74] presented a working mechanism for detecting anomalies in BGP indirect BGP anomaly worm attacks: Nimda, Code Red, link failure, Florida blackout, and East Coast by using C4.5 ML algorithms and extracting 35 features every minute. However, The study fails to adequately address the issue of anomaly identification, while it does successfully detect various other types of anomalies. Xianbo et al. [75] presented a method for BGP anomaly detection based on (SVM-BGPAD); They also used Fisher Linear and Markov Random algorithms to select and select features. Thus, three types of Nimda, Slammer, and Code-Red Worms Attacks are made by extracting ten features to build BGP update messages. This method does not locate the attacker who caused the anomaly and cannot detect other types of anomalies.

Edwards et al. [76] present machine learning methods for BGP detecting anomalies. It used clustering techniques, including K-means and DBSCAN. Which can detect some anomalies of historical events, such as TMNet BGP misconfiguration, Moscow blackout hardware failure, and Worm attacks such as (slammer and Nimda, and TTNET BGP misconfiguration), by selecting 13 features to build BGP update messages; this method did not identify the site that caused the anomaly, nor could it detect other types of anomalies.

Ding et al. [18] The mRMR algorithm was used to extract relevant features, classify the anomalies in BGP, and then apply the SVM algorithm to compare the accuracy. Ten features from BGP update message data were chosen to detect Worm assaults like slammer, Nimda, and Code Red I, which were not concerned with locating the attacker or detecting other irregularities. Al-Rousan and Trajković [36] use ten-volume features BGP update message data for anomaly detection, using it in algorithms (SVM) and (HMMs) to design an anomaly detection method BGP. Worm attacks such as Slammer, Nimda, and Code Red I were the subject of this study and did not care about locating the attacker or identifying other types of anomalies.

Allahdadi et al. [29] The Support Vector Machine (SVM) technique is employed to identify Border Gateway Protocol (BGP) anomalies. Before this, 18 features are extracted, and a novel connection and link among these features are established. Subsequently, the BGP behavior is analyzed to detect two specific forms of anomalies, namely Worm assaults and power supply outages. The study fails to adequately address the issue of anomaly identification, as it solely focuses on the detection of a different type of anomaly. McGlynn et al. [77] The key premise of this research is that good path updates have qualities that bad updates lack. As a result, using a deep learning method (Auto-Encoder) that effectively encodes good path updates, good properties can be learned. Even yet, there is a significant discrepancy between the input and output with random or malicious updates. Named attack data was not used by the company. For this effort, adversarial updates ensure their IP addresses are not tied to the AS path's origin. They cannot pinpoint the source of the anomaly since their AS origin is not next to the AS origin. The following Table 3 summarizes BGP anomaly detection based on machine-learning techniques.

Table 3. Shows a summary of BGP anomaly detection based on ML techniques.

Ref	Data source	Features	Used Techniques	Anomaly Detection	Identify locating sources of anomaly	Detection accuracy
[19]	RIPE NCC and RouteViews	extracting 37 features classified as AS-Path and Volume features	RNN	slammer, WannaCrypt, Moscow Blackout	did not address the issue of anomaly locating, and the work was unable to detect direct anomalies in the BGP	93.24%
[70]	RIPE NCC and RouteViews	extracting 8 features to build BGP update messages	Neural Network Classifier	Nimda, Slammer, and Code-Red Worms	did not solve the problem of locating the attacker that was anomaly-causing, nor the detection of direct anomalies	98%

(*continued*)

Table 3. (*continued*)

Ref	Data source	Features	Used Techniques	Anomaly Detection	Identify locating sources of anomaly	Detection accuracy
[71]	RIPE NCC	AWS Route Leak 15 AS-Path and Volume features were extracted over 5 days	ANN	Malaysian Telecom AS 4788, Indosat Routing Table Leak, and Routing Leak AS9121	cannot detect all other types of anomalies except for Link Failures	AS9121 RTL = 0.99375,
[72]	RIPE NCC	extracting 14 features every five minutes	SVM, MLP, DT, RF, NB	specifically, BGP path leaks	cannot identify the source of the anomaly	AWS RTL = 0.99431,
[73]	SNMP-MIB Big Dataset	extracting 8 features to classify the data	SNMP-MIB Big Data Set uses Random Forest, J48 (Decision Tree), and REP Tree classifiers	DOS	does not identify the source of the anomaly and cannot detect direct anomalies	Malaysian Telecom RTL = 0.9925,
[74]	RIPE NCC	extracting 35 features every single minute	C4.5	Nimda, CodeRed, link failure, Florida blackout, and East Coast	does not address the problem of identifying the anomaly. It was able to detect other kinds of anomalies	Indosat RTL0.93056,
[75]	RIPE NCC and BCNET	extracting 10 features to build BGP update messages	SVM-BGPAD	Nimda, Slammer, and Code-Red Worms Attacks	does not address the problem of identifying the anomaly; it wasn't able to detect other kinds of anomalies	91.36%
[76]	RIPE NCC	selecting 13 features to build BGP update messages	clustering techniques, including K-means and DBSCAN	TMNet BGP misconfiguration, Moscow blackout hardware failure, Worm attacks such as slammer and Nimda, and TTNET BGP misconfiguration	did not identify the site that caused the anomaly, nor could it detect other types of anomalies	silhouette score of 0.652 using 14 clusters
[18]	RIPE NCC and BCNET	10 features were selected from the BGP update message data	SVM and LSTM algorithms	Slammer, Nimda, and Code Red are examples of worm attacks	was not concerned with locating the attacker nor identifying other types of anomalies	68.60%

(*continued*)

Table 3. (*continued*)

Ref	Data source	Features	Used Techniques	Anomaly Detection	Identify locating sources of anomaly	Detection accuracy
[36]	RIPE NCC and BCNET	10 volume features BGP update message data	SVM and HMM algorithms	Slammer, Nimda, and Code Red I are examples of worm attacks	did not care about locating the attacker nor identifying other types of anomalies	SVM = 86.1%
[29]	RIPE NCC and RouteViews	extracting 18 features and building a new correlation and relationship between them	SVM algorithm	Worm attacks and power supply outgas	the work does not address the problem of identifying the anomaly; it was able to detect other kinds of anomalies	92.16%
[77]	RouteViews	The fundamental aspects of input data are learned using an auto-encoder	Auto-Encoder	Adversarial updates are created for this work	cannot identify the source of the anomaly	60%

7 Conclusions

This writing audit is expected to give a hypothetical comprehension of the irregularity identification issue and its various perspectives. It also intended to investigate different methods to solve this issue and present various existing solutions to these challenges. A conversation is given about what the anomaly is and identifying its simplest signs. Anomaly detection is the most widely known identification because it is dynamic and detects known and obscure anomalies. This investigation has evaluated numerous papers to give an expanded view of what was accomplished in detecting anomalies and what can be enhanced. A wide range of outcomes can be imagined from direct methods of agreeing with complicated frameworks to solve the anomaly detection problem. However, each method has its points of downsides and interests; there are still many open problems to enhance the viability and anomaly detection possibility, yet interestingly, a significant number of promising rules for specialists to continue additional examinations on anomaly detection topics. This research aimed to compare and contrast the various machine learning models for BGP anomaly detection and assess their utility.

References

1. Al-Musawi, B., Branch, P., Armitage, G.: Recurrence behaviour of BGP traffic. In: 2017 27th International Telecommunication Networks and Applications Conference (ITNAC), pp. 1–7. IEEE (2017)
2. Al-Musawi, B., Branch, P., Armitage, G.: BGP anomaly detection techniques: a survey. IEEE Commun. Surv. Tutor. **19**, 377–396 (2017). https://doi.org/10.1109/COMST.2016.2622240

3. Sriram, K., Borchert, O., Kim, O., et al.: A comparative analysis of bgp anomaly detection and robustness algorithms. In: 2009 Cybersecurity Applications & Technology Conference for Homeland Security, pp. 25–38. IEEE (2009)
4. Sermpezis, P., Kotronis, V., Gigis, P., et al.: ARTEMIS: neutralizing BGP hijacking within a minute arXiv:1801.01085 (2018)
5. Bush, R., Austein, R.: The resource public key infrastructure (RPKI) to router protocol, version 1. RFC 8210 (2017). https://doi.org/10.17487/RFC8210
6. Hassan Muosa, A., Mohan Hamed, A.: Remote monitoring and smart control system for greenhouse environmental and automation irrigations based on WSNs and GSM module. In: IOP Conference Series: Materials Science and Engineering, vol. 928, p. 032037 (2020). https://doi.org/10.1088/1757-899X/928/3/032037
7. Muosa, A.H., Ali, A.H.: Internet routing anomaly detection using LSTM based autoencoder. In: 2022 International Conference on Computer Science and Software Engineering (CSASE), pp. 319–324. IEEE (2022). https://doi.org/10.1109/CSASE51777.2022.9759613
8. Muosa, A.H.: Comparison of energy efficient routing for multipath protocols in mobile ad-hoc network. Int. J. Tech. Phys. Probl. Eng. (IJTPE) **15**(1), 218–223 (2023)
9. Testart, C., Richter, P., King, A., et al.: Profiling BGP serial hijackers: capturing persistent misbehavior in the global routing table. In: Proceedings of the ACM SIGCOMM Internet Measurement Conference, IMC, pp. 420–434. ACM, New York (2019)
10. Mahajan, R., Wetherall, D., Anderson, T.: Understanding BGP misconfiguration. In: Computer Communication Review, Pittsburgh, Pennsylvania, USA, pp. 3–16 (2002)
11. Al-Musawi, B., Branch, P., Armitage, G.: Detecting BGP instability using recurrence quantification analysis (RQA). In: 2015 IEEE 34th International Performance Computing and Communications Conference (IPCCC). IEEE, pp. 1–8 (2015)
12. Shi, X., Xiang, Y., Wang, Z., et al.: Detecting prefix hijackings in the internet with argus. In: 2012 Proceedings of the ACM SIGCOMM Internet Measurement Conference, IMC, New York, NY, USA, pp. 15–28 (2012)
13. Deshpande, S., Thottan, M., Ho, T.K., Sikdar, B.: An online mechanism for bgp instability detection and analysis. IEEE Trans. Comput. **58**, 1470–1484 (2009). https://doi.org/10.1109/TC.2009.91
14. Xu, M., Li, X.: BGP anomaly detection based on automatic feature extraction by neural network. In: 2020 IEEE 5th Information Technology and Mechatronics Engineering Conference (ITOEC), pp. 46–50. IEEE, Chongqing (2020). https://doi.org/10.1109/ITOEC49072.2020.9141762
15. Ding, Q., Li, Z., Haeri, S., Trajković, L.: Application of machine learning techniques to detecting anomalies in communication networks: datasets and feature selection algorithms. In: Dehghantanha, A., Conti, M., Dargahi, T. (eds.) Cyber Threat Intelligence. Advances in Information Security, vol. 70, pp. 47–70. Springer, Cham (2018). https://doi.org/10.1007/978-3-319-73951-9_3
16. Fonseca, P., Mota, E.S., Bennesby, R., Passito, A.: BGP dataset generation and feature extraction for anomaly detection. In: 2019 IEEE Symposium on Computers and Communications (ISCC), pp. 1–6. IEEE, Barcelona (2019). https://doi.org/10.1109/ISCC47284.2019.8969619
17. Muosa, A.H.: An improvement routing algorithm based on leach protocol. Int. J. Tech. Phys. Probl. Eng. (IJTPE) **14**(3), 66–72 (2022)
18. Ding, Q., Li, Z., Batta, P., Trajkovic, L.: Detecting BGP anomalies using machine learning techniques. In: 2016 IEEE International Conference on Systems, Man, and Cybernetics (SMC), pp. 003352–003355. IEEE (2016)
19. Li, Z., Rios, A.L.G., Trajkovic, L.: Detecting internet worms, ransomware, and blackouts using recurrent neural networks. In: 2020 IEEE International Conference on Systems, Man, and Cybernetics (SMC), pp. 2165–2172. IEEE, Toronto (2020). https://doi.org/10.1109/SMC42975.2020.9283472

20. Cheng, M., Xu, Q., Jianming, L.V., et al.: MS-LSTM: a multi-scale LSTM model for BGP anomaly detection. In: 2016 IEEE 24th International Conference on Network Protocols (ICNP), pp. 1–6. IEEE, Singapore (2016). https://doi.org/10.1109/ICNP.2016.7785326
21. Batta, P., Singh, M., Li, Z., et al.: Evaluation of support vector machine kernels for detecting network anomalies. In: 2018 IEEE International Symposium on Circuits and Systems (ISCAS), pp. 1–4. IEEE, Florence (2018). https://doi.org/10.1109/ISCAS.2018.8351647
22. Moustafa, N., Creech, G., Slay, J.: Anomaly detection system using beta mixture models and outlier detection. In: Pattnaik, P., Rautaray, S., Das, H., Nayak, J. (eds.) Progress in Computing, Analytics and Networking. Advances in Intelligent Systems and Computing, vol. 710, pp. 125–135. Springer, Singapore (2018). https://doi.org/10.1007/978-981-10-7871-2_13
23. Roy, B., Cheung, H.: A deep learning approach for intrusion detection in internet of things using bi-directional long short-term memory recurrent neural network. In: 2018 28th International Telecommunication Networks and Applications Conference (ITNAC), pp. 1–6. IEEE, Sydney (2018)
24. Samir, A., El Ioini, N., Fronza, I., et al.: Anomaly detection and analysis for reliability management clustered container architectures. Int. J. Adv. Syst. Meas. **12**, 247–264 (2020)
25. Lo, O., Buchanan, W.J., Griffiths, P., Macfarlane, R.: Distance measurement methods for improved insider threat detection. Secur. Commun. Netw. **2018**, 1–18 (2018). https://doi.org/10.1155/2018/5906368
26. Sermpezis, P., Kotronis, V., Dainotti, A., Dimitropoulos, X.: A survey among network operators on bgp prefix hijacking. ACM SIGCOMM Comput. Commun. Rev. **48**, 64–69 (2018). https://doi.org/10.1145/3211852.3211862
27. Mitseva, A., Panchenko, A., Engel, T.: The state of affairs in BGP security: a survey of attacks and defenses. Comput. Commun. **124**, 45–60 (2018). https://doi.org/10.1016/j.comcom.2018.04.013
28. Cho, S., Fontugne, R., Cho, K., et al.: BGP hijacking classification. In: TMA 2019 - Proceedings of the 3rd Network Traffic Measurement and Analysis Conference, pp. 25–32 (2019)
29. Allahdadi, A., Morla, R., Prior, R.: A framework for BGP abnormal events detection. abs/1708.03453 (2017)
30. Oberheide, J.: Pybgpdump. https://jon.oberheide.org/pybgpdump/. Accessed 13 Apr 2021
31. RIPE NCC RIS Projec. libbgpdump. http://ris.ripe.net/source/bgpdump/. Accessed 13 Apr 2021
32. Ćosović, M., Obradović, S., Trajković, L.: Using databases for bgp data analysis. In: Proceedings of the International Scientific Conference, UNITECH, pp. 367–370 (2014)
33. Ćosović, M., Obradović, S., Trajković, L.: Performance evaluation of BGP anomaly classifiers. In: 2015 3rd International Conference on Digital Information, Networking, and Wireless Communications, DINWC 2015, Moscow, Russia, pp. 115–120 (2015)
34. Yan, H., Oliveira, R., Burnett, K., et al.: BGPmon: A real-time, scalable, extensible monitoring system. In: Proceedings - Cybersecurity Applications and Technology Conference for Homeland Security, CATCH 2009, Washington, DC, USA, pp. 212–223 (2009)
35. Chi, Y.-J., Oliveira, R., Zhang, L.: Cyclops: the AS-level connectivity observatory. In: ACM SIGCOMM Computer Communication Review, p. 5 (2008)
36. Al-Rousan, N.M., Trajkovic, L.: Machine learning models for classification of BGP anomalies. In: 2012 IEEE 13th International Conference on High Performance Switching and Routing, pp. 103–108. IEEE, Belgrade (2012)
37. Li, Y., Xing, H.J., Hua, Q., et al.: Classification of BGP anomalies using decision trees and fuzzy rough sets. In: Conference Proceedings - IEEE International Conference on Systems, Man and Cybernetics, pp. 1312–1317. IEEE, San Diego (2014)

38. Al-Rousan, N., Haeri, S., Trajković, L.: Feature selection for classification of BGP anomalies using Bayesian models. In: Proceedings - International Conference on Machine Learning and Cybernetics, Xi'an, China, pp. 140–147 (2012)
39. Gu, Q., Li, Z., Han, J.: Generalized fisher score for feature selection. In: Proceedings of the 27th Conference on Uncertainty in Artificial Intelligence, UAI 2011, Barcelona, Spain, pp. 266–273 (2011)
40. Li, Z., Gonzalez Rios, A.L., Trajkovic, L.: Border gateway protocol (BGP) routing records from route views. IEEE Dataport (2021). https://doi.org/10.21227/wpph-ex74
41. Peng, H., Long, F., Ding, C.: Feature selection based on mutual information criteria of max-dependency, max-relevance, and min-redundancy. IEEE Trans. Pattern Anal. Mach. Intell. **27**, 1226–1238 (2005). https://doi.org/10.1109/TPAMI.2005.159
42. Liu, H., Lang, B.: Machine learning and deep learning methods for intrusion detection systems: a survey. Appl. Sci. **9**, 4396 (2019). https://doi.org/10.3390/app9204396
43. Yuanyan, L., Xuehui, D., Yi, S.: Data streams anomaly detection algorithm based on self-set threshold. In: Proceedings of the 4th International Conference on Communication and Information Processing, pp. 18–26. ACM, New York (2018). https://doi.org/10.1145/3290420.3290451
44. Jonker, M., Pras, A., Dainotti, A., Sperotto, A.: A first joint look at DoS attacks and BGP blackholing in the wild. In: Proceedings of the ACM SIGCOMM Internet Measurement Conference, IMC, Boston, USA, pp. 457–463 (2018)
45. Miller, S., Curran, K., Lunney, T.: Detection of anonymising proxies using machine learning (2021)
46. Al-Kasassbeh, M., Al-Naymat, G., Al-Hawari, E.: Towards generating realistic SNMP-MIB dataset for network anomaly detection. Int. J. Comput. Sci. Inf. Secur. **14**, 1162 (2016)
47. Al-Naymat, G., Al-Kasassbeh, M., Al-Hawari, E.: Using machine learning methods for detecting network anomalies within SNMP-MIB dataset. Int. J. Wireless Mobile Comput. **15**, 67–76 (2018). https://doi.org/10.1504/IJWMC.2018.094644
48. Aljawarneh, S., Aldwairi, M., Yassein, M.B.: Anomaly-based intrusion detection system through feature selection analysis and building hybrid efficient model. J. Comput. Sc. **25**, 152–160 (2018). https://doi.org/10.1016/j.jocs.2017.03.006
49. Belavagi, M.C., Muniyal, B.: Performance evaluation of supervised machine learning algorithms for intrusion detection. In: Procedia Computer Science, pp. 117–123 (2016)
50. Rezaei, S., Liu, X.: Deep learning for encrypted traffic classification: an overview. IEEE Commun. Mag. **57**, 76–81 (2019). https://doi.org/10.1109/MCOM.2019.1800819
51. Yun, X., Wang, Y., Zhang, Y., Zhou, Y.: A semantics-aware approach to the automated network protocol identification. IEEE/ACM Trans. Netw. **24**, 583–595 (2016). https://doi.org/10.1109/TNET.2014.2381230
52. Berthier, R., Sanders, W.H., Khurana, H.: Intrusion detection for advanced metering infrastructures: requirements and architectural directions, pp. 350–355. IEEE, Dresden (2010)
53. Maamar, A., Benahmed, K.: A hybrid model for anomalies detection in ami system combining K-means clustering and deep neural network. Comput. Mater. Continua **60**, 15–39 (2019). https://doi.org/10.32604/cmc.2019.06497
54. Wang, Z., Gong, G., Wen, Y.: Anomaly diagnosis analysis for running meter based on BP neural network. In: Proceedings of the 2016 International Conference on Communications, Information Management and Network Security, Gold Coast, Australia (2016)
55. Chen, Y., Tao, J., Zhang, Q., et al.: Saliency detection via the improved hierarchical principal component analysis method. Wirel. Commun. Mob. Comput. **2020**, 1–12 (2020). https://doi.org/10.1155/2020/8822777
56. Farooq, M.S., Khalid, H., Arooj, A., et al.: A conceptual multi-layer framework for the detection of nighttime pedestrian in autonomous vehicles using deep reinforcement learning. Entropy **25**, 135 (2023). https://doi.org/10.3390/e25010135

57. Tahir, T., Gencel, C., Rasool, G., et al.: Early software defects density prediction: training the international software benchmarking cross projects data using supervised learning. IEEE Access **11**, 141965–141986 (2023). https://doi.org/10.1109/ACCESS.2023.3339994
58. Ashfaq, R.A.R., Wang, X.Z., Huang, J.Z., et al.: Fuzziness based semi-supervised learning approach for intrusion detection system. Inf. Sci. **378**, 484–497 (2017). https://doi.org/10.1016/j.ins.2016.04.019
59. Fernandes, G., Rodrigues, J.J.P.C., Carvalho, L.F., et al.: A comprehensive survey on network anomaly detection. Telecommun. Syst. **70**, 447–489 (2019). https://doi.org/10.1007/s11235-018-0475-8
60. Wang, W., Sheng, Y., Wang, J., et al.: HAST-IDS: learning hierarchical spatial-temporal features using deep neural networks to improve intrusion detection. IEEE Access **6**, 1792–1806 (2017). https://doi.org/10.1109/ACCESS.2017.2780250
61. Erfani, S.M., Rajasegarar, S., Karunasekera, S., Leckie, C.: High-dimensional and large-scale anomaly detection using a linear one-class SVM with deep learning. Pattern Recogn. **58**, 121–134 (2016). https://doi.org/10.1016/j.patcog.2016.03.028
62. Wang, Y., Zhou, H., Feng, H., et al.: Network traffic classification method basing on CNN. Tongxin Xuebao/J. Commun. **39**, 14–23 (2018). https://doi.org/10.11959/j.issn.1000-436x.2018018
63. Kaur, S., Singh, M.: Hybrid intrusion detection and signature generation using deep recurrent neural networks. Neural Comput. Appl. **32**, 7859–7877 (2020). https://doi.org/10.1007/s00521-019-04187-9
64. Swarnkar, M., Hubballi, N.: OCPAD: One class Naive Bayes classifier for payload based anomaly detection. Expert Syst. Appl. **64**, 330–339 (2016). https://doi.org/10.1016/j.eswa.2016.07.036
65. Van Efferen, L., Ali-Eldin, A.M.T.: A multi-layer perceptron approach for flow-based anomaly detection. In: 2017 International Symposium on Networks, Computers and Communications (ISNCC), pp. 1–6. IEEE, Marrakech (2017)
66. Buczak, A.L., Guven, E.: A survey of data mining and machine learning methods for cyber security intrusion detection. IEEE Commun. Surv. Tutor. **18**, 1153–1176 (2016). https://doi.org/10.1109/COMST.2015.2494502
67. Kaya, ŞM., İşler, B., Abu-Mahfouz, A.M., et al.: An intelligent anomaly detection approach for accurate and reliable weather forecasting at IoT edges: a case study. Sensors **23**, 2426 (2023). https://doi.org/10.3390/s23052426
68. Al-Yaseen, W.L., Othman, Z.A., Nazri, M.Z.A.: Multi-level hybrid support vector machine and extreme learning machine based on modified K-means for intrusion detection system. Expert Syst. Appl. **67**, 296–303 (2017). https://doi.org/10.1016/j.eswa.2016.09.041
69. Arasteh, B., Seyyedabbasi, A., Rasheed, J., Abu-Mahfouz, A.M.: Program source-code re-modularization using a discretized and modified sand cat swarm optimization algorithm. Symmetry **15**, 401 (2023). https://doi.org/10.3390/sym15020401
70. Karimi, M., Jahanshahi, A., Mazloumi, A., Sabzi, H.Z.: Border gateway protocol anomaly detection using neural network. In: 2019 IEEE International Conference on Big Data (Big Data), pp. 6092–6094. IEEE, Los Angeles (2019). https://doi.org/10.1109/BigData47090.2019.9006201
71. Cosovic, M., Obradovic, S., Junuz, E.: Deep learning for detection of BGP anomalies. In: Rojas, I., Pomares, H., Valenzuela, O. (eds.) Time Series Analysis and Forecasting, pp. 95–113 (2018). https://doi.org/10.1007/978-3-319-96944-2_7
72. Sanchez, O.R., Ferlin, S., Pelsser, C., Bush, R.: Comparing machine learning algorithms for BGP anomaly detection using graph features. In: Proceedings of the 3rd ACM CoNEXT Workshop on Big DAta, Machine Learning and Artificial Intelligence for Data Communication Networks, pp. 35–41. ACM, New York (2019)

73. Manna, A., Alkasassbeh, M.: Detecting network anomalies using machine learning and SNMP-MIB dataset with IP group. In: 2019 2nd International Conference on new Trends in Computing Sciences (ICTCS), pp. 1–5. IEEE Amman (2019). https://doi.org/10.1109/ICTCS.2019.8923043
74. Li, J., Dou, D., Wu, Z., et al.: An internet routing forensics framework for discovering rules of abnormal BGP events. In: Computer Communication Review, pp. 55–66 (2005)
75. Dai, X., Wang, N., Wang, W.: Application of machine learning in BGP anomaly detection. J. Phys.: Conf. Ser. **1176**, 032015 (2019). https://doi.org/10.1088/1742-6596/1176/3/032015
76. Edwards, P., Cheng, L., Kadam, G., et al.: Border gateway protocol anomaly detection using machine learning techniques. In: SMU Data Science Review, p. 5 (2019)
77. McGlynn, K., Acharya, H.B., Kwon, M.: Detecting BGP route anomalies with deep learning. In: IEEE INFOCOM 2019 - IEEE Conference on Computer Communications Workshops (INFOCOM WKSHPS), pp. 1039–1040. IEEE (2019)

A Survey of Next Words Prediction Models

Mortadha Adnan Abood(✉) and Suhad Malallah Kadhem

Computer Science Department, University of Technology, Baghdad, Iraq
cs.21.16@grad.uotechnology.edu.iq,
suhad.m.kadhem@uotechnology.edu.iq

Abstract. Word prediction, also called language modeling, is one field of natural language processing that can help predict the next word. Next-word prediction technology has emerged as a crucial tool for enhancing text input efficiency and reducing spelling errors. Globally, people are spending a cumulative amount of time on their mobile devices, laptops, desktops, etc. for messaging, sending emails, banking, interacting through social media, and all other activities. It's necessary to cut down on the time spent typing through these devices. This paper aims to provide a comprehensive overview of prior research in the field of word prediction, specifically focusing on the utilization of deep learning (DL), statistical approaches, and hybrid methodologies. The primary objective is to compare the outcomes achieved through these various techniques. This study concludes that the word prediction system can be categorized into four models, which are statistical, DL, pre-trained, and hybrid models, and concludes that the selection of an appropriate model for word prediction depends on several factors, such as the nature of the application, the availability of data, computational resources, and accuracy requirements.

Keywords: Word Prediction · Natural Language Processing (NLP) · Long Short-Term Memory (LSTM) · Bi-directional LSTM · BERT

1 Introduction

One of the most important branches of artificial intelligence, natural language processing (NLP) aims to teach computers to interpret human language almost like humans. Deep Learning (DL) approaches, which combine rules-based linguistics with statistical and deep learning models, have been crucial in the rapid advancement of natural language processing (NLP). These techniques are currently used extensively in a wide range of NLP applications, including text creation, machine translation, and speech recognition. Improving typing productivity and text coherence are two main goals of these improvements, which rely on accurately predicting the next words in a sequence [1]. Constantly improving to generate more complicated and natural language interactions, these predictive capabilities have broader applications beyond just improving typing speed and accuracy through user-specific learning, such as content creation, automated answering systems, and personalized recommendation engines [2].

J. Rasheed et al. (Eds.): FoNeS-AIoT 2024, LNNS 1035, pp. 165–185, 2024.
https://doi.org/10.1007/978-3-031-62871-9_14

Individuals with physical, perceptual, or intellectual impairments are receiving crucial support from the rapidly expanding science of natural language processing (NLP), particularly next-word prediction, which is transforming assistive devices. More effective communication is made possible by these systems' predictive text, which adjusts to user patterns of use; commonly used terms are quickly recommended, making typing easier and saving the user time [3, 4].

This feature notably enhances typing efficiency by reducing keystrokes, minimizing time investment, and mitigating spelling errors. The software dynamically generates word suggestions as users type, updating the options with each new input. When the desired word surfaces within the suggested list, users can seamlessly insert it into the text with a single keystroke. Typically, these word suggestions are accompanied by numeric references, facilitating user selection by corresponding numbers. In cases where the desired word isn't predicted, manual entry remains an option [5].

But several challenges must be overcome before next-word prediction can be fully assimilated into the field of NLP, each of which has unique implications:

- Language Complexity: Accurate word prediction has substantial challenges, especially when applied to languages with a variety of linguistic patterns and limited resources due to their complex nature [3, 6, 7].
- Model Selection: To provide optimal prediction accuracy, the choice of an appropriate model architecture, loss functions, and activation functions must be carefully considered [3, 8].
- Privacy and Data Sharing: Balancing the collaborative nature of training models with data privacy considerations, particularly in federated learning setups, raises complexities in securely aggregating model updates [3, 9, 10].
- Data Quality and Ambiguity: It can be difficult to improve model accuracy in the face of popular influences [3, 6, 7, 11].
- Linguistic nuance: Capturing nuanced context and keeping contextual awareness are key to making correct forecasts and guaranteeing user delight [3, 12].
- Cross-Domain Adaptation: The continual difficulty is in adapting models across diverse domains and languages while maintaining relevance and performance across varied situations [3].

The remainder of the sections of this paper are as follows: Sect. 2 presents a survey of research papers and studies related to word prediction; Sect. 3 explains the different techniques and models used in word prediction; Sect. 4 discusses and compares the results; and Sect. 5 concludes the importance of next-word prediction technology in improving text input and reducing errors.

2 Literature Survey

Predicting the next word in a sequence has given rise to several techniques with broad implications for natural language processing. This survey paper provides an insight into these methodologies, offering a comprehensive view of their usage and effectiveness. Based on the types of models employed in the research, We categorized the survey responses as follows:

- Statistical models
- Deep Learning Models
- Pre-Trained Models
- Hybrid Models.

2.1 Probabilistic Models

Probabilistic models in word prediction use statistical patterns to anticipate the next word, improving typing speed by suggesting contextually relevant options based on word frequencies. These models estimate word likelihood, aiding users in faster and more accurate text input. Here is a survey of papers in which researchers used probabilistic models in their research.

Ms. Niti Shah in [13] proposed a novel approach for Hindi word prediction using "Syntactic N-Grams" (Sn-Grams), which marked a considerable advancement over traditional N-Grams. This approach successfully utilized the intricacy of Hindi, making a substantial contribution to linguistic technology for Indian languages. Additionally, it emphasized the development of a user-friendly Hindi keyboard for Android, improving its accessibility. Nevertheless, the emphasis on Hindi limited its broader applicability, and the integration of Sn-Grams presented technological difficulties. Moreover, there was a potential for proposing grammatically accurate but contextually inappropriate words, and the research report would have been enhanced by incorporating more thorough evaluation measures, such as error rates, user satisfaction, and comparative analysis with a wider range of datasets.

Bhuyan et al. [14] investigated the impact of prediction list length on the accuracy of word prediction in Assamese using n-gram models. The major results indicated that the highest accuracies achieved were 81.39% for texts within the same domain and 63.81% for texts from other domains. These accuracies were obtained when using longer prediction lists, ranging from 1 to 10 words. The study's strengths encompassed a meticulous analysis of the correlation between forecast length and accuracy, as well as thorough testing across various list lengths and domains. Nevertheless, the n-gram model encountered constraints arising from its inability to handle unfamiliar words and its heightened computational requirements for larger n-grams. Moreover, the emphasis on the Assamese language could restrict the applicability of its conclusions. The research underscored a balance between achieving high accuracy and maintaining user-friendly system design.

Yazdani et al. [15] presented an innovative N-gram model for word prediction in electronic health records (EHR). The model's effectiveness was demonstrated through tests conducted on real clinical reports, resulting in notable reductions in both time and keystrokes. The research emphasized its novel methodology and effectiveness, although its implementation was restricted to particular medical records, and it lacked a comparison evaluation with alternative systems. There has to be additional research and development into the model's complexity, wide applicability, and usability, as the study highlighted its promise in clinical settings but also highlighted these issues.

Hamarashid et al. [8] presented an analysis of text prediction systems. The study achieved an accuracy rate of 88.2% on the first dataset, while the findings on the second dataset, which included phonetic transcription, varied. Notable strengths encompass a comprehensive exploration of techniques ranging from statistical to AI models, as well as their practical implementations in assistive technologies and games. An effectiveness assessment was conducted using measures such as the keystroke saving rate. Nevertheless, the technical complexity of the content may have posed difficulties for readers without expertise in the subject matter, and it also lacked comprehensive case studies and empirical data. Furthermore, it failed to thoroughly address forthcoming patterns in text prediction, which may have been rectified to achieve a more comprehensive analysis.

Table 1 includes information on the reference used, the dataset, accuracy, strengths, and limitations of each model. The models discussed in the table employ N-gram and trigram approaches, and they demonstrate relevance to Indian languages, comprehensive analysis, extensive testing, and real-world applications. However, limitations such as limited scope, domain specificity, and concerns regarding user experience are also mentioned.

Table 1. Summary of Probabilistic Models research papers.

Ref.	Algorithm	Dataset	Accuracy	Strength points	Limitations
[13]	Syntactic N-Gram	Hindi language	–	Novel approach Relevance to Indian language User-friendly interface	Limited scope Technological challenge Potential for over-prediction Evaluation metrics
[14]	N-gram-based language models	In-domain and mixed-domain text data	Max 81.39%/63.81%	Comprehensive analysis Extensive testing High accuracy	Model limitation Domain specificity User experience concerns
[15]	Trigram Language Model	Various medical reports	–	Novel approach Empirical testing Time &efficiency saving	Scope of data Complexity and usability
[8]	N-gram Model	random sample of nearly 22,000 queries	88.2% (1st dataset) 90% (2nd dataset)	Comprehensive coverage Real-world application Effectiveness assessment	Technical Depth Case studies and data Future trends analysis

2.2 Deep Learning (DL) Models

DL models for word prediction use neural networks to comprehend complex language patterns from large datasets, improving their capacity to correctly anticipate the next word in text, making them useful for applications like auto-correction and recommendation systems. Here is an overview of research publications in which DL models were applied.

Barman et al. [7] introduced an LSTM recurrent neural network (RNN) model for predicting the next word in Assamese. The model was trained using datasets derived from the novel "Bhanumati" and its corresponding phonetic transcription. The original text achieved an accuracy of 88.20%, while the transcription earned an accuracy of 72.10%. The model excelled in efficiently managing the intricacies of the Assamese language and utilizing LSTM for acquiring knowledge of long-term dependencies. Nevertheless, a significant drawback was its dependence on a restricted dataset from a sole source and the lack of comparative studies with alternative models. This research made a noteworthy contribution to the field of computational linguistics, particularly for languages that have received less attention, such as Assamese.

Ramya et al. [16] provided a comprehensive comparison of RNN, stacked RNN, LSTM, and bidirectional LSTM models. The reported accuracy ranged from 60% to 72%, based on the utilization of Kaggle's news dataset. The study's robustness was derived from its comprehensive model range and its practical implementation using real-world datasets. Nevertheless, it failed to provide a comprehensive examination of the elements influencing model performance and the underlying causes of the varying accuracies, which could have contributed to a deeper comprehension of the strengths and shortcomings of each architecture. Abujar et al. [17] presented a method that utilized a bi-directional RNN, namely LSTM cells, for processing the Bengali language. This methodology achieved a remarkable accuracy of 98.766% with a minimal loss of 0.0430. Notable strengths encompass proficient utilization of LSTM cells and thorough data pretreatment employing Word2Vec embedding. Nevertheless, the system had constraints arising from its self-gathered dataset, which may have lacked diversity, and its narrow concentration on Bengali, hence restricting its applicability to other languages.

Parihar et al. [18] utilized a DL approach, specifically employing LSTM networks and phonetics, to accurately estimate the complexity of words. The model attained a precision of 97.5% and underwent training using a dataset including 1800 words. The robustness of the system was derived from the efficient use of phonetics and LSTM, which improved its prediction skills. Nevertheless, the system's principal weaknesses were attributed to its dependence on a limited dataset and its exclusive emphasis on word-level difficulty without taking into account broader contextual linguistic complexity. This methodology provided a hopeful yet relatively limited approach to analyzing linguistic complexity. Mahmud et al. [6] introduced a model based on long-short-term memory (LSTM) for predicting the next word in a sequence of text. The model attained a precision of 75%, but a substantial loss rate of 55% highlighted areas that require enhancement. The dataset, which specifically emphasizes Indonesian tourism destinations, restricts the model's broader use. The use of LSTM was appropriate for text data, although its performance may have been hampered by the small dataset size and diversity. In general, the study showcased a robust methodology, while it would have been advantageous to have a more varied dataset and use optimizations to minimize loss.

Rianti et al. [19] introduced a model based on LSTM to forecast the next word in a sequence of text. The model demonstrated 75% accuracy, but a significant loss rate of 55% highlighted areas that require enhancement. The dataset, which specifically targets

Indonesian tourism destinations, restricts the model's wider application. The selection of LSTM was suitable for textual data, but the constrained, limited sample size and lack of diversity perhaps hindered its efficacy. In general, the study showcased a robust methodology, while it would have been advantageous to have a more varied dataset and use optimizations to minimize loss. Naulla et al. [20] aimed at the Sinhala language, with a particular emphasis on Sri Lankan cricket news. The study included datasets from six news websites, encompassing stories published between 2016 and 2019. The model successfully attained a training accuracy of around 73% by effectively employing RNNs for the intricate and inflected Sinhala language. An evident drawback was the elevated rates of validation and testing loss, suggesting the possibility of overfitting and a constraint in the model's ability to apply to a broader range of data due to its dataset being specific to a certain domain. Nevertheless, the study's strength lay in its ability to showcase the practicality of precise word anticipation in Sinhala, marking a substantial advancement in language processing for languages with complex inflection patterns. This study has implications beyond cricket news; it highlights the need to use larger and more varied datasets to improve accuracy and decrease overfitting.

Nayak et al. [21] presented an LSTM model as a means to improve text prediction, surpassing the performance of traditional RNNs. It successfully handled dependencies that lasted for a long time, with an accuracy range of 92–96%. The paper's main advantage was its emphasis on utilizing LSTM for enhanced syntax prediction and operational efficiency. Nevertheless, it failed to adequately address the datasets and the model's constraints, which are crucial for practical implementation. This research was a significant breakthrough in the domain of text prediction technology. Endalie et al. [11] introduced a bi-directional LSTM-Gated Recurrent Unit (BLSTM-GRU) model designed for predicting the next word in Amharic. Using a dataset generated from the Amharic Bible, it attained a remarkable accuracy of 78.6%. The main advantages were its novel integration of BLSTM and GRU, its applicability to Amharic language processing, and the development of a fresh dataset. The weaknesses encompassed a restricted dataset domain, the possibility of high processing demands, and the risk of overfitting. This study was notable for its contribution to the advancement of language technology in locations with limited resources. Trigreisian et al. [22] utilized the Bi-LSTM algorithm to develop a model for word prediction in Indonesian book titles, resulting in an accuracy rate of 81.82%. The Bi-LSTM model's proficiency with sequential text was showcased by utilizing a particular dataset of Indonesian book titles, highlighting its strength. Nevertheless, the study's scope was constrained by the dataset's exclusive emphasis on the Indonesian language, impacting its broader relevance. Consequently, additional investigation was required, using several datasets to encompass a wider range of languages and contexts.

Table 2 provides a summary of various research publications that utilized different algorithms and datasets to achieve specific accuracy levels in NLP tasks. For example, the use of Bi-LSTM on a dataset of Indonesian book titles resulted in 81.82% accuracy, demonstrating the efficiency of this approach for language-specific and domain-specific tasks. However, limitations such as dataset specificity and limited dataset diversity were also observed in some studies.

Table 2. A summary of DL models research papers.

Ref.	Algorithm	Dataset	Accuracy	Strength points	Limitations
[7]	LSTM	Assamese text	88.20%	Novel approach Handling the complexity of the Assamese language	Small-single source dataset Lack of comparative study
[16]	RNN Stacked RNN LSTM Bi-LSTM	English language	60–72%	Comprehensive model comparison Real-world dataset application	Insufficient performance analysis Unexpected accuracy differences
[17]	LSTM	Dataset contains a few sorts of Bengali posts	98.766%	Effective use of LSTM cells Pre-processing and Word embedding	Dataset limitation Specificity to Bengali
[18]	LSTM networks and phonetics	A dataset comprising 1800 words	97.50%	efficient use of phonetics and LSTM	Limited dataset focused exclusively on word-level difficulty
[6]	Bi-LSTM and Bi-GRU	Bangla newspapers, Wikipedia, etc	99.43%	High accuracy Diverse data source Comprehensive application	Limited dataset Exclusive focus on the Bangla language
[19]	LSTM	Web-scraped Indonesian destinations	75%	Appropriate LSTM model	Limited dataset diversity
[20]	RNN	Gathering from Six Sinhala news websites datasets	–	Targeted Sinhala language Effective RNN use	increased validation and testing loss possibility of overfitting limited dataset scope
[21]	LSTM	–	92 -96%	LSTM efficiency syntax improvement	Insufficient dataset analysis model constraints

(*continued*)

Table 2. (*continued*)

Ref.	Algorithm	Dataset	Accuracy	Strength points	Limitations
[11]	BLST-GRU	63,300 Amharic sentences	78.60%	Novel BLSTM, GRU integration Fresh dataset development	Restricted dataset domain Possible high processing demands Risk of model overfitting
[22]	Bi-LSTM	Dataset of Indonesian book titles with 5618 data	81.82%	Efficiency use of Bi-LSTM Specific Dataset Utilization	language-specific (Indonesian) and domain-specific (book titles) in dataset

2.3 Pre-trained Models

The term "pre-trained" is used in the field of word prediction to describe a language model that has already been trained on a large dataset to recognize language patterns and context and that has been further customized or fine-tuned to predict the next word in a given context. Before making predictions for a specific use case, the model may draw on a broader understanding of language thanks to this pre-training. Here is an overview of research publications in which pre-trained models were applied.

Please note that the first paragraph of a section or subsection is not indented. The first paragraphs that follows a table, figure, equation etc. does not have an indent, either.

Subsequent paragraphs, however, are indented.

Qu et al. [23] focused on the application of BERT and GPT-2 models for text generation. This was achieved by utilizing previously collected BaiduBaike and LLKT corpora. The text emphasized GPT-2's proficiency in generating lengthy sentences and BERT's effectiveness in predicting words. The key findings revealed a high level of accuracy, while there were some concerns with coherence and repetition in the generated content. The strengths of the approach included the proficient utilization of sophisticated models for Chinese text. However, the shortcomings were evident in the presence of inconsistencies in the training data, which had an impact on the coherence of sentences. This study indicated potential advancements in data standardization and the utilization of models. Ma et al. [24] presented a new pre-training technique called PROP, which improved the efficiency of ad-hoc retrieval. PROP demonstrated substantial enhancements in retrieval tasks and showed strong performance in low-resource environments when applied to English Wikipedia and the MS MARCO dataset. The main advantage of this is its ability to be applied universally and its strong foundation in traditional information retrieval methods. Nevertheless, the extent to which the latest language model developments are effective in comparison and their applicability in various real-world scenarios have not been thoroughly investigated. This analysis provided a thorough and inclusive summary, emphasizing the method's groundbreaking methodology and prospective avenues for future exploration.

Agarwal et al. [25] utilized BERT and MLM models to predict Hindi text. The study made use of a dataset obtained from IIT Bombay. The primary emphasis was on Hindi, specifically addressing the necessity for inclusive language representation in NLP. Nevertheless, it was deficient in comprehensive performance measures, hence rendering its efficacy uncertain. The utilization of advanced models was a notable strength; however, the shortcomings included the restricted breadth of the dataset and the lack of precise accuracy estimates. This approach was noteworthy for its emphasis on an underrepresented language but necessitated more comprehensive evaluation metrics for practical applicability. Kim et al. [26] presented KM-BERT, a Korean medical language model, which greatly enhanced the processing of Korean medical literature. It demonstrated superior performance compared to baseline models such as M-BERT and KR-BERT, namely in the field of Korean Medical Named Entity Recognition, resulting in a higher F1-score. By utilizing a rich collection of clinical and health books, the model successfully tackled linguistic and domain-specific difficulties. Nevertheless, the enhanced vocabulary version of the model did not consistently surpass the performance of the original model, and the limitations of computational resources restricted the pre-training to the initial weights of KR-BERT. The highlighted characteristics of KM-BERT demonstrate its robustness in comprehension and practicality, as well as identify potential areas for future improvement.

Sirrianni et al. [27] investigated the performance of GPT-2 and GPT-Neo models in predicting medical text based on dental clinical notes. The study revealed that GPT-2 exhibited superior accuracy, achieving a rate of 76%, in contrast to GPT-Neo, which had a lower accuracy of 53%. The effective prediction of names and abbreviations in GPT-2 involves robust characteristics. The weaknesses of both approaches encompass difficulties in handling patient-specific data and specialized language within specialized domains. This study emphasized the possibilities and constraints of employing pre-trained models in medical charting, indicating the need for additional progress and improvement.

Table 3 provides a summary of the different pre-trained models used in research publications. It includes information about the algorithms used, datasets employed, accuracy, strengths, and limitations of each model. This summary helps researchers understand the effectiveness and applicability of these models in handling complex tasks, improving retrieval tasks, and making effective predictions in specific domains such as medical literature and dental clinical notes.

Table 3. A summary of pre-trained models research publication.

Ref.	Algorithms	Dataset	Accuracy	Strength points	Limitations
[23]	GPT-2 and BERT	BaiduBaike and LLKT Datasets	–	Effectively use of BERT and GPT2 Handling complex tasks	Phrase coherence and repetition Diverse training data
[24]	PROP	English Wikipedia the MS MARCO dataset	–	Novel pre-trained model Use of dataset Improved retrieval task	Comparative effectiveness Real-world applicability
[25]	BERT and Masked Language Model	Collected by IIT Bombay at ILTC	–	Use advanced models Focused on the Hindi language	Limited dataset scope Lack of explicit performance metrics
[26]	KM-BERT	Three types of datasets Clinical research articles Health information articles Medical textbooks	–	Introduction of KM-BERT High F1-score Utilization of rich collection	Inconsistent performance Computational limitations
[27]	GPT-2 and GPT-Neo	A large dataset of dental clinical notes	76%/ 53%	Use of GPT-2 and GPT-Neo Effective prediction	Struggles with patient-specific data Domain-specific terminology

2.4 Hybrid Models

Word prediction has witnessed the introduction of several modern and novel methods. Some are brand-new findings, while others result from blending several techniques. We'll look at several publications that have used these modern methods in the following summary; in Table 4, we will see a summary of hybrid models.

Rakib et al. [28] discussed a study on Bangla word prediction utilizing GRU, an advanced variant of RNN, with a specific emphasis on n-gram models. The accuracy of the models varied significantly, with 5-g models achieving a peak accuracy of 99.70% and uni-gram models dropping to 32.17%. The primary advantages consisted of the successful implementation of GRU to enhance conventional RNN capabilities and impressive performance in greater n-gram models. Nevertheless, simpler n-gram models exhibited poorer accuracy, and achieving ideal outcomes depended on the quality and size of the dataset. This study was remarkable for its contribution to the advancement of Bangla language processing. Yang et al. [29] introduced MCNN-ReMGU, a previous method for predicting natural language words. It combines multi-window convolution with a residual-connected Minimal Gated Unit network. This model demonstrated improved precision when evaluated on the Penn Treebank (PTB) and WikiText-2 (WT2) datasets. The notable characteristics of this system encompass enhanced feature extraction and the resolution of the vanishing gradient problem. Nevertheless, it encountered possible overfitting concerns and had restricted evaluation in real-life situations. MCNN-ReMGU presented a promising, although incompletely validated, approach to several NLP problems.

Habib et al. [30] investigated the creation of a telemedicine prediction system for Arabic text using DL techniques. The exhibition featured models such as LSTM, BiLSTM, and CONV1D, with CONV1D reaching remarkable accuracy. The main advantages were its innovative approach to processing Arabic text in telemedicine and comprehensive evaluations of models. The dataset's diversity was limited due to its reliance on Altibbi's database and the difficulties of the Arabic language, which affected its real-world applicability. The intricate nature of DL models presented difficulties in terms of deploying them in real time and ensuring computing efficiency. Shakhovska et al. in [12] presented an innovative approach for Ukrainian language next-word prediction, utilizing LSTM, Markov chains, and a hybrid model trained on a unique dataset of Ukrainian poems. The LSTM model attained an accuracy of around 75%. The study's strengths included the innovative approach used for a language with limited resources and the unique selection of the dataset. Nevertheless, the study revealed limitations in the small sample size, which could potentially compromise its generalizability. Additionally, the reliance on a poetry-based dataset may restrict its broader applicability. Proposed enhancements for the model include refining punctuation prediction and word form suggestions, hence augmenting the overall efficacy of the program.

Table 4 summarizes the word prediction model algorithm and dataset research articles. GRU-based RNNs trained on Bangla language data, MCNN-ReMGU models trained on PTB and WT2 datasets, LSTM, BiLSTM, CONV1D, and LSTM-CONV1D models applied to Altibbi's medical recommendation database, LSTM models with Markov chains for Ukrainian poetry datasets, and federated learning with pre-trained methods using Stack Overflow and Shakespeare's works datasets. Each approach's accuracy, merits, and weaknesses are summarized, emphasizing dataset quality, model complexity, and application.

Table 4. A summary of the Hybrid Models research papers.

Ref.	Algorithm	Dataset	Accuracy	Strength points	Limitations
[28]	GRU Based RNN	Collected from different sources in the Bangla language	99.70%	Use of GRU technology in RNN High accuracy in 5-g model	Low accuracy in uni-gram model Reliance on dataset quality and size
[29]	MCNN-ReMGU	Penn Treebank (PTB) and WikiText-2 (WT2) datasets	–	Improved feature extraction Improved addressing of the vanishing gradient problem	Overfitting issues Limited testing in real-world
[30]	LSTM, BiLSTM, CONV1D, and LSTM-CONV1D	Altibbi's medical recommendation DB	–	Innovative approach Comprehensive model evaluation Practical application	Dataset diversity Complexity of Arabic language Model complexity
[12]	LSTM Markov Chains Hybrid model	Ukrainian poems dataset	–	Innovative approach Usage of a unique dataset	Limited dataset The reliance on a poetry-based dataset

3 Techniques

The next-word prediction techniques involve suggesting the next or most probable words based on user input or a predefined set of phrases. Various approaches have been employed in the creation of diverse next-word prediction systems, such as statistical approaches DL, pre-trained models, and hybrid models, which are a combination of two models or more. Based on the findings of this study, it has been observed that word prediction models may be categorized into four distinct parts, as seen in Fig. 1.

3.1 Statistical Models

Word prediction in statistical methods is based on the probabilities of words occurring in a specific text corpus. The probabilistic method is a popular choice. The Markov Chain assumption provides a foundation for word prediction in statistical approaches. According to this concept, the next word is heavily influenced by the context of the previous work, which in this case is the previous (N-1) word. Therefore, the N-gram model is another name for this method [8]. The following are examples of statistical methodologies that frequently use different methods:

Markov Chain Model. Markov chains are fundamental components of stochastic processes; in the Markov chain model, the likelihood of the text determines the prediction

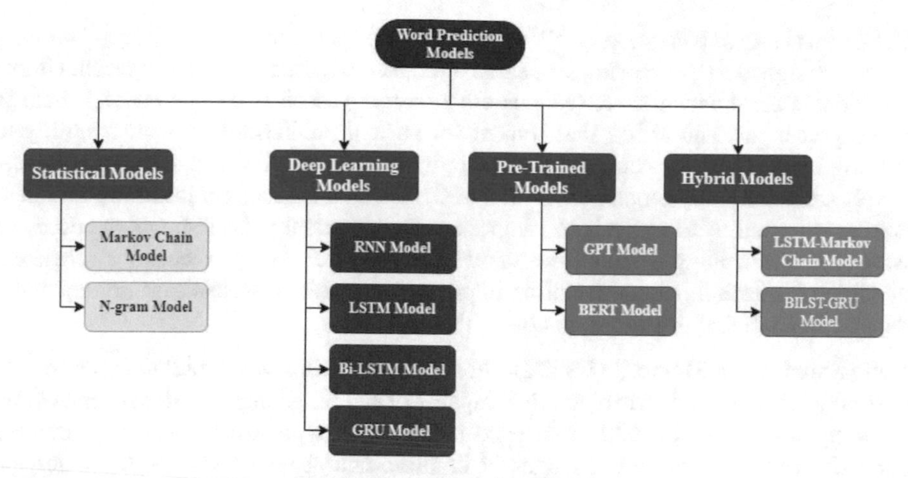

Fig. 1. Word Prediction Models.

of the following word. They serve several functions in a wide range of fields of study. Since the past and the future are unknowable based on knowledge of the present, Markov chains are a type of stochastic process that satisfies the Markov characteristics. Knowing the current state of a process is sufficient for making accurate predictions about its future behavior. Markov chains are used to estimate the outcome of a certain event when we only have access to the event's prior state and the state it is now in at the time of estimation. By "state", we mean the current climatic and political circumstances. If we have sequential variables for each state s1, s2,..., si in a more rigorous model. The Markov model will incorporate the Markov model's premise about the list of probabilities for sequences, i.e., that we need only know the present to make predictions [23, 25].

N-gram Model. The N-gram model is a fundamental approach in computer science's NLP field. It involves analyzing sequences of n items (phonemes, syllables, letters, words, and base pairs) to predict the next word. N-grams are derived from text/speech corpora, employing a Markov chain model to measure word occurrence probability. They predict the next word based on the preceding N-1 words. Commonly, N is 2 (bigrams) or 3 (trigrams), leveraging prior context for word generation. N-gram model types' include; a) Unigram model (considers the current word in isolation), b) Bigram model (accounts for the current and preceding words), and c) Trigram model (incorporates the current word and two preceding words). These models find applications in speech recognition, machine translation, and text classification, and underpin advanced language models like neural language models. N-grams power auto-completions (e.g., Gmail), spell correction, grammar checking, etc. [5, 23].

3.2 Deep Learning Models

DL is a part of AI that focuses on algorithms that work like the human brain to process data, or it is a network that can learn from data; it is also commonly named a deep neural network [8]. The following subsections illustrate some of the DL algorithms.

Recurrent Neural Networks (RNNs). RNNs are a specialized class of Deep Learning models designed for processing sequential data like text, time series, and speech. Unlike traditional neural networks, RNNs possess a memory mechanism that enables them to retain past inputs and utilize that context for current predictions. This adaptability to varying input lengths is achieved by employing the same set of weights for all inputs. RNNs are extensively employed in NLP tasks such as language translation, text generation, and sentiment analysis. An important RNN variant is LSTM, which addresses issues like vanishing gradients in standard RNNs. By iteratively executing a consistent operation for each input and factoring in prior computations, RNNs capture sequential dependencies, a feat distinct from non-RNNs [22, 24].

Long Short-Term Memory (LSTM). In 1997, Hochreiter & Schmidhuber introduced LSTMs as an advanced form of an RNN capable of handling long-term dependencies and retaining information [31]. LSTMs, a specialized RNN type, excel at learning extended dependencies. To address challenges in iterative neural networks, LSTMs emerged. Unlike RNNs, LSTMs have 2-input gates, forget gates, and output gates that interact specially, as shown in Fig. 2. These gates are integrated into repeating neural network cells. The input gate manages cell state updates, determining data significance (1) or insignificance (0) via sigmoid processing. The tanh activation function regulates network adjustments within the -1 to 1 range. Multiplying sigmoid and tanh outputs guides data update decisions. Conversely, the exit gate governs input (ht + 1) for the next cell. It combines past data and current input using sigmoid processing, followed by tanh processing for existing cell state data. The multiplication of these results dictates the input (ht + 1) for the subsequent cell. Following gate operations in the current cell, decisions about transferring cell state and hidden state (ht) data to the next cell occur [2].

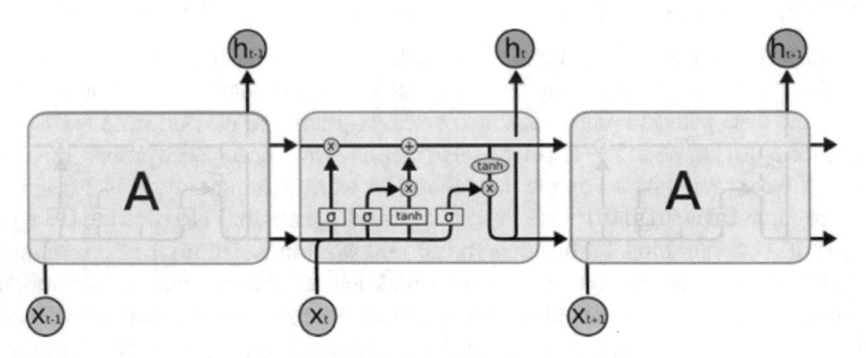

Fig. 2. Structure of LSTM network [2].

Gated Recurrent Units (GRU). GRU offers a simplified variant of the LSTM model in RNNs, focusing on resolving short-term memory issues. GRU features a single state vector and two gate vectors, namely the reset and update gates. This configuration equips GRU to undertake tasks such as natural language processing, speech signal modeling, and music modeling, much like LSTM. Notably, GRU streamlines its structure by combining the input and forget gates into a single gate. In scenarios involving smaller datasets,

GRU exhibits superior performance compared to LSTM. The crux of the distinction lies in GRU's utilization of only two gates—the update and reset gates—facilitating seamless information transfer through its hidden state. Unlike LSTM, GRU dispenses with a separate cell state. These characteristics collectively position GRU as an efficient alternative for enhancing short-term memory management within neural networks [1, 26].

Bidirectional LSTM. A bidirectional LSTM (Bi-LSTM), commonly known as a bidirectional LSTM, represents a significant advancement in the realm of natural language processing, particularly in the domain of next-word prediction. It serves as an enhanced iteration of the RNN, meticulously crafted to elevate the accuracy and precision of models when predicting the next word in a sequence of text. In this innovative approach, the input sequence is subjected to a unique dual processing by two LSTM units, diverging from the traditional LSTM's solitary operation. The first LSTM analyzes the preceding words in the sequence, while concurrently, the second LSTM scrutinizes the subsequent words, effectively looking both forward and backward in the text to predict the next word with greater context awareness, as illustrated in Fig. 3.

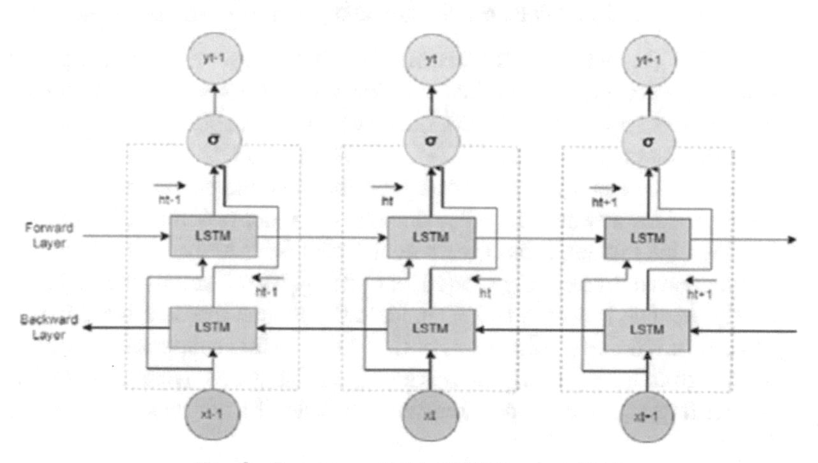

Fig. 3. Structure of Bi-LSTM Network [22].

The key distinction that sets a bi-LSTM apart from a conventional LSTM is how it manages the flow of information to excel in next-word prediction. While a unidirectional LSTM retains information solely from the words that have previously traversed its hidden state, a bidirectional LSTM operates with a broader perspective. It processes inputs in both directions, from left to right (past to future) and from right to left (future to past), simultaneously. Consequently, a backward-running LSTM can only consider information from upcoming words, limiting its predictive capabilities. In contrast, a bidirectional LSTM, thanks to its fusion of two hidden states, can effectively leverage knowledge from both historical and forthcoming words, greatly improving its accuracy in predicting the next word [32]. Generally, in a normal LSTM network, we take output directly, as shown in Fig. 2, but in a bidirectional LSTM network, the output of the forward and backward layers at each stage is given to the activation layer, which is a neural network, and the output of this activation layer is considered. This output contains information about the relationship between past and future words.

3.3 Pre-trained Models

Deep neural networks that have been pre-trained on large-scale unlabeled corpora and which may then be fine-tuned for various downstream tasks are what we call pre-trained language models (PLMs) [33]. PLMs have been proven to encode a substantial amount of linguistic information into their voluminous quantities of parameters [34, 35]. As a result, using PLMs to better comprehend language and provide better-quality output is an exciting prospect. As a result of Transformer's widespread adoption, it serves as the foundation for virtually every PLM. The GPT [35] and the BERT [36]are two common PLMs that initially rely on a transformer decoder and an encoder, respectively. The PLMs XLNet [37], RoBERTa [38], ERNIE [39], T5, and BART [40] followed GPT and BERT. T5 and BART are encoder-decoder-based PLMs, whereas XLNet, RoBERTa, and ERNIE are built on the BERT architecture. The OpenAI GPT-2 and BERT models are currently widely used language models for text generation and prediction.

The OpenAI GPT (Generative Pre-Training). GPT is a state-of-the-art natural language processing technique that employs a series of transformer decoder blocks [35]. It alters the conventional transformer by doing away with the encoder-less multihead attention module that uses cross-attention. A feedforward network for processing, positional encoding, and masked multi-head self-attention make up the decoders. The successor, GPT-2, likewise takes advantage of the decoder-only construction of a transformer. Assuming a beginning token, it iteratively generates tokens and adds them to the input for future predictions until reaching a length or end flag. By repeatedly feeding the model with the newly predicted word, the GPT model may predict the next word in a phrase. Its ability to train on massive volumes of online data and its unsupervised nature makes it versatile enough to handle many categories of text, such as psychology. In this way, each given text fragment can be coherently extended until the process finds a logical finish [23, 41].

BERT (Bidirectional Encoder Representations from Transformers). The BERT model has been designed to pre-train deep bidirectional representations from unannotated text. This is achieved by simultaneously considering the contextual information from both the left and right sides in all layers of the model. Consequently, the pre-trained BERT model may be effectively fine-tuned by incorporating a single supplementary output layer. This approach enables the creation of cutting-edge models for many tasks, including question-answering and language inference, without the need for significant alterations to the task-specific architecture [36]. BERT's technique is based on masked language modeling. The model hides 15% of the words inside the input document or corpus, prompting the model to generate predictions on the missing words. Each transformer encoding block in the stack is given a sentence with a missing word and charged with anticipating the missing word. In the above phrases, 15% of words are missing, with each missing word assigned to every encoder block inside the stack. The technique used in this context is unsupervised, which means it does not rely on outside guidance or supervision. Each word inside a sentence has the potential to be hidden, and the goal is to predict which word has been masked. Because the job is unsupervised and there are no labeling restrictions, the large amount of textual data available on the internet may be used to train the BERT model. This entails picking words at random to be masked during the training process. It is critical to recognize that BERT is a bidirectional model that learns to predict missing words by considering both previous and following words via attention processes. As a result, BERT evaluates both the prior (left) and subsequent (right) contexts of each word at the same time [37, 42–49].

3.4 Hybrid Techniques

To increase effectiveness or accuracy, hybrid methods are implemented. These approaches integrate several different techniques, such as a combination of statistical and DL techniques or methods that utilize a single technique [8].

BILST-GRU. The Bi-LSTM-GRU model is an advanced deep learning algorithm used for word prediction and contextual understanding in natural language processing (NLP) applications. This sophisticated model may readily incorporate Bidirectional Long-Term Memory (BLSTM) and Gated Recurrent Unit (GRU) layers to effectively understand complex textual connections. The BLSTM component effectively captures important features at each phase by utilizing a dual-directional method, hence mitigating the problem of overfitting. The GRU layer, renowned for its straightforward traversal approach, retrieves pertinent content attributes from incoming data. The model's structure has a distinct hierarchy, starting with communication and culminating with digital conversion. The text employs a bidirectional LSTM layer to scrutinize its many intricacies. Ultimately, a GRU layer is employed to retrieve precise context. This seamless integration ensures optimal performance and accurate interpretation of textual information in many business languages. This approach assists businesses in enhancing language difficulties by employing word prediction. [11].

4 Discussion and Future Work

The inclusion of word prediction technology has significant promise to assist those with physical, perceptual, or intellectual disabilities who struggle with typing. The poll utilized many methods, such as DL, statistical techniques, and a combination of numerous methodologies, to predict the next word. Certain researchers have utilized modern methods, such as the implementation of pre-trained models. The methods mentioned above were employed by several studies, resulting in a notable level of accuracy. To improve the effectiveness and performance of prediction systems, researchers must make substantial efforts to tackle issues such as language complexity or dataset quality, among others. The following proposals are suggested as possible remedies to tackle current obstacles and improve the accuracy of forecasting.

- Employ contemporary language modeling techniques such as BERT and its variants, GPT-3, or other state-of-the-art models, if accessible. These models possess enhanced comprehension of intricate linguistic systems, resulting in more precise predictions, offering a potential resolution to the issue of language complexity.

- The objective is to construct and curate a robust dataset including high-quality sources for languages spoken globally while minimizing the presence of noisy data and addressing the scarcity of datasets for some languages. The resolution of this issue will contribute to the enhancement of predictive accuracy about the term.

5 Conclusion

Various word prediction algorithms have been devised to accomplish multiple objectives, such as speeding typing, enhancing communication rates, and reducing the cognitive strain involved in text input. At first, these methods were conceived as assistance for people who suffered from motor limitations to improve their overall quality of life. However, these tools are not just for people with disabilities; others without disabilities can also use them to feel more comfortable and improve their spelling.

The objective of this study is to conduct a complete examination of the application of next-word prediction technology within the field of NLP. The main emphasis lies on the potential of this technology to improve the efficiency of text entry, reduce spelling errors, and aid users who encounter language or physical difficulties. This paper looks at different prediction models like N-grams, RNNs, LSTMs, bi-LSTMs, and hybrid techniques like BILST-GRU. It also looks at how they can be used and their effects.

This work elucidated the categorization of prediction models for the next words into four distinct groups, including statistical, DL, and hybrid models, and a novel classification termed pre-trained models. In recent years, there has been extensive utilization of pre-trained models in the field of language modeling. This has facilitated the incorporation of word prediction systems into several domains, such as next-sentence prediction, missing word prediction, question answering, and various other applications. It is also deduced that no model exhibits superior performance over another, contingent upon several elements inside the used system. The selection of an acceptable model for the prediction of the following word system may be made. Several factors are involved, namely:

- The model selection is heavily influenced by the nature of the application. Some examples include mobile keyboards, assistive technologies, and content-creation tools.
- The availability of data: The quantity and quality of accessible training data significantly impacts the selection of a model.
- Computational Resources: The selection of a model is influenced by the computational resources available. Models like BERT or GPT-3, which operate on a large scale, need substantial computational resources and may necessitate the utilization of high-performance GPUs or TPUs for both training and inference processes.
- Accuracy Requirements: The level of accuracy necessary for the word prediction job is an important aspect. In situations where achieving a high level of prediction accuracy is crucial, it may be advantageous to employ more intricate models that possess the ability to capture nuanced language patterns. Conversely, in situations where a modest level of accuracy is deemed sufficient, less complex models may be deemed satisfactory.

References

1. ChhayaGudade, S.B.: Models of word prediction .pdf. Int. J. Sci. Res. **12**(7), 120–127 (2023). https://doi.org/10.21275/SR23630151700
2. Atçılı, A., Özkaraca, O., Sarıman, G., Patrut, B.: Next word prediction with deep learning models. In: Hemanth, D.J., Kose, U., Watada, J., Patrut, B. (eds.) ICAIAME 2021. Engineering Cyber-Physical Systems and Critical Infrastructures, vol. 1, pp. 523–531. Springer, Cham (2023). https://doi.org/10.1007/978-3-031-09753-9_38
3. Narula, K.: A critical review on next word prediction. Int. J. Adv. Res. Sci. Commun. Technol. (IJARSCT) **3**(1) (2023). www.ijarsct.co.in
4. Makkar, R., Kaur, M., Sharma, D.V.: Word prediction systems: a survey. Adv. Comput. Sci. Inf. Technol. (ACSIT) **2**(2), 177–180 (2015). http://www.krishisanskriti.org/acsit.html
5. HaCohen-Kerner, Y., Applebaum, A., Bitterman, J.: Improved language models for word prediction and completion with application to Hebrew. Appl. Artif. Intell. **31**(3), 232–250 (2017). https://doi.org/10.1080/08839514.2017.1315503
6. Mahmud, A., Rony, M.N.H., Bhowmik, D.D., Kuri, R., Rana, A.R.M.M.H.: An RNN based approach to predict next word in Bangla language. In: Satu, M.S., Moni, M.A., Kaiser, M.S., Arefin, M.S. (eds.) MIET 2022. LNICST, vol. 490, pp. 551–565. Springer, Cham (2022). https://doi.org/10.1007/978-3-031-34619-4_43
7. Barman, P.P., Boruah, A.: A RNN based approach for next word prediction in assamese phonetic transcription. Procedia Comput. Sci. **143**, 117–123 (2018)
8. Hamarashid, H.K., Saeed, S.A., Rashid, T.A.: A comprehensive review and evaluation on text predictive and entertainment systems. Soft. Comput. **26**(4), 1541–1562 (2022). https://doi.org/10.1007/s00500-021-06691-4
9. Hard, A., et al.: Federated learning for mobile keyboard prediction. arXiv Preprint arXiv: 1811.03604 (2018)
10. Stremmel, J., Singh, A.: Pretraining federated text models for next word prediction. In: Arai, K. (ed.) FICC 2021. AISC, vol. 1364, pp. 477–488. Springer, Cham (2021). https://doi.org/10.1007/978-3-030-73103-8_34
11. Endalie, D., Haile, G., Taye, W.: Bi-directional long short term memory-gated recurrent unit model for Amharic next word prediction. PLoS ONE **17**(8), e0273156 (2022)
12. Shakhovska, K., Dumyn, I., Kryvinska, N., Kagita, M.K.: An approach for a next-word prediction for Ukrainian language. Wirel. Commun. Mob. Compu. **2021** (2021). https://doi.org/10.1155/2021/5886119
13. Niti Shah, M.N.K.: Syntactic_Word_Prediction_for_Hindi.pdf. IJSART **3**(3) (2017). ISSN 2395-1052
14. Bhuyan, M.P., Sarma, S.K.: Effects of prediction-length on accuracy in automatic Assamese word prediction. In: Proceedings of 2019 3rd IEEE International Conference on Electrical, Computer and Communication Technologies, ICECCT 2019 (2019). https://doi.org/10.1109/ICECCT.2019.8869431
15. Yazdani, A., Safdari, R., Golkar, A., Niakan Kalhori, S.R.: Words prediction based on N-gram model for free-text entry in electronic health records. Heal. Inf. Sci. Syst. **7**(1), 6 (2019). https://doi.org/10.1007/s13755-019-0065-5
16. Ramya, M.S., Selvi, D.C.S.K.: Recurrent neural network based models for word prediction. Int. J. Recent Technol. Eng. **8**(4), 7433–7437 (2019). https://doi.org/10.35940/ijrte.D5313.118419
17. Abujar, S., Masum, A.K.M., Chowdhury, S.M.M.H., Hasan, M., Hossain, S.A.: Bengali text generation using bi-directional RNN. In: 2019 10th International Conference on Computing, Communication and Networking Technologies, ICCCNT 2019 (2019). https://doi.org/10.1109/ICCCNT45670.2019.8944784

18. Parihar, S., Miradwal, S., Panse, A., Patel, R.: Word difficulty level prediction system using deep learning approach. In: Ethics and Information Technology, pp. 109–112. Volkson Press (2020). https://doi.org/10.26480/etit.02.2020.109.112
19. Rianti, A., Widodo, S., Ayuningtyas, A.D., Hermawan, F.B.: Next word prediction using LSTM. J. Inf. Technol. Util. **5**(1), 432033 (2022)
20. Naulla, N.T.K., Fernando, T.G.I.: Predicting the next word of a Sinhala word series using recurrent neural networks. In: 2022 2nd International Conference on Advanced Research in Computing (ICARC), pp. 13–18 (2022). https://doi.org/10.1109/ICARC54489.2022.9754174
21. Nayak, C., Kumar, A.: Next word prediction using machine learning techniques, vol. 54, no. 02 (2022)
22. Trigreisian, A.A., Harani, N.H., Andarsyah, R.: Next word prediction for book title search using bi-LSTM algorithm. Indones. J. Comput. Sci. **12**(3) (2023). https://doi.org/10.33022/ijcs.v12i3.3233
23. Qu, Y., Liu, P., Song, W., Liu, L., Cheng, M.: A text generation and prediction system: pre-training on new corpora using BERT and GPT-2. In: ICEIEC 2020 - Proceedings 2020 IEEE 10th International Conference on Electronics Information and Emergency Communication, pp. 323–326 (2020). https://doi.org/10.1109/ICEIEC49280.2020.9152352
24. Ma, X., Guo, J., Zhang, R., Fan, Y., Ji, X., Cheng, X.: Prop: Pre-training with representative words prediction for ad-hoc retrieval. In: Proceedings of the 14th ACM International Conference on Web Search and Data Mining, pp. 283–291 (2021)
25. Agarwal, S., Sukritin, Sharma, A., Mishra, A.: Next word prediction using Hindi language. In: Hu, Y.C., Tiwari, S., Trivedi, M.C., Mishra, K.K. (eds.) Ambient Communications and Computer Systems. Lecture Notes in Networks and Systems, vol. 356, pp. 99–108. Springer, Singapore (2022). https://doi.org/10.1007/978-981-16-7952-0_10
26. Kim, Y., et al.: A pre-trained BERT for Korean medical natural language processing. Sci. Rep. **12**(1), 13847 (2022)
27. Sirrianni, J., Sezgin, E., Claman, D., Linwood, S.L.: Medical text prediction and suggestion using generative pretrained transformer models with dental medical notes. Methods Inf. Med. **61**(05/06), 195–200 (2022)
28. Rakib, O.F., Akter, S., Khan, M.A., Das, A.K., Habibullah, K.M.: Bangla word prediction and sentence completion using GRU: an extended version of RNN on N-gram language model. In: 2019 International Conference on Sustainable Technologies for Industry 4.0, STI 2019 (2019). https://doi.org/10.1109/STI47673.2019.9068063
29. Yang, J., Wang, H., Guo, K.: Natural language word prediction model based on multi-window convolution and residual network. IEEE Access **8**, 188036–188043 (2020)
30. Habib, M., Faris, M., Qaddoura, R., Alomari, A., Faris, H.: A predictive text system for medical recommendations in telemedicine: a deep learning approach in the Arabic context. IEEE Access **9**, 85690–85708 (2021). https://doi.org/10.1109/ACCESS.2021.3087593
31. Tiwari, A., Sengar, N., Yadav, V.: Next word prediction using deep learning. In: 2022 IEEE Global Conference on Computing, Power and Communication Technologies, GlobConPT 2022 (2022). https://doi.org/10.1109/GlobConPT57482.2022.9938153
32. Raza, M.F.A., Naeem, M.A.: Saraiki language word prediction and spell correction framework. In: 2022 24th International Multitopic Conference (INMIC), pp. 1–6 (2022). https://doi.org/10.1109/INMIC56986.2022.9972938
33. Almutiri, T., Nadeem, F.: Markov models applications in natural language processing: a survey. Int. J. Inf. Technol. Comput. Sci **2**, 1–16 (2022). https://doi.org/10.5815/ijitcs.2022.02.01

34. Zhong, J., Li, W.: Detecting customer churn signals for telecommunication industry through analyzing phone call transcripts with recurrent neural networks detecting customer churn signals for telecommunication industry through analyzing phone call transcripts with recurrent, no. February (2020)
35. Hochreiter, S., Schmidhuber, J.: Long short-term memory. Neural Comput. **9**(8), 1735–1780 (1997). https://doi.org/10.1162/neco.1997.9.8.1735
36. Khalid, M.F.: Text prediction using machine learning, p. 45 (2022). diva2: 1632760
37. Soam, M., Thakur, S.: Next word prediction using deep learning: a comparative study. In: Proceedings of the Confluence 2022 - 12th International Conference on Cloud Computing, Data Science and Engineering, pp. 653–658. Institute of Electrical and Electronics Engineers Inc. (2022). https://doi.org/10.1109/Confluence52989.2022.9734151
38. Li, J., Tang, T., Zhao, W.X., Wen, J.R.: Pre-trained language models for text generation: a survey. In: IJCAI International Joint Conference on Artificial Intelligence, pp. 4492–4499 (2021). https://doi.org/10.24963/ijcai.2021/612
39. Ribeiro, L.F.R., Schmitt, M., Schütze, H., Gurevych, I.: Investigating pre-trained language models for graph-to-text generation. In: NLP Conversational AI, NLP4ConvAI 2021 – Proceedings of the 3rd Workshop, pp. 211–227 (2021). https://doi.org/10.18653/v1/2021.nlp4convai-1.20
40. Vaswani, A., et al.: Attention is all you need. Adv. Neural Inf. Process. Syst. **30** (2017)
41. Radford, A., Narasimhan, K., Salimans, T., Sutskever, I.: Improving language understanding by generative pre-training (2018)
42. Devlin, J., Chang, M. W., Lee, K., Toutanova, K.: BERT: pre-training of deep bidirectional transformers for language understanding. arXiv Preprint arXiv:1810.04805 (2018)
43. Yang, Z., Dai, Z., Yang, Y., Carbonell, J., Salakhutdinov, R.R., Le, Q.V.: XLNet: generalized autoregressive pre-training for language understanding. Adv. Neural Inf. Process. Syst. **32** (2019)
44. Liu, Y., et al.: RoBERTa: a robustly optimized bert pretraining approach. arXiv Preprint arXiv: 1907.11692 (2019)
45. Zhang, Z., Han, X., Liu, Z., Jiang, X., Sun, M., Liu, Q.: ERNIE: enhanced language representation with informative entities. arXiv Preprint arXiv:1905.07129 (2019)
46. Lewis, M., et al.: BART: denoising sequence-to-sequence pre-training for natural language generation, translation, and comprehension. arXiv Preprint arXiv:1910.13461 (2019)
47. Ghojogh, B., Ghodsi, A.: Attention mechanism, transformers, BERT, and GPT: tutorial and survey (2020)
48. Sun, K., Luo, X., Luo, M.Y.: A survey of pretrained language models. In: Memmi, G., Yang, B., Kong, L., Zhang, T., Qiu, M. (eds.) KSEM 2022. LNCS, vol. 13369, pp. 442–456. Springer, Cham (2022). https://doi.org/10.1007/978-3-031-10986-7_36
49. Mahdi, A.F., Farhan, R.N., Al-Rawi, S.S.: Improvement of semantic-BERT model with bidirectional LSTM for Arabic language. In: 2022 8th International Conference on Contemporary Information Technology and Mathematics (ICCITM), pp. 202–207. IEEE (2022)

Email Spam Detection by Machine Learning Approaches: A Review

Mohammad Talib Hadi[1](✉) and Salwa Shakir Baawi[2]

[1] Department of Computer Science, College of Computer Science and Information Technology, University of Al-Qadisiyah, Babylon, Iraq
it.mast.23.16@qu.edu.iq

[2] Department of Computer Information Systems, College of Computer Science and Information Technology, University of Al-Qadisiyah, Diwanyah, Iraq

Abstract. Currently, technology has exhibited substantial advancement, resulting in the improvement of communication. Emails are often regarded as the most effective method for both informal and formal communication. Furthermore, individuals utilize email as a means to save and distribute significant data, encompassing textual content, images, documents, and various other things. Due to emails' simple and easy-to-use nature, some people abuse this mode of communication by sending an excessive amount of unwanted emails, usually referred to as spam emails. The spam emails may include malicious content that is disguised as attachments or URLs, posing a risk of security breaches to the host system and potential theft of sensitive information such as credit card data. These days, spam detection poses serious and massive challenges to email and IoT service providers. Various previous studies have concentrated on machine-learning methods to detect spam emails in the mailbox. The primary aim of this work is to provide a comprehensive examination and comparative evaluation of machine learning techniques utilized in the detection of email spam. Also, it highlights the main challenges that face spam email detection. Furthermore, a thorough evaluation of various strategies is conducted, taking into account metrics such as accuracy, precision, recall, and F1-score. Finally, a thorough analysis and potential areas for future research are also examined.

Keywords: Email · Spam · Non-Spam · Spam Detection · Machine Learning

1 Introduction

In the age of information technology, the process of distributing information has become more efficient and rapid [1]. Various platforms exist for individuals to disseminate knowledge globally. Email is the most straightforward, cost-effective, and expedient means of transmitting information globally among all communication platforms. However, because of their uncomplicated nature, emails are susceptible to several forms of attacks, with the most prevalent and perilous being spam. Unwanted emails that are unrelated to one's interests are undesirable as they consume the recipient's time and resources. In addition, these emails may contain harmful content that is concealed as attachments or URLs, which can result in security breaches to the host system [2].

J. Rasheed et al. (Eds.): FoNeS-AIoT 2024, LNNS 1035, pp. 186–204, 2024.
https://doi.org/10.1007/978-3-031-62871-9_15

Email spam is a pervasive global problem that affects all email users. The growth of bulk mailing tools has led to a significant increase in the volume of email spam. This has caused increasing frustration among recipients and placed Internet service providers (ISPs) under constant strain, as they grapple with the issue of unsolicited email communications [3]. Therefore, the escalating volume of data available on the Internet, encompassing diverse novel attributes, poses a significant obstacle for spam detection systems. In other words, the volume of spam emails will probably increase as more people utilize the internet.

Technological innovation is being misused for unethical and unlawful activities such as phishing and scamming. The detection of online spammers has become a prominent social concern because of the threat they pose to internet security [4]. Accurately detecting spam emails is important, and there are many different tools and methods available for this purpose. There are three types of spam detection models: deep learning (DL), machine learning (ML), and computational models [5].

Machine Learning (ML) refers to the systematic procedures designed to enable computing devices to perform tasks without the need for explicit programming [6]. Machine Learning is highly advantageous in addressing the spam problem due to its capacity to adapt and optimize itself over time, thereby overcoming a significant limitation present in other types of spam detection methods known as 'Concept Drift'. Concept Drift, as identified by researchers, refers to the phenomena where the structure and content of spam emails evolve, rendering current tactics ineffective in the future. This is due to the constant changes in the content and operating mechanism of spam emails [7].

Therefore, the email system necessitates a substantial need for security [8]. Many email providers allow their users to make keyword-based rules that automatically filter emails. Still, this approach is not very useful because it is difficult, and users do not want to customize their emails, due to which spammers attack their email accounts.

The main objective of this study is to conduct a thorough investigation and comparative analysis of machine learning methods employed in the detection of email spam. The primary aim of this study is to evaluate the performance and effectiveness of several machine-learning approaches in the context of email spam detection. Additionally, it aims to emphasize the main challenges encountered in email spam detection. Our goal is to assess the precision, recall, accuracy, and F-Score measures of different machine learning algorithms using datasets like Enron, Ling Spam, Spam Assassin, and Spambase. By analyzing performance indicators, it is possible to identify algorithms that demonstrate exceptional accuracy and reliability in detection.

The remained of this paper is organized as follows. In Sect. 2, the performance evaluation metrics are presented. Section 3, provides a related work of email spam detection. The discussion is presented in Sect. 4. Section 5, offers challenges in email spam detection. Concluding remarks are presented in Sect. 6.

2 Performance Evaluation Metrics

In the age of information technology, the process of distributing information has become more efficient and rapid. Various platforms exist for individuals to disseminate knowledge globally. Email is the most straightforward, cost-effective, and expedient means of transmitting information globally among all communication platforms. However, because of their uncomplicated nature, emails are susceptible to several forms of attacks, with the most prevalent and perilous being spam. Unwanted emails that are unrelated to one's interests are undesirable as they consume the recipient's time and resources. In addition, these emails may contain harmful content that is concealed as attachments or URLs, which can result in security breaches to the host system [2].

The literature review has revealed several commonly employed metrics, such as accuracy, recall, precision, and F-score, that are pertinent to the present inquiry. The four primary-level properties outlined below form the foundation for measuring accuracy, recall, and precision. The F-score, in turn, is a calculated metric that relies on recall and precision. There are four primary characteristics at the core level:

- True positive (TP): A legitimate email that has been appropriately identified as a legitimate email.
- True negative (TN): A spam email that has been appropriately identified as spam.
- False positive (FP): A legitimate email that was mistakenly identified as spam.
- False negative (FN): A spam email that was mistakenly identified as a legitimate email [9].

Accuracy refers to the percentage of accurately anticipated instances out of all cases. It is mathematically represented by the following ratio:

$$Accuracy = \frac{TP + TN}{TP + TN + FP + FN} \tag{1}$$

Precision, which is the ratio of true positives (TP) to all positive predictions, can be mathematically represented as:

$$Precession = \frac{TP}{TP + FP} \tag{2}$$

As a Recall: The mathematical expression for the ratio of true positives (TP) to the total number of actual positive cases is as follows:

$$Recall = \frac{TP}{TP + FN} \tag{3}$$

The F1-score is a mathematical expression that represents the harmonic mean of Precision and Recall. It may be calculated using the following formula [10]:

$$F1score = 2 \times \frac{Precessioin \times Recall}{Precession + Recall} \tag{4}$$

Table 1 displays the relationships between the true positive (TP), true negative (TN), false positive (FP), and false negative (FN).

Table 1. The relationship of TP, TN, FP, and FN.

Actual value	Predicted value	
	Spam	Ham
Spam	True positive (TP)	False positive (FP)
Ham	False negative (FN)	True negative (TN)

3 Related Work

In recent years, the research community has focused a lot of attention on content-based spam detection, and several algorithms have been developed. We offer an overview of these spam detection methods. Many supervised machine learning algorithms have been examined in the literature to establish an email detection system. For instance, in [11], Sai Charan Lanka et al. conducted a comprehensive comparison of various machine-learning methods employed to detect Email spam. The approaches include Naive Bayes (NB), Support Vector Machines (SVM), Decision Trees (DT), Random Forests (RF), Bagging, and AdaBoost. The dataset used for this study, spam.csv, came from Kaggle and consists of 5573 records. As a result of the comparison, it was found that Naive Bayes produced the most accurate predictions, achieving almost 98% accuracy due to its minimal false positives. Nevertheless, the program had a limited scope when it came to real-time applications.

In [12], Lakshman Narayana Vejendla et al. introduced a model that enables the classification of emails as either spam or ham, while also providing the ability to calculate the proportion of each category within the Enron1 dataset. This study introduced the SVM model and conducted a comparative analysis with the K-Nearest Neighbor (KNN) model in terms of various performance metrics including accuracy, training time, and feature extraction and selection time. The suggested model demonstrated high efficacy across all indicated performance parameters and obtained a 98% accuracy rate in email spam detection, owing to its robust generalizability and performance.

Archana Saini et al. have performed a comparative analysis of four machine-learning methods for identifying spam [13]. The evaluated techniques are Naive Bayes, K*, J48, and Random Forest (RF). The performance of the four classifiers was evaluated based on four criteria: accuracy, precision, recall, and F1 measure. The Spambase dataset used for predicting email spam was acquired from the UCI repository. The findings indicated that the Random Forest classifier outperformed all other models in terms of all evaluation criteria, with an accuracy rate of 95.48%. Conversely, the Naive Bayes classifier exhibited the lowest performance, with an accuracy rate of 79.29%. Additional machine-learning techniques can be employed and augment the dataset, thereby improving the accuracy of future predictions.

Arka Ghosh et al. [14], employed an ensemble learning technique for spam detection instead of relying on a single model to enhance the overall accuracy of the results. The Spam mail dataset from Kaggle was utilized, and exhaustive testing was conducted on all conceivable combinations of four classifiers: SVM, Multinomial Naïve Bayes (MNB), RF, and DT. The computation time for the combination of SVM, MNB, and DT was very low, with an accuracy of 98%. Nevertheless, this study had the drawback of not incorporating the manual classification of emails as spam or the identification of the sender's IP addresses as malicious.

In their study [15], Prazwal Thakur et al. (2022) conducted a comprehensive analysis of various machine-learning methods to detect email spam. The performance of several approaches (SVM, KNN, Logistic Regression (LR), NB, and DT) has been evaluated using accuracy and precision as the criteria. The analysis utilized the data set named "spam.csv". The findings indicated that the SVM classifier outperformed the other employed machine learning techniques, with an accuracy of 98%. In contrast, the DT methodology exhibited the lowest accuracy of 84%. Nevertheless, the precision of the spam detection models can fluctuate based on the dataset employed and the particular features chosen for the models.

Rodica Paula Cota and Daniel Zinca In their study conducted a comparison of several supervised machine-learning algorithms, namely SVM, RF, LR, MNB, and Gaussian Naive Bayes (GNB), for spam identification [16]. The experiments were carried out on two datasets, namely emails.csv and Input. The evaluation of performance was based on metrics such as accuracy, precision, recall, and F-score. During the test on the data set (emails.csv), the MNB classifier had a maximum accuracy of 99%. However, while testing on the (Input) dataset, the Ran-dom Forest classifier outperformed other methods with an accuracy of 85%. The decline in the accuracy of the methods observed throughout the evaluation of the two datasets can be attributed to the relatively limited amount of training data available for the second dataset in comparison to the first. Furthermore, the second dataset includes a field labeled "hard_ham" that comprises emails that are challenging to differentiate from spam emails.

Aryan Rawat and his coworkers employed a pre-trained model that utilizes SVM technology. This model also employs word vectorization to encode the data [17]. Furthermore, the incorporation of multilingual embedding for spam detection can be applied across several languages, including Arabic, Chinese, Dutch, English, French, German, Italian, Korean, Polish, Portuguese, Russian, Spanish, and Turkish. The findings indicated that the proposed model achieved a 98% accuracy rate. However, when the model was implemented in real-time using the GUI application, the accuracy dropped to 70%. The suggested approach undergoes testing on a restricted range of languages, and the outcomes may not apply to other languages.

Ajay Chakravarty and V. Manikandan conducted a study on six widely used machine-learning algorithms [18] to examine their effectiveness in detecting spam. The dataset (Spam Assassin) consisting of 7000 emails was used to test various classifiers including NB, SVM, Artificial Neural Network (ANN), K-NN, Artificial Immune System (AIS), and Rough sets (RS) algorithms. The results showed that the NB method was the highest of the six algorithms in terms of accuracy, reaching 99.48%, while the least-performing algorithm was K-NN with an accuracy of 97.22%. Regarding the remaining classifiers, their performance was satisfactory, although inferior to the NB approach. To enhance existing work, it is advisable to broaden its scope to encompass other industries, including platforms centered around job profiles, e-commerce, and other platforms where the dissemination of false information is widespread.

V. Sasikala et al. conducted a thorough examination of several machine learning filters employed in spam detection, as outlined in [19]. The algorithms (Multilayer Perceptron (MLP), NB, and J48) were assessed for their performance in text categorization for spam detection, using accuracy as the evaluation criterion. The results indicated that the MLP classifier demonstrated the highest degree of performance, attaining an accuracy rate of 99%. The NB method achieved the second-highest ranking, exhibiting an accuracy rate of 98.6%. In contrast, the J48 algorithm performed less well, with an accuracy rate of 96.6%. The research is deficient in providing details on the precise dimensions and diversity of the dataset used for training and testing the models.

In [20], P. Vishnu Raja et al. (2022) conducted a study that utilized the CRISP-DM methodology to enhance the accuracy of spam categorization, without depending on optimization strategies. The suggested system employed a combination of two classifiers SVM and NB, to analyze a dataset consisting of 5728 emails from the Kaggle website, specifically the Spam filter dataset. The findings indicated that the combined use of SVM and NB classifiers yielded the highest performance among other methods, with an accuracy rate of 99%. The authors have omitted any discussion of the computational complexity and scalability of the suggested system, which could pose a problem in real-world scenarios involving substantial amounts of email data.

To detect spam, Tasnia Toma et al. used several classification techniques, including MNB, Bernoulli Naïve Bayes (BNB), GNB, RF, and SVM [21]. Using a SpamAssassian dataset from the Kaggle website, these algorithms were tested, and the results indicated that accuracy depends on the dataset used. Based on all approved evaluation criteria, the classifier MNB was the best with an accuracy of 98.8%, while the classifier GNB was the lowest with an accuracy of 91.5%. The inequality in the distribution of training and testing data was the reason for the high accuracy of the used classifiers, as the dataset contained a small percentage of spam compared to non-spam.

In [22], Riya et al. investigated the effectiveness of various supervised email spam detection algorithms, including SVM, KNN, RF, and NB. They specifically focused on measures such as Accuracy, Precision, Recall, and F1-score to assess the effectiveness of these algorithms. The analysis results indicated that the NB classifier achieved the highest accuracy rate of 98%, followed by the RF technique achieving an accuracy rate of 97%. The KNN classifier ranked third with an accuracy rate of 93%, while the SVM approach had the lowest accuracy rate of 86%. In comparison to other machine-learning techniques, the NB method proved to be the most efficient and suitable for categorizing texts. No information is given regarding the source, quantity, and quality of the email dataset used in these tests.

Doaa Mohammed et al. [23] have introduced a model that incorporates feature selection and ensemble classification techniques. The technique of information gain was employed to pick features from the Spambase dataset to decrease the number of characteristics, hence enhancing the efficiency of the training and testing procedures and reducing memory consumption. To assess the effectiveness, several machine-learning techniques were employed in the identification of email spam, including Naive Bayes, Decision Trees, Ensemble methods, and Hybrid Ensembles. The Hybrid Ensemble algorithm achieved the highest performance among the algorithms utilized in this study, with an accuracy of 94.41%. This can be attributed to the implementation of a feature selection approach, which significantly enhanced the classification results.

Argha Ghosh et al. [24] introduced a comparative analysis of the traditional Naive Bayes (NB) and the modified NB models, as described in [22]. Their analysis focused on comparing the accuracy of both models in detecting email spam. To conduct the testing on both models, two datasets were utilized, namely the spam corpus and spam base. The feature selection approach for the improved NB classifier employed NB-based embedded incremental wrapper subset selection as the search method and wrapper-based feature subset selection as the evaluator. During the implementation stage on the spam corpus dataset, the modified NB classifier demonstrated a higher accuracy of 93.51%, compared to the accuracy of 87.63% achieved by the traditional NB classifier. Similarly, when conducting tests on the spam-based dataset, the modified NB classifier achieved an accuracy of 89.12%, while the traditional NB classifier achieved an accuracy of 79.56%.

Veysel Aslantas et al. [25] discovered that the reduced accuracy of email spam categorization and the heightened complexity in feature selection processes are closely linked to the existing obstacles faced by email spam detection methods. The Firefly algorithm was introduced to decrease feature dimensionality and enhance the accuracy of the Naive Bayes classifier. Experiments were conducted on the Spambase dataset, which consists of 4601 emails and 57 features. The proposed method was evaluated against several established classifiers, including NB, SVM, and KNN, which utilized all features in the dataset. In contrast, the proposed algorithm identified a subset of 21 features that significantly improved the accuracy of the NB classifier. Specifically, the accuracy increased from 79.6% when using all features to 95.14% with the selected subset.

S. Sharma and C. Azad [26] developed a hybrid approach to enhance the feature selection strategy, aiming to decrease the time and space complexity while simultaneously improving the accuracy of the model employed in spam classification. The hybrid method utilized two optimization techniques. The first strategy employed the chi2 select best method as a global optimization technique. The second methodology employed a tree-based method to discover features as a local optimization tool. The first technology's results are regarded as inputs for the second technology. The study involved conducting experiments on two datasets, namely Spambase and Enron, using four established classifiers: SVM, RF, LR, and Ensemble. The results indicated that the suggested method achieved a 98.81% accuracy when applied to the Enron dataset using the SVM classifier. Similarly, when applied to the Spambase dataset, the RF classifier achieved the best accuracy of 95.95%.

Chirag Bansal and his colleagues [27] introduced a hybrid approach to identify email spam by combining Term Frequency-Inverse Document Frequency (TF-IDF) with an artificial neural network. TF-IDF is a method employed to transform words into numerical representations. The researchers conducted experiments on the Enron1 dataset. They compared the results with various algorithms often used for email spam classification, including NB and eXtreme Gradient Boosting (XGboost). The evaluation of performance was based on several parameters, including Accuracy, Precision, Recall, and F1-score. The experimental results demonstrated that the suggested hybrid approach attained exceptional performance across all assessment parameters, with an accuracy of

97.50%, precision of 95.50%, recall of 94.50%, and F1-Score of 95.50%. The computational resources and time required for executing the proposed approach are not addressed in this paper, potentially affecting its practicality and scalability.

A hybrid approach for spam detection was proposed by A Ahmed I. Taloba and Safaa S.I. Ismail [28]. This GADT hybrid technology incorporates a genetic algorithm to enhance the performance of the conventional decision tree technique. It also employs PCA technology, which is effective in eliminating inappropriate features and decreasing processing requirements. The experiments were carried out using the spam-based dataset, comprising 4,601 emails. The hybrid method has been compared to several classifiers often employed for spam categorization, including NB, SVM, KNN, and J-48. The findings indicated that the hybrid approach GADT, employing the PCA technique, achieved the highest level of accuracy at 95.5% in comparison to the other classifiers.

In [29], Hadeel M. Saleh proposed a hybrid approach that combines the artificial bee colony (ABC) algorithm with the particle swarm optimization (PSO) algorithm. The objective of this approach is to reduce the dimensionality of data and choose the most optimal subset of original features. By doing so, the accuracy of email spam classifiers can be improved. To assess the effectiveness of the suggested method, we conducted a comparative analysis with several established techniques employed in email spam categorization, including NB, SVM, KNN, SVM with Ant Colony Optimization (SVM&ACO), Naive Bayes with Genetic Algorithm (NB&GA), and Naive Bayes with Ant Colony Optimization (NB&ACO). The dataset used in this study is Spambase, which consists of 4601 emails and 57 features. The test results demonstrated that the previously described hybrid approach, in conjunction with the NB classifier, attained a commendable performance of 91.26%. This performance is deemed superior when compared to the individual spam classifiers.

In [30], Z. Hassani et al. introduced a hybrid approach to identify the most effective features from a dataset, which greatly enhances the accuracy of email spam detection. This strategy combines optimization algorithms (Binary Whale Optimization (BWO) & Binary Grey Wolf Optimization (BGWO)) with classification approaches (KNN & Fuzzy K-Nearest Neighbour (FKNN)). The experiments were performed on the Spambase dataset, which comprises 4601 emails and 57 features. The performance of the proposed hybrid method was evaluated by comparing it with several classifiers commonly employed for email spam detection. The experimental results demonstrated that when employing the BWO algorithm in conjunction with the FKNN classifier using 25 features, the accuracy achieved was 97.28%. Similarly, when utilizing the BGWO algorithm with the FKNN classifier using 19 features, the accuracy reached 97.61%. This work lacks an analysis of the applicability of the suggested approach to various datasets.

Nandan Parmar and colleagues in [31] introduced a comprehensive approach for identifying email spam by utilizing the NB classifier and PSO algorithm. The PSO methodology is a type of stochastic optimization method [32, 33] that was employed to enhance the parameters of the NB classifier, which is afterward utilized for the classification of emails into spam or ham categories. The correlation-based feature selection method was utilized to decrease the dimensions and choose a pertinent subset of the data in the long Spam dataset, which comprises 1500 emails. To assess the effectiveness of the suggested approach, it was contrasted with the conventional NB classifier. The test outcomes demonstrated that the new method outperformed the standard NB classifier, achieving an accuracy rate of 95.50% compared to the latter's accuracy of only 87.75%. The inadequacies of employing Naïve Bayes and Particle Swarm Optimization for email spam detection are insufficiently addressed, encompassing prospective difficulties in managing intricate and evolving spamming methodologies.

Uma Bhardwaj and Priti Sharma [34] employed an ensemble-based boosting technique, specifically Adaboost, in conjunction with machine learning classifiers (NB and Decision Tree J48), to detect email spam. The proposed strategy utilized Adaboost to overcome the limitations of the classifiers. The experiments were conducted on the Lingspam dataset, which consists of two classes: bare and Lemm. The proposed boosting technique produced an accuracy of 92.07% and a sensitivity of 79.38% when evaluating the results on the bare category of the dataset. On the other hand, when using the Lemm category, the accuracy reached 97.24% and the sensitivity was 88.66%. The assessment of the suggested ensemble technique solely relies on accuracy and sensitivity parameters, without considering any other evaluation metrics or performance measurements.

Ismail B. Mustapha et al. [35] introduced an enhanced model that surpassed prior efforts in email spam detection by utilizing XGBoost. To carry out realistic tests, the researchers utilized the spam base data set, which included 1813 spam and 2788 non-spam instances. To assess the effectiveness of the proposed model, it was compared to several well-known email spam detection algorithms, including SVM, PSO, LR, and J48. The XGBoost classifier demonstrated its superiority over other algorithms, with an accuracy of 96.88% according to the final results. The research does not discuss the utilization of any feature selection or extraction techniques on the dataset, which has the potential to enhance the effectiveness of the spam detection model.

Tsehay Admassu Assegie conducted a comparative analysis of various supervised machine-learning algorithms commonly employed in spam categorization, including SVM, RF, KNN, DT, NB, LR, GB, and XGB [36]. This comparison is conducted using a predefined set of evaluation criteria, including Accuracy, Precision, Recall, F1-score, and AUC-ROC. To carry out practical experiments, we utilized the spam base dataset, which comprises 4601 emails and 57 distinct characteristics. The test results indicate that the RF classifier attained the highest classification accuracy of 96.6%, followed by the XGboost classifier with an accuracy of 96.2%. Conversely, the DT classifier ranked last on the list with an accuracy of 89.8%. In terms of the criterion (AUC-ROC), the XGboost classifier attained the maximum accuracy rate of 99.4%, whereas the DT classifier achieved the lowest value of 95.8%. There remains potential for the model to undergo more testing with additional training data in the future, enabling its practical implementation in real-time scenarios.

Table 2 displays other relevant works that researchers have done in spam detection, listing the methods, dataset used, the findings, and the limitations.

Table 2. Summary of relevant work done on spam detection

Ref	Dataset	Method	Finding	Limitations
[11]	Spam	NB, SVM, DT, RF, Bagging, and AdaBoost classifiers	Naïve Bayes gave better results: Accuracy = 98% Precision = 94% F1-Score = 94%	The implementation is rarely used in real-time applications
[12]	Enron1	SVM, KNN	SVM gave the better result: Accuracy = 98% Precision = 98%	The biggest problem is that too much information demands too much time and memory
[13]	Spambase	Naïve Bayes, K*, J48, RF	Random Forest gave better results: Accuracy = 95.48% Precision = 0.955 Recall = 0.955 F1-Measure = 0.955	The research does not address false positives and negatives, which can influence spam detection model effectiveness
[14]	Spam Mails	Ensemble learning among all of the models: SVM, MNB, RF, DT	The combination of SVM, MNB, and DT gave the computational time very less and: Accuracy = 98%	Manual spam marking and malicious IP address marking are not included
[15]	Spam	SVM, KNN, LR, NB, DT	SVM gave the better result: Accuracy = 98% Precision = 96%	Datasets and features determine spam detection accuracy. Only machine-learning models are compared in the report, not other spam detection methods
[16]	emails.csv & Input	SVM, RF, LR, MNB, and GNB	* Were emails.csv dataset used: MNB gave the better result with Accuracy = 99 * Were Input datasets used: RF gave the better result with Accuracy = 85%	For larger datasets or real-world email systems, the research does not consider algorithm scalability
[17]	Spam	A pre-trained SVM for email spam detection in different languages	Accuracy = 98% While in real-time Accuracy = 70%	The proposed approach is tested on a few languages, thus the results may not apply to others

(*continued*)

Table 2. (*continued*)

Ref	Dataset	Method	Finding	Limitations
[18]	Spam Assassin	NB, SVM, ANN, K-NN, AIS, and RS classifiers	Naïve Bayes gave better results: Accuracy = 99.48% While K-NN gave the lowest result: Accuracy = 97.22%	By expanding to employment profile sites, e-commerce, and other fake news sites, existing efforts can be improved
[19]		Multilayer Perceptron, NB, and J48	MLP gave better results: Accuracy = 99.2% with False positive rate = 2%	Only one model is compared to the proposed model in the paper. The study does not specify the dataset size and diversity for model training and testing
[20]	Spam filter	SVM & NB	Improved SVM& NB gave the better Accuracy = 99%	In real-world applications with enormous email data, the proposed system's computational complexity and scalability may be a limitation
[21]	Spam Assassin	MNB, BNB, GNB, RF, and SVM	MNB gave the better results: Accuracy = 98.8 Precision = 0.96 Recall = 0.94 F1-Score = 0.95	The article doesn't examine machine learning's ethical implications for spam email detection. If email content is studied without consent, privacy concerns may arise
[22]		SVM, KNN, RF, and NB	NB gave the better results: Accuracy = 98% Precision = 0.99 Recall = 0.99 F1-Score = 0.99	The research does not describe the email dataset utilized in the tests, including its size, source, and quality
[23]	Spambase	NB, DT, Ensemble, Hybrid Ensemble	Hybrid Ensemble gave better results: Accuracy = 94.41 Precision = 0.944 Recall = 0.944 F1-Score = 0.944	The paper does not compare the suggested spam email classification system to state-of-the-art approaches, which may restrict its usefulness

(*continued*)

Table 2. (*continued*)

Ref	Dataset	Method	Finding	Limitations
[24]	Spam corpus & Spambase	NB, and modified NB	Modified NB gave better results: In spam corpus dataset: Accuracy = 93.51% In the spam base dataset: Accuracy = 89.12%	The research exclusively examines the Naïve Bayes classifier, not comparing its performance against other algorithms
[25]	Spambase	SVM, KNN, NB, and NB with Firefly algorithm	NB with Firefly algorithm gave better results: Accuracy = 95.14%	
[26]	Spambase & Enron1	SVM, RF, LR, and Ensemble. With hybrid FS techniques (chi2 select best &Tree-based FS)	* Were Spambase dataset used: RF gave the better result with Accuracy = 95.95% * Were Enron1 dataset used: SVM gave the better result with Accuracy = 98.81%	
[27]	Enron1	NB, XGboost, Hybrid ANN & TF-IDF	ANN with TF-IDF gave the better results: Accuracy = 97.50% Precision = 95.50% Recall = 94.50% F1-Score = 95.50%	The research does not explore the computational resources or time needed to implement the proposed strategy, which could affect its practicality and scalability
[28]	Spambase	NB, SVM, KNN, J-48, and hybrid GA & DT (GADT)	GADT gave the better results: Accuracy = 95.5% Precision = 95.5% Recall = 97.2% F-measure = 96.3%	The hybrid technique's computational complexity, which may limit its scalability to larger datasets, is not discussed in the research
[29]	Spambase	NB, SVM, KNN, SVM&ACO, NB&GA, NB&ACO, and NB& ABC-PSO	NB & ABC-PSO gave the better results: Accuracy = 91.26%	The research does not discuss any drawbacks or limitations of using Chaotic Particle Swarm Optimization (PSO) and Artificial Bees Colony (ABC) for feature selection

(*continued*)

Table 2. (*continued*)

Ref	Dataset	Method	Finding	Limitations
[30]	Spambase	NB, Rotation Forest, Random Forest, Random committee, and (BWOA-FKNN& BGWOA-FKNN)	BWOA-FKNN gave the: Accuracy = 97.28% BGWOA-FKNN gave the: Accuracy = 97.61%	The research does not discuss how the proposed strategy applies to other datasets or email spam scenarios
[31]	Ling spam	NB, and Integrated NB with PSO	Integrated NB with PSO gave the better results: Accuracy = 95.50 Precision = 96.42 Recall = 94.50 F-measure = 95.45	The constraints of Naïve Bayes and Particle Swarm Optimization for email spam detection are not fully addressed, including potential difficulties in managing sophisticated and dynamic spamming strategies
[34]	Ling spam	NB, DT J48, and Proposed boosting technique	* With the Bare category Proposed boosting gave the: Accuracy = 92.07% Sensitivity = 79.38% * With Lemm category Proposed boosting gave the: Accuracy = 97.24% Sensitivity = 88.66%	The suggested ensemble technique for email spam detection is evaluated based on accuracy and sensitivity parameters, but no other metrics or performance measurements are included
[35]	Spambase	SVM, PSO, LR, J48, and Proposed XGboost	XGboost gave the better results: Accuracy = 96.88% Precision = 96.47% Recall = 95.59% F1-Score = 96.03%	No feature selection or extraction procedures are described for the dataset, which could improve the spam detection model
[36]	Spambase	SVM, RF, KNN, DT, NB, LR, GB, and XGB	RF gave the better results: Accuracy = 96.6% Precision = 96.7% Recall = 96.6% F1-Score = 96.6% But in the AUC-ROC measure, XGB gave the better = 99.4%	Future testing on more training data could make the model useful in real-time applications

4 Discussion

Accuracy is a fundamental metric used to assess the effectiveness of spam detection techniques. Several spam detection models have been proposed and tested in literature but still, the reported accuracy begs for more work in this direction to achieve better accuracy. By referring to Table 2, we can observe several techniques that have been implemented, leading to the most favorable outcomes, as indicated in [11]. These strategies have yielded an accuracy rate of 98%, a precision rate of 94%, and an F1-Score of 94%. In addition, the study reported a minimum accuracy rate of 84% [13]. The analysis in [11, 15], and [17] used the spam.csv dataset. Certain methods necessitate distinct datasets for implementation, such as the Enron dataset mentioned in [12, 26], and [27], as well as the Spam Mails dataset used in reference [14]. The studies [12, 14], and [17] achieved the highest level of accuracy, with a rate of 98%. The accuracy of the model, as stated in [17], decreased to 70% when it was applied in real-time using the GUI application. The technique suggested in [26] attained the highest outcome, with an accuracy of 98.81%. Additionally, [26] achieved the greatest accuracy of 95.95% while utilizing the Spambase dataset. On the other side, the experiments [16] were carried out on two datasets, namely emails.csv and input. During the test on the dataset (emails.csv), the MNB classifier had a maximum accuracy of 99%.

Table 2 also presents several techniques that have been implemented, leading to the most favorable outcomes, as indicated in [13]. These strategies have yielded the highest accuracy rate of 95.48%, and exhibited the lowest performance, with an accuracy rate of 79.29%. The findings of [25] have yielded an accuracy increase from 79.6% when using all features to 95.14% with the selected subset. In [29] the finding of accuracy is 91.26%. While the accuracy achieved was 97.28% in the [30]. Similarly, when utilizing the 19 features, the accuracy reached 97.61%.

As seen in Table 2, [20] used a Spam filter dataset and proved the performed best, with 99% accuracy. The authors have not discussed the computational complexity and scalability of the proposed system. On the other hand, [18] and exploited 7000 Spam Assassin emails. The best accuracy of the six algorithms reached 99.48%, while it had the lowest at 97.22%. While [21] provided the most accuracy at 98.8% and the lowest at 91.5% based on all evaluation criteria. The study presented in [19] used an unmentioned dataset and several classifiers. The results indicated that the MLP classifier demonstrated the highest degree of performance, attaining an accuracy rate of 99%. The NB method achieved the second-highest ranking, exhibiting an accuracy rate of 98.6%. In contrast, the J48 algorithm performed less well, with an accuracy rate of 96.6%. In [23] and [24], information gain was employed to pick features from the Spambase dataset. The study [23] achieved an accuracy of 94.41%. [24] when conducting tests on the spam base dataset, the modified NB classifier achieved an accuracy of 89.12%, while the traditional NB classifier achieved an accuracy of 79.56%.

In Table 2, several researchers such as [27, 31], and [34], utilized the Ling spam dataset to detect spam emails by employing various classifiers. Researchers in [27] had 97.50% accuracy, 95.50% precision, 94.50% recall, and 95.50% F1-Score.

However, this study does not examine how computational resources and time may affect the suggested approach's feasibility and scalability. While [31] selected a 1500-email Spam dataset subset with reduced dimensions. The results of the technique outperformed with 95.50% accuracy versus 87.75% after testing. In [34] the proposed boosting approach had 92.07% accuracy and 79.38% sensitivity on the dataset's basic category. The result was 97.24% accurate and 88.66% sensitive. [28] used a spam-based collection that contains 4,601 emails. The proposed classifier is provided the most accurate at 95.5% and in [22], investigated the effectiveness of various supervised email spam detection algorithms. The analysis results indicated that the NB classifier achieved the highest accuracy rate of 98%, followed by the RF technique achieving an accuracy rate of 97%. The KNN classifier ranked third with an accuracy rate of 93%, while the SVM approach had the lowest accuracy rate of 86%.

The study [35] utilized the spam base data set, which included 1813 spam and 2788 non-spam instances. The proposed XGboost classifier provided an accuracy of 96.88%. [36] employ different classifiers and achieved the highest classification accuracy of 96.6%, followed by the XGboost classifier with an accuracy of 96.2%. Conversely, the DT classifier ranked last on the list with an accuracy of 89.8%. In terms of the AUC-ROC, the XGboost classifier attained the maximum value of 99.4%, whereas the DT classifier achieved the lowest value of 95.8%.

This paper provides a thorough analysis of the most recent technology designed to detect email spamming and assesses its effectiveness. This research suggests a possibly beneficial alternative strategy that might be applied to get better results and lower risks. The researcher's objective is to develop cutting-edge spam detection algorithms capable of efficiently analyzing multimedia content and detecting spam emails. Hybrid algorithms are expected to replace the existing supervised and unsupervised learning approaches used for spam detection to improve accuracy and precision. Ultimately, we offer a succinct evaluation of the accuracy of the proposed approaches and a thorough summary of the many strategies used for spam detection. All spam detection technologies were deemed effective. While some individuals are attempting to use different approaches to improve accuracy, others have achieved great achievements. Although all of them possess usefulness, specific spam detection algorithms still display shortcomings, which is the main source of concern among researchers.

In the Future may enhance the method of extracting features. Email spam can be detected more effectively with clustering. Machine learning is not the only approach available for spam detection. To construct classification models, future research should prioritize acquiring readily available datasets with standardized labels. In addition, spam detection models must include attacker IP addresses and/or locations. Researchers examined the header, subject, and content of an email before classifying it as spam.

5 Spam Detection Challenges

Some challenges must be overcome in the spam detection process:

- The expanding volume of data on the Internet, along with different new features, poses a significant challenge to spam detection systems.

- It is also difficult for spam filters to evaluate features from several dimensions such as chronological, writing styles, semantic, and statistical [2].
- Spammers constantly evolve their techniques to bypass detection systems. They may use obfuscation, encryption, and other tactics to make spam emails look more like legitimate ones.
- Balancing the detection of spam without flagging legitimate emails (false positives) or allowing spam to pass through undetected (false negatives) is a persistent challenge. Striking the right balance is crucial to avoid inconveniencing users or compromising security.
- Spam emails can take various forms, including text, images, attachments, and hyperlinks. Detecting spam in diverse content formats requires sophisticated algorithms capable of analyzing different types of data.

6 Conclusion

This paper examined various machine-learning techniques and approaches proposed by researchers to detect email spam. The study's findings indicated that most existing approaches for identifying spam emails are based on supervised machine learning methodologies. And It also offered a comprehensive analysis of how it validated spam emails, taking multiple perspectives into account.

Different algorithms exhibit varying levels of accuracy and efficiency, with each possessing its own unique set of strengths and weaknesses. Additionally, there exists a disparity in the datasets employed by each method in the study. The duration of processing will also vary based on the algorithm's capabilities and level of adaptability.

This work has contributed to the assurance that researchers can select suitable approaches and, if required, integrate them to acquire optimal findings, through the examination of diverse methodologies. Besides that, this study highlighted some challenges that must be overcome in the spam detection process.

Acknowledgment. Special thanks to my advisor for the valuable insight provided regarding the topic at hand.

References

1. Cevik, T., Cevik, N., Rasheed, J., Asuroglu, T., Alsubai, S., Turan, M.: Reversible logic-based hexel value differencing—a spatial domain steganography method for hexagonal image processing. IEEE Access **11**, 118186–118203 (2023). https://doi.org/10.1109/ACCESS.2023.3326857
2. Ahmed, N., Amin, R., Aldabbas, H., Koundal, D., Alouffi, B., Shah, T.: Machine learning techniques for spam detection in email and iot platforms: analysis and research challenges. Secur. Commun. Netw. **2022** (2022). https://doi.org/10.1155/2022/1862888
3. Idris, I., Selamat, A.: Improved email spam detection model with negative selection algorithm and particle swarm optimization. Appl. Soft Comput. J. **22**, 11–27 (2014). https://doi.org/10.1016/j.asoc.2014.05.002

4. Guo, Z., Tang, L., Guo, T., Yu, K., Alazab, M., Shalaginov, A.: Deep graph neural network-based spammer detection under the perspective of heterogeneous cyberspace. Futur. Gener. Comput. Syst. **117**, 205–218 (2021). https://doi.org/10.1016/j.future.2020.11.028
5. Bagui, S., Nandi, D., Bagui, S., White, R.J.: Machine learning and deep learning for phishing email classification using one-hot encoding. J. Comput. Sci. **17**(7), 610–623 (2021). https://doi.org/10.3844/jcssp.2021.610.623
6. Tahir, T., et al.: Early software defects density prediction: training the international software benchmarking cross projects data using supervised learning. IEEE Access **11**, 141965–141986 (2023). https://doi.org/10.1109/ACCESS.2023.3339994
7. Karim, A., Azam, S., Shanmugam, B., Kannoorpatti, K., Alazab, M.: A comprehensive survey for intelligent spam email detection. IEEE Access **7**, 168261–168295 (2019). https://doi.org/10.1109/ACCESS.2019.2954791
8. Olatunji, S.O.: Extreme Learning machines and Support Vector Machines models for email spam detection. In: Canadian Conference on Electrical and Computer Engineering, pp. 1–6 (2017). https://doi.org/10.1109/CCECE.2017.7946806
9. Khan, S.A., Iqbal, K., Mohammad, N., Akbar, R., Ali, S.S.A., Siddiqui, A.A.: A novel fuzzy-logic-based multi-criteria metric for performance evaluation of spam email detection algorithms. Appl. Sci. **12**(14) (2022). https://doi.org/10.3390/app12147043
10. Mathur, S., Purohit, A.: Performance evaluation of machine learning algorithms on textual datasets for spam email classification. Int. J. Res. Appl. Sci. Eng. Technol. **10**(7), 4726–4734 (2022). https://doi.org/10.22214/ijraset.2022.46072
11. Lanka, S.C., Akhila, K., Pujita, K., Sagar, P.V., Mondal, S., Bulla, S.: Spam based email identification and detection using machine learning techniques. In: 2nd International Conference on Sustainable Computing and Data Communication Systems, ICSCDS 2023 - Proceedings, pp. 69–74 (2023). https://doi.org/10.1109/ICSCDS56580.2023.10104659
12. Vejendla, L.N., Bysani, B., Mundru, A., Setty, M., Kunta, V.J.: Score based support vector machine for spam mail detection. In: 7th International Conference on Trends in Electronics and Informatics, ICOEI 2023 - Proceedings, no. Icoei, pp. 915–920 (2023). https://doi.org/10.1109/ICOEI56765.2023.10125718
13. Saini, A., Guleria, K., Sharma, S.: Machine learning approaches for an automatic email spam detection. In: 2023 International Conference on Artificial Intelligence and Applications (ICAIA 2023) Alliance Technology Conference (ATCON-1 2023) - Proceeding, pp. 1–5 (2023). https://doi.org/10.1109/ICAIA57370.2023.10169201
14. Ghosh, A., Das, R., Dey, S., Mahapatra, G.: Ensemble learning and its application in spam detection. In: ICCECE 2023 - International Conference on Computer, Electrical & Communication Engineering, pp. 1–6 (2023). https://doi.org/10.1109/ICCECE51049.2023.10085378
15. Thakur, P., Joshi, K., Thakral, P., Jain, S.: Detection of email spam using machine learning algorithms: a comparative study. In: 2022 8th International Conference on Signal Processing and Communication, ICSC 2022, pp. 349–352 (2022). https://doi.org/10.1109/ICSC56524.2022.10009149
16. Cota, R.P., Zinca, D.: Comparative results of spam email detection using machine learning algorithms. In: 14th International Conference on Communications, COMM 2022 - Proceedings, pp. 4–8 (2022). https://doi.org/10.1109/COMM54429.2022.9817305
17. Rawat, A., Behera, S., Rajaram, V.: Email spam classification using supervised learning in different languages. In: 2022 1st International Conference on Computer, Power and Communiction, ICCPC 2022 - Proceedings, pp. 294–298 (2022). https://doi.org/10.1109/ICCPC55978.2022.10072054

18. Chakravarty, A., Manikandan, V.: An intelligent model of email spam classification. In: 4th International Conference on Emerging Research in Electronics, Computer Science and Technology, ICERECT 2022, pp. 1–6 (2022). https://doi.org/10.1109/ICERECT56837.2022.10059620
19. Sasikala, V., Mounika, K., Sravya Tulasi, Y., Gayathri, D., Anjani, M.: Performance evaluation of spam and non-spam E-mail detection using machine learning algorithms. In: Proceedings of the International Conference on Electronics and Renewable Systems, ICEARS 2022, no. Icears, pp. 1359–1365 (2022). https://doi.org/10.1109/ICEARS53579.2022.9752202
20. Raja, P.V., Sangeetha, K., Suganthakumar, G., Madesh, R.V., Vimal Prakash, N.K.K.: Email spam classification using machine learning algorithms. In: Proceedings of the 2nd International Conference on Artificial Intelligence and Smart Energy, ICAIS 2022, pp. 343–348 (2022). https://doi.org/10.1109/ICAIS53314.2022.9743033
21. Toma, T., Hassan, S., Arifuzzaman, M.: An analysis of supervised machine learning algorithms for spam email detection. In: 2021 International Conference on Automation, Control and Mechatronics for Industry 4.0, ACMI 2021, no. July, pp. 1–5 (2021). https://doi.org/10.1109/ACMI53878.2021.9528108
22. Riya, Gupta, S., Vishvashdeep, Kumar, V.: Performance metrices of different machine learning algorithms. In: Proceedings - 2021 3rd International Conference on Advances in Computing, Communication Control and Networking, ICAC3N 2021, pp. 262–264 (2021). https://doi.org/10.1109/ICAC3N53548.2021.9725404
23. Ablel-Rheem, D.M.: Hybrid feature selection and ensemble learning method for spam email classification. Int. J. Adv. Trends Comput. Sci. Eng. **9**(1.4), 217–223 (2020). https://doi.org/10.30534/ijatcse/2020/3291.42020
24. Ghosh, A., Senthilrajan, A.: A modified naïve bayes classifier for detecting spam E-mails based on feature selection. In: Proceedings - 2022 6th International Conference on Intelligent Computing and Control Systems, ICICCS 2022, no. May, pp. 1634–1641 (2022). https://doi.org/10.1109/ICICCS53718.2022.9788340
25. Ahmed, B.: Wrapper feature selection approach based on binary firefly algorithm for spam E-mail filtering. J. Soft Comput. Data Min. **2**(1), 44–52 (2020)
26. Sharma, S., Azad, C.: A hybrid approach for feature selection based on global and local optimization for email spam detection. In: 2021 12th International Conference on Computing Communication and Networking Technologies, ICCCNT 2021, pp. 1–6 (2021). https://doi.org/10.1109/ICCCNT51525.2021.9580038
27. Bansal, C., Sidhu, B.: Machine learning based hybrid approach for email spam detection. In: 2021 9th International Conference on Reliability, Infocom Technologies and Optimization (Trends and Future Directions ICRITO 2021), pp. 1–4 (2021). https://doi.org/10.1109/ICRITO51393.2021.9596149
28. Taloba, A.I., Ismail, S.S.I.: An intelligent hybrid technique of decision tree and genetic algorithm for e-mail spam detection. In: Proceedings - 2019 IEEE 9th International Conference on Intelligent Computing and Information Systems, ICICIS 2019, pp. 99–104 (2019). https://doi.org/10.1109/ICICIS46948.2019.9014756
29. Saleh, H.M.: An Efficient feature selection algorithm for the spam email classification. Period. Eng. Nat. Sci. **9**(3), 520–531 (2021). https://doi.org/10.21533/pen.v9i3.2202
30. Hassani, Z., Hajihashemi, V., Borna, K., Sahraei Dehmajnoonie, I.: A classification method for e-mail spam using a hybrid approach for feature selection optimization. J. Sci. Islam. Repub. Iran **31**(2), 165–173 (2020). https://doi.org/10.22059/JSCIENCES.2020.288729.1007444
31. Agarwal, K., Kumar, T.: Email spam detection using integrated approach of naïve bayes and particle swarm optimization. In: Proceedings of the 2nd International Conference on Intelligent Computing and Control Systems, ICICCS 2018, no. March, pp. 685–690 (2019). https://doi.org/10.1109/ICCONS.2018.8662957

32. Tavakol Aghaei, V., SeyyedAbbasi, A., Rasheed, J., Abu-Mahfouz, A.M.: Sand cat swarm optimization-based feedback controller design for nonlinear systems. Heliyon **9**(3), e13885 (2023). https://doi.org/10.1016/j.heliyon.2023.e13885
33. Arasteh, B., Seyyedabbasi, A., Rasheed, J., Abu-Mahfouz, A.M.: Program source-code re-modularization using a discretized and modified sand cat swarm optimization algorithm. Symmetry **15**(2), 401 (2023). https://doi.org/10.3390/sym15020401
34. Bhardwaj, U., Sharma, P.: Detection of email spam using an ensemble based boosting technique. Int. J. Innov. Technol. Explor. Eng. **8**(11), 403–408 (2019). https://doi.org/10.35940/ijitee.K1365.0981119
35. Mustapha, I.B., Hasan, S., Olatunji, S.O., Shamsuddin, S.M., Kazeem, A.: Effective email spam detection system using extreme gradient boosting (2020). http://arxiv.org/abs/2012.14430
36. Assegie, T.A.: Evaluation of supervised learning models for automatic spam email detection, pp. 1–10 (2023)

Implementing Cyclical Learning Rates in Deep Learning Models for Data Classification

Hussein A. A. Al-Khamees[1] (✉), Mehdi Ebady Manaa[2,3], Zahraa Hazim Obaid[1], and Noor Abdalkarem Mohammedali[1]

[1] Computer Techniques Engineering Department, College of Engineering and Technologies, Al-Mustaqbal University, 51001 Hillah, Babil, Iraq
Hussein.Alkhamees@uomus.edu.iq
[2] Artificial Intelligence Science Department, College of Science, Al-Mustaqbal University, 51001 Hillah, Babil, Iraq
[3] Department of Information Networks, College of IT, University of Babylon, Hillah, Babil, Iraq

Abstract. Neural networks are effectively used in a variety of applications including data mining. The neural network can realize different complex nonlinear functions by making them attractive to identify a system. One of the most important issues of classifying datasets through neural networks is the formation of an ideal network, that consists of many successive steps like set parameters. Perhaps the most prominent parameter is the learning rate. Indeed, choosing an appropriate learning rate value is one of the things that greatly helps to control the overall network performance. In contrast, any inappropriate value for the learning rate negatively affects the classification model and can therefore destabilize the model's performance and thus seriously deteriorate its quality. This paper presents a new model by adopting a cyclical learning rate instead of using a constant value for training deep neural networks by Multi-Layer Perceptron (MLP) architecture. This model is tested on various real-world datasets; Electricity, NSL- KDD, and four sub-datasets of HuGaDB (HuGaDB-01-01, HuGaDB-05- 12, HuGaDB-13-11, and HuGaDB-14-05). The proposed model achieves an accuracy of, 89.57%, 99.12%, 99.2%, 97.83%, 96.19%, and 99.85% for these datasets respectively. Accordingly, the proposed model outperforms many previous models. As a result, the deep neural network models can be more effective when they adopt an appropriate value for the learning rate.

Keywords: Deep neural networks · Data mining · Cyclical Learning Rate · Multi-Layer Perceptron (MLP)

1 Introduction

In recent years, machine learning techniques have been widely used in various fields and neural networks (NNs) are a sub-field of these techniques [1]. It has been significantly applied in many areas, this is due to several reasons including, its simple and excellent self-learning ability, high accuracy when it comes to mapping complex non-linear relationships, and hence, modeling these relationships [2]. Accordingly, neural networks are useful tools with great results [3].

J. Rasheed et al. (Eds.): FoNeS-AIoT 2024, LNNS 1035, pp. 205–215, 2024.
https://doi.org/10.1007/978-3-031-62871-9_16

Neural networks simulate the human brain and consist of input, hidden, and output layers. Every layer has a different number of neurons that can receive input (s), process it, and deliver a single output [4]. Moreover, neural networks can realize different complex nonlinear functions by making them attractive to identify a system [5, 6].

Neural networks can be classified into two types that are; shallow and deep. The deep neural networks use the deep learning techniques to learn the data [7]. The deep learning is based significantly on neural networks that are originally inspired by neurons in the human brain [8]. Multilayer Perceptron is an important and effective architectural type of deep learning [9, 10].

The training algorithm represents the core of any deep neural network that requires adjusting different parameters initially. Some of these parameters are the number of hidden layers, the neuron numbers per layer, the weights, the learning rate, and so forth. The learning rate (or step-wise) is arguably the most important parameter to adjust when initializing or training a neural network [11].

In the same context, the slow convergence and longer training times are the main drawbacks of deep neural network training [12]. Moreover, in most training algorithms, the learning rate value is still constant (does not change) for the weights in a layer throughout the training phase [13].

Choosing the appropriate learning rate value contributes and helps greatly in controlling the overall network performance. In contrast, any inappropriate value for the learning rate negatively affects the model and can therefore destabilize the model's performance, and network oscillatory, and thus seriously deteriorate its quality [4].

Thereupon, in the deep learning community, researchers are starting to find alternative methods to the traditional method that ensure the learning rate remains constant. One of these methods is to rely on a cyclic learning rate [14]. The application of the cyclical learning rates has become more popular and successfully used. In practice, the cyclic learning rate typically requires fewer steps for converging [15, 16].

This study presents a suggested framework based on the cyclic learning rate strategy that ensures the learning rate is changed cyclically and continuously between boundaries. This framework can train the network through the Multi-Layer Perceptron (MLP) technique.

Because of the importance of learning rate in training deep networks, this study adopts a new strategy based on the value of cyclical learning rate to reduce the computational effort of training neural networks as well as reduce training time. This represents the main contribution of this study. Whereas in training most traditional neural networks, the value of the learning rate cannot be modified or adjusted during processing. On the other hand, the proposed framework (a cyclic learning rate adjustment algorithm) overcomes the traditional models of constant learning rates.

2 Related Work

Since the learning rate indicates the prominent parameter in any network, the related work section discusses different previous papers and frameworks that are based on an effective learning rate to train deep networks. It displays the method of selecting the learning rate value.

The authors [17] suggested a deep neural network based on a fit cycle policy of learning rate to classify the white blood cells (WBCs). Also, this strategy is used for diagnosing specific diseases like liver disease and cancer. They compared the results of the fixed learning rate with the results of the cyclical learning rate to prove their strategy is better than others. This strategy is more accurate and needs fewer cycles to train the network since it is based on a fit cycle policy.

According to [18], the authors proposed a framework for finding the parameters which able to minimize the cost function built through data learning. They proved the learning rate is an effective parameter to do this task and it can be changed during the working of a network according to the network weights. Thus, it initialized the learning rate value with 0.001 and then it decreased at each network update. The results proved the learning rate is considered an important parameter in optimizing the networks.

The authors in [4] proposed a developed deep network based on dynamic (non-constant) learning rate value instead of applying the fixed value. This step aims to obtain an optimal learning rate value that can reduce the error in every iteration. The main structure is done by deriving a new parameter which is either added to or subtracted from the learning rate value. The paper classifies the data stream samples into different classes by the above steps.

The study [1] presented firstly the idea of a cyclical learning rate. In this paper, instead of using the classic learning rate value to train the deep network, we applied the idea of changing this value during the training step. However, for best balancing the training speed, the cyclic learning rate is presented that varies the learning rate between the maximum and minimum value and sufficiently improves the model accuracy.

3 Materials and Proposed Method

3.1 Learning Rate

In many training algorithms, the learning rate is constant during the processing steps. Since manually determining the exact learning rate is still a cumbersome step, an adaptive learning rate algorithm aims to solve this problem in various methods (arguably). Within this realm, the learning rate algorithms can be classified into two types, constant learning rate algorithms and adaptive learning rate algorithms [19].

In the same context, the slow convergence and longer training times are the main drawbacks of neural network training [12]. Moreover, in most training algorithms, the learning rate value is still constant (does not change) for the weights in a layer [17].

When the learning rate value is very small, the network is overfitting. On the other hand, if it is very large, the training can be normalized [20].

Accordingly, one possible solution is to make appropriate adjustments to the learning rate parameter during the training process [21]. A small learning rate leads to small changes in network weights. In contrast, when the learning rate is large, the training process is done speed up. Thus, the network results may cause instability, and sometimes then, the network becomes oscillatory [18].

3.2 Finding Learning Rate

To determine an accurate learning rate, the most important and common methods used are as follows [18, 22, 23]:

- Reduce learning rate: In this method, the learning rate is decreasing by a set fraction that can mainly achieve network stability.
- Finder learning rate: It is based on selecting a randomly weighted value and then, determining the learning rate value. These steps are repeated and every time the learning rate is recomputed.
- Cyclic learning rate: One of the latest methods to find an ideal learning rate, and it does not require traditional experimental research for the best learning rate value. In reality, the cyclic learning rate method allows the learning rate to vary periodically between boundaries.

3.3 Cyclical Learning Rate

The cyclical learning rate addresses the learning rate issue through repeated cycles of linearly increasing and decreasing the learning rates, constructing the triangle policy for every cycle [24]. In other words, by applying the cyclical learning rate, the network forgets both fixed or exponentially reducing value. This observation leads to the idea of letting the learning rate vary within a range of values rather than adopting a step-wise fixed or exponentially decreasing value (arguably). The cyclical learning rate can lead to better training results than a constant learning rate, which will be reflected in the results of the testing data, as shown in Fig. 1 [17].

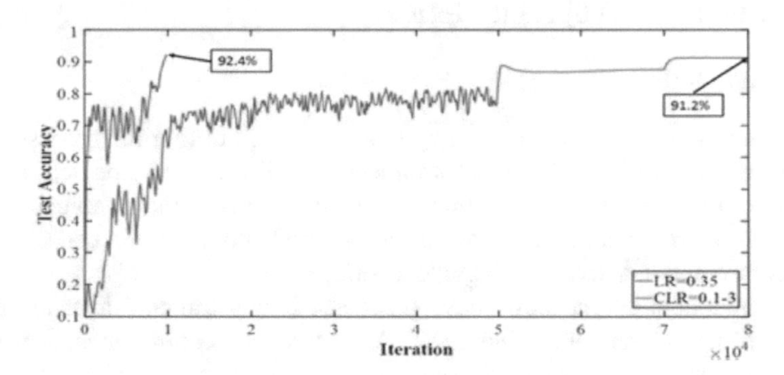

Fig. 1. The test accuracy with one cycle of training [17]

The strategy of cyclical learning rate is used for adjusting the learning rate value through the cycling step between the boundaries (the upper and lower bounds) [25]. More specifically, a lower bound called min_lr and an upper bound known. as max_lr. The input parameter step size is the iteration number in half of the cycle. More deeply, the step size represents a half one cycle perfectly as displayed in Fig. 2.

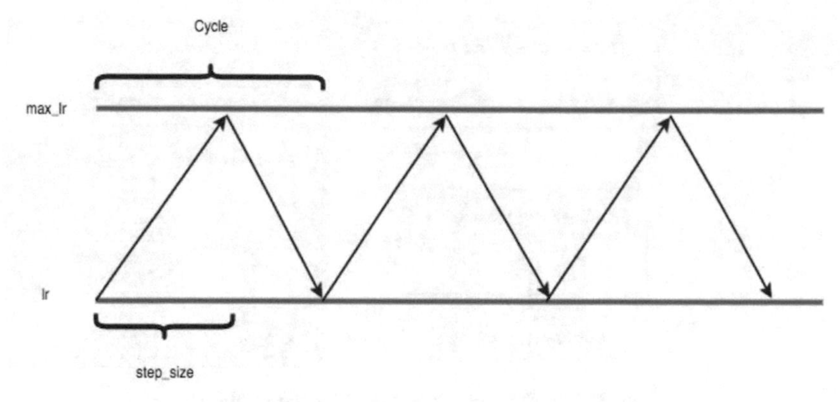

Fig. 2. A half one cycle for training data [1]

3.4 Learning Rate Taxonomy

In addition to the previous taxonomies in the last section, there is another taxonomy of selecting learning rate that is categorized into [26]:

1. Fixed learning rate.
2. Decaying learning rate.
3. Cyclic learning rate.

More deeply, the cyclic learning rate is partitioned into two classes that are: cyclic adaptive learning rate (CALR), and traditional cyclic learning rate strategy [27]. Furthermore, the learning rate can also be constant, linear decay, and exponential decay learning rate. The linear decay learning rate can be either with time-based or with drop-based [28].

3.5 The Proposed Method

The proposed method adopts the training of deep networks through the Multi-Layer Perceptron (MLP) technique.

In reality, it is, however, different methods and frameworks that have been proposed previously to train the network depending on a fixed learning rate. Thus, the current paper suggested a training method based on a cyclic learning rate that will be between two boundaries that are, max_lr as an upper bound and min_lr as a lower bound. In other words, by applying a cyclical learning rate, the network abandons the idea of relying on a constant learning rate value or one that decreases exponentially.

Instead, the learning rate is varied within a range of values. Figure 3 explains the framework of the proposed method in more detail. However, the proposed method consists of four phases that are:

Pre-processing Phase. This phase has two steps. The first one is the normalization and the min-max technique is used. The second step of this phase is dataset splitting where all datasets are split into an equal ratio that is, 80% for the training phase and 20% for the testing phase.

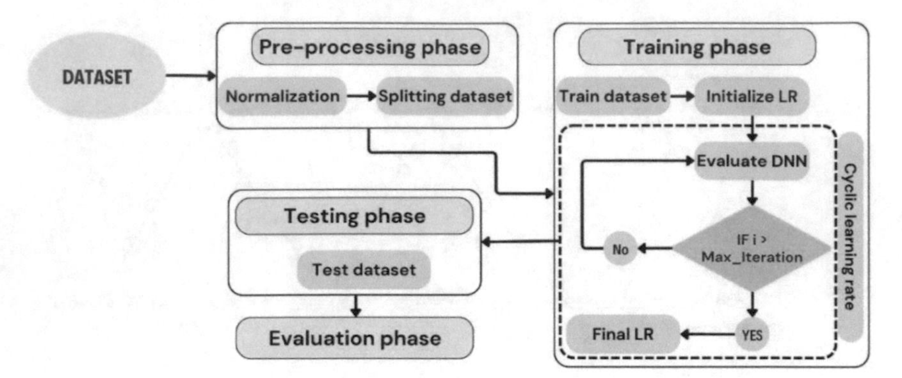

Fig. 3. The framework of the proposed method

The Training Phase. It represents the core of the proposed method. This phase has many steps such as:

- Firstly, the training data is used for this phase.
- Initialize the learning rate value for training the network. The value 0.0001 is selected for all datasets as an initial value.
- After that, the network results are evaluated according to this value to obtain the best results and then the best learning rate value until the method gets the best learning rate value. Then, all the rest dataset samples are trained with this value.
- The repeated evaluation step above is done if the current iteration (i) is less than the max iteration. Otherwise, it returns the final learning rate value.
- In this phase, the max learning rate (max_lr) is 0.1, while the min learning rate (min_lr) is 0.0.
- The final learning rate value is carried forward to the next phase (the testing phase).

The Testing Phase. During this phase, the final learning rate that is obtained from the previous phase is used.

The Evaluation Phase. It is the last phase in this method. The results are evaluated (in this phase) by four measurements which are accuracy, precision, recall, and F1-score.

4 Results

First of all, the proposed method is tested by six datasets. Table 1 describes these datasets and the number of samples, features, and classes for each dataset [29].

Firstly, the network is trained by MLP with a fixed (not change) learning rate value meanwhile, this value differs from one dataset to another. Thereafter, the proposed method with a cyclic learning rate is applied to train the network. Table 2 displays the learning rate value for every tested dataset. However, this value is constant for the whole processing steps for every dataset. While Table 3 illustrates the cyclic learning rate value for each dataset. Furthermore, Table 4 demonstrates the four measurements

Table 1. The details of datasets.

Dataset	No. of Samples	No. of Features	No. of Classes
Electricity	45,312	8	2
NSL-KDD	148,517	41	5
HuGaDB-v2-various-01-01	2,435	39	4
HuGaDB-v2-various-05-12	4,393	39	3
HuGaDB-v2-various-13-11	5,272	39	3
HuGaDB-v2-various-14-05	2,392	39	2

(accuracy, precision, recall, and F1-score) of the model with a fixed learning rate value. Alternatively, Table 5 displays these measurements with the proposed method by the cyclic learning rate. It should be noted that the accuracy as well as the other three measurements increase when the network is trained on the proposed method with a cyclic learning rate than when these measurements were with a constant learning rate. For all of the above reasons, it becomes clear that the proposed method outperforms the one based on a constant learning rate.

Table 2. The constant learning rate value for all datasets.

Dataset	Learning Rate Value
Electricity	0.0012
NSL-KDD	0.0016
HuGaDB-v2-various-01-01	0.0015
HuGaDB-v2-various-05-12	0.00092
HuGaDB-v2-various-13-11	0.0017
HuGaDB-v2-various-14-05	0.0009

The results compare with the other previous related works as shown in Table 6 where the best accuracy as well as the other measurements result of the proposed method are shown.

Table 3. The cyclic learning rate value for all datasets.

Dataset	Cyclic Learning Rate Value
Electricity	0.00322
NSL-KDD	0.0182
HuGaDB-v2-various-01-01	0.042
HuGaDB-v2-various-05-12	0.0075
HuGaDB-v2-various-13-11	0.049
HuGaDB-v2-various-14-05	0.0071

Table 4. The four measurements with a fixed learning rate value for all datasets.

Dataset	Accuracy (%)	Precision (%)	Recall (%)	F1-score (%)
Electricity	88.16	88.37	88.16	88.27
NSL-KDD	98.67	98.67	98.67	98.67
HuGaDB-v2-various-01-01	98.63	98.64	98.63	98.63
HuGaDB-v2-various-05-12	97.11	97.12	97.11	97.12
HuGaDB-v2-various-13-11	95.06	95.11	95.06	95.09
HuGaDB-v2-various-14-05	99.72	99.72	99.72	99.72

Table 5. The four measurements with a cyclic learning rate value for all datasets.

Dataset	Accuracy (%)	Precision (%)	Recall (%)	F1-score (%)
Electricity	89.57	89.05	89.57	89.00
NSL-KDD	99.12	98.92	99.24	99.08
HuGaDB-v2-various-01-01	99.2	99.12	99.22	99.16
HuGaDB-v2-various-05-12	97.83	97.57	97.83	97.52
HuGaDB-v2-various-13-11	96.19	96.03	96.21	96.1
HuGaDB-v2-various-14-05	99.85	99.79	99.83	99.8

Table 6. The comparison results with other related works.

Ref	Dataset	Accuracy (%)	Proposed Method Accuracy (%)
[30]	Electricity	81.06	89.57
[31]	NSL-KDD	97.05	99.12
[32]	NSL-KDD	97.97	99.12
[33]	HuGaDB	92.5	96.19
[34]	HuGaDB	88.0	96.19

5 Conclusion

In the machine learning community, neural networks are considered an indispensable cornerstone. Moreover, these networks are of two types, shallow and deep. Deep neural networks are the most difficult and complex in formation, application, and training. (MLP) is considered one of the most important architectures used to train this type of network, which has always relied on a set of traditional parameters, including the learning rate. This paper presents a new architecture for training a deep network based on a cyclic learning rate instead of a constant learning rate. This method was tested on six data sets which are; Electricity, NSL-KDD, and four sub datasets of the main dataset HuGaDB (HuGaDB-01-01, HuGaDB-05-12, HuGaDB-13-11, and HuGaDB-14-05), and the proposed method achieves an accuracy of 89.57%, 99.12%, 99.2%, 97.83%, 96.19%, and 99.85% for these datasets respectively. This method outperformed different previous methods and for the same data sets.

Acknowledgments. The authors would like to express our deepest gratitude to Al-Mustaqbal University for their invaluable support in the publication of this paper.

Disclosure of Interests. No potential conflict of interest relevant to this paper was reported.

References

1. Smith, L.N.: Cyclical learning rates for training neural networks. In: 2017 IEEE Winter Conference on Applications of Computer Vision (WACV), pp. 464–472. IEEE, Santa Rosa, CA, USA (2017)
2. Waziry, S., Wardak, A.B., Rasheed, J., Shubair, R.M., Rajab, K., Shaikh, A.: Performance comparison of machine learning driven approaches for classification of complex noises in quick response code images. Heliyon **9**(4), e15108 (2023)
3. Xu, A., Chang, H., Xu, Y., Li, R., Li, X., Zhao, Y.: Applying artificial neural networks (ANNs) to solve solid waste-related issues: a critical review. Waste Manag. **124**, 385–402 (2021)
4. Al-Khamees, H.A.A., Al-A'araji, N., Al-Shamery, E.S.: Enhancing the stability of the deep neural network using a non-constant learning rate for data stream. Int. J. Electr. Comput. Eng. (IJECE) **13**(5), 2123–2130 (2023)
5. Kaya, E.: A new neural network training algorithm based on artificial bee colony algorithm for nonlinear system identification. Mathematics **10**(19), 3487 (2022)

6. Rasheed, J., Alsubai, S.: A hybrid deep fused learning approach to segregate infectious diseases. Comput. Mater. Contin. **74**(2), 4239–4259 (2023)
7. Al-Khamees, H.A.A., Al-A'araji, N., Al-Shamery, E.S.: Classifying the human activities of sensor data using deep neural network. In: Bennour, A., Ensari, T., Kessentini, Y., Eom, S. (eds.) Intelligent Systems and Pattern Recognition. ISPR 2022. CCIS, vol. 1589. Springer, Cham (2022). https://doi.org/10.1007/978-3-031-08277-1_9
8. Sarker, I.H.: Machine learning: algorithms, real-world applications and research directions. SN Comput. Sci. **2**(3), 160 (2021)
9. Zhang, J., Li, C., Yin, Y., Zhang, J., Grzegorzek, M.: Applications of artificial neural networks in microorganism image analysis: a comprehensive review from conventional multilayer perceptron to popular convolutional neural network and potential visual transformer. Artif. Intell. Rev. **56**(2), 1013–1070 (2023)
10. Farooq, M.S., et al.: A conceptual multi-layer framework for the detection of nighttime pedestrian in autonomous vehicles using deep reinforcement learning. Entropy **25**(1), 135 (2023)
11. Kaveh, M., Mesgari, M.S.: Application of meta-heuristic algorithms for training neural networks and deep learning architectures: a comprehensive review. Neural. Process. Lett. **55**(4), 4519–4622 (2023)
12. Marza, N.H., Manaa, M.E., Lafta, H.A.: Classification of spam emails using deep learning. In: 2021 1st Babylon International Conference on Information Technology and Science (BICITS), pp. 63–68. IEEE, Babil, Iraq (2021)
13. Arasteh, B., Seyyedabbasi, A., Rasheed, J., Abu-Mahfouz, A.M.: Program source-code re-modularization using a discretized and modified sand cat swarm optimization algorithm. Symmetry **15**(2), 401 (2023)
14. L. N. Smith: General Cyclical Training of Neural Networks. arXiv preprint arXiv:2202.08835 (2022)
15. Xiao, J., Fan, Y., Sun, R., Wang, J., Lou, Z.: Stability Analysis and generalization bounds of adversarial training. Adv. Neural. Inf. Process. Syst. **35**, 15446–15459 (2022)
16. Wang, W., Lee, C.M., Liu, J., Colakoglu, T., Peng, W.: An empirical study of cyclical learning rate on neural machine translation. Nat. Lang. Eng. **29**(2), 316–336 (2023)
17. Houssein, E.H., et al.: Using deep DenseNet with cyclical learning rate to classify leukocytes for leukemia identification. Front. Oncol. **13**, 1230434 (2023)
18. Park, J., Yi, D., Ji, S.: A novel learning rate schedule in optimization for neural networks and it's convergence. Symmetry **12**(4), 660 (2020)
19. Xie, Z., Wang, X., Zhang, H., Sato, I., Sugiyama, M.: disentangling the effects of adaptive learning rate and momentum, pp. 24430–24459, PMLR (2022)
20. Ziouzios, D., Tsiktsiris, D., Baras, N., Dasygenis, M.: A distributed architecture for smart recycling using machine learning. Future Internet **12**(9), 141 (2021)
21. Usmani, I.A., Qadri, M.T., Zia, R., Alrayes, F.S., Saidani, O., Dashtipour, K.: Interactive effect of learning rate and batch size to implement transfer learning for brain tumor classification. Electronics **12**(4) (2023)
22. Anwar, T., Zakir, S.: Deep learning based diagnosis of COVID-19 using chest CT-scan images. In: 2020 IEEE 23rd international multitopic conference (INMIC), pp. 1–5. IEEE, Bahawalpur, Pakistan (2020)
23. You, K., Long, M., Wang, J., Jordan, M.I.: How Does Learning Rate Decay Help Modern Neural Networks? arXiv preprint arXiv:1908.01878 (2019)
24. Lee, C.M., Liu, J., Peng, W.: Applying cyclical learning rate to neural machine translation. arXiv preprint arXiv:2004.02401 (2020)
25. Alyafi, B., Tushar, F.I., Toshpulatov, Z.: In: Jmd in medical image analysis and applications pattern recognition module, pp. 1–4, Cassino, Italy (2018)

26. Wu, Y., Liu, L.: Selecting and composing learning rate policies for deep neural networks. ACM Trans. Intell. Syst. Technol. **14**(2), 1–25 (2023)
27. Gao, Y., Lu, G., Gao, J., Li, J.: A high-performance federated learning aggregation algorithm based on learning rate adjustment and client sampling. Mathematics **11**(20), 4344 (2023)
28. Yu, T., Zhu, H.: Hyper-parameter optimization: a review of algorithms and applications. arXiv preprint arXiv:2003.05689 (2020)
29. Al-Khamees, H.A.A., Al-A'araji, N., Al-Shamery, E.S.: Data stream: statistics, challenges, concept drift detector methods, applications and datasets. Int. J. Comput. Digit. Syst. **13**(1), 717–728 (2023)
30. JALIL: DElStream: an ensemble learning approach for concept drift detection in dynamic social big data stream learning. IEEE Access **9**, 66408–66419 (2021)
31. Kyatham, A.S., Nichal, M.A., Deore, B.S.: A novel approach for network intrusion detection using probability parameter to ensemble machine learning models. In: 2020 Fourth International Conference on Computing Methodologies and Communication (ICCMC), pp. 608–613. IEEE, Erode, India (2020)
32. Ahanger, A.S., Khan, S.M., Masoodi, F.: an effective intrusion detection system using supervised machine learning techniques. In: 2021 5th International Conference on Computing Methodologies and Communication (ICCMC), pp. 1639–1644. IEEE, Erode, India (2021)
33. Javeed, M., Gochoo, M., Jalal, A., Kim, K.: HF-SPHR: hybrid features for sustainable physical healthcare pattern recognition using deep belief networks. Sustainability **13**(4), 1699 (2021)
34. Sun, Y., Yang, G., Lo, B.: An artificial neural network framework for lower limb motion signal estimation with foot-mounted inertial sensors. In: 2018 IEEE 15th International Conference on Wearable and Implantable Body Sensor Networks (BSN), pp. 132–135. IEEE, Las Vegas, NV, USA (2018)

Multimodal ML Strategies for Wind Turbine Condition Monitoring in Heterogeneous IoT Data Environments

Syed Shahryar Jameel[1], Syed Muhammad Khaliq-ur-Rahman Raazi[1], and Syed Muslim Jameel[2](✉)

[1] Muhammad Ali Jinnah University (MAJU), Karachi, Pakistan
[2] Atlantic Technological University, Galway, Republic of Ireland
muslimjameel.syed@atu.ie

Abstract. Addressing the pressing need for efficient wind turbine monitoring in the sustainable energy sector, this paper begins with an extensive literature review focused on condition monitoring techniques specific to wind turbines. This foundational review uncovers significant gaps, particularly in managing diverse and voluminous data streams that are characteristic of wind turbine operations. The core objective of this research is to thoroughly analyze the unique characteristics of heterogeneous data environments in wind turbine monitoring, tackling challenges like data diversity, volume, and reliability, where their effectiveness in interpreting complex data is scrutinized. This analysis provides critical insights into the applicability of these models in practical monitoring situations. Further, the research broadens its scope to assess the implications of these findings within the Artificial Intelligence of Things (AIoT) domain. It highlights the potential of AI and IoT integration in revolutionizing wind turbine monitoring, leading to smarter, more resilient renewable energy systems. The study sets a foundation for future advancements in AIoT, especially in enhancing the efficiency and intelligence of renewable energy infrastructures. It paves the way for the development of more sophisticated AI-driven tools for energy management, envisioning a future where renewable energy systems are managed with greater efficiency and intelligence.

Keywords: Artificial Intelligence of Things · Renewable Energy · Wind Turbines · Condition Monitoring · Multi-Modal ML

1 Introduction and Background

The landscape of wind energy research has witnessed significant advancements in the areas of operation, maintenance, and monitoring. A study from 2020 [1] delves into the application of condition monitoring on wind turbine blades. The research highlights the critical role of Bayesian decision analysis frameworks in optimizing maintenance costs. This study demonstrates the importance of reliability and the consequential decisions derived from effective condition monitoring.

In a 2017 study [2] on the condition monitoring of wind turbines, the focus is on health assessment and fault detection. The study introduces a novel methodology that employs

J. Rasheed et al. (Eds.): FoNeS-AIoT 2024, LNNS 1035, pp. 216–228, 2024.
https://doi.org/10.1007/978-3-031-62871-9_17

co-integration analysis of SCADA data. The innovative approach involves analyzing nonlinear data trends for wind turbines, proving practical and feasible in real-world settings. However, it's worth noting that this study lacks proper validation, and there is room for improvement in this regard. Within the domain of offshore wind turbines, another study [3] examines the current status and future trends, providing a comprehensive overview of condition-based maintenance, deterioration models, and fault diagnosis. This research sheds light on the evolving landscape of offshore wind turbine maintenance strategies. Additionally, a study [4] presented in the context of condition monitoring for wind turbine blades proposes a methodology focused on damage tolerance. This methodology integrates material properties, monitoring systems, and modeling, emphasizing the potential for efficient wind turbine operation through proactive monitoring and maintenance. These studies collectively contribute to advancing the field of wind energy research, offering insights into cutting-edge methodologies and technologies for optimizing the performance and reliability of wind turbines.

2 Literature Analysis and Deduction

2.1 Traditional Approaches for Condition Monitoring for Wind Turbines

Over the past twenty years, many researchers have developed an interest in the field of condition monitoring through the use of damage detection in composite materials or carbon fiber reinforced polymers. The literature mostly focuses on applications within the wind turbine domain, among other domains. Many classical techniques (see below) have made a significant contribution to literature.

- Statistical methods [5–7]
- Vibration-based trend analysis [8, 9]
- Filtering methods [10]
- Cepstrum analysis [11]
- Time-synchronous averaging [12]
- Fast Fourier transform [13]
- Amplitude demodulation [14]
- Order analysis [15]
- Wavelet transforms [16]

2.2 Machine Learning Approaches for Condition Monitoring for Wind Turbines

To understand the current state of the testing/sensor data, machine learning (ML) techniques are employed to analyze the data and extract useful information. Although methods enable quantifiable computations for condition monitoring, these computations are not necessarily reliable and are subject to bias or human error. Data-driven/machine learning-based approaches are becoming more and more popular as a means of getting around the drawbacks of old methodologies [17]. A collection of algorithms known as machine learning (ML) can automatically identify hidden patterns inside massive data sets. When there is enough experimental data, researchers have discovered that condition or predictive monitoring (using ML) performs more accurately in terms of damage detection, damage initiation, harm accumulation/growth, damage interaction, and others [18]. For instance, a study [19] that employs decision trees, neural networks, and machine learning-based quadratic discriminant analysis to identify features for damage assessment achieves 91.5% accuracy before wind turbine blades. Only a small number of previous research used linear discrimination analysis and machine learning (ML) -based statistical pattern recognition for blade performance monitoring. Researchers have also combined fatigue/stress testing data with machine learning in numerous studies [20, 21]. Cycling loading and fatigue can cause cracks to propagate, which in turn can induce fractures over the turbine blades and other components [22]. To locate the impending bond degradation, for instance, a predictive monitoring strategy utilizing artificial neural networks is suggested [22]. An early study predicts the longitudinal split growth of carbon-epoxy notched specimens under tension-dominated fatigue testing using a neural network. For split growth, it is discovered that the suggested ML-based method is more effective than the power-law model [23]. The load cycle and shallow learning techniques—including support vector machines, random forests, decision trees, AdaBoost, ensemble learning, and others—were employed in these investigations. To lower the high error rate, these methods need a sufficient amount of training and testing data. Many studies [24–27] have suggested applying deep learning models to condition monitor wind blades in recent years. Specifically, large data and image processing/classification techniques (e.g., photographs of both damaged and undamaged wind turbine blades, CT-Scan/x-ray images, ultrasonic wave images, and stress/fatigue data in wave frequency transform images) are used for testing and training. Using deep neural networks (DNNs), particularly region-based convolutional neural networks (CNNs) like the Fast R-CNN, Faster R-CNN, or similar variants, is the most popular deep learning technique for automatically detecting, localizing, and estimating the severity of surface fractures [28]. For instance, a study [17] demonstrated how deep learning (DL) can speed up composite design optimization by using CNN to predict material attributes like stiffness, strength, and toughness that are derived using the finite element approach. The data visualization stage was carried out in a different study using time-series signals that were converted into picture vectors and piecewise plotted in grayscale images for model training. Because of this, a deep neural network or cluster of deep neural networks is trained using methods known as stacked autoencoders and greedy layer-wise training [29]. The training dataset is time-series signal-based images that are composed of randomly selected and manually labeled image vectors.

Following a thorough evaluation of the literature, it is also possible to conclude that, while the machine learning techniques currently in use are more successful than their traditional counterparts, more advancements are still needed to ensure decision-making trust. Furthermore, real-time monitoring of structural health in a harsh marine environment remains a difficulty for the scientific community. For instance, there is no interoperability method to foster confidence in the decision-making process of the present ML and DL approaches, and their decision functionality is concealed [30].

Dealing with uncertainty in real time presents another difficulty. Examples of such elements include weather, maritime environmental conditions, operational and environmental unpredictability, and other contributing factors [31, 32]. Published research, however, indicates a notable difference between the real system environment and the conducted experimental scenarios/conditions. This is a result of academics' published work ignoring the wide variety of stochastic environmental variables. For example, the existing ML- or DL-based condition monitoring techniques are static in nature and neglect to take these presumptions into account [33]. On the other hand, while working in a real-time setting, these presumptions must be taken into account [32], Ignoring such presumptions may cause the condition monitoring system to eventually deteriorate or become outdated [34]. The adaptive or continuous learning mechanism is used in literature to address these assumptions; however, the current adaptive mechanism is reserved for other uses [35–37]. To handle the uncertain feature arrival during the real-time condition monitoring of the wind blades, it is crucial to create a Multi-Modal system with adaptive capabilities. A comprehensive ICT framework must also be created for implementation purposes, beginning with the collecting of sensor data and continuing through its dependable transmission, processing, storage, and other stages. Survey research [38] describes the current ICT frameworks for condition monitoring, including data routing tactics, system implementation, and abstract perspectives of Internet of Things (IoT) technologies.

3 Contextual Overview and Problem Definition

3.1 Typical Machine Learning Reflection on Condition Monitoring

There are two main categories of typical Machine Learning (ML) approaches: Non-Imaging and imaging data [39]. The fundamental idea behind both situations is that a machine learning model may be trained on previously labeled data, and then used on fresh, unlabeled data to generate predictions or evaluations, as demonstrated in Fig. 1.

This method is essential to condition monitoring, which aims to preserve infrastructure safety and integrity via early detection of possible problems. For instance, machine learning (ML) for non-imaging data uses historical data in the form of numerical values or time-series data, such as strain measurements, vibration patterns, and other sensor data that has been labeled with pertinent structural health indicators [40]. A machine learning model is trained using this data, and it is trained to correlate particular data patterns with structural health labels. The trained model evaluates the input data, assigns labels to the structure, or projects future conditions. This allows it to assess the structure's current state of health and maybe pinpoint any possible problem areas. On the other hand, ML for picture data starts with a dataset made up of previously identified photos. These labels could indicate any number of characteristics or states that are important to a structure's health, such as corrosion, deformation, or fissures. A machine learning model is trained using the labeled dataset. During the training phase, the model learns to identify and comprehend the characteristics and patterns connected to each label. The model can capture new photos of unlabeled structures after it has been trained. Based on its knowledge from previous data, the trained model analyzes the incoming image and produces labels, classifications, or categories that indicate the structural problems or health state. In both cases, developing two separate models is necessary, requiring more processing power or capabilities, as one ML model is unable to manage the variety of data.

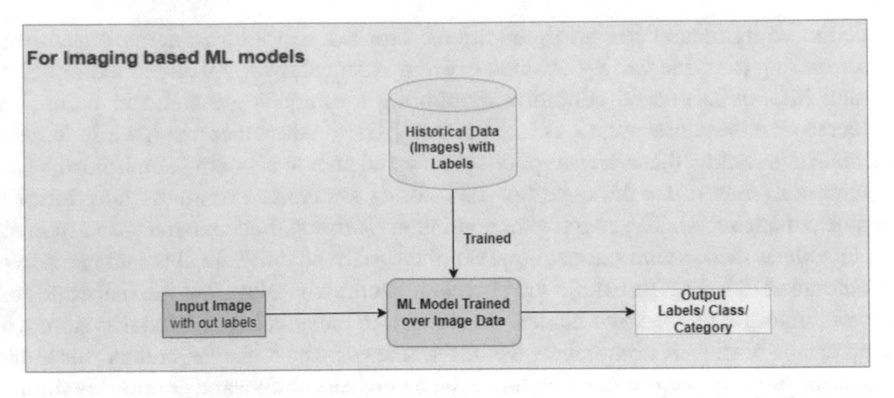

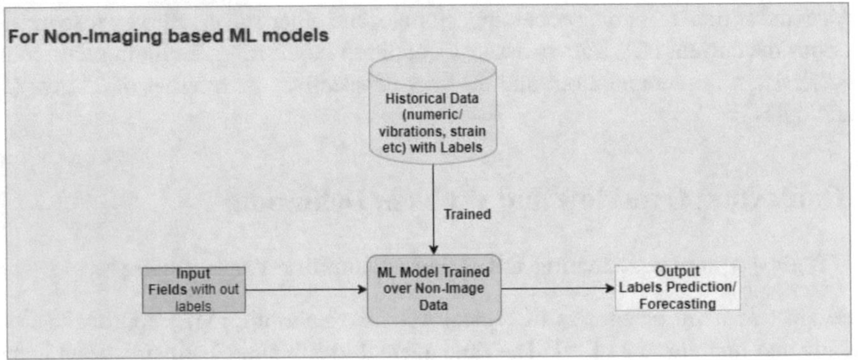

Fig. 1. The typical ML approaches deal with non-imaging and imaging data.

3.2 Define the Concept of Multimodal ML and Its Different Scenarios

Some foundational publications, such as Multimodal Machine Learning: A Survey and Taxonomy, offer a noteworthy consideration of multimodal to define the multimodal ML approach. A thorough introduction to multimodal machine learning techniques and applications is given in [41], which also defines the terms "data source diversity" and "joint feature learning." The most recent developments in deep multimodal learning are reviewed in a survey [42], which also includes comments on dependency learning and cross-modal interactions. Understanding how many modalities can be combined and processed in a machine-learning model is important. One study addresses multimodal fusion in multimedia systems [43]. Together, these studies lay the groundwork for the noteworthy examination of multimodal machine learning and offer theoretical and practical insights that complement the justification offered. If the study satisfies any or all of the following requirements, it can be used to support the case for a multimodal approach in heterogeneous data environments.

Leveraging diverse data types.

- Joint feature representation
- Adaptability and Customization
- Improve complex decision-making of framework.

Based on the noteworthy examination of multimodal, Fig. 2 illustrates a potential scenario 1 in which several data sources input their data into a single machine learning model and obtain distinct outputs according to the type of data involved. As seen in Fig. 2, another potential scenario 2 is when several data sources provide their data to a single machine learning model, which produces a single enhanced output. Both situations meet the important requirements of the multimodal ML technique, including 1) data source diversity (using a variety of data sources or kinds). Both of the previously described scenarios (scenario 1 and scenario 2) include the system ingesting data from many sources (such as audio, visual, and thermal). 2) Cooperative feature representation and learning. Joint feature learning or representation can occur during the learning process even in cases where, as in scenario 1, the outputs are distinct for every type of data.

This implies that while one modality may contribute more than the other, the model may learn to extract and represent features from the combined input in a way that partially takes use of the linkages and correlations across modalities. Nevertheless, scenario 2

Multi-Model ML Approaches for Structure Health Monitoring

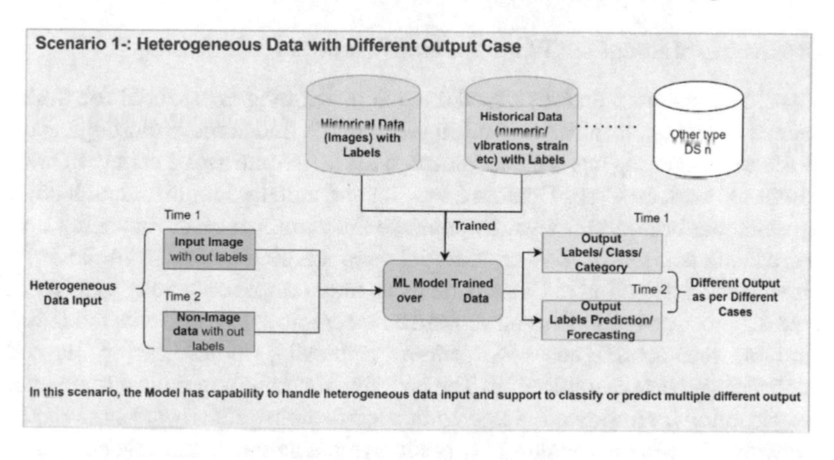

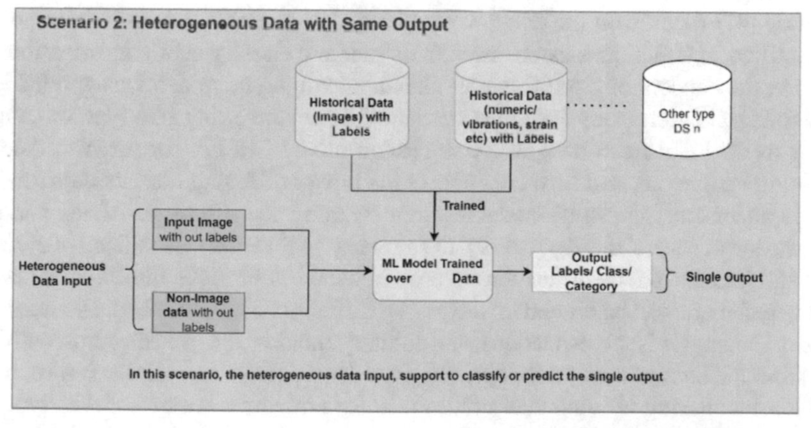

Fig. 2. The representation of multimodal machine modal two distinct scenarios

itself is a reflection of the joint feature representation and learning. 3) Customization and adaptation enable flexibility and one important component of multimodal learning is customization to particular types of inputs. It recognizes that while various data types may benefit from being a part of an integrated learning system, they may also require different processing and output formats, which both scenarios fulfill. 4) Enhance difficult decision-making by taking into account a variety of data points. In addition to improving the overall efficiency of the framework, even though the outputs are distinct, the decision-making process gains from a holistic picture that takes into account various aspects of the data environment, which is essential to multimodal learning.

Defining a multimodal is justified by its adherence to the noteworthy considerations of 1) diversity of data types, 2) joint feature learning, 3) adaptability and customization, and 4) leveraging these for enhanced analysis and decision-making, according to the thorough literature analysis. As a result, both of the scenarios shown in Fig. 1—with separate (scenario 1) or single outputs (scenario 2)—satisfy the important factor of multimodal nature. They also offer a customized method for managing the nuances of each data source within a single learning framework.

3.3 Use the Multimodal for Wind Turbine Condition Monitoring

It has been demonstrated that multimodal machine learning is essential for working in heterogeneous data environments, resulting in analysis that is more thorough and accurate. Multimodal learning has been documented in the literature for a variety of purposes. For example, a work on Cross Domain Learning and unpaired multi-modal learning for deep cross-modal hashing retrieval emphasizes the significance of managing a variety of unpaired data sources [44]. A multimodal deep learning approach for MGMT promoter methylation detection in GBM patients is one example of a study that focused on improved diagnostic capabilities and showed how combining several data modalities can result in more accurate diagnoses [45]. More specifically, studies like [46] have shown that it enhances picture classification. The benefits of combining conventional methods with deep learning for improved categorization are demonstrated by the suggested SIFT-CNN strategy. Emotion recognition for predictive maintenance has recently advanced. The study [47] focuses on the emotion categorization of English speech and illustrates how multimodal techniques can be used to extract and classify subtle information. Furthermore, the concept of a multi-modal machine learning approach was applied in the sound-to-imagination study [48] to demonstrate how converting one type of data into another might yield fresh insights. A suggested model EFAFN for sarcasm detection combining text, audio, and face cues shows the power of fusing distinct data kinds for nuanced understanding. Multi-modal machine learning algorithms have been published for the fusion of several data types [49]. In summary, to find patterns indicative of turbine health and improve the identification of flaws in wind turbines, the multimodal machine learning technique can be crucial in integrating different unstructured data sources, such as sensor readings, visual inspections, and auditory monitoring. We can confidently conclude from the literature that this approach provides a rich framework for improving the condition monitoring of wind energy turbines. By utilizing a variety of data types and sophisticated analytical techniques, it is possible to achieve more accurate, early detection of potential issues, which can lead to better maintenance strategies and longer turbine lifespans. By combining data from several sources or types, multimodal machine learning provides a more thorough understanding of intricate systems, such as wind turbines. Therefore, it will be beneficial to improve fault detection. For instance, certain fault types may be better identified by different data modalities; for example, thermal imaging may be more efficient in detecting overheating difficulties, while sound sensors may immediately identify a mechanical failure. By combining these data sources, possible issues can be identified more precisely and beforehand. Additionally, it enhances predictive maintenance. Through the analysis of data from many sources, machine learning models can predict maintenance needs more precisely, which minimizes downtime and increases turbine lifespan. Similar to this, in an environment with heterogeneous data, several data sources can feed into a single machine learning model and provide distinct outputs depending on the kind of data that is also valuable in wind turbine situations. Furthermore, because wind turbines operate in a variety of contexts, multimodal data can be used to better understand how various geographic locations or weather patterns affect the wear and performance of the turbines. In wind turbine condition monitoring, using a multimodal machine learning method can result in more precise, effective, and economical maintenance plans. This strategy is in line with the larger trend in artificial intelligence and machine learning, which is the growing recognition that using a variety of data sources can lead to deeper insights and more reliable solutions across a range of fields.

4 Experimental Results of Preliminary Simulations

In this research, we address class imbalance in a wind turbine surface defect dataset used for preventative maintenance via deep learning-based computer vision. The dataset comprises six classes of frequent physical damages: Crack, Erosion, Good, Mechanical Damage, Paint Off, and Scratch. Due to the limited number of samples for certain classes, we employed image augmentation techniques, expanding the dataset tenfold. Subsequent balancing was achieved through down-sampling the augmented instances of prevalent classes and up-sampling the scarce ones, standardizing all class sizes to 700 samples. The dataset sample represents diverse surface defects on wind turbine components. It includes clear instances of ‘crack’ and ‘erosion’, showing distinct damage patterns. The ‘good’ class displays unblemished turbine surfaces, contrasting with the ‘mechanical damage’ class, where the damage is severe and structural. ‘Paint off’ shows sections of missing paint, and ‘scratch’ depicts surface lines indicative of abrasion. The ‘blade’ class images illustrate turbine blades from varying perspectives, while the ‘turbine’ class provides broader views, including the entire turbine structure. The variability within these samples underlines the complexity of automated damage classification and the importance of a well-trained model capable of discerning subtle yet critical differences in wind turbine surface conditions (Fig. 3).

The resulting confusion matrix, as depicted in Fig. 4, from our model’s predictions exhibits flawless classification accuracy across all classes, with no observed misclassifications. While such perfection is rare and often indicative of overfitting, the balanced nature of the augmented dataset suggests a high degree of model reliability. Nevertheless, further validations such as cross-validation and external test set evaluations are essential

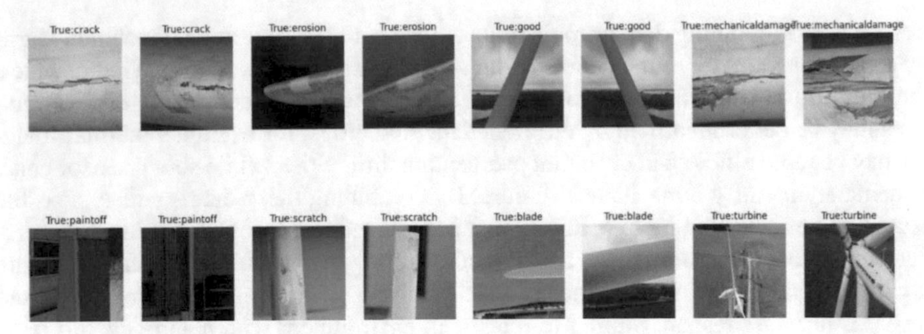

Fig. 3. The dataset samples for the wind turbine blade dataset.

to ensure the model's generalization capabilities. This study's approach demonstrates a rigorous methodology to mitigate potential biases due to imbalanced data, ensuring a robust predictive maintenance model for wind turbines. The updated confusion matrix for the balanced wind turbine surface defect dataset reveals a high classification accuracy, with most classes exhibiting strong true positive rates. However, certain inter-class confusions are evident, notably between the 'blade' and 'good' classes. Specifically, 'blade' was misclassified as 'good' 57 times, indicating a challenge in differentiating between intact blades and those with subtle defects. Similarly, 'good' was occasionally mistaken for 'blade', with 74 instances suggesting an overlap in feature representation for these conditions. Despite this, the classes 'paint off' and 'scratch' show remarkable discriminative performance with no recorded misclassifications, underscoring the model's effectiveness in recognizing these defects. A solitary misclassification occurred

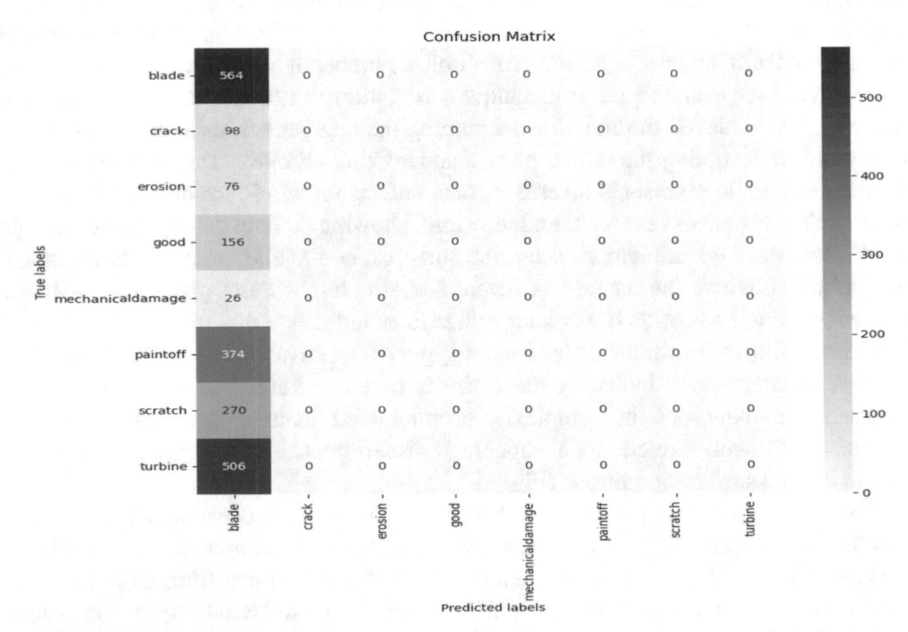

Fig. 4. The confusion matrix on the wind turbine data when the classes are imbalanced.

between 'turbine' and 'blade', which is negligible but should not be overlooked. This matrix underscores the necessity for further refinement in feature extraction and model training, particularly to enhance the distinction between 'blade' and 'good', ensuring that subtle defect features are captured with greater fidelity, thereby improving the overall robustness and reliability of the predictive maintenance system, as depicted in Fig. 5.

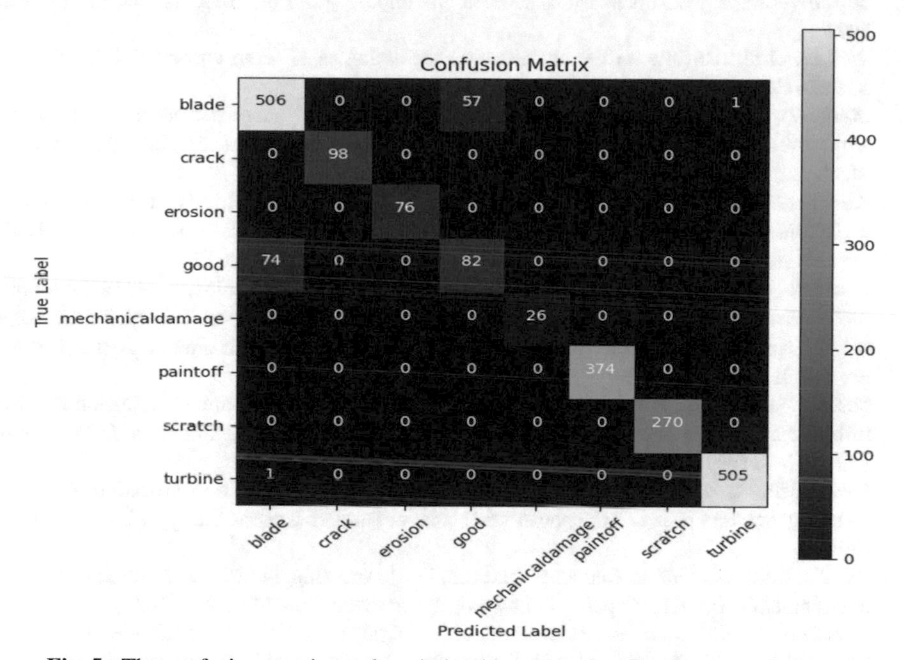

Fig. 5. The confusion matrix on the wind turbine data when the classes are balanced.

5 Conclusion and Future Consideration

This research is a rich literature review, that identifies the current state of the art in condition monitoring, focusing on the ML approach for condition monitoring of wind turbines. Another contribution is to showcase the importance of multimodal data integration scenarios to enhance detection accuracy in wind energy turbine condition monitoring. Reinforcing the role of AI in the sustainable operation of wind turbines.

Later, this research marks a significant stride in wind turbine defect classification, addressing class imbalances with advanced deep learning techniques. In the future, this research will lead towards integrating the diverse streams of IoT data into the machine learning models and will help enhance the decision-making processes for condition maintenance of wind energy turbines or similar applications. To develop an AI-driven, real-time monitoring system that ensures the optimal performance and longevity of wind turbines, contributing to the sustainability and reliability of wind energy in the future concern of this study.

References

1. [Climate at a Glance: Global Mapping. https://www.ncdc.noaa.gov/cag/. Accessed on 12 Oct 2021
2. Farrok, O., et al.: Electrical power generation from the oceanic wave for sustainable advancement in renewable energy technologies. Sustainability **12**(6), 2178 (2020)
3. https://www.epa.gov/ghgemissions/sources-greenhouse-gas-emissions. Accessed 12 Oct 2021
4. Wei, L., et al.: Review on the blade design technologies of wind current turbine. Renew. Sustain. Energy Rev. **63**, 414–422 (2016)
5. Marks, R., Gillam, C., Clarke, A., Armstrong, J., Pullin, R.: Damage detection in a composite wind turbine blade using 3D scanning laser vibrometry. Proc. Inst. Mech. Eng. Part C: J. Mech. Eng. Sci. **231**(16), 3024–41 (2017)
6. Kirchner-Bossi, N., Porté-Agel, F.: Wind farm layout optimization through a crossover-elitist evolutionary algorithm performed over a high performing analytical wake model. In: EGU General Assembly Conference Abstracts, vol. 19, p. 18074, April 2017
7. Desvaux, M., Multon, B., Ahmed, H.B., Sire, S., Fasquelle, A., Laloy, D.: Gear ratio optimization of a full magnetic indirect drive chain for wind turbine applications. In: Ecological Vehicles and Renewable Energies (EVER), 2017 Twelfth International Conference on, pp. 1–9. IEEE, 11 April 2017
8. Sarrafi, A., Mao, Z., Niezrecki, C., Poozesh, P.: Vibration-based damage detection in wind turbine blades using phase-based motion estimation and motion magnification. J. Sound Vib. 421, 300–18 (2018). 67
9. Farajzadeh, S., Ramezani, M.H., Nielsen, P., Nadimi, E.S.: A testing procedure for wind turbine generators based on the power grid statistical model. Renew. Energy **1**(116), 136–144 (2018)
10. Yu, X., Dan, D.: Online frequency and amplitude tracking in structural vibrations under environment using APES spectrum postprocessing and Kalman filtering (2021)
11. Hemmati, A., Khorasanchi, M., Barltrop, N.: Analysis of offshore wind turbine foundations with soil damping models. In: ASME 2017 36th International Conference on Ocean, Offshore and Arctic Engineering 2017, pp. V07BT06A037–V07BT06A037. American Society of Mechanical Engineers, 25 June 2017
12. Reder, M., Yürüşen, N.Y., Melero, J.J.: Data-driven learning framework for associating weather conditions and wind turbine failures. Reliab. Eng. Syst. Saf. **1**(169), 554–569 (2018)
13. Pliego, A., Ruiz de la Hermosa, R., García Márquez, F.P.: Big data and wind turbines maintenance management. In: García Márquez, F., Karyotakis, A., Papaelias, M. (eds.) Renewable Energies, LNCS, pp. 111–125. Springer, Cham (2018). https://doi.org/10.1007/978-3-319-45364-4_8
14. Chandrasekhar, K.: On the structural health monitoring of operational wind turbine blades. Lasers Eng. **49**(12), 1361–1371 (2017)
15. Gan, L.K., Shek, J.K., Mueller, M.A.: Modeling and characterization of downwind tower shadow effects using a wind turbine emulator. IEEE Trans. Ind. Electron. **64**(9), 7087–7097 (2017)
16. Kankanamge, Y., Hu, Y., Shao, X.: Application of wavelet transform in structural health monitoring. Earthq. Eng. Eng. Vib. **19**(2), 515–532 (2020)
17. Tibaduiza, D., Torres-Arredondo, M. Á., Vitola, J., Anaya, M., Pozo, F.: A damage classification approach for structural health monitoring using machine learning. Complexity **2018** (2018)
18. Azimi, M., Eslamlou, A.D., Pekcan, G.: Data-driven structural health monitoring and damage detection through deep learning: state-of-the-art review. Sensors **20**(10), 2778 (2020). https://doi.org/10.3390/s20102778

19. Chen, T.: A Machine Learning Based Framework for Load Forecasting and Optimal Operation of Power Systems with Distributed Generation
20. Pervez, M., Kamal, T., Fernández-Ramírez, L.M.: A novel switched model predictive control of wind turbines using artificial neural network-Markov chains prediction with load mitigation. Ain Shams Eng. J. **13**(2), 101577 (2022)
21. Luna, J., Falkenberg, O., Gros, S., Schild, A.: Wind turbine fatigue reduction based on economic-tracking NMPC with direct ANN fatigue estimation. Renew. Energy **147**, 1632–1641 (2020)
22. Gupta, S., Krishnan, S., Sundaresan, V.: Structural health monitoring of composite structures via machine learning of mechanoluminescence. In: ASME 2019 Conference on Smart Materials, Adaptive Structures and Intelligent Systems SMASIS 2019 (2019). https://doi.org/10.1115/SMASIS2019-5697
23. Choi, S.W., Song, E.-J., Hahn, H.T.: Prediction of fatigue damage growth in notched composite laminates using an artificial neural network (2002)
24. Zhang, Y.M., Wang, H., Wan, H.P., Mao, J.X., Xu, Y.C.: Anomaly detection of structural health monitoring data using the maximum likelihood estimation-based Bayesian dynamic linear model. Struct. Health Monit. **20**(6), 2936–2952 (2021)
25. Park, J.-H., Huynh, T.-C., Choi, S.-H., Kim, J.-T.: Vision-based technique for bolt-loosening detection in wind turbine tower. Wind Struct. Int. J. **21**, 709–726 (2015). [CrossRef]
26. Wang, L., Zhang, Z. Automatic detection of wind turbine blade surface cracks based on UAV-taken images. IEEE Trans. Ind. Electron. **64**, 7293–7303 (2017). [CrossRef]
27. Reddy, A., Indragandhi, V., Ravi, L., Subramaniyaswamy, V.: Detection of cracks and damage in wind turbine blades using artificial intelligence-based image analytics. Measurement **147**, 106823 (2019). [CrossRef]
28. Civera, M., Surace, C.: Non-destructive techniques for the condition and structural health monitoring of wind turbines: a literature review of the last 20 years. Sensors **22**(4), 1627 (2022)
29. Bao, Y., Tang, Z., Li, H., Zhang, Y.: Computer vision and deep learning–based data anomaly detection method for structural health monitoring. Struct. Health Monit. **18**(2), 401–421 (2019)
30. Molnar, C.: Interpretable Machine Learning. A Guide for Making Black Box Models Explainable. Book, p. 247 (2019)
31. Pozo, F., Tibaduiza, D.A., Vidal, Y.: Sensors for structural health monitoring and health assessment. Sensors **21**(5), 1558 (2021)
32. Sankararaman, S., Goebel, K.: Uncertainty in prognostics and systems health management. Int. J. Prong. Health Manag. **6**(4) (2015)
33. Flah, M., Nunez, I., Ben Chaabene, W., Nehdi, M.L.: Machine learning algorithms in civil structural health monitoring: a systematic review. Arch. Comput. Methods Eng. **28**(4), 2621–2643 (2021)
34. Hashmani, M.A., Jameel, S.M., Rehman, M., Inoue, A.: Concept drift evolution in machine learning approaches: a systematic literature review (2019)
35. Azeem, A., Ismail, I., Jameel, S.M., Harindran, V.R.: Electrical load forecasting models for different generation modalities: a review. IEEE Access (2021)
36. Zenisek, J., Holzinger, F., Affenzeller, M.: Machine learning based concept drift detection for predictive maintenance. Comput. Ind. Eng. **137**, 106031 (2019). https://doi.org/10.1016/j.cie.2019.106031
37. Disabato, S., Roveri, M.: Tiny machine learning for concept drift (2021). arXiv preprint arXiv:2107.14759
38. Tokognon, C.A., Gao, B., Tian, G.Y., Yan, Y.: Structural health monitoring framework based on Internet of Things: a survey. IEEE Internet Things J. **4**(3), 619–635 (2017)

39. Waziry, S., Wardak, A.B., Rasheed, J., et al.: Performance comparison of machine learning driven approaches for classification of complex noises in quick response code images. Heliyon **9**, e15108 (2023). https://doi.org/10.1016/j.heliyon.2023.e15108
40. Rasheed, J., Jamil, A., Hameed, A.A., et al.: A survey on artificial intelligence approaches in supporting frontline workers and decision makers for the COVID-19 pandemic. Chaos Solitons Fractals **141**, 110337 (2020). https://doi.org/10.1016/j.chaos.2020.110337
41. Baltrusaitis, T., Ahuja, C., Morency, L.-P.: Multimodal machine learning: a survey and taxonomy. IEEE Trans. Pattern Anal. Mach. Intell. **41**(2), 423–443 (2019). https://doi.org/10.1109/tpami.2018.2798607
42. Ramachandram, D., Taylor, G.W.: Deep multimodal learning: a survey on recent advances and trends. IEEE Signal Process. Mag. **34**(6), 96–108 (2017). https://doi.org/10.1109/msp.2017.2738401
43. Atrey, P.K., Hossain, M.A., El Saddik, A., Kankanhalli, M.S.: Multimodal fusion for multimedia analysis: a survey. Multimed. Syst. **16**(6), 345–379 (2010). https://doi.org/10.1007/s00530-010-0182-0
44. Williams-Lekuona, M., Cosma, G., Phillips, I.: A framework for enabling unpaired multimodal learning for deep cross-modal hashing retrieval. J. Imaging **8**(12), 328 (2022). https://doi.org/10.3390/jimaging8120328
45. Capuozzo, S., Gravina, M., Gatta, G., Marrone, S., Sansone, C.: A multimodal Knowledge-Based deep learning approach for MGMT promoter methylation identification. J. Imaging **8**(12), 321 (2022). https://doi.org/10.3390/jimaging8120321
46. Tsourounis, D., Kastaniotis, D., Theoharatos, C., Kazantzidis, A., Economou, G.: SIFT-CNN: when convolutional neural networks meet dense SIFT descriptors for image and sequence classification. J. Imaging **8**(10), 256 (2022). https://doi.org/10.3390/jimaging8100256
47. Yue, L., Hu, P., Chu, S.-C., Pan, J.: English speech emotion classification based on multi-objective differential evolution. Appl. Sci. **13**(22), 12262 (2023). https://doi.org/10.3390/app132212262
48. Fanzeres, L.A., Nadeu, C.: Sound-to-imagination: an exploratory study on cross-modal translation using diverse audiovisual data. Appl. Sci. **13**(19), 10833 (2023). https://doi.org/10.3390/app131910833
49. Qu, Z., Han, T., Yi, T.: MFFAMM: a small object detection with multi-scale feature fusion and attention mechanism module. Appl. Sci. **12**(18), 8940 (2022). https://doi.org/10.3390/app12188940

A Cybersecurity Procedure to Vulnerabilities Classification of Windows OS Based on Feature Selection and Machine Learning

Noor Alhuda Abdul Hasan Al-Sarray(✉) and Sait Demir

Department of Computer Engineering, Karabuk University, Karabuk, Turkey
noorhassan336655@gmail.com, 2128150013@ogrenci.karabuk.edu.tr, saitdemir@karabuk.edu.tr

Abstract. The fast advancement of systems and technology has led to problems in several areas, with data protection and information security being particularly significant concerns. Cybersecurity is a recently developed field that comprises many strategies aimed at safeguarding sensitive data. By utilizing machine learning techniques, this study has improved the security of the Windows operating system. The categorization of Windows system vulnerabilities was performed using machine learning techniques such as Random Forest, Logistic Regression, Naive Bayes, K-Nearest Neighbors, and Support Vector Machine. The dataset was compiled using materials from the National Institute of Standards and Technology (NIST) and exploit-deb, Care was taken to ensure that the data is real and exists. The parameters that were assessed include accuracy, precision, recall, f1-score, and ROC AUC Score. The Random Forest approach yielded the most precise findings, with an accuracy rate of 97%. The findings indicated that the Random Forest methodology was successful in identifying vulnerabilities in security.

Keywords: Cybersecurity · Vulnerability · Feature Selection · Random Forest · Logistic Regression · Naive Bayes · K-Nearest Neighbors · SVM

1 Introduction

One of the most significant issues that software faces is vulnerabilities; when the malicious hacker discovers vulnerabilities, then they can sabotage and disrupt, the hacker can stop the service or even take complete control of the system or view essential files that may be confidential in operating systems (OS) and software [1, 2]. Vulnerabilities have a high-scale impact and cause economic damage and problems to individuals, governments, and companies [3]. The Windows system is one of the most popular OS used for computers, so it is the main target for most hackers. Vulnerabilities must be discovered periodically and fixed before the hacker discovers them.

The action of protecting information and communication technology (ICT) systems from various cyber-threats or hackers has come to be known as 'cybersecurity'; numerous sides are linked with cybersecurity: measurement to protect ICT; the raw data and

J. Rasheed et al. (Eds.): FoNeS-AIoT 2024, LNNS 1035, pp. 229–243, 2024.
https://doi.org/10.1007/978-3-031-62871-9_18

information it includes and their transmitting and processing; The system's virtual and physical components, the level of protection achieved through their implementation, and ultimately the related professional field. The main concern in cybersecurity is comprehending various cyber-attacks and creating defense strategies that safeguard multiple properties, In general, three basic factors must be present for data to be protected, which are confidentiality, availability, and integrity, and it is called the CIA triangle [4].

Cybersecurity is a determination of procedures, technologies, programs, etc., that aim to protect networks, computers, systems, and data from unauthorized access and malicious hackers; due to technical advancements, there has been an increase in vulnerabilities, attacks, and unauthorized access, therefore, to meet the increasing demand for cybersecurity solutions, it is crucial to develop procedures to enhance security, via the improvement of machine learning (ML) capabilities [5].

ML is already a significant factor in the development of current and future information systems, and a significant proportion of this use is being generated in other areas of work that incorporate ML. Despite this, the utilization of ML in cybersecurity is still in its early stages, highlighting a significant divergence between research and practice. The fundamental reason for the discrepancy is the current state of the art, which hinders our understanding of the role of ML in cybersecurity. ML will always be fully realized without the knowledge of a widespread understanding of its implications and advantages, as the methods employed at the time do not fully benefit the whole system [6].

The dataset in ML is essential to find good results, we claim the data set via two of the most famous global vulnerability data sites. We focused on selecting the relevant features, directly determining the severity of the vulnerability. Therefore, we will try to use ML algorithms to classify vulnerabilities as a cybersecurity procedure to increase the protection of the Windows system.

2 Related Work

Vulnerabilities present significant challenges in terms of detection and prevention. Despite extensive research and application of ML models, current defense mechanisms struggle to provide complete protection. Over the past period, there have been many studies using ML to detect vulnerabilities and malware.

Islam, Rejwana, et al. [7] utilized ensemble ML and optimal feature selection to classify Android malware. Utilizing dynamic feature analysis, the researchers employed the widely-used ensemble ML technique of weighted voting for multi-classification on the CCCS-CIC-AndMal-2020 dataset. This dataset comprises a substantial collection of Android app and malware samples. Random Forest, K-nearest Neighbors, Decision Trees, SVM, Logistic Regression, and Multi-Level Perceptrons are among the methods employed to analyze the ensemble model. By combining weighted voting based on ML classifiers' R2 scores with careful data preparation, we achieve a 95.0% accuracy rate. By removing 60.2% of the features of the ML classifiers, it is still possible to get a 95.0% accuracy rate in the predictions.

This study utilized ML approaches to predict potential security vulnerabilities in the source code of the software. Mandal, Dilek, and İrfan KÖsesoy [1], by using the Decision Tree, Logistic Regression, and SVM algorithms devised techniques to forecast security flaws in source code. The dataset for the test was sourced from the OWASP Benchmark Test pocket. The Java code set was utilized to train many ML models, such as RF, Decision Tree, SVM, K-Nearest Neighbors, and Logistic Regression. The TFIDF and Doc2Vec algorithms were employed to extract feature vectors from the source code. An experimental investigation was undertaken to reach the highest prediction accuracy (0.97).

Rakesh Verma, Employing ML methodologies to uncover software vulnerabilities [8]. This work exemplifies a significant degree of inaccuracy. An ML classifier is employed to analyze the text functions extracted from the C source code. The extracted functions derived from the basic features, such as character count and character diversity, exhibited a very high accuracy rate of 64%. In contrast, the accuracy rate for the complex features was 69%.

Alan, Umaru, and Adamu [9] investigated several ML techniques for detecting ransomware. They subsequently employ these techniques to choose a specific feature and develop an SVM approach. This method achieves a Root Mean Square Error (RMSE) of 0.179 and accurately classifies ransomware with an accuracy rate of 88.2%. The researchers report a corresponding RMSE score of 0.179 in their publication. The SVM surpasses most ML classifiers in terms of ransomware classification.

Ying Xue, [10] The essay mostly focuses on ML approaches. A preliminary overview of the network vulnerabilities is offered. To proceed, the vulnerability data has to be prepared for extraction and matching using a Convolutional Neural Network (CNN) + Long-Short-Term-Memory (LSTM) approach, while considering the national vulnerability. The study revealed that the CNN-LSTM approach outperformed the CNN-LSTM vector machine in terms of accuracy, recall, F1 score, and Mathews correlation coefficient (MCC). The training precision is also rather impressive. To classify the various susceptibility groups, detecting methods were devised with assistance and further examination. The F1 and MCC values, beyond 0.85 and reaching 0.88 and 0.97 respectively, were used for evaluation. The results provide as evidence of the reliability of the CNN-LSTM approach for vulnerability detection. It is feasible to utilize the CNN-LSTM technique

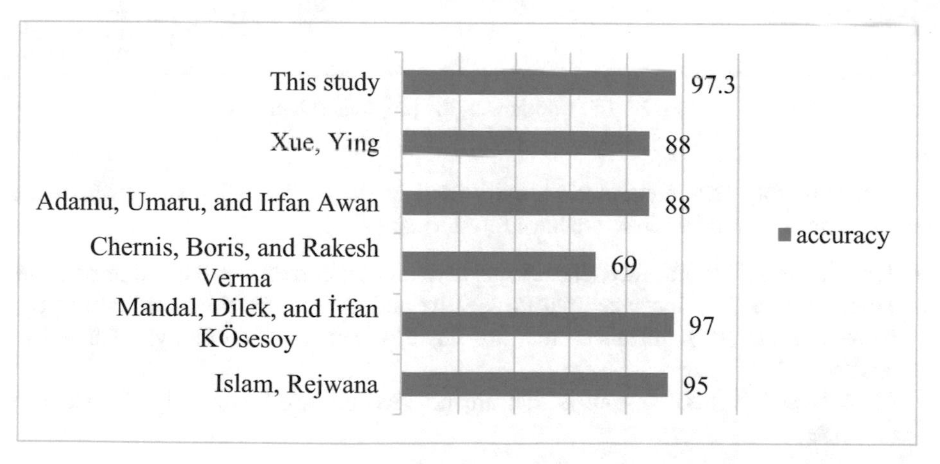

Fig. 1. A comparison chart with the prior studies.

on real-world networks. Refer to (see Fig. 1.) To assess these studies about the present one.

3 Methodology

In this section, we will see the steps through which we conducted the methodology. First, we collected data on Windows vulnerabilities over the past six years. Based on real data and actual vulnerabilities, we collected 12 characteristics for each vulnerability. Then we selected the most critical characteristics directly related to the strength of the vulnerabilities. The data was then classified into 4 groups according to the strength of the risk for each vulnerability. After that, we tested the performance of five ML algorithms to reach the highest possible accuracy. Figure 2 explains the workflow of the proposed methodology.

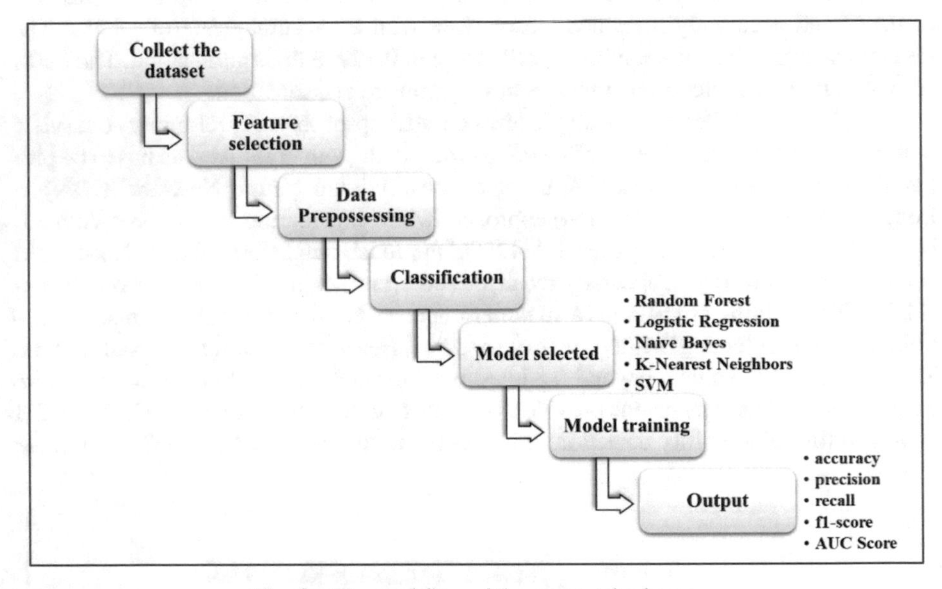

Fig. 2. The workflow of the proposed scheme.

These performance metrics are usually used in classification tasks to evaluate the performance of a model. First, we have four terms:

- True Positive (TP): Instances that are positive and are correctly predicted as positive.
- True Negative (TN): Instances that are negative and are correctly predicted as negative.
- False Positive (FP): Instances that are negative but are incorrectly predicted as positive.
- False Negative (FN): Instances that are positive but are incorrectly predicted as negative.

To calculate the performance metrics, we utilized the following equations [11, 12]:

$$Accuracy = \frac{TP + TN}{TP + TN + FP + FN} \quad (1)$$

$$Precession = \frac{TP}{TP + FP} \quad (2)$$

$$Recall = \frac{TP}{TP + FN} \quad (3)$$

$$F1score = 2 \times \frac{Precessioin \times Recall}{Precession + Recall} \quad (4)$$

4 Dataset

4.1 Dataset Collection

In this study, data was collected from Exploit-deb[1], and we also relied on NIST (National Institute of Standards and Technology)[2] to take some features of the data, considering the period between 2018 and 2023, we collected (189) vulnerabilities on the Windows OS in a CSV file, then the data was pre-processed as we will explain in the next steps.

4.2 Dataset Details

The dataset has 12 features, each feature has a different value and different data type (see Table. 1).

4.3 Feature Selection

There are many features for vulnerabilities, and not all of these features directly affect the severity of the vulnerability, so we need the actual features that have a direct impact on the strength of the vulnerability, and we do this by using the feature selection method. After implementing the feature in the selection process, the selected features can be seen in Table 2.

[1] https://www.exploit-db.com/ (Last accessed: October 03, 2023).
[2] https://nvd.nist.gov/ (Last accessed: October 01, 2023).

Table 1. Description of the dataset.

Feature	Description	Type
EDB_ID	This is the ID of vulnerability in the ExploitDB website	Integer
CVE	The Common Vulnerabilities and Exposures (CVE), is a general identifier for vulnerabilities	Integer
CVSS	A Common Vulnerability Scoring System (CVSS) or a custom severity scale, is a general measure of the strength of a vulnerability	Integer
Title	The title of vulnerability	String
EDB_Verified	This Feature indicates whether the vulnerability has been verified, exploited, or tested, this column has two values (0,1)	Integer
Type	The type of vulnerability (e. remote, local, DOS, etc.), the categorizing vulnerabilities by type can be useful for analysis	String
Platform	The platform means in which platform this vulnerability can work, (like. Windows 10, Windows 11, etc.)	String
Date	The date of vulnerability published	Date
Version	The version of the vulnerability	String
Author	The name of the author of this vulnerability	String
Description	Description of the vulnerability, like security issues and more details	String
Source	The link of the vulnerability source	String

Table 2. Feature selection for the dataset.

Feature	Description	Type
EDB_ID	The ID used to differentiate between each of the vulnerabilities, as each vulnerability has a different ID	Integer
CVSS	We are classifying vulnerabilities into four classes based on their strength level, based on this value which ranges from 0 to 10	Integer
Type	Represents the type of vulnerability	String
Platform	The platform on which the vulnerability runs	String

4.4 Dataset Preprocessing

We note that in the previous step, which is referred to as feature selection, we have four features, two of which are data types of strings (type, platform). ML technologies cannot directly process text data. To proceed, it is necessary to have numerical data. We employed the one-hot encoding technique to assign numerical values to these attributes. By utilizing a sparse vector where only one element is assigned a value of 1 and all other elements are assigned a value of 0, one-hot encoding can effectively do its task. This is a frequently used method of representing strings that have a limited number of possible values. Utilizing one-hot encoding results in the creation of feature vectors with a large cardinality. Nevertheless, due to its simplicity, one-hot encoding is frequently employed. Onehot encoding is effective for brief, non-repetitive phrases or tweets when working with models that possess robust smoothing capabilities. Neural networks that need input inside the discrete range of [0, 1] or [–1, 1] often utilize one-hot encoding [13–15]. A uniform vector is a 1 × N matrix (vector) where all cells but one have zeros, and the unique identifying cell contains a value of one. One-hot encoding allows for a more expressive representation of categorical data. Four of these encodings would represent various words, including [DOS, Local, REMOTE, WEBAPPS] (see Fig. 3).

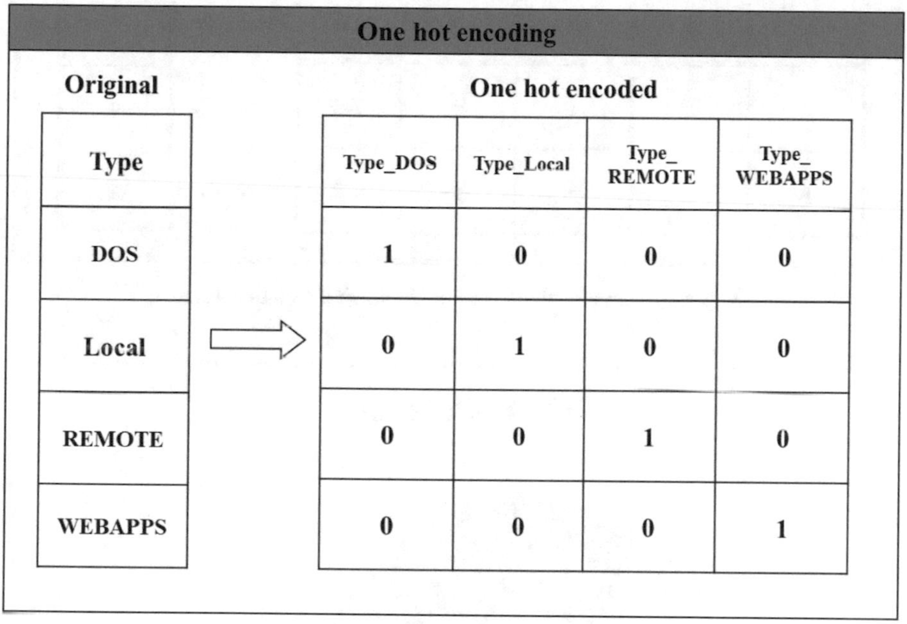

One hot encoding

Original

Type
DOS
Local
REMOTE
WEBAPPS

One hot encoded

Type_DOS	Type_Local	Type_ REMOTE	Type_ WEBAPPS
1	0	0	0
0	1	0	0
0	0	1	0
0	0	0	1

Fig. 3. A sample of one-hot encoding technique.

4.5 Classification

To do classification we add the 'class' column to the dataset file, this column contains four values (–1, 0, 1, 2), –1 means that it is low risk, 0 means that the vulnerability risk is medium, 1 means that it is high risk, and 2 means the risk is critical based on CVSS values [16]. We have four different classes of vulnerabilities based on the risk level (low, medium, high, critical). Figure 4 depicts the addition process of adding the class column, whereas Fig. 5 represents the dataset details with respect to class categories.

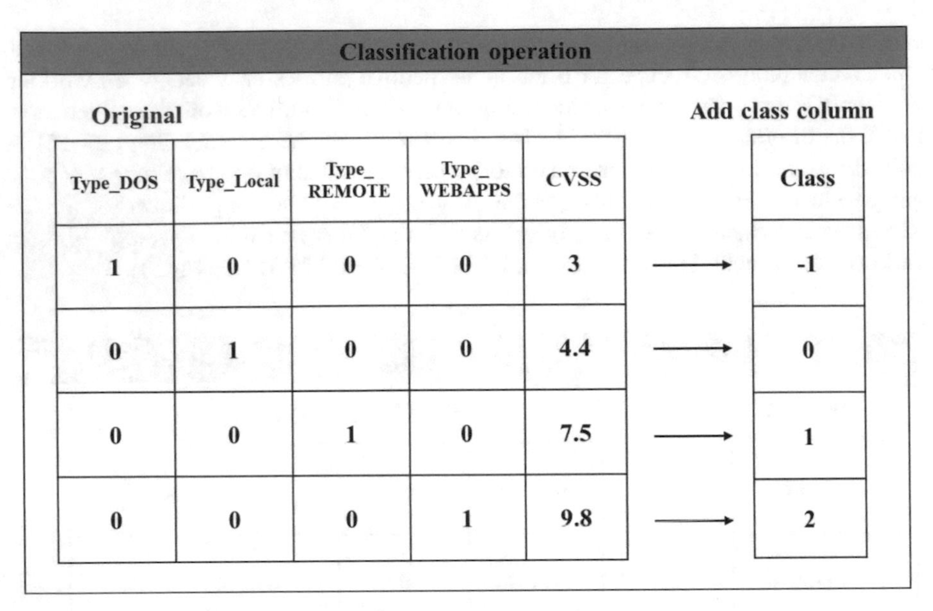

Fig. 4. A classification operation by adding a class column.

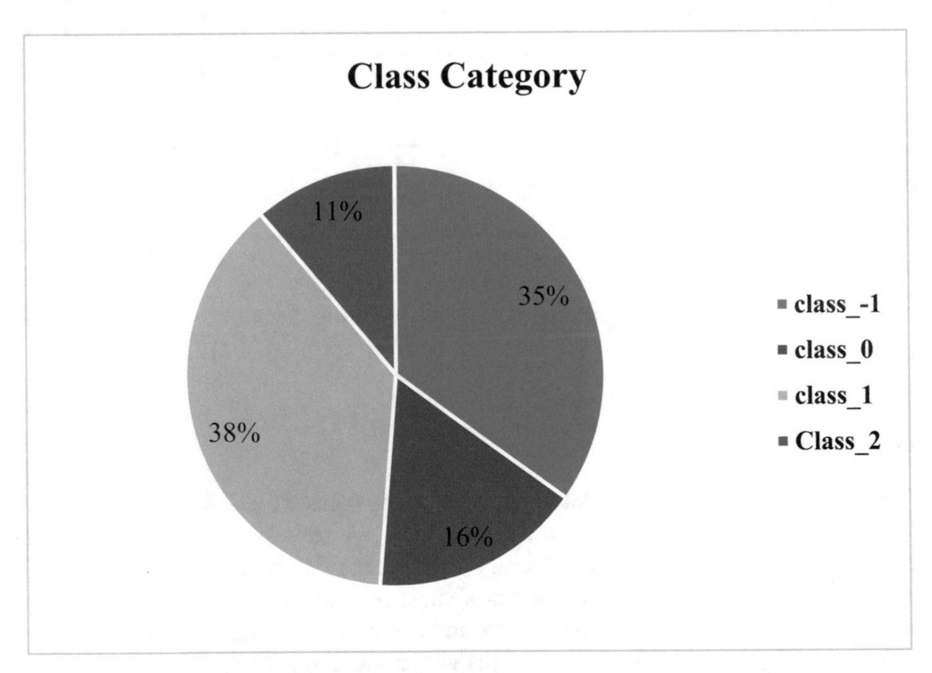

Fig. 5. A dataset details with respect to class categories.

5 Exploited Methods

After preparing the data, five distinct strategies were used to train the model and achieve the desired outcomes. SVMs, Naive Bayes, Random Forest, and K-Nearest Neighbors are the algorithms being referred to.

5.1 Random Forest

Random Forest is a renowned ML technique created by Leo Breiman and Adele Cutler. It involves aggregating the outcomes of several decision trees. Due to its versatility and ease of use, it has gained extensive popularity for both regression and classification applications [17]. Before commencing training, RF algorithms need the definition of three primary hyperparameters. These factors encompass the number of items analyzed, the dimensions of the nodes, and the count of trees. The RF classifier may be utilized to address issues related to classification and regression. The Random Forest methodology uses a bootstrap sample, which is a subset of the training set, to train a sequence of decision trees. One-third of the training sample has been put aside specifically for testing. This subset is referred to as the "out-of-bag sample" or "oob". By employing feature clustering, we may introduce more unpredictability to the dataset, so enhancing its variety and reducing the correlation among decision trees. The definition of prediction might vary depending on the sort of situation. In contrast to regression tasks, which involve averaging decisions from several decision trees, classification issues like as Volume employ majority decision-making. Frequently occurring category variable: The anticipated class is generated. In order to complete the prediction, the out-of-bag sample is utilized for cross-validation. It attained a precision of 0.97 and outperformed all other algorithms in the study.

5.2 Logistic Regression

For classification and prediction, the preferred statistical model is often the logit model. Logistic regression calculates the likelihood of an occurrence, such as making a choice or not making a choice, based on a group of independent variables [18]. The dependent variable is bounded between zero and one due to its probabilistic nature. Logistic regression utilizes the logit transformation of probabilities, which represents the ratio of the probability of success to the probability of failure.

The logit of pi is the dependent or response variable, while x is the independent variable. The study produced an algorithmic performance accuracy of 0.81.

5.3 Naïve Bayes

The Naive Bayes classifier is a supervised ML algorithm commonly used to address classification issues, including text categorization. As a generative learning algorithm, it also aims to replicate the input data distribution of a certain class or category. Unlike discriminative classifiers such as logistic regression, it does not acquire knowledge about the most crucial characteristics for distinguishing across classes. Due to its versatility in addressing different issues or categories, this study's technique exhibited the lowest performance accuracy among the alternatives, reaching a maximum accuracy of 0.68.

5.4 K-Nearest Neighbors

The k-Nearest Neighbors (KNN) approach, sometimes referred to as KNN or k-NN, is a nonparametric supervised learning classifier that uses proximity to produce predictions or classifications for a single data point in clustering. While it may be used for regression or classification problems, its primary application is as a classification method that relies on the premise that comparable data points are located close to each other. While this method attained a commendable accuracy of 0.89, other algorithms outperform it in classification tasks.

5.5 Support Vector Machine

The categorizing method, known as Simple Vs. Complex is considered one of the most straightforward and aesthetically pleasing techniques. All objects requiring classification are positioned as points in the last dimension of space, with designated coordinates. SVMs utilize a hybrid method to address the classification difficulty. In other words, a hyperplane refers to a two-dimensional line or three-dimensional plane that separates points into two categories, with all points on one side belonging to one category and all points on the other side not belonging to that category. Despite the existence of other hyperplanes, SVMs strive to find the most effective one for separating the two cats. In other words, the outcome represents the greatest distance to points within each category. This space has been designated as a margin. Support vectors are precisely located near the border. Initially, we locate this hyperplane. The SVM depends on a training set, which consists of a collection of locations that have been previously assigned accurate categorical labels. Thus, SVM may be classified as a form of supervised learning. In order to optimize this margin, SVM is utilized to solve convex optimization issues. The restrictions specify that the points in each category must be located on the right side of the hyperplane. One of the key advantages of SVMs is their simplicity, making them straightforward to comprehend, execute, and interpret [19]. The algorithm demonstrated high efficiency, achieving an accuracy of 0.94, which was the second-best result recorded in this experiment.

6 Results and Discussion

Hence, we calculated the following metrics: F1-score, ROC AUC score, recall, accuracy, precision, and a convolution matrix plot. The findings demonstrated that the RF algorithm attained the best level of accuracy (0.97%) when compared to the other methods. Table 3 and Fig. 6 demonstrate that performance exhibited variability, ranging from a minimum accuracy of 0.68 to a maximum accuracy of 0.97. Furthermore, Fig. 7 depicts the confusion matrices of these models.

To construct an ROC curve, the initial step involves computing its values by tracking the specificities and sensitivities of a continuous test measure at various time intervals. Upon completion of the process, you will have a compilation of potential test results, along with their respective specificity and sensitivity levels. The ROC curve in ROC charts displays a summary of the data using the sensitivity (true positive rate) and 1-specificity (false positive rate) [20]. Ultimately, the y-axis values corresponding to the various factors are utilized to generate a visual representation of the ROC curve. Figure 8 shows the ROC curves obtained for each model.

Table 3. The performance analysis of the exploited machine learning algorithms.

Method	Accuracy	Precision	Recall	F1-score	ROC AUC Score
Random Forest	0.97368	0.9737	0.9759	0.9738	0.9993
Logistic Regression	0.8157	0.8158	0.8285	0.7987	0.9697
Naïve Bayes	0.6843	0.6842	0.5417	0.5942	0.8825
K-Nearest Neighbors	0.8948	0.8947	0.9049	0.8965	0.9658
Support Vector Machine	0.9473	0.9474	0.9555	0.9479	0.9622

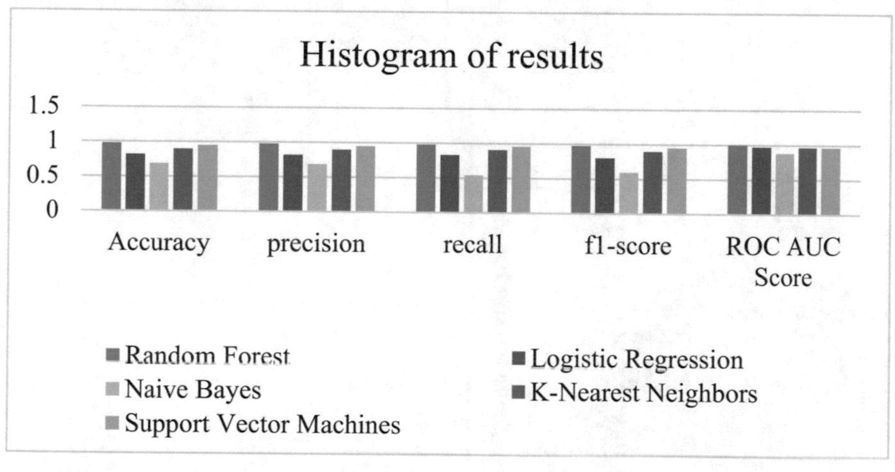

Fig. 6. A performance comparison of exploited machine learning algorithms.

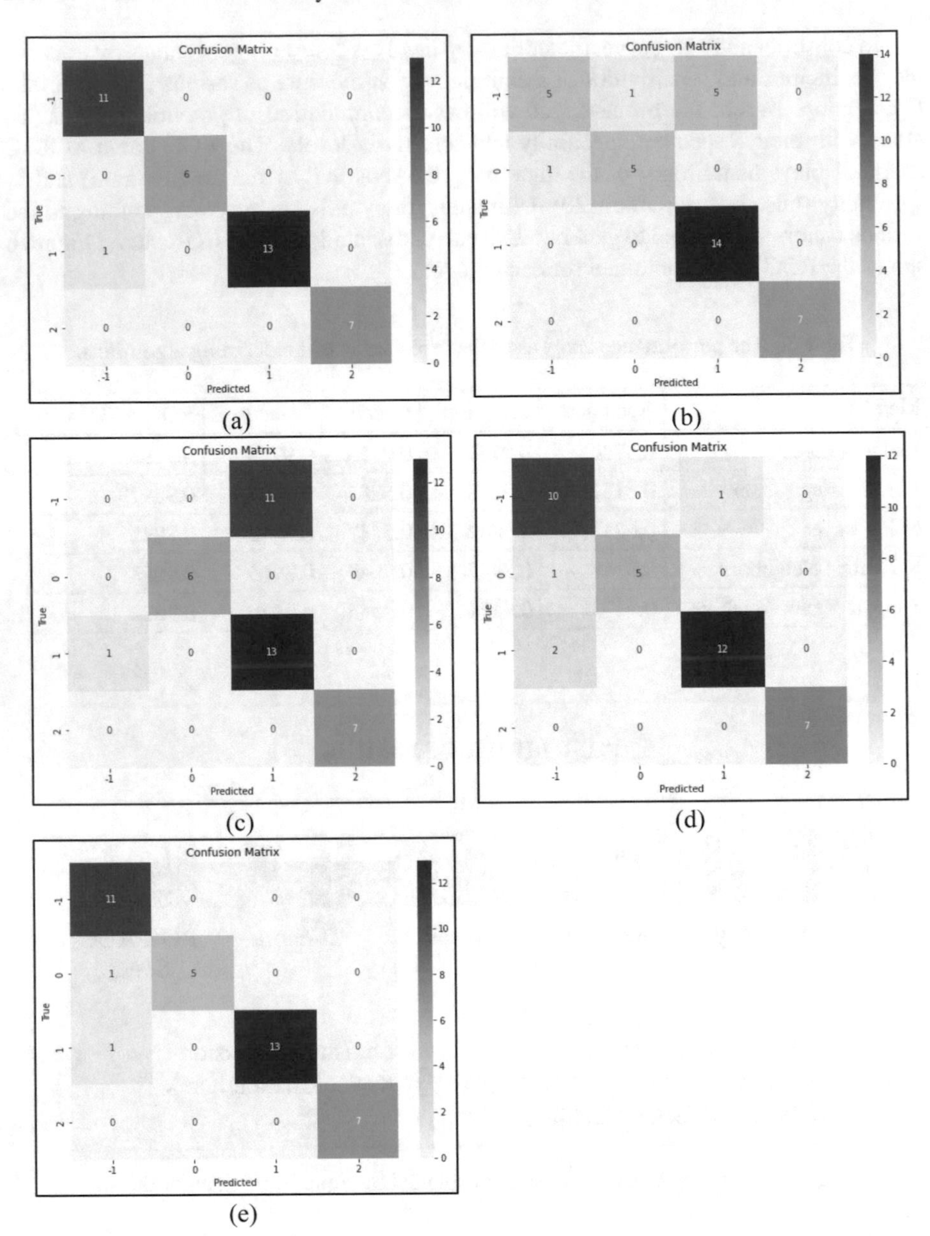

Fig. 7. The confusion matrix of various exploited models, (a) random forest, (b) logistic regression, (c) Naïve Bayes, (d) k-nearest neighbors, and (e) support vector machine.

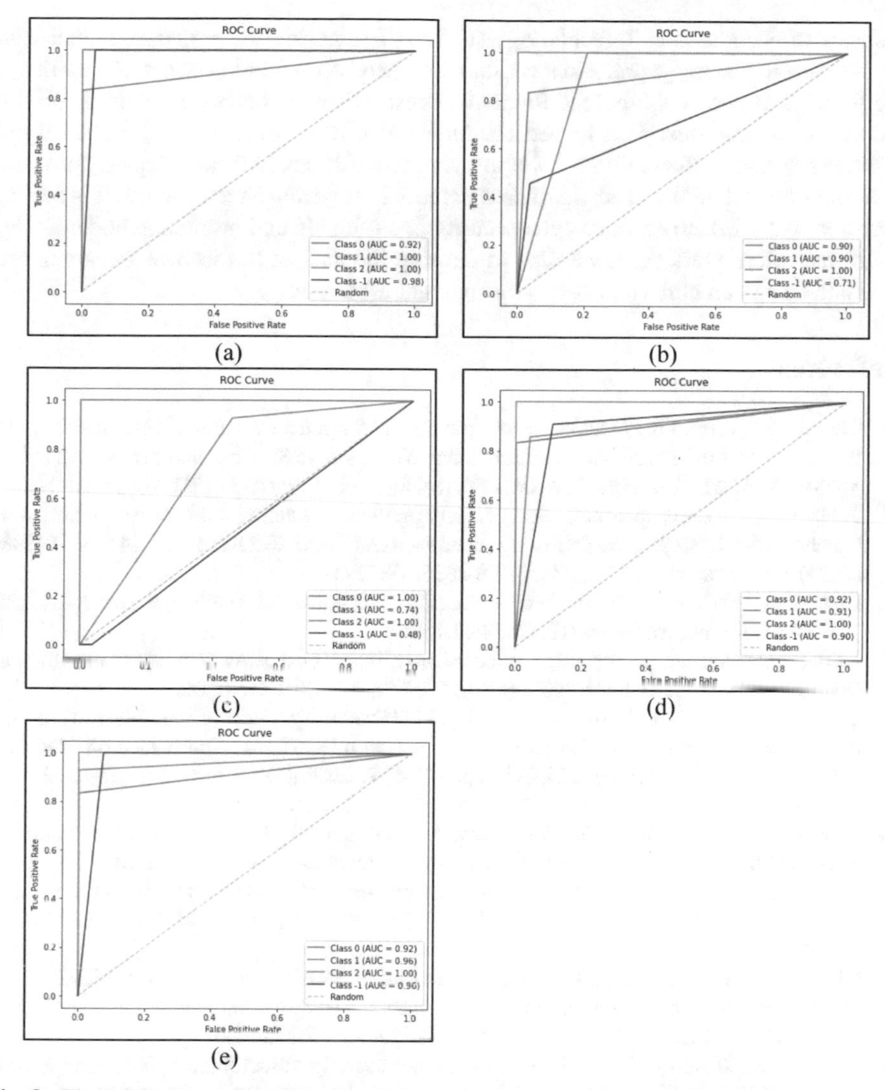

Fig. 8. The ROC curves of exploited models, (a) random forest, (b) logistic regression, (c) Naïve Bayes, (d) k-nearest neighbors, and (e) support vector machine.

7 Conclusion

The fast advancement of systems and technology has resulted in several challenges in the fields of data security and information protection. As a result, the discipline of cybersecurity has developed several tactics and procedures to ensure data protection. We used ML methodologies to bolster the security of the Windows OS in our study. To classify the vulnerabilities in the Windows system, we utilized five ML classification techniques: Support Vector Machine (SVM), Random Forest, Logistic Regression, and Naive Bayes. In addition to exploit-deb, the dataset was obtained from the National

Institute of Standards and Technology (NIST). The performance measures that were assessed were accuracy, precision, recall, ROC AUC score, and f1-score. According to the findings, it is evident that the Random Forest technique outperforms in identifying security vulnerabilities. We achieved a commendable accuracy rate of 97% in classifying Windows system vulnerabilities by employing five different ML techniques. SVM and Random Forest demonstrated significant potential for performance. Hence, it is evident that employing ML to enhance cybersecurity is a valuable and promising undertaking.

In the future work plan, we plan to develop the dataset to discover threats in real time and design an anti-virus system using machine learning.

References

1. Chernis, B., Verma, R.: Machine learning methods for software vulnerability detection. In: Proceedings of the Fourth ACM International Workshop on Security and Privacy Analytics, pp. 31–39. ACM, New York, NY, USA (2018). https://doi.org/10.1145/3180445.3180453
2. Tahir, T., et al.: Early software defects density prediction: training the international software benchmarking cross projects data using supervised learning. IEEE Access **11**, 141965–141986 (2023). https://doi.org/10.1109/ACCESS.2023.3339994
3. Bharadiya, J.: Machine learning in cybersecurity: techniques and challenges. Eur. J. Technol. **7**, 1–14 (2023). https://doi.org/10.47672/ejt.1486
4. Kim, L.: Cybersecurity: Ensuring Confidentiality, Integrity, and Availability of Information. Presented at the (2022). https://doi.org/10.1007/978-3-030-91237-6_26
5. Yesiltepe, M., Jamil, A., Rasheed, J., Kurulay, M.: Hexadecimal hash value hiding in a message in web service against deep learning. In: 2019 11th International Conference on Electrical and Electronics Engineering (ELECO), pp. 693–697. IEEE (2019). https://doi.org/10.23919/ELECO47770.2019.8990576
6. Mahmood, N.Z., Ahmed, S.R., Al-Hayaly, A.F., Algburi, S., Rasheed, J.: The evolution of administrative information systems: assessing the revolutionary impact of artificial intelligence. In: 2023 7th International Symposium on Multidisciplinary Studies and Innovative Technologies (ISMSIT), pp. 1–7. IEEE (2023). https://doi.org/10.1109/ISMSIT58785.2023.10304973
7. Islam, R., Sayed, M.I., Saha, S., Hossain, M.J., Masud, M.A.: Android malware classification using optimum feature selection and ensemble machine learning. Internet Things Cyber-Phys. Syst. **3**, 100–111 (2023). https://doi.org/10.1016/j.iotcps.2023.03.001
8. Mandal, D., KÖsesoy, İ.: Prediction of software security vulnerabilities from source code using machine learning methods. In: 2023 Innovations in Intelligent Systems and Applications Conference (ASYU), pp. 1–6. IEEE (2023). https://doi.org/10.1109/ASYU58738.2023.10296747
9. Adamu, U., Awan, I.: Ransomware prediction using supervised learning algorithms. In: 2019 7th International Conference on Future Internet of Things and Cloud (FiCloud), pp. 57–63. IEEE (2019). https://doi.org/10.1109/FiCloud.2019.00016
10. Xue, Y.: Machine learning: research on detection of network security vulnerabilities by extracting and matching features. J. Cyber Secur. Mobil. (2023). https://doi.org/10.13052/jcsm2245-1439.1254
11. Waziry, S., Wardak, A.B., Rasheed, J., Shubair, R.M., Yahyaoui, A.: Intelligent facemask coverage detector in a world of chaos. Processes **10**, 1710 (2022). https://doi.org/10.3390/pr10091710
12. Churcher, A., et al.: An experimental analysis of attack classification using machine learning in IoT networks. Sensors **21**, 446 (2021). https://doi.org/10.3390/s21020446

13. Cerda, P., Varoquaux, G., Kégl, B.: Similarity encoding for learning with dirty categorical variables. Mach. Learn. **107**, 1477–1494 (2018). https://doi.org/10.1007/s10994-018-5724-2
14. Davis, M.J.: Contrast coding in multiple regression analysis: strengths, weaknesses, and utility of popular coding structures. J. Data Sci. **8**, 61–73 (2021). https://doi.org/10.6339/JDS.2010.08(1).563
15. Bagui, S., Nandi, D., Bagui, S., White, R.J.: Machine learning and deep learning for phishing email classification using one-hot encoding. J. Comput. Sci. **17**, 610–623 (2021). https://doi.org/10.3844/jcssp.2021.610.623
16. Bitton, R., Maman, N., Singh, I., Momiyama, S., Elovici, Y., Shabtai, A.: Evaluating the cybersecurity risk of real-world, machine learning production systems. ACM Comput. Surv. **55**, 1–36 (2023). https://doi.org/10.1145/3559104
17. Waziry, S., Wardak, A.B., Rasheed, J., Shubair, R.M., Rajab, K., Shaikh, A.: Performance comparison of machine learning driven approaches for classification of complex noises in quick response code images. Heliyon **9**, e15108 (2023). https://doi.org/10.1016/j.heliyon.2023.e15108
18. Kaya, ŞM., İşler, B., Abu-Mahfouz, A.M., Rasheed, J., AlShammari, A.: An intelligent anomaly detection approach for accurate and reliable weather forecasting at ioT edges: a case study. Sensors **23**, 2426 (2023). https://doi.org/10.3390/s23052426
19. Yahyaoui, A., et al.: Performance comparison of deep and machine learning approaches toward COVID-19 detection. Intell. Autom. Soft Comput. **37**, 2247–2261 (2023). https://doi.org/10.32604/iasc.2023.036840
20. Hoo, Z.H., Candlish, J., Teare, D.: What is an ROC curve? Emerg. Med. J. **34**, 357–359 (2017). https://doi.org/10.1136/emermed-2017-206735

Cyberbullying Detection for Urdu Language Using Machine Learning

Hamza Mustafa(✉) and Kashif Zafar

National University of Computer and Emerging Sciences, Lahore, Pakistan
hamza.mustafa@lhr.nu.edu.pk, kashif.zafar@nu.edu.pk

Abstract. As social media users are rapidly growing, cyberbullying detection has become an increasingly essential topic for research. Cyberbullying is described as the sending or posting of text or photos designed to hurt or portray shame to another person or group of people using the Internet, cell phones, video game systems, or other technology. Cyberbullying is the use of social media to hurt or embarrass any other person. In the past, research in cyberbullying detection in English has been carried out, and cyberbullying detection involving the Urdu language has been ignored mostly because the Urdu language lacks resources. Urdu is a widely spoken language, especially in some parts of South Asia. It is the National Language of Pakistan. In this research, machine learning-based approaches are used for cyberbullying detection. (Dataset collection and source). The dataset is labeled by several different native speakers. The majority voting scheme is used for assigning a final label to a Tweet. Three main feature extraction techniques used for the detection of cyberbullying are TF-IDF, BOW, and Glove, and different machine learning algorithms are implemented. After doing multiple experiments, it has been proven that the Extra Tree Classifier (ETC) with TF-IDF outperformed other algorithms having a 79% accuracy score. However, the proposed approach performed better than the reported approaches that are based on machine learning for the Urdu language on sentiment analysis. Our experimental results also performed better on the sentiment dataset.

Keywords: Cyberbullying Detection · Urdu language · Machine Learning

1 Introduction

1.1 Cyberbullying

Cyberbullying is defined as the sending or posting of textual content or photos to hurt or disgrace every other person or organization through the usage of the Internet, cell telephones, online game structures, or other eras (Haidar et al., 2017). Most cyberbullying scholars recall the definition of cyberbullying. Cyberbullying is described as "willful and repetitive harm delivered through the medium of electronic conversation". The majority of advanced research on cyberbullying focused on the consequences of cyberbullying. Cyberbullying on social media, in contrast to conventional bullying, which normally occurs at school through face-to-face verbal exchange, can occur anywhere at any time.

J. Rasheed et al. (Eds.): FoNeS-AIoT 2024, LNNS 1035, pp. 244–257, 2024.
https://doi.org/10.1007/978-3-031-62871-9_19

Bullies have no qualms approximately hurting their peers' sentiments because they do not have to confront everyone and may cover behind the Internet. Because everyone, especially children, is continuously connected to the Internet or social media, victims have without difficulty uncovering harassment. Open-source network (OSN) websites (which include Twitter and Facebook) have ended up quintessential equipment in the lives of customers.

As a result, these OSNs have come to be the most commonplace for cyberbullying harassment and victimization (Dinakar et al., 2012) and their popularity and fast increase have expanded in phrases of cyberbullying incidences over the preceding several years. Cyberbullying is bullying that happens on the web and can happen at different destinations where individuals read, take an interest, or give content. Little exertion is being made to foster mechanical strategies to identify and dis-turb continuous cyberbullying, or to forestall cyberbullying from happening in any case, (Deb Barma et al., 2020). Cyberbullying is a growing problem that needs to be addressed quickly and effectively.

1.2 Social Media and Cyberbullying

Online media, according to the author (Kini et al., 2020), is a set of web-based applications that developed from the e functional and technological basis of Web 2.0. These tools make it easier to create and share user-generated content. People can gain access to a wealth of knowledge, as well as a convenient communication experience, through social media. However, social media can have bad consequences, such as cyberbullying, which can negatively impact people's lives, particularly children and teenagers. Cyberbullying in theory is the use of harsh language on the Internet, which has become a big issue for people of all ages. The issue of automatically detecting objectionable language in social media programs, websites, and blogs is challenging yet critical. Because of the widespread availability and popularity of the Internet, laptops, tablets, and cell phones in the present era, cyberbullying may occur at any time and from any location, making it a major problem. As a result, we require an automated method for detecting cyberbullying conduct quickly and effectively.

In Pakistan, the number of people utilizing YouTube has increased dramatically in recent years. YouTube is Pakistan's second most popular website (Alexa - Top Sites in Pakistan - Alexa, n.d.). YouTube is used by people of various religions, cultures, and educational backgrounds. People sometimes submit movies that are offensive to other cultures or religions, which can lead to verbal assaults in comments due to differing viewpoints. Such actions have the potential to destabilize the country's law and order situation. People have the right to free expression; therefore, they can comment on any video on YouTube, which can lead to harsh language, racial comments, religious hatred, and even threats. Hate speech (HS) has a negative influence on civilization and harms the public's intellectual health, hence it must be stopped.

1.3 Effects of Cyberbullying

Both the bully (predator) and the effector go through mental and bodily penalties as a result of bullying and cyberbullying. Cyberbullying (CB) is more serious than physical bullying since it is more widespread, and public, and the victim has no way of escaping. Cyberbully fatalities described emotional, focus, and behavioral problems, as well as difficulties interacting with their peers. Common headaches, persistent stomach discomfort, and difficulties sleeping were more common among these victims. One out of every four pupils reported feeling uncomfortable at school. They were likewise bound to be hyperactive, have conduct issues, drink unreasonably, or use cigarettes. Cyberbullying has a catastrophic effect on this group, resulting in victims having low self-esteem. Bullying can have a variety of harmful consequences, including poor effects on mental and physical health (Pawar & Raje, 2019), sadness and anxiety, and suicidal thoughts (Klomek et al. 2010).

1.4 Categories of Cyberbullying

According to (Haidar et al., 2017), there are numerous types of cyberbullying:

- Flaming is the act of starting an online brawl.
- Masquerade: When a bully disguises himself as someone else to carry out nasty intentions.
- Denigration is the act of transmitting or distributing rumors to harm someone's reputation.
- Impersonation is the act of pretending to be someone else and emailing or uploading something to bring that person into trouble or harm their reputation or friendships.
- Harassment: Sending vulgar and nasty communications regularly.
- Publication of someone's embarrassing facts, photographs, or secrets is known as an outing.
- Trickery is when you persuade someone to expose secrets or humiliate information to share them online.
- Exclusion is defined as the deliberate and brutal exclusion of somebody from an online assemblage.
- Cyberstalking is defined as persistent, severe aggravation and denigration that comprises intimidations or instills substantial anxiety in the target.

2 Literature Review

Urdu is a resource-poor language (Khattak et al., 2020), so cyberbullying detection of Urdu text is challenging compared to other languages. There is a lot of work already done in English, but in Urdu, some people contribute to sentiment analysis, not to cyberbullying detection. Urdu language research for cyberbullying detection is not made publicly available. This section contains previous work done on cyberbullying detection in different languages and sentiment analysis for the Urdu language.

2.1 Cyberbullying Detection in Different Languages

In this section, we discuss the research that detects cyberbullying in different languages. There are different languages like Arabic, Hindi, Spanish (Mercado et al., 2018), Turkish (Özel et al., 2017) and Dutch in which research has been performed and researchers have also done some experimentation for the detection and some researchers have achieved good accuracy results. However, none of the research has been publicly available for the detection of cyberbullying in the Urdu language although some research has been performed on the Urdu language for sentiment analysis that we have also discussed below.

Arabic. They worked on information readiness before demonstrating preparation (Alhumoud et al., 2015). WEKA was utilized to complete fundamental handling. Vector Filter for Tweet to Senti Strength Feature, change of character units into word vectors, and normalization among past handling strategies utilized. The framework was then prepared utilizing SVM. The method involved with preparing and testing the model consumes a large chunk of the day (it requires around 8 h). In this review, they observed that 31245 peaceful episodes were accurately recognized utilizing the Nave Bayes model. 1832 and misleading allegations are recorded as intruding. As far as occurrences of viciousness, 801 were precisely grouped, while 1395 were thought of as peaceful. Eventual outcomes of WEKA in the wake of preparing and model testing. In the two models, Nave Bayes and SVM, precision, review, genuine great rating, bogus rate, and F-Measure are shown. The general consequences of accuracy and review for Naıve Bayes are 0.901% and 0.909% individually. Furthermore, the general aftereffects of accuracy and review for the SVM model are 0.934% and 0.941% individually. They offer a strategy to distinguish cyberbullying in Arabic Twitter streams continuously in (Mouheb, Abushamleh, et al., 2019 a, b).

In (Mouheb, Ismail, et al., 2019, a b) Arabic language online media remarks from YouTube and Twitter were additionally dissected utilizing vocabulary approach, in view of corpus of cyberbullying and forceful words.

Hindi and Marathi. In (Pawar & Raje, 2019) ML-based arrangement prototypes are utilized for distinguishing cyberbullying. ML is essentially ordered into three classes: I) Administered Learning: in this methodology, the numerical model is fabricated dependent on information which encompasses both arrangement of information sources and wanted results; Unsupervised Erudition: in this methodology, the model accepts set of information as information, and attempt to discover erection (e.g., gathering or bunching of the information); and iii) Reinforcement Knowledge: this methodology is worried about making appropriate moves to expand the award specifically circumstance. They accumulated this informational collection from different sources which incorporate tweets, paper surveys, and traveler audits. For the Hindi-associated investigations, they accumulated and utilized 621 audits. For the Marathi study, they gathered 810 audits. Because of setting the affectability of Indian dialects, and to guarantee the right marking of wry messages, they physically named the posts in both the Hindi and Marathi datasets. They directed every one of the trials utilizing the MNB, Logistics Regression (LR), and Stochastics Gradient Descent (SGD) calculations. Moreover, they completed investigations with and without the incorporated informational collection. Results, with the integrated dataset, show that the expansion of more information to our dataset extra develops the F1 score. This demonstrates that their model is summed up and executes better on both the classes (i.e., tormenting and non-harassing) than when it is made with the extreme (i.e. genuine) dataset. In (Hussain et al., 2019), they identify oppressive Bangla remarks which are gathered from different social locales where individuals share their feeling, sentiments, sees and so forth.

Dutch and English. (Van Hee et al., 2018) worked on customized cyberbullying disclosure in electronic media text by showing the posts formed by dangers, losses, and observers of online bothering. They focus on modified cyberbullying acknowledgment in online media text by showing posts made by hazards, setbacks, and spectators of web irritation. They utilized direct help vector machines (SVM) and played out a progression of trials on a two-fold grouping to decide programmed cyberbullying identification. English and Dutch corpus were made after gathering information from ASKfm. For the programmed location of cyberbullying, double grouping tests utilizing SVM executed in LIBLINEAR were performed by working with Scikit-learn, and the AI library for Python. A set of paired order tests was directed to investigate the achievability of programmed cyberbullying identification via web-based media. Ten times cross-approval was acted in a thorough lattice search over various component types and hyper-boundary mixes. The outcomes report AUC scores are heartier to information irregularity than review, accuracy, and F1-score. After highlighting and hyper-boundary enhancement of our models, a most extreme F1 score of 64.32% and 58.72% was acquired for English and Dutch, separately.

2.2 Urdu Language Sentiment Analysis

Bilal proposed a supervised machine learning technique to classify the Roman Urdu text by applying various machine learning classifiers, namely KNN, decision tree, and Naïve Bayes. In experiments, the authors proved that the naive Bayes algorithm delivered better results based on different assessment parameters such as precision, accuracy, recall, and F-measure. They stated that the main drawback is the small size of the dataset, which has not significantly improved the performance (Bilal et al., 2016). Usman determined text classification using machine learning approaches and with the maximum majority voting to gain satisfactory accuracy. There are several steps in the process. The information is first gathered from news websites and saved in a text file. Second, data is preprocessed, including tokenization, stop word removal, and stemming. In space, features are handled as a vector. Health, sports, science, culture, strange, entertainment, and business are divided into five categories. The link between classes and documents is established. The record is mapped and validated in the course once the papers appear. Following the classification of the classes by five classification techniques, majority voting was implemented in the class (Cover & Hart et al., 1967).

Finally, the results of the algorithms are gathered, and the mod of each document is calculated to predict the relevant category (Usman et al., 2016). Sentiment analysis is performed at the sentence level in this paper (Mukhtar & Khan, 2018). Khan developed the Urdu sentiment lexicon with the help of the existing English sentiment lexicon. They also developed the dataset from Twitter consisting of 1000 preprocessed tweets. The research contains Lexicon-based and Machine learning-based experiments and achieved 60% accuracy in both approaches. They mentioned that due to negations in their dataset, model performance was average. When we checked the dataset, it contained taunts, sarcasm, and negations (Khan et al., 2017).

3 Proposed Methodology

3.1 Data Collection

First, we choose a platform from which we can download the Urdu data. We choose Twitter. Twitter is a microblogging and long-range interpersonal communication stage established in the United States that permits clients to post tweets or short messages. Enrolled clients can make, like, and retweet tweets, however, unregistered clients can just see tweets. Twitter is a main informal community stage where a great many clients communicate their thoughts day by day as a large number of short instant messages (up to 140 people) known as tweets. For cyberbullying detection, Twitter is the best available choice. However, it's not possible to manually download random tweets, even 500 tweets. I collect the Dataset from (Khan et al., 2017) containing 1K Urdu tweets. They used this dataset for measuring sentiment analysis and attained an accuracy of around 70%. Each tweet in the dataset was labeled into two labels, whether it is positive (P), or negative (N).

3.2 Data Preprocessing

Data Classification. The dataset is classified into six levels. Following are the descriptions of the levels:

- Level 1 is to tell positive or negative tweets.
- Level 2 is to tell if it is offensive or not.
- Level 3 is to tell if it is a targeted insult or an untargeted insult.
- Level 4 is to tell if it is individual cyberbullying or grouped.
- Level 5 is Merging other labels with hate speech and predicting binary.
- Level 6 is Merging other labels with cyberbullying and predicting binary.

Data Labeling by Majority Voting Scheme. First, the downloaded data set is labeled according to our needs then the file with just labels according to the classification of cyberbullying. Then the next step is to identify other people to label this file too so that we can compile this file based on majority voting. The downloaded dataset contains only sentiment-labeled files means only positive, and negative classifications have been done according to tweets.

The original file of the dataset contains 510 negative values and 486 positive values. The logic behind the labeling is that only negative tweets can proceed for classification because positive tweets cannot be classified further. The final dataset file contains 249 offensive and 237 non- offensive which were derived from negative tweets. And 145 target insult and 92 classify as not target. Out of 145 target tweets, 52 tweets are cyberbullying, 48 tweets are hating speech, and 28 tweets are others.

Stop Words Removal. Stop words are disputes that in any language do not pay abundant meaning to a catchphrase. They can be safely ignored without affecting the sense of the phrase. Articles, prepositions, pronouns, conjunctions, and so on are the most prevalent terms in any language, and they offer nothing to the text (Figs. 1 and 2).

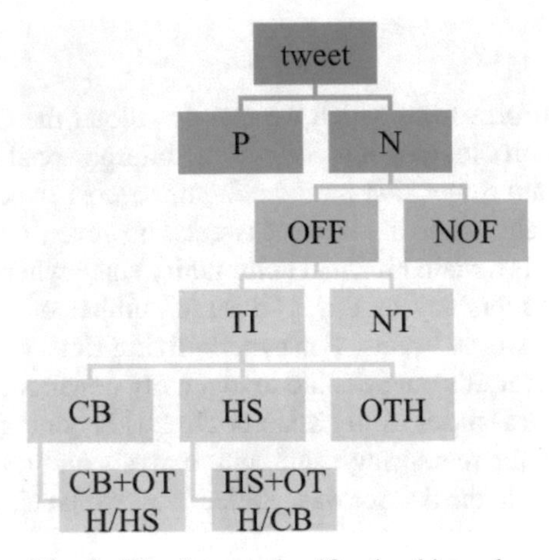

Fig. 1. The dataset classification hierarchy.

آئی	آئے	آتا
آئیں	آتی	آتے
آنا	آنی	آپ
آہ	آیا	اب
ابھی	تجھی	تجھے
ترا	تمام	تمہاری
آگے	تمہارے	تک
تو	تھی	تھا
تری	تم	تمہارا

Fig. 2. Sample list of stop words.

Tokenization. When it comes to working with text data, tokenization is one of the most typical processes. But precisely a token is the name given to each of these smaller components. There are several Urdu tokenizers, such as Urdu hack, Spacy, and Inltk, however, we chose Spacy in this study since it is more efficient than the others.

Data Cleaning. To render Urdu text usable for machine learning tasks, it must first be normalized. This module contains the following features (Fig. 3):

- Normalizing Single Characters
- Normalizing Combine Characters
- Put Spaces Before and After Digits
- Put Spaces After Urdu Punctuations
- Put Spaces Before and After English Words
- Removal of Diacritics from Urdu Text

Index	Tweets	After preprocessing
1	نواز شریف کے استعفے کے بغیر کنٹینر : سے نہیں نکلوں گا - عمران خان (جھوٹے ہر روز نکلتا ہے بنی گالا بھی جاتا ہے	نواز شریف استعفے کنٹینر نکلوں عمران خان جھوٹے روز نکلتا بنی گالا
2	بریکنگ :- دبئی ٹیسٹ، دوسری اننگز، یونس خان کی آسٹریلیا کے خلاف سنچری مکمل	بریکنگ دبئی ٹیسٹ اننگز یونس خان آسٹریلیا خلاف سنچری مکمل
3	چندے سے انقلاب نہیں لایا جاسکتا اور نہ بندے اُٹھانے سے انہیں چُپ کروایا جا سکتا ہے	چندے انقلاب جاسکتا بندے اٹھانے چپ کروایا
4	کس نے کھیل کھیلا ہے ، کس نے ہجر جھیلا ہے اب گزر گیا جاناں اس سوال کا موسم	کھیل کھیلا ہجر جھیلا گزر جاناں سوال موسم
5	جس کام کے بارے میں آپ سوچنا نہیں چھوڑ سکتے ،اس کام کو پایۂ تکمیل تک پہنچانے کیلئے کوشش کرنا بھی نہ چھوڑیئے -	کام سوچنا چھوڑ کام پایہ تکمیل پہنچانے کیلئے کوشش چھوڑیئے

Fig. 3. Sample tweets before and after preprocessing.

4 Experiments and Results

There are two approaches, the first one is the multi-label approach and the second one is the multi-class approach. Different machine learning models are implemented for comparison such as Logistic Regression, Naïve Bayes, Decision Tree, Random Forest,

Support Vector Machine, Extra tree Classifier, KNN Classifier, Ada Boost, and Bagging Classifier (Freund & Schapire et al,.1996).

4.1 Multi-label Approach

Multi-label classification (MLC) each data instance is associated with multiple binary class variables. There are different methods to solve a multi-label classification problem (Table 1 and Fig. 4).

Table 1. Accuracies with multi-label approaches.

Algorithm	Problem Transformation			Ensemble Classifiers
	Binary Relevance	Classifier Chains	Label Power Set	
SVM	45.33	45.66	55.85	55.51
KN	40.66	38.0	52.17	52.17
NB	36.33	36.33	41.13	38.46
DT	34.66	46.33	43.81	42.80
LR	48.0	48.0	55.18	55.18
RF	42.66	44.66	50.50	51.50
AdaBoost	44.66	47.0	54.51	55.18
BgC	44.66	47.33	51.17	50.83
ETC	40.66	42.0	44.14	32.44

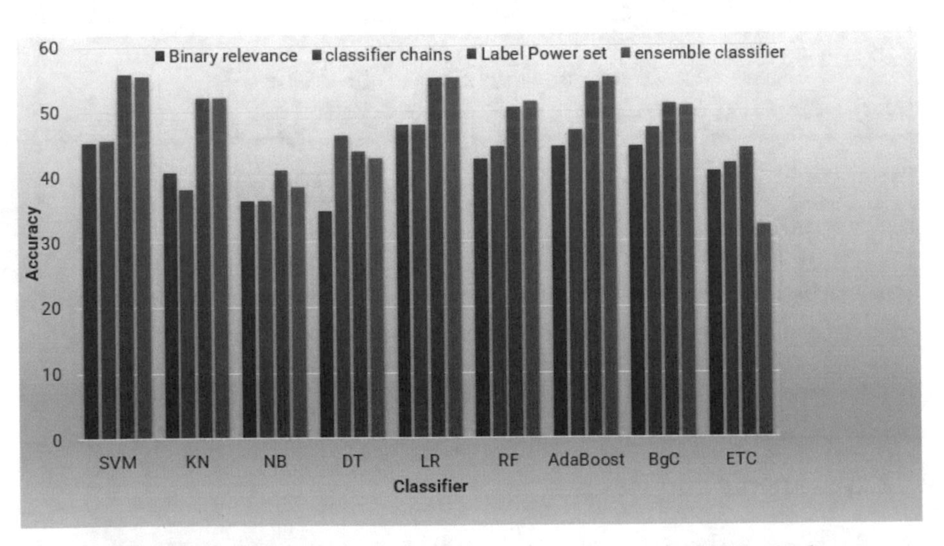

Fig. 4. The performance analysis of the models using multi-label approaches.

4.2 Multi-class Approach

Multi-class classification is a popular problem in supervised ML. It is a type of classification in which the machine should classify an instance as only one of three classes or more. The classified dataset for multi-class classification has entries as follows:

- HS 48
- CB 51
- OTH 28

4.3 Implementation Approach

In this section, we discuss abbot implementation approaches that are TF-IDF with 10-fold-cross validation, 5-fold-cross validation, and splitting data into training and testing data.

TF-IDF (Term Frequency-Inverse Document Frequency). The TF-IDF (term recurrence converse report recurrence) measurement evaluates a word's significance to a record in a bunch of archives. This is finished by duplicating two measurements: the occasions a word shows up in a report and the word's converse archive recurrence across a gathering of records. It has a wide scope of utilizations, including programmed text investigation and word scoring in Natural Language Processing AI strategies (NLP).

K-fold Cross-Validation. Cross-approval is a resampling strategy for assessing AI models on a little example of information. The cycle incorporates just a single boundary, k, which indicates the quantity 42 of gatherings into which a given information test ought to be separated. Thus, the interaction is much of the time alluded to as K overlap cross-approval. At the point when an exact incentive for k is indicated, it very well may be filled in for k in the model's reference, for instance, k = 10 for 10-crease cross-approval. In this approach, all labels are compared according to machine learning algorithms. P/N label, OFF/NOF label, TI/NT label, and CB/HS/OTH achieve the highest accuracy in the Naïve base, CB + OTH/HS label achieves the highest accuracy in SVM, and the last label HS + OTH/CB achieves the highest accuracy in extra tree classifier (Figs. 5 and 6).

Train Test Split Dataset. The train-test split is a methodology for assessing the exhibition of an AI calculation. It very well may be utilized for arrangement or relapse errands and can be utilized for any directed learning strategy. The strategy involves parting a dataset into two subgroups. The underlying subset used to fit the model is the preparation dataset. The dataset's feedback component is given to the model, which then, at that point, makes forecasts and analyzes them to the anticipated qualities. The subsequent subset isn't utilized to prepare the model; all things being equal, the dataset's feedback component is given to the model, which then, at that point, makes expectations and analyzes them to the anticipated qualities (Figs. 7 and 8).

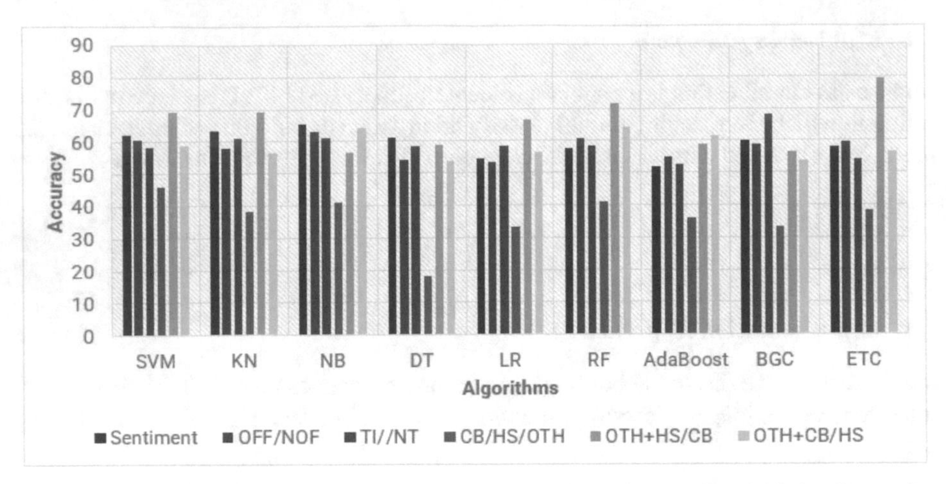

Fig. 5. The performance analysis of the models using TF-IDF (Train Test Splitting Dataset).

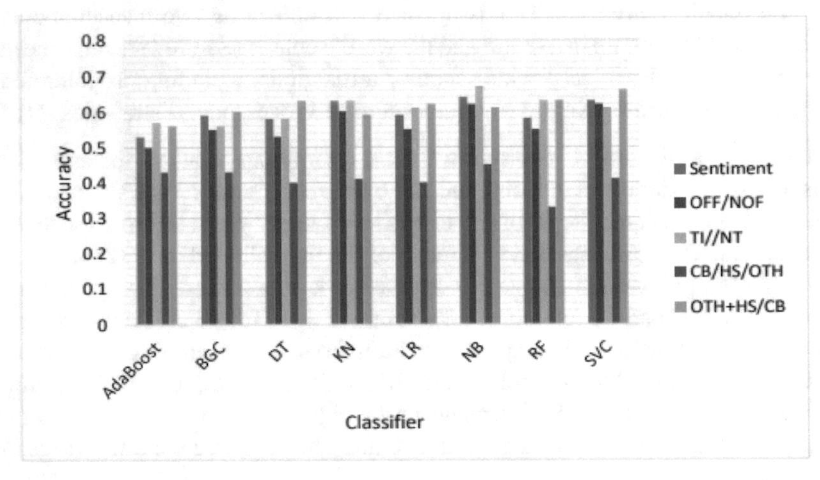

Fig. 6. The performance analysis of the models with 10 K-fold using TF-IDF.

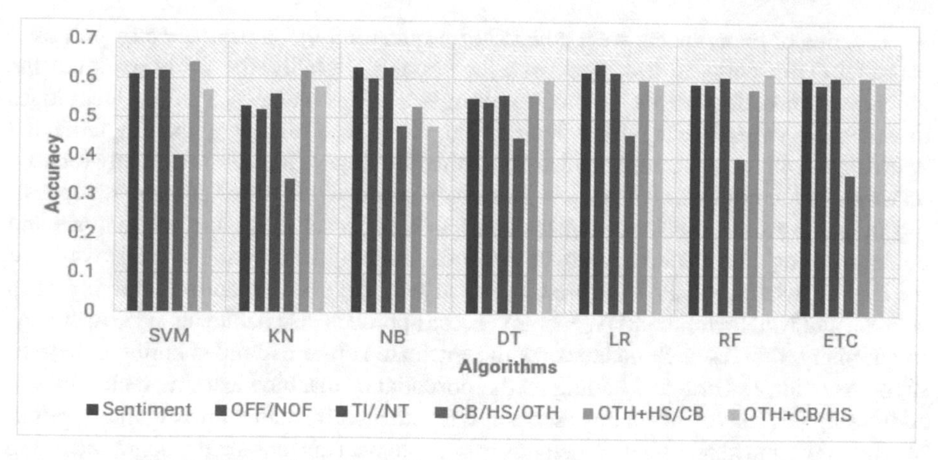

Fig. 7. The performance analysis of the models using Bag of Words.

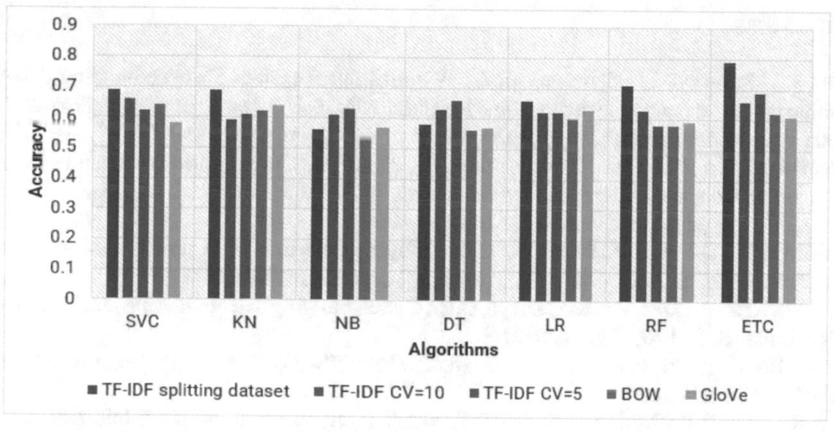

Fig. 8. The comparison between current implementation approaches.

5 Conclusion

In conclusion, the implementation of manual cyberbullying detection in the Urdu language presents distinct challenges, primarily due to the constraints of time and resources, as well as the scarcity of available research and tools. Unlike English, where various libraries for cyberbullying detection exist, Urdu lacks such resources, making the task more intricate. Within the realm of social media cyberbullying, particularly on the Twitter platform, this research leveraged Machine Learning techniques to address the issue. The dataset was prepared in two versions, one with sentiment labels (positive and negative) and the other with more intricate labels encompassing different aspects of cyberbullying. Through a collaborative labeling process and subsequent preprocessing, the data was refined for analysis. Employing Python, a range of Machine Learning algorithms and diverse feature extraction methods were employed.

A series of experiments were conducted, comparing the performance of proposed classifiers with classical Machine Learning models. Notably, the study explored the efficacy of three feature vector models alongside seven machine learning techniques in detecting cyberbullying from both Urdu datasets. Comparing the implemented approaches with previous sentiment analysis methodologies, this research solely focused on machine learning-based detection. The study revealed additional prospects, suggesting potential extensions involving lexicon-based approaches for further analysis and comparison against machine learning methods. Among the classifiers, ETC exhibited the highest accuracy at 79% followed by SVM with the GloVe feature vector approach at 64%, and KNN with the BOW feature vector approach also achieving 64% accuracy. In summary, this research underscores the intricate nature of implementing cyberbullying detection in Urdu and highlights the potential of machine learning techniques in addressing this issue. The findings serve as a foundation for future investigations, possibly incorporating lexicon-based methods to complement and augment machine-learning approaches.

References

Haidar, B., Chamoun, M., Serhrouchni, A.: A multilingual system for cyberbullying detection: arabic content detection using machine learning. Adv. Sci. Technol. Eng. Syst. **2**(6), 275–284 (2017). https://doi.org/10.25046/aj020634

Dinakar, K., Jones, B., Havasi, C., Lieberman, H., Picard, R.: Common sense reasoning for detection, prevention, and mitigation of cyberbullying. ACM Trans. Interact. Intell. Syst. **2**(3) (2012). https://doi.org/10.1145/2362394.2362400

Deb Barma, D., Dwvedi, V., Ballal, A.: Cyberbullying detection prevention on social networks. Int. J. Adv. Res. **8**(7), 87–90 (2020). https://doi.org/10.21474/ijar01/11266

Kini, P.M., Keni, A., Deepika, K.V, Divya, C.H.: Cyber-bullying detection using machine learning algorithms. **8**(8), 1966–1972 (2020)

Alexa—Top Sites in Pakistan—Alexa. (n.d.). https://www.alexa.com/topsites/countries/PK. Accessed 28 July 2020

Pawar, R., Raje, R.R.: Multilingual cyberbullying detection system. In: IEEE International Conference on Electro Information Technology, 2019-May (October), pp. 040–044 (2019). https://doi.org/10.1109/EIT.2019.8833846

Klomek, A.B., Sourander, A., Gould, M.: The association of suicide and bullying in childhood to young adulthood: a review of cross-sectional and longitudinal research findings. Can. J. Psychiat. **55**(5), 282–288 (2010)

Khattak, A., Asghar, M.Z., Saeed, A., Hameed, I.A., Asif Hassan, S., Ahmad, S.: A survey on sentiment analysis in Urdu: A resource-poor language. Egypt. Inform. J. (2020). https://doi.org/10.1016/j.eij.2020.04.003

Mercado, R.N.M., Chuctaya, H.F.C., Gutierrez, E.G.C.: Automatic cyberbullying detection in Spanish-language social networks using sentiment analysis techniques. Int. J. Adv. Comput. Sci. Appl. **9**(7), 228–235 (2018)

Ozel, S.A., Akdemir, S., Sarac¸, E., Aksu, H.: Detection of cyberbullying on social media messages in Turkish. In: 2nd International Conference on Computer Science and Engineering, UBMK 2017, pp. 366–370 (2017)

Mouheb, D., Abushamleh, M.H., Abushamleh, M.H., Al Aghbari, Z., Kamel, I.: Real-time detection of cyberbullying in Arabic twitter streams. In: 2019 10th IFIP International Conference on New Technologies, Mobility and Security, NTMS 2019 - Proceedings and Workshop, pp. 1–5 (2019a)

Mouheb, D., Ismail, R., Al Qaraghuli, S., Al Aghbari, Z., Kamel, I.: Detection of offensive messages in Arabic social media communications. In: Proceedings of the 2018 13th International Conference on Innovations in Information Technology, IIT 2018, pp. 24–29 (2019b)

Hussain, M.G., Al Mahmud, T., Akthar, W.: An approach to detect abusive Bangla text. In: 2018 International Conference on Innovation in Engineering and Technology, ICIET 2018, pp. 1–5 (2019). https://doi.org/10.1109/CIET.2018.8660863

Van Hee, C., et al.: Automatic detection of cyberbullying in social media text. PLoS ONE **13**(10), 1–22 (2018). https://doi.org/10.1371/journal.pone.0203794

Bilal, M., Israr, H., Shahid, M., Khan, A.: Sentiment classification of Roman Urdu opinions using Naïve Bayesian, decision tree and KNN classification techniques. J. King Saud Univ. Comput. Inf. Sci. **28**(3), 330–344 (2016). https://doi.org/10.1016/j.jksuci.2015.11.003

Cover, T., Hart, P.: Nearest neighbor pattern classification. IEEE Trans. Inf. Theory **13**(1), 21–27 (1967). https://doi.org/10.1109/TIT.1967.1053964

Usman, M., Shafique, Z., Ayub, S., Malik, K.: Urdu text classification using majority voting. Int. J. Adv. Comput. Sci. Appl. **7**(8), 265–273 (2016). https://doi.org/10.14569/ijacsa.2016.070836

Mukhtar, N., Khan, M.A.: Urdu sentiment analysis using supervised machine learning approach. Int. J. Pattern Recognit Artif Intell. **32**(2), 1–15 (2018). https://doi.org/10.1142/S0218001418510011

Khan, M.Y., Emad-ud-din, S.M., Junejo, K.N.: Harnessing English sentiment lexicons for polarity detection in Urdu tweets: a baseline approach. In: Proceedings - IEEE 11th International Conference on Semantic Computing, ICSC 2017, pp. 242–249 (2017)

Freund, Y., Schapire, R.E.: Experiments with a new boosting algorithm. In: Proceedings of the Thirteenth International Conference on International Conference on Machine Learning (ICML '96), pp. 148–156. Morgan Kaufmann Publishers Inc., San Francisco, CA, USA (1996)

Alhumoud, S., AlBuhairi, T., Altuwaijri, M.: Arabic sentiment analysis using WEKA a hybrid learning approach (2015). https://doi.org/10.5220/0005616004020408

Human Activity Recognition Using Convolutional Neural Networks

Omer Fawzi Awad[1], Saadaldeen Rashid Ahmed[2,3](✉), Atheel Sabih Shaker[4], Duaa A. Majeed[5], Abadal-Salam T. Hussain[6], and Taha A. Taha[7]

[1] Department of Surgery, College of Medicine, Tikrit University, Tikrit, Salahalden, Iraq
omer.fawzi@tu.edu.iq
[2] Artificial Intelligence Engineering Department, College of Engineering, Alayan University, Nasiriyah, Iraq
saadaldeen.ahmed@alayen.edu.iq, saadaldeen.aljanabi@bnu.edu.iq
[3] Computer Science Department, Bayan University, Erbil, Kurdistan, Iraq
[4] Computer Engineering Techniques, Baghdad College of Economic Sciences University, Baghdad, Iraq
atheel.sabih@baghdadcollege.edu.iq
[5] Aeronautical Engineering Department, Baghdad University, Baghdad, Iraq
[6] Department of Medical Instrumentation Techniques Engineering, Technical Engineering College, Al-Kitab University, Altun Kupri, Kirkuk, Iraq
[7] Unit of Renewable Energy, Northern Technical University, Kirkuk, Iraq

Abstract. This research addresses human activity recognition (HAR) using a deep learning framework. Particularly convolutional neural networks (CNNs) to identify and categorize human actions using sensor data. Employing a complete dataset containing numerous actions. It features reclining, sitting, standing, and different walking modes. The study applies CNN models to attain the best precision and accuracy. The models' strong performance in these CNN. Constraints in dataset diversity and size could impact real-world applicability. This article asks for more investigation and design. It combines additional sensors and concentrates emphasis on real-time HAR systems. The paper illustrates the potential of deep learning models in HAR.

Keywords: CNN · Classification · Real-time · Resnet101 · HAR

1 Introduction

Human Activity Recognition (HAR) is a key topic in current technology. Its purpose is to classify and explain human behavior. It can recognize behavior by applying computational activity. The relevance of HAR lies in its varied approach to the digital world. Encompassing a wide range of uses. It includes healthcare and fitness monitoring. as well as smart surroundings and ubiquitous computing.

J. Rasheed et al. (Eds.): FoNeS-AIoT 2024, LNNS 1035, pp. 258–274, 2024.
https://doi.org/10.1007/978-3-031-62871-9_20

1.1 Context and Importance

The notion of HAR has gained substantial interest owing to its capability. It can alter various industries [1]. HAR systems have been augmented using machine learning methods. It can accurately detect and comprehend human activities. With a surprising degree of precision [2]. These systems employ datasets that consist of sensors. This dataset can be read from many sources. Illustrations of accelerometers and other sensors that may be used in the environment.

1.2 The Uses of HAR Are Manifold

HAR contributes greatly to patient monitoring. It can be used for rehabilitation tracking in the healthcare sector [3]. It supports adaptive systems by identifying user behaviors and improving resource consumption. HAR assists in performance analysis, training optimization, and injury avoidance in the sports industry [4].

1.3 Research Problems and Objectives

In classical machine learning algorithms for HAR, there are significant hurdles in attaining increased accuracy, robustness, and scalability [5]. When presented with complicated real-world events and various activity patterns [6]. This research aims to solve these constraints by applying deep learning approaches.

1.4 The Significance of Deep Learning in HAR

Deep learning provides a paradigm shift in HAR due to its intrinsic power to automatically extract complicated hierarchical patterns. It can represent raw sensor data. The application of neural network topologies, such as convolutional neural networks (CNNs) and recurrent neural networks (RNNs). It promotes increased identification and understanding of human behaviors.

2 Literature Review

HAR has undergone major breakthroughs. It was spurred by the incorporation of deep learning technologies. This section critically analyzes current research initiatives in HAR. This section focuses on the application of different deep learning models and approaches. It addresses evaluating their strengths, weaknesses, and extant gaps in the literature. A research study fully examining many elements of HAR highlighted the growth of deep-learning neural networks for identifying human behaviors [7]. The author highlights their usefulness in managing sensor data from cell phones. Utilizing deep learning neural networks, particularly those designed for smartphone sensors [8]. This study exhibits greater accuracy in activity recognition. One research study did a detailed survey primarily concentrating on deep learning in HAR [9]. Their investigation examined and contrasted multiple deep learning architectures of convolutional neural networks (CNNs). They also considered recurrent networks (RNNs) when simulating temporal Mainly, they focused on geographical connections within activity data. They accepted these models while noting their limits in addressing complicated real-world settings.

Deep learning architectures combining CNNs, and Long Short-Term Memory (LSTM) networks have received attention. One of the researchers presented a CNN-LSTM hybrid model [10]. They showed a greater ability to spot sequential patterns in HAR datasets. Attention-based models helped with sophisticated activity detection by concentrating on important characteristics within the data [11]. The investigation of sensor-based datasets for HAR remains an important feature, offering a thorough review of the literature [12–14]. They stressed the availability and diversity of sensor-based datasets for training and assessing HAR models and highlighted issues in ensuring the robustness and generalizability of models trained on such datasets due to data-gathering procedures [15].

Recent investigations have gone into real-time HAR applications machine learning and deep learning-based techniques [16–21]. They are concentrating on real-time human activity identification. They underlined the necessity for lightweight and efficient models for real-world deployment. The author examines the trade-offs between accuracy and computing efficiency. Some gaps abound in the literature. Limited research has been undertaken on developing machine-learning algorithms for HAR [22, 23]. It remains an area warranting additional exploration.

This section presents a complete summary of recent literature. It demonstrates the growth of deep learning techniques. Discussed applications and the present research landscape in HAR. The following sections delve deeper into the methodology and experimentation. Undertaken in this study, aiming to address existing gaps and challenges identified in the reviewed literature. As shown in Table 1.

Table 1. Compressive study relative to literature review.

Ref	Domain	Method	Findings	Limitations
Jobanputra et al. [7]	HAR	Survey	Evolution of deep learning in HAR, smartphone sensor utilization, overall efficacy	The general overview lacks a specific model analysis
Ronao et al. [8]	Smartphone sensors	Deep learning neural networks	Tailored models for smartphone sensors, improved accuracy in activity recognition	Limited scope, focus on smartphone-specific data

(*continued*)

Table 1. *(continued)*

Ref	Domain	Method	Findings	Limitations
Gu et al. [9]	Deep learning in HAR	Comparative analysis of deep learning architectures	Effectiveness of CNNs, RNNs, limitations in complex real-world scenarios	Lack of exploration in real-world scenarios, model scalability issues
Mutegeki et al. [10]	CNN-LSTM architecture	A hybrid model combining CNNs and LSTMs	Superiority in recognizing sequential patterns in HAR datasets	Lack of scalability, potential overfitting
Murahari et al. [11]	Attention-based models	Utilization of attention mechanisms	Improved recognition by focusing on salient features in the data	Complexity in model implementation, computation overhead
Hayat et al. [15]	Sensor-based datasets	Literature review	Diversity and availability of sensor-based datasets, challenges in robustness	Variability in sensor quality, issues in generalization
Kulsoom et al. [24]	Real-time HAR	Deep learning-based models	Focus on efficiency for deployment, real-time applications	Challenges in achieving high accuracy, and scalability concerns
Ramasamy Ramamurthy et al. [2]	Evolving ML techniques	Research on dynamic environments	Limited research in evolving techniques for dynamic environments	Lack of adaptability to dynamic changes, scalability issues
De-La-Hoz-Franco et al. [25]	Elderly care	Deep learning in elderly care	Applications in elderly care, need for robust models	Lack of robustness in diverse elderly settings, limited datasets

3 Methodology

The technique employed for this work includes a systematic strategy to leverage deep learning technologies for HAR. Utilizing human activity recognition data. To guarantee extensive analysis and effective modeling. The process may be separated into numerous crucial components.

3.1 Dataset Description

The study employed an enormous Human Activity Recognition (HAR) dataset [26]. This dataset has X samples produced by a thorough collection method. It comprised 30 participants aged between 19 and 48 years. Each participant completed six unique activities (WALKING, WALKING_UPSTAIRS, WALKING_DOWNSTAIRS, SITTING, STANDING, LAYING). These women are wearing a Samsung Galaxy S II smartphone attached to their waist. The integrated accelerometer and gyroscope sensors gathered 3-axial linear acceleration and 3-axial angular velocity. These devices have a steady rate of 50 Hz. The dataset was partitioned randomly into two subsets: 70% for training data creation and 30% for testing purposes. It ensures a full model assessment as shown in Fig. 1.

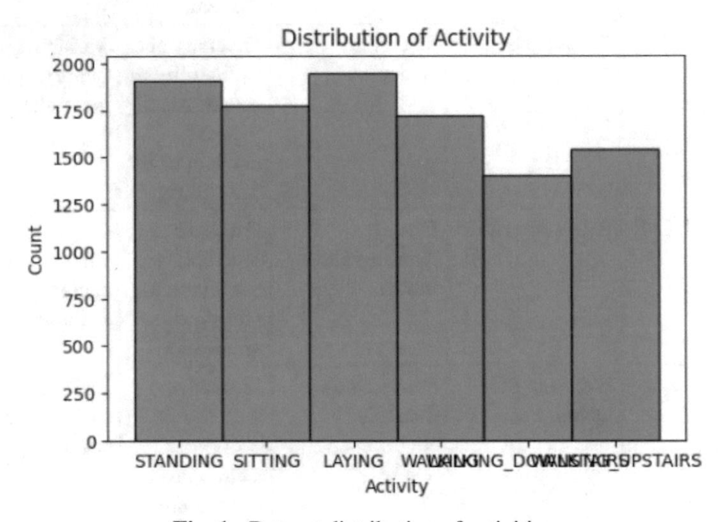

Fig. 1. Dataset distribution of activities.

3.2 Data Collection and Preprocessing

The sensor signals underwent thorough preprocessing processes, including:

- Noise Filtering: Applying noise filters to ensure signal clarity and accuracy.
- Fixed-width Sliding Windows: Segmentation into fixed-width sliding windows of 2.56 s with a 50% overlap. It has 128 readings per window.

- Signal Separation: Using a Butterworth low pass filter to distinguish gravity and body motion. These components are within the acceleration signal, resulting in body acceleration.
- Feature Extraction: Calculating a variety of time and frequency domain variables from the signals. It generates a vector of informative characteristics.

3.3 Experimental Analysis with PCA

The dataset features were derived predominantly from the accelerometer. This dataset gyroscope 3-axial raw signals. tAcc-XYZ and tGyro-XYZ were acquired at a constant rate of 50 Hz. These signals underwent a series of transformations:

- Filtering: Median and low-pass Butterworth filters (3rd order) with a corner frequency of 20 Hz. This was applied to remove noise.
- Signal Separation: Differentiating the body's linear acceleration and angular velocity signals. It is used for additional low-pass Butterworth filters.
- Derivation: Transforming linear acceleration and angular velocity into Jerk signals (tBodyAccJerk-XYZ and tBodyGyroJerk-XYZ). It calculates their magnitudes (tBodyAccMag, tGravityAccMag, tBodyAccJerkMag, tBodyGyroMag, tBodyGyroJerkMag).
- Frequency Domain Conversion: Employing the Fast Fourier Transform (FFT). It converts selected signals into frequency domain representations (fBodyAcc-XYZ, fBodyAccJerk-XYZ, fBodyGyro XYZ, fBodyAccJerkMag, fBodyGyroMag, and fBodyGyroJerkMag).

This comprehensive dataset encapsulates a diverse range of features capturing both temporal and spatial aspects. It detects human activities and offers a robust foundation for model development and evaluation in HAR.

3.4 Deep Convolutional Neural Networks for HAR

The deep convolutional neural network architecture is applied for HAR. It involves a succession of convolutional layers and fully linked layers as shown in Fig. 2. Here's a thorough summary of the methodology:

The mathematical representation for HAR in the context of deep learning. It involves various equations and operations used within neural network architectures. One common mathematical representation using a neural network for HAR can be denoted as:

Given an input data set, $X = \{x_1, x_2, \ldots\ldots\ldots, x_N\}$, where N represents the number of features. a set of corresponding activity labels $Y = \{y_1, y_2, \ldots\ldots\ldots\ldots, y_N\}$. It can be denoting different human activities. The goal is to learn mapping functions. Here, f that relates X to Y.

This operation of a neural network can be represented as:

$$z = W \cdot X + b \tag{1}$$

where X shows the input features, W denotes the weights matrix, b is the bias vector, while z represents the output of the weighted. And the sum of operations.

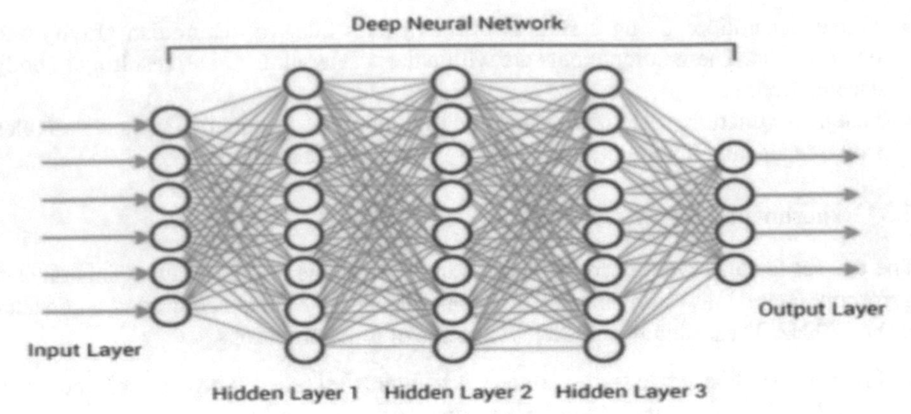

Fig. 2. Deep Learning for human activity recognition.

This operation is followed by an activation function.

$$a = \sigma(z) \tag{2}$$

where a denotes the output. Here, after passing through the activation function.

For classification tasks in HAR, the SoftMax function is commonly used at the output layer:

$$P(Y = j|X) = \frac{e^{zj}}{\sum_{k=1}^{k} e^{zk}} \tag{3}$$

where $P(Y = j|X)$ represents the probability that the input X belongs to class j given the learned parameters, zj represents the output of class j, and k denotes the total number of classes.

The model is trained by optimizing the parameters (weights and biases) using techniques like backpropagation. Gradient descent to minimize the error between predicted and actual labels.

This mathematical representation provides a foundation for how neural networks learn and classify human activities. Specific architectures and variations may include additional layers and activations for HAR tasks.

Convolutional Layer Operations. The network starts with convolutional layers that execute convolutional operations on incoming sensor data. It extracts features using a collection of learnable filters and kernels. The output of the convolutional layer is generated by applying an activation function. Include the weighted total of input values and biases. This process continues via additional convolutional layers. It builds hierarchical feature representations.

Pooling Layer Operation. Max-pooling procedures are undertaken to summarize neighboring outputs by picking the maximum value. It depends on the set of inputs. This pooling procedure minimizes spatial dimensions. This assists in feature selection and extraction.

Hierarchical Feature Extraction. The layered convolutional layers constitute a hierarchical feature extractor. This hierarchical representation enables the network to detect progressively more complicated. It has discriminative properties important to the HAR challenge as shown in Fig. 3.

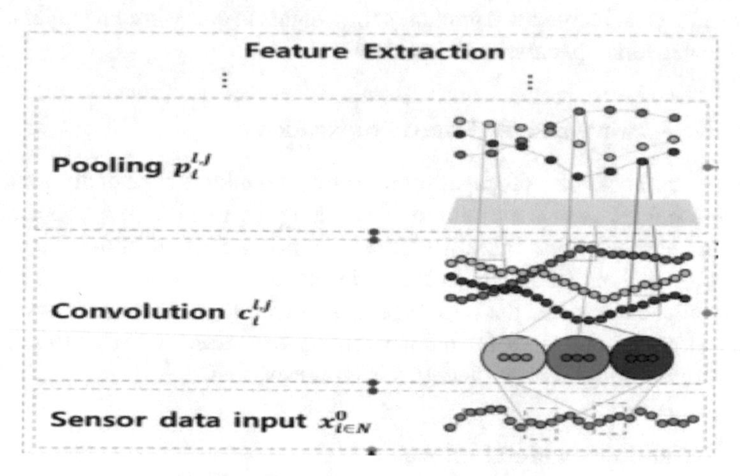

Fig. 3. Hierarchical Feature Extraction.

Fully Connected and SoftMax Layers. Flattened feature vectors derived from the pooling layers are fed into fully linked layers. These layers further process the characteristics using weights and biases. It creates intermediate outputs. The topmost layer, a SoftMax classifier, gives inferred activity classifications. And it applies the Soft-Max function to the last layer's outputs as shown in Fig. 4.

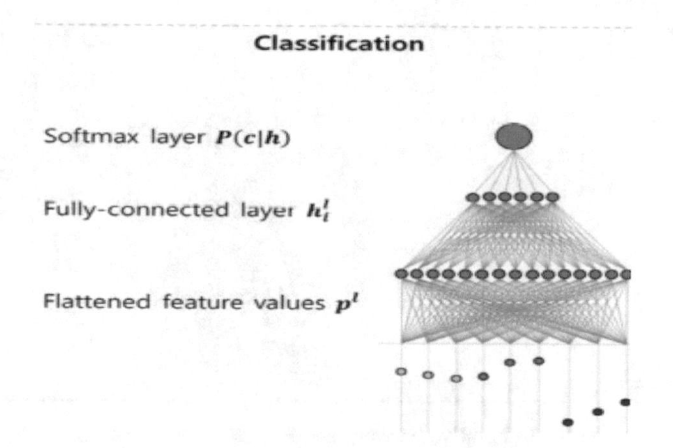

Fig. 4. Fully Connected and SoftMax Layers.

3.5 Training and Optimization

The network performs forward propagation using derived equations. It offers incorrect values. Stochastic gradient descent (SGD) lowers error costs through weight changes. in their mini-batches of sensor training data. Backpropagation modifies weights in completely linked layers. It computes gradients for weights in convolutional layers. That can optimize the network's parameters iteratively.

3.6 Iterative Forward and Backward Propagation

The forward and backward propagation procedures continue repeatedly until a preset stopping threshold is met. An example of attaining the maximum number of epochs. This repeated approach fine-tunes the network's parameters. It boosts its capacity to perceive human actions efficiently as shown in Fig. 5.

This technique elucidates the various processes and stages needed in deploying deep convolutional neural networks for human activity detection. It stresses the network's power to learn meaningful characteristics from sensor data.

3.7 Training and Evaluation Protocol

Training and Evaluation Protocol. The study implemented a meticulous training and evaluation protocol harnessing a CNN-based architecture. It is tailored for HAR. The models underwent rigorous training, optimizing hyperparameters to attain optimal performance.

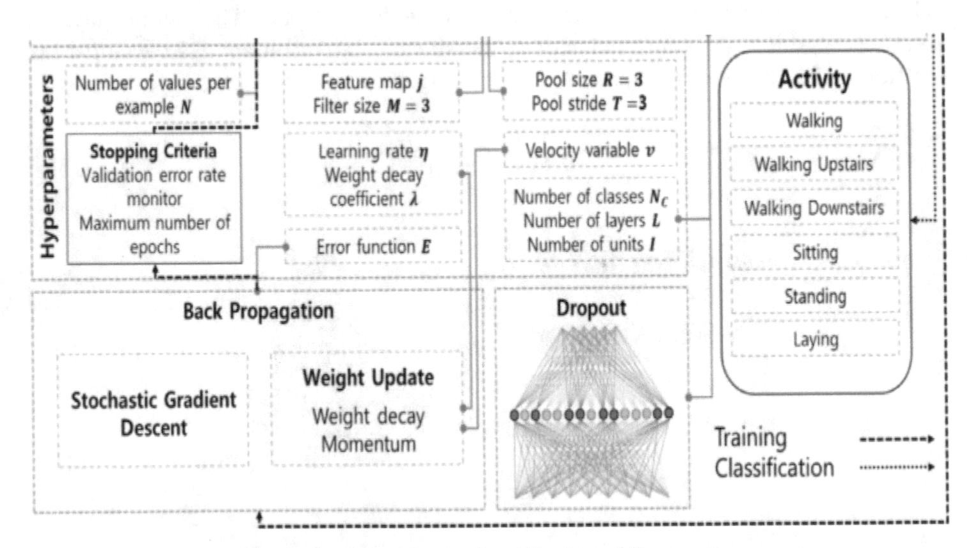

Fig. 5. Iterative Forward and Backward Propagation

Model Training. The CNN models were systematically trained on a meticulously curated training dataset. It can leverage an array of convolutional layers for feature extraction. These models were trained iteratively with optimized hyperparameters. It ensures the convergence of the training process. Regular validation is included. While training to prevent overfitting. It ensures the generalization of unseen data as shown in Fig. 6.

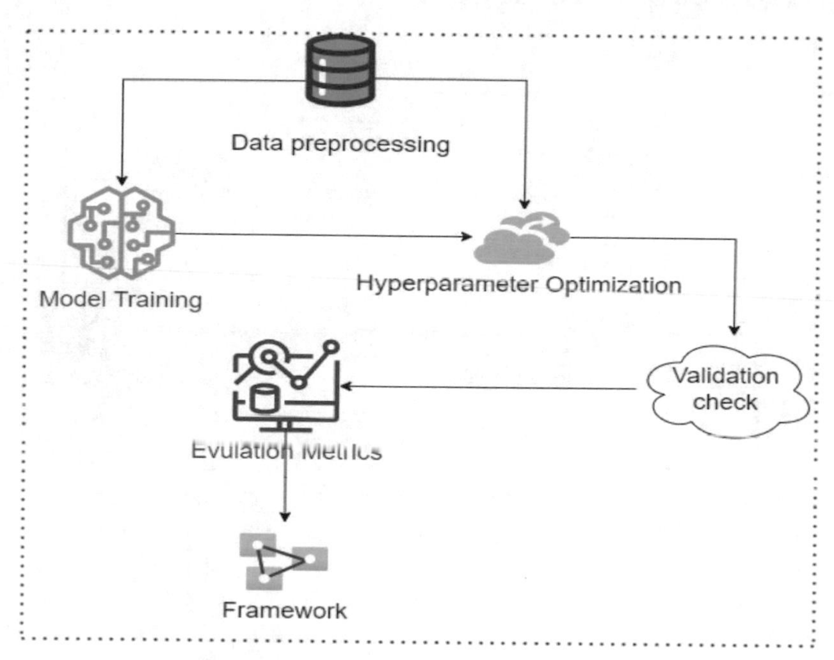

Fig. 6. Training and Evaluation Protocol Workflow for HAR.

3.7.1 Evaluation Metrics

Comprehensive assessment metrics were used. When it comes to analyzing the efficacy of the trained models appropriately. The performance of the models was tested using different measures. Including:

Accuracy: It assesses the overall accuracy of the forecasts.
Precision: It indicates properly anticipated positive cases.
Recall: Recall captures genuine positive instances.
F1-score: offering a harmonic meaning between accuracy and recall. It can balance the trade-off between them.

Performance Analysis. These assessment criteria jointly permitted an in-depth investigation of varied human activities. It can be obtained from sensor data. The combination of various measurements produced a sophisticated picture. There are problems in precisely recognizing diverse activities.

3.8 Proposed Framework for HAR

This is a well-constructed technique. It contains optimized hyperparameters. It is a regular validation. It gives extensive assessment criteria. It can fortify a robust architecture for HAR. It leverages deep learning. The CNN-based architecture functioned as a formidable instrument. It draws intricate patterns from sensor data. It offers a complete comprehension of human action (see Fig. 7).

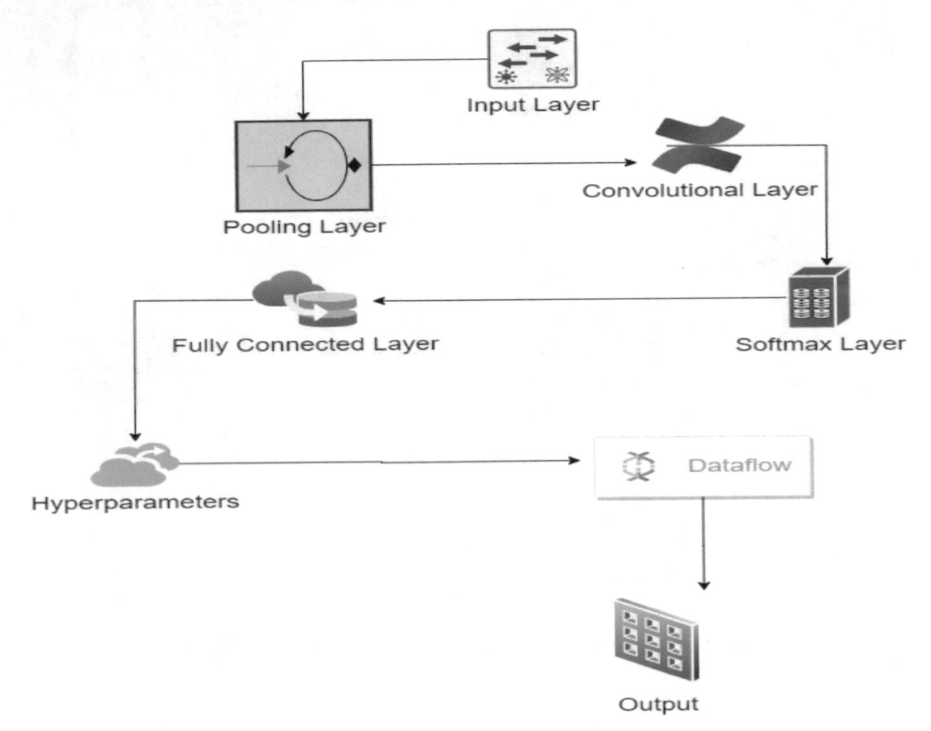

Fig. 7. Convolutional Neural Network (CNN) Framework for Human Activity Recognition.

This organized approach: 1. Combining optimum hyperparameters; 2. Rigorous model training; 3. Detailed assessment metrics; 4. Emphasizing the resilience of CNNs in interpreting complicated patterns within sensor-derived data for human activity detection. The systematic methodology provides deep insights regarding CNN performance across multiple activity categories. It can indicate both strengths and limits. This thorough technique highlights the usefulness of deep learning. Models in capturing the intricacies of human actions, increasing the area of human activity identification.

4 Result and Discussion

Performance assessment is especially related to convolutional neural network (CNN) models. There are various displays with unique tendencies. In Fig. 8, the cumulative distribution across datasets displays the distinctive behavior of CNNs. These models display great variability in peak performance. Between the best-performing CNN technique and the least-performing DNN, there's about a 15% mean f1-score difference and 12% on DG. The premier CNN model and b-LSTM-S outperform the present state-of-the-art by 4%. mean.

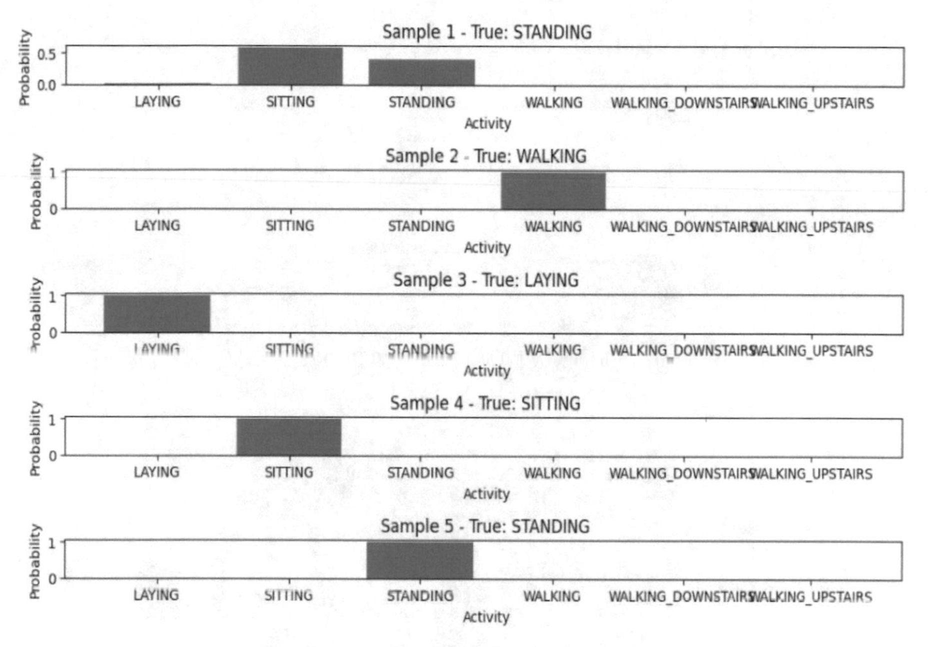

Fig. 8. Sample probability True Standing.

F1-score (1% weighted f1-score) on OPP, showcasing a substantial improvement over prior CNN-based models reported in the literature. It scores more than 5% mean f1-score and weighted f1-score. The effectiveness of CNNs in recognizing detailed patterns from raw sensor data gives exciting potential for real-time HAR applications without the necessity for data segmentation.

Figures 9, 10, 11 and 12 vividly display the varied performance distributions among datasets. It affirms CNN models' diverse performance profiles. The observed difference between different CNN designs shows Certain configurations can demonstrate limited efficacy (e.g., 20% on PAMAP2), while others indicate reasonably constant performance. The shorter CNN designs tend to outperform the deeper ones. It shows potential benefits in researching generative pretraining approaches for deeper networks. These findings underscore the importance of fine-tuning parameters, especially in CNN architectures, to enhance recognition performance.

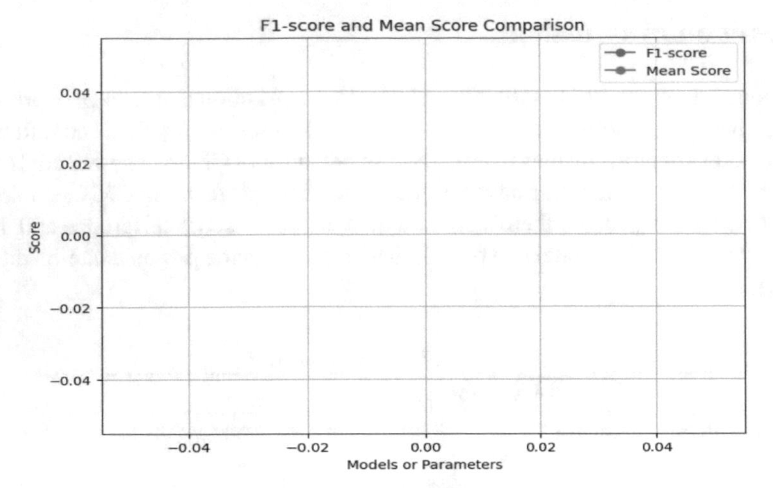

Fig. 9. F1-score and Mean Score Comparison

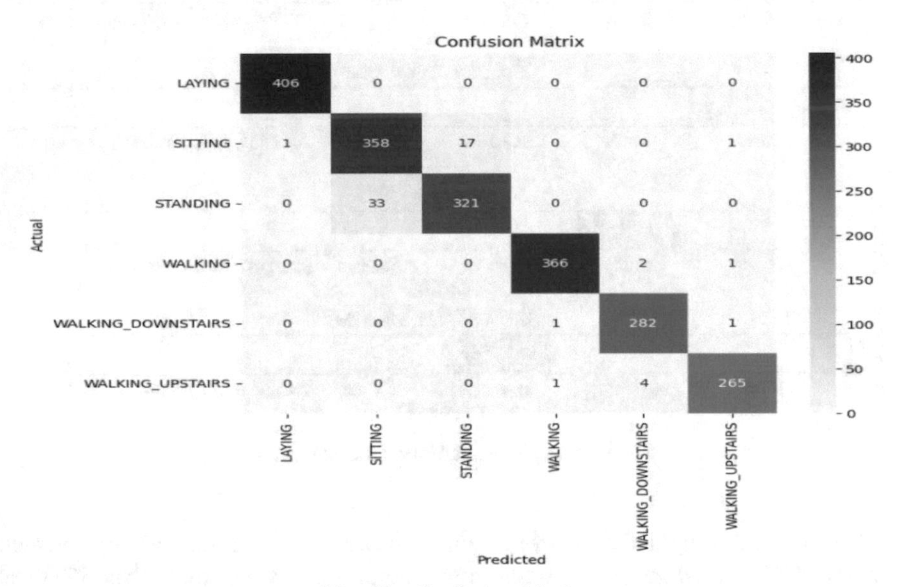

Fig. 10. Confusion Matrix

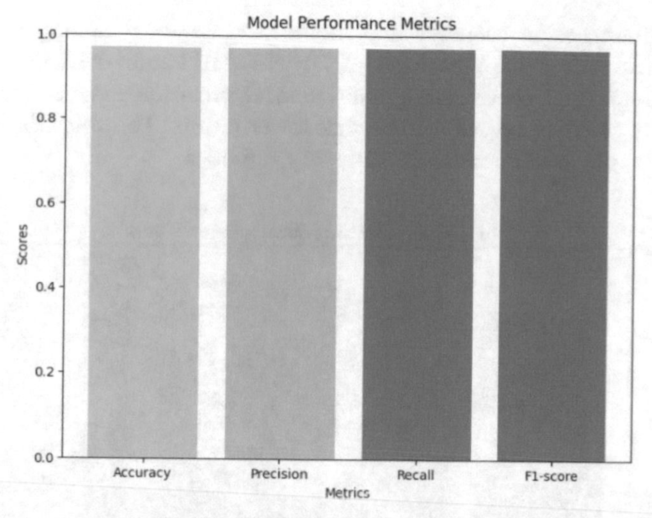

Fig. 11. Performance distribution.

Class	Precision	Recall	F1-Score	Support
LAYING	1.00	1.00	1.00	406
SITTING	0.92	0.95	0.93	377
STANDING	0.95	0.91	0.93	354
WALKING	0.99	0.99	0.99	369
WALKING_DOWNSTAIRS	0.98	0.99	0.99	284
WALKING_UPSTAIRS	0.99	0.98	0.99	270
Accuracy			0.97	2060
Macro Average	0.97	0.97	0.97	2060
Weighted Average	0.97	0.97	0.97	2060

Fig. 12. Class-wise classification Report with an accuracy of 97%.

5 Discussion

The observed findings reveal remarkable performance in identifying human actions. It uses a convolutional neural network (CNN)-based model. The classification report exhibits strong accuracy and recall across multiple activities. Showing the resilience of the model in discriminating activities. Examples of LAYING, SITTING, STANDING, WALKING, WALKING_DOWNSTAIRS, and WALKING UPSTAIRS.

The acquired accuracy of 97% reflects the model's competency. It can correctly classify the activities based on the sensor data. Actions like laying and walking displayed remarkable accuracy, recall, and F1 scores. It is consistently reaching or approaching 1.00 (see Fig. 13). This demonstrates the model's ability to properly detect these actions without misunderstanding or misclassification.

Minor differences in accuracy and memory were noted among various activities. Whereas behaviors like WALKING_DOWNSTAIRS and WALKING_UPSTAIRS demonstrated strong F1-scores. It suggests balanced precision and recollection in tasks. An example of standing revealed somewhat lower results. This disparity could reflect inherent challenges in differentiating stationary activities.

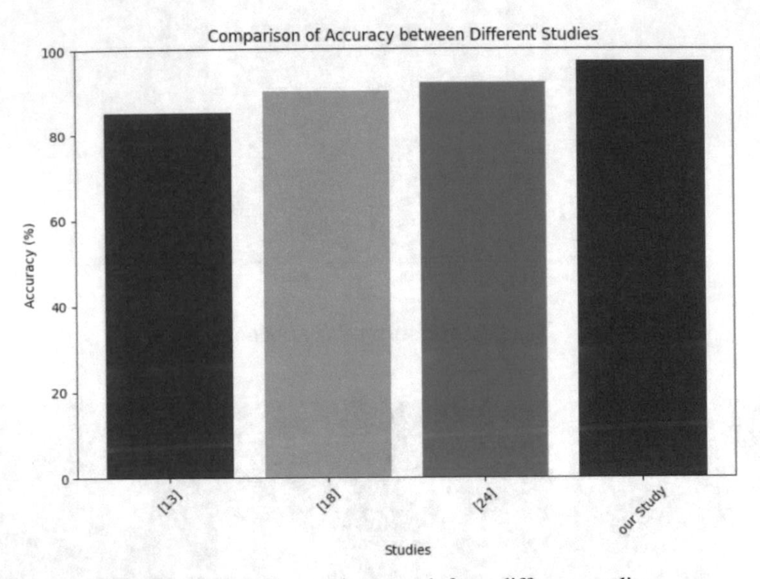

Fig. 13. Result Comparison graph from different studies.

The macro and weighted averages for accuracy, recall, and F1 scores are at 0.97. It signified a harmonious balance across all pursuits. This balanced performance demonstrates the model's generalizability to accurately recognize a varied variety of human activities based on sensor data.

The CNN-based model for properly identifying and categorizing diverse human actions. More study and fine-tuning could be essential to handling complex differences in accuracy and memory among different activity types.

6 Conclusion

This research went into the application of deep learning-based CNN models for human activity recognition (HAR). It showcases their commendable performance in accurately categorizing diverse activities using sensor data. The acquired high precision and accuracy across actions including laying, sitting, standing, and various types of walking exceeded past standards, verifying the developments in HAR technology. Constraints in dataset diversity and size might impair the model's real-world usefulness. Future initiatives in HAR might examine hybrid model architectures, incorporate new sensors, and go into real-time application deployment. It aims to create more adaptable and efficient recognition systems for practical use. This work emphasizes the potential of deep learning in HAR. It can advocate for continuous research to strengthen these systems for greater real-world deployment.

References

1. Zheng, X., Wang, M., Ordieres-Meré, J.: Comparison of data preprocessing approaches for applying deep learning to human activity recognition in the context of industry 4.0. Sensors **18**(7), 2146 (2018)
2. Ramasamy Ramamurthy, S., Roy, N.: Recent trends in machine learning for human activity recognition—a survey. Wiley Interdiscip. Rev. Data Min. Knowl. Discov. **8**(4), e1254 (2018)
3. Tang, C.I., Perez-Pozuelo, I., Spathis, D., Mascolo, C.: Exploring contrastive learning in human activity recognition for healthcare (2020). arXiv preprint arXiv:2011.11542
4. Host, K., Ivašić-Kos, M.: An overview of human action recognition in sports based on computer vision. Heliyon **8**(6), e09633 (2022)
5. Jegham, I., Khalifa, A.B., Alouani, I., Mahjoub, M.A.: Vision-based human action recognition: an overview and real world challenges. Forensic Sci. Int. Digit. Investig. **32**, 200901 (2020)
6. Yadav, S.K., Tiwari, K., Pandey, H.M., Akbar, S.A.: A review of multimodal human activity recognition with special emphasis on classification, applications, challenges and future directions. Knowl.-Based Syst. **223**, 106970 (2021)
7. Jobanputra, C., Bavishi, J., Doshi, N.: Human activity recognition: a survey. Procedia Comput. Sci. **155**, 698–703 (2019)
8. Ronao, C.A., Cho, S.B.: Human activity recognition with smartphone sensors using deep learning neural networks. Expert Syst. Appl. **59**, 235–244 (2016)
9. Gu, F., Chung, M.H., Chignell, M., Valaee, S., Zhou, B., Liu, X.: A survey on deep learning for human activity recognition. ACM Comput. Surv. (CSUR) **54**(8), 1–34 (2021)
10. Mutegeki, R., Han, D.S.: A CNN-LSTM approach to human activity recognition. In: 2020 International Conference on Artificial Intelligence in Information and Communication (ICAIIC), pp. 362–366. IEEE, February 2020
11. Murahari, V.S., Plötz, T.: On attention models for human activity recognition. In: Proceedings of the 2018 ACM International Symposium on Wearable Computers, pp. 100–103, October (
12. Beddiar, D.R., Nini, B., Sabokrou, M., Hadid, A.: Vision-based human activity recognition: a survey. Multimed. Tools Appl. **79**(41–42), 30509–30555 (2020)
13. Sousa Lima, W., Souto, E., El-Khatib, K., Jalali, R., Gama, J.: Human activity recognition using inertial sensors in a smartphone: an overview. Sensors **19**(14), 3213 (2019)
14. Gupta, S.: Deep learning based human activity recognition (HAR) using wearable sensor data. Int. J. Inf. Manag. Data Insights **1**(2), 100046 (2021)
15. Hayat, A., Morgado-Dias, F., Bhuyan, B.P., Tomar, R.: Human activity recognition for elderly people using machine and deep learning approaches. Information **13**(6), 275 (2022)
16. Xu, C., Chai, D., He, J., Zhang, X., Duan, S.: InnoHAR: a deep neural network for complex human activity recognition. IEEE Access **7**, 9893–9902 (2019)
17. Murad, A., Pyun, J.Y.: Deep recurrent neural networks for human activity recognition. Sensors **17**(11), 2556 (2017)
18. Hammerla, N.Y., Halloran, S., Plötz, T.: Deep, convolutional, and recurrent models for human activity recognition using wearables (2016). arXiv preprint arXiv:1604.08880
19. Thu, N.T.H., Han, D.S.: HiHAR: a hierarchical hybrid deep learning architecture for wearable sensor-based human activity recognition. IEEE Access **9**, 145271–145281 (2021)
20. San, P.P., Kakar, P., Li, X.L., Krishnaswamy, S., Yang, J.B., Nguyen, M.N.: Deep learning for human activity recognition. In Big Data Analytics for Sensor-Network Collected Intelligence, pp. 186–204. Academic Press (2017)

21. Arasteh, B., Seyyedabbasi, A., Rasheed, J., Abu-Mahfouz, A.M.: Program source-code re-modularization using a discretized and modified sand cat swarm optimization algorithm. Symmetry **15**(2), 401 (2023)
22. Kumar, P., Chauhan, S.: Human activity recognition with deep learning: overview, challenges & possibilities. CCF Trans. Pervasive Comput. Interact. **339**(3), 1–29 (2021)
23. Farooq, M.S., et al.: A conceptual multi-layer framework for the detection of nighttime pedestrian in autonomous vehicles using deep reinforcement learning. Entropy **25**(1), 135 (2023)
24. Kulsoom, F., Narejo, S., Mehmood, Z., Chaudhry, H.N., Butt, A., Bashir, K.A.: A review of machine learning-based human activity recognition for diverse applications. Neural Comput. Appl. **34**, 18289–18324 (2022)
25. García-Restrepo, J.K., Ariza-Colpas, P.P., Butt-Aziz, S., Piñeres-Melo, M.A., Naz, S., De-la-hoz-Franco, E.: Evaluating techniques based on supervised learning methods in casas Kyoto dataset for human activity recognition. In: Saeed, K., Dvorský, J., Nishiuchi, N., Fukumoto, M. (eds.) Computer Information Systems and Industrial Management. CISIM 2023. LNCS, vol. 14164, pp. 253–269. Springer, Cham (2023). https://doi.org/10.1007/978-3-031-42823-4_19
26. Human Activity Recognition Dataset. https://www.kaggle.com/datasets/arunasivapragasam/human-activity-recognition-dataset/data. Accessed 10 Nov 2023

Robot Hand-Controlled by Gyroscope Sensor Using Arduino

Fatima Ghali[1](✉) and Atheer Y. Ouda[2,3]

[1] Computer Techniques Engineering Department, Imam Alkadhim University College, Nasiriyah, Thi-Qar, Iraq
lecdhi97@alkadhum-col.edu.iq

[2] Department of Computer Sciences, College of Education for Pure Science, University of Thi-Qar, Nasiriyah, Iraq

[3] Information and Communication Technology Research Group, Scientific Research Center, Al-Ayen University, Nasiriyah, Thi-Qar, Iraq
atheer@alayen.edu.iq

Abstract. The robotic arm's movements are primarily guided by an accelerometer sensor coupled with an artificial intelligence algorithm. This paper proposes a gesture recognition-based 6DOF robotic arm controller that utilizes a gyrometer and accelerometer to enhance stability and detect human arm rotational gestures. Equipped with object grasping capabilities, the arm utilizes a sensor fusion approach, combining data from a low-cost MEMS chip housing a 3-axis accelerometer and 3-axis gyroscope, to determine the angular position of objects. This cost-effective solution effectively detects human arm gestures and precisely identifies object orientation. The gyroscope provides gesture orientation data to assess dynamic gesture behavior. An artificial intelligence algorithm evaluates the gesture data, guiding the robotic arm's training process. The widely used Kalman filter accurately pinpoints the human arm's position. Wireless communication between the human hand and the robotic arm is facilitated through the IEEE standard Zigbee protocol interface. This enables seamless interaction, with the robotic arm's movements mirroring human arm gestures in real-time, akin to a shadow mode. The artificial arm's response to human arm gestures is remarkably swift. This control strategy offers a more intuitive and straightforward approach compared to traditional methods like joystick control, making it well-suited for industrial applications. Developed on the Arduino IDE platform, this robotic arm's versatility extends to various platforms, including embedded systems and intelligent peripherals. Thus, extensive testing of the robotic arm yielded promising results.

Keywords: Voltage Regulator · Motor Sheild · MPU6050 Sensor · DC Motor · RF Transmitter

---

The original version of the chapter has been revised. The affiliation of Fatima Ghali has been corrected. A correction to this chapter can be found at
https://doi.org/10.1007/978-3-031-62871-9_36

J. Rasheed et al. (Eds.): FoNeS-AIoT 2024, LNNS 1035, pp. 275–285, 2024.
https://doi.org/10.1007/978-3-031-62871-9_21

1 Introduction

Controlling robotic arms is a complex task that requires a deep understanding of robotics, control systems, and software engineering. In many industrial settings, robotic arm control is still done manually, which is time-consuming and error-prone. This is why many universities and researchers are working on developing new technologies to make robotic arm control simpler and smarter [1]. Human-robot interaction (HRI) is a unique and increasingly important field of research and development. HRI encompasses a wide range of interactions between humans and robots, from simple tasks like opening doors to complex collaborations in manufacturing and healthcare [2]. The nature of HRI varies depending on the specific purpose and application [3].

Here are a few examples of potential HRI applications inspired by the sci-fi TV channel Robot Combat League and the augmentation of traditional robotic surgery (Da Vinci system). Indeed, the interaction with machines has increased day by day with the evolution of technology [4]. From simple tasks like switching on lights from a smartphone to complex ones like controlling vehicles, machines have become an integral part of our lives. Traditional methods of human-machine interaction (HMI) like remote controls and joysticks have limitations, such as requiring physical contact with the device and being limited to certain types of interactions. Gesture control, on the other hand, offers a more natural and intuitive way to interact with machines, breaking down language barriers and expanding the possibilities for HMI [5, 6]. The Arduino Uno is a microcontroller board that can be used to read data from sensors and control devices. In this case, the Arduino Uno is being used to detect motion and transmit data via transceivers [7, 8].

The gesture-controlled robot manipulator system concept was shown [9]. To manipulate the robotic arm, they employed a band-style device that attaches to the elbow and thumb. They concluded that the robotic arm was rotated to a specific angle using the elbow and that the thumb handled the gripping movement.

The system of gesture-controlled image processing was designed [10]. They were assigning tasks to specific photos using MATLAB. They converted the webcam-captured image into a binary format before performing the specified tasks [11]. Similar work is done in [12]. However, our study uses a Gyro-accelerometer with an MPU6050 sensor that is not only efficient but also cost-effective.

2 Circuit Components

The primary objective of this project is to develop a Gyro-accelerometer-based control system for a robotic hand utilizing the MPU6050 sensor.

The components of the practical circuit are as follows:

2.1 Arduino Uno Board

2.2 The Structure of the Robotic Hand

It is a hand with five fingers. Arduino L293.

2.3 Tools

- 3D printer (or access to one)
- Soldering iron
- Screwdriver and other basic workshop tools
- Super glue and/or epoxy
- Drill and pliers for clean support structure
- Wire strippers/cutter

2.4 Parts

- 3D printing filament (black and white PLA was used for this project, but ABS is also a suitable option.
- At least 8 SG90 servos, as some may be damaged during the modification process.
- Fishing line
- Thin elastic cord
- Four springs
- Heat shrink tube
- Spare wire
- Braided nylon sleeve
- Arduino Uno
- Feed sources: Two batteries, one for feeding the Arduino with a voltage of 9 V and the other for feeding the voltage and volts of the sensor.

2.5 Motor Sheild

Arduino shields are one of the most amazing things because they are easy to use when dealing with installation or programming. We find that DC motors consume a very high current that the microcontroller (Arduino) cannot the motor is directly supplied with power because its current is very limited, so we resort to a circuit called a driving circuit the function of the motors is to provide sufficient current for the motor to operate utilizing commands coming from the controller, and thus, we have electrical insulation to protect the controller from damage It also protects the controller from currents in reverse, it contains a diode operating in the forward biased state only and prevents any currents in the case of reverse bias. in this paper, we used the L293 circuit called (Motor Driver L293). Backed by a very powerful library that allows this Shield to handle DC motors, Servo motors, and stepper motors.

2.6 MPU6050 Senso

It is a Micro-Electro-Mechanical System (MEMS) device that integrates a 3-axis accelerometer and a 3-axis gyroscope, enabling the measurement of acceleration, velocity, direction, displacement, and various other motion parameters as in Fig. 1 and 2.

Fig. 1. MPU6050 sensor

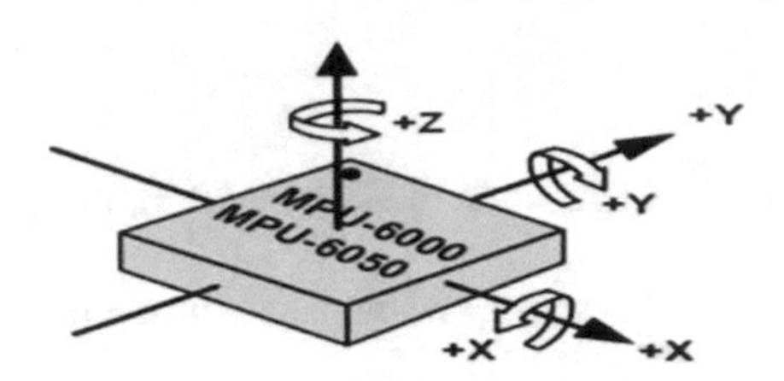

Fig. 2. MPU6050 sensors work.

2.7 DC Motor

Direct Current (DC) motors are electromechanical devices that convert electrical energy into mechanical energy, specifically rotational motion. They operate solely on DC power supplies and are characterized by their high speed-to-size ratio and ability to rotate in both clockwise and counterclockwise directions.

2.8 Servo Motor

It is a DC Motor equipped with an electronic circuit to precisely control the direction of rotation of a shaft, the motor, its placement, and a gearbox. There are two types of servo motors: The first is the standard servo in which the motor can be rotated from (0–180) in both directions clockwise and counterclockwise. The second is a continuous servo motor and the motor can rotate from (0–360) It also rotates in both directions. In this research, we used a type 1 servo motor to control the motion-sensitive Ultrasonic.

2.9 Voltage Regulator

It is an electronic piece made of semiconductors that is used to supply voltage to the circuit fixed by varying the resistance according to the load, these pieces have a common designation. It is 78XX as a positive linear voltage regulator and 79XX as a negative linear voltage regulator specified the output with the last two numbers, for example, 7805 outputs + 5 V, and 7905 outputs −5 V.

3 Methodology and Results

The idea of the Robot hand controlled by a gyroscope sensor could help patients open and close objects using computers, smartphones, or anything that requires precise grasping. To achieve this, the tilt sensor, which is a sensor that measures the surface movement of the user's body, is connected to the Arduino controller that houses the robotic hand actuators (see Fig. 3).

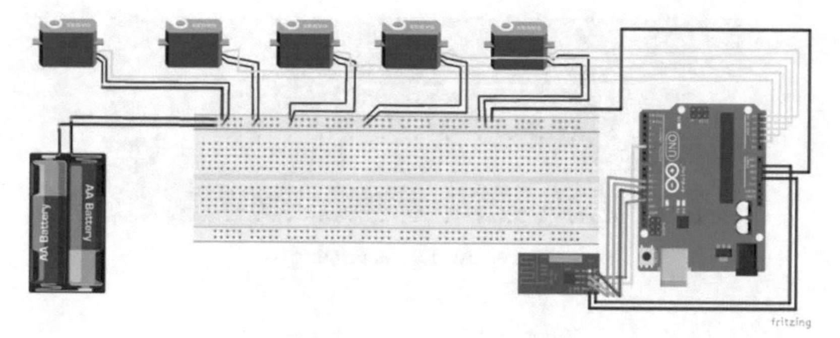

Fig. 3. Circuit setup for the robotic hand

It is an interesting project that can be of great benefit. It needs an Arduino and a tilt sensor, and the sensor will be used to measure the bending force. To start the design, we determine the number of fingers needed to design the robot, then draw a design for it using CAD software and take the necessary dimensions. 3D printers can be used to produce the various parts of the hand. Servo motors can then be used to activate the 3D-printed hand, with each motor attached to a part of the hand. After that, the inclination sensor is installed (this can be done by attaching the sensor to the inner side of the joint to determine the degree of extension of the fingers) and the other end of the sensor is connected to the Arduino. Download a program for Arduino, where the program contains code that displays and measures the number of degrees of rotation of the robotic mechanism. This signal will automatically adjust the degree of curvature of each finger in the robot based on the values transmitted from the sensor. In the meantime, other robots will have to do some programming related to achieving this adaptation for sensing and control. When the required steps are completed, test the bot and make sure everything is working properly. We also work on improving the design when considering more factors such as weight, balance, level of motion, and other factors that may be important.

3.1 Project Simulation Design Using the Software TINKERCAD

Step 1: Arm Parts Printing. The hand's design was created using Fusion 360. To 3D print a hand, first import the design file into Fusion 360. Then, export each component as an STL file, which is a standard as in Fig. 4, format for 3D printing. Next, prepare the fingertips for printing using dual extrusion, a technique that allows printing with two different materials simultaneously. This ensures that the black fingertips are not too thin and print successfully. Alternatively, if there is a single extrusion printer, consider designing the black fingertips to be thicker. Moreover, it can use either PLA or ABS plastic but print with the support structure. To reduce printing time, consider printing the larger components like the palm and forearm at a 0.3 mm layer height, while printing the smaller, more intricate pieces like the fingers at a 0.1 mm layer height. Ultimately, the choice of layer height depends on the desired level of detail and the specific capabilities of your 3D printer.

Fig. 4. Arm Parts Printing

Step 2: Hacking the Servos (part 1) – Continuous Rotation Servo. The rotational range of servos is constrained by the limitations of the potentiomcters they employ. These potentiometers, devices that exhibit variable resistance as they rotate, can only rotate up to 180 degrees, effectively limiting the servo's rotational range to the same extent. To circumvent the rotational limitation imposed by the potentiometer, the servo's casing can be carefully opened and the plastic stopper located on the topmost gear can be removed. After that, a hole should be drilled along the gear's axis to enable it to rotate freely without being hindered by the D-pin mount of the potentiometer, as depicted in the fourth image above. Regular servos typically have a limited rotational range of approximately 180 degrees due to the constraints of the potentiometer they employ. However, by modifying the servo's internal components, it is possible to convert it into a continuous rotation servo, enabling unrestricted rotation in both directions (as seen in Fig. 5).

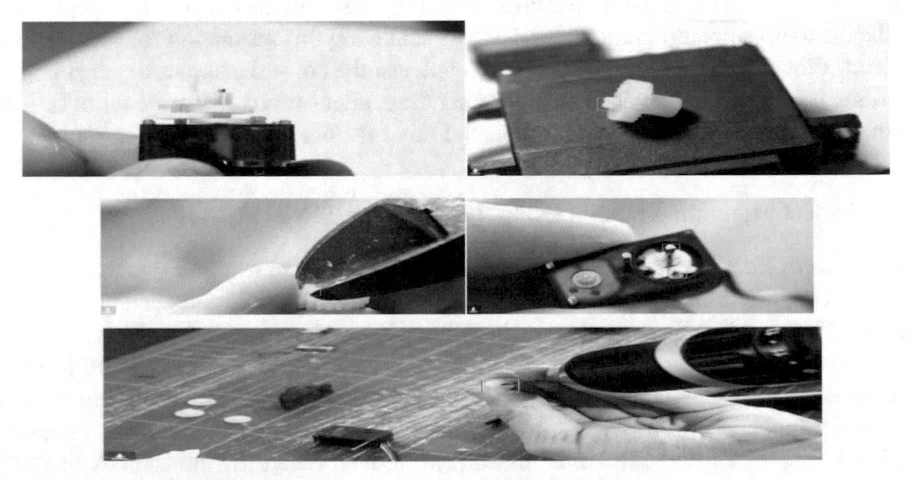

Fig. 5. Hacking the servos (part 1) – continuous rotation servo

3.2 Learn About the Interface of the Program

At the top of the screen, there is the largest menu bar in the screen called the editing window, which acts as a drawing window where it is placed and delivered ingredients. The small space on the left of the screen is called the view window and displays the view General drawing Completely.

Charting. Firstly, we must do is get the parts of the libraries we need in the Chart. Secondly, the components are placed in the placing components. Thirdly, the components are connected to form the desired circle. Lastly, the program runs for the best results.

Conventional servos typically have a restricted rotational range of about 180 degrees due to the limitations imposed by their internal potentiometers. However, this limitation can be overcome by modifying the servo's internal components, converting it into a continuous rotation servo capable of unrestricted rotation in both directions.

Step 3: Modifying Servos for Continuous Rotation – Integrating. Figure 6 shows the compact DC Motors (Part 2).

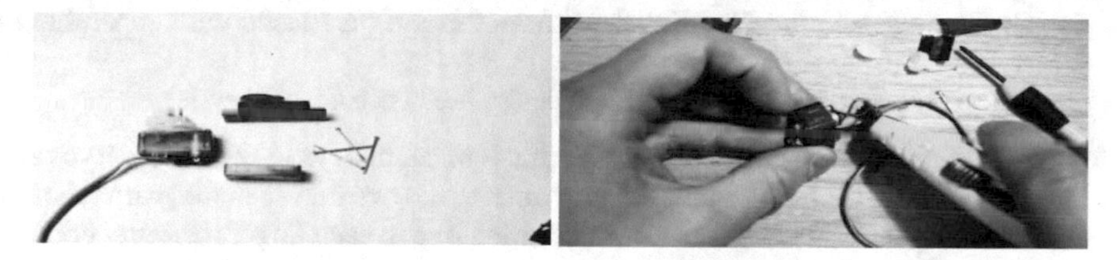

Fig. 6. Hacking the servos (part 2), 2-in-1 compact DC motors.

Step 4: Servo Steps.

1. Open the servos and remove the potentiometer and chip.
2. The steps on how to cut two of the blue plastic case components and remove the stopper on the top gear.
3. Reassemble the case as shown in the image.
4. Cut scrap rods for the gears with no axle.
5. Repeat steps for the second servo.
6. Glue the servos together.
7. The solder extension wires to the DC motors are a relatively straightforward process, but it is important to follow the steps carefully to avoid damaging the motors or the wire.
8. Test the motors.
9. Use the saved potentiometers in the knuckles for reading.
10. Repurposing a potentiometer and a DC motor as a servo using Arduino involves modifying the existing connections and incorporating Arduino code to control the motor's position based on potentiometer readings.

Step 5: Hacking the Servos (Part 3) - The Remaining Servos. Two servos will be dedicated to controlling the wrist's rotational and pitch movements, with the remaining servos serving as backups in case of failure. The black potentiometer is the only component that needs to be retained; the rest can be set aside.

Step 6: Assemble the Fingers.
Before proceeding, use a drill to clear out the support structure and ensure that the string can pass through each hole.

1. Apply glue to the finger segments and attach them. Cut the elastic to the desired length. The length of the elastic will depend on the size of the object you are attaching it to.
2. For testing, thread the fishing line through the holes on the underside of a finger for testing purposes and remove it afterward.
3. Repeat for the main 4 fingers.
4. Solder extension wires to the 3 potentiometers saved earlier and differentiate each wire with a color code.
5. Gently turn each of the pots to a 90° angle and place one pot between each of your index, middle, and ring fingers.
6. Attach the fingers to the 3D-printed palm and fasten the elastic with superglue or epoxy.
7. Carefully route all the wires through the designated hole near the wrist.

Step 7: Build the Hand – Shorten Step. Attach black pulleys to modified servos and tie the fishing line to each of the Pulleys. Next, screw the two halves of the palm together to secure the fingers in place. Then, attach a fishing line to each finger for better control.

Step 8: Make the Wrist.

1. Secure all four springs using eight screws, ensuring the screw heads are wide enough to prevent them from slipping through the holes.
2. Attach each spring to the forearm and palm by carefully screwing in both sides.
3. Put the 3D-printed sphere or ping pong ball in between the springs to reduce friction.
4. The steps on how to connect the armatures to the two SG90 servos and rotate both servos to a 90° angle, along with images:
5. Affix two lengths of fishing line to each servo motor, ensuring that one end of each line is securely fastened to an extremity of the armature.
6. With the servos securely fastened to the base, proceed to meticulously guide the fishing line through each designated aperture shown in Fig. 7.

Step 9: Forearm and Wiring. Instead of saying "Route the wires carefully through the designated tube located on the side of the first forearm piece", we could simply say "Thread wires through the tube on the first forearm piece". Also, instead of listing all the pictures and their descriptions, we summarize the main steps and just mention relevant pictures as needed. For example, we might say "Wire the modified servos as shown in Fig. 6 or connect the power plug from the battery to the corresponding socket on the forearm".

Figure 8 and 9 show the resultant images of hand closed and hand open, respectively.

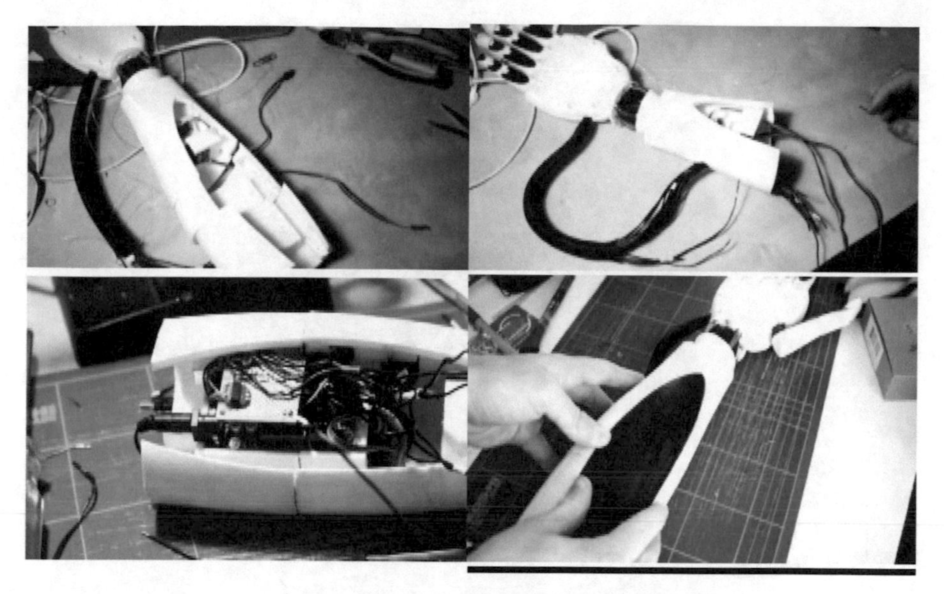

Fig. 7. Make the wrist.

Fig. 8. Hand closed.

Fig. 9. Hand open.

4 Conclusion

It can be concluded that the idea of graduation research is to manufacture a robotic hand that works with a gyroscope sensor and Arduino. The study involves developing a robot capable of manipulating and controlling a human hand. The robot is likely to require advanced sensors and software to accurately identify and replicate the movements of the human hand. The project could have applications in a variety of industries, from manufacturing and construction to healthcare and entertainment. The development of such a robot could lead to significant advances in human-robot interaction and collaboration. As with any robotics project, there will likely be challenges around cost, safety, and ethical considerations that must be carefully considered and addressed throughout the development process.

References

1. Mihara, I., Yamauchi, Y., Doi, M.: A real-time vision-based interface using motion processor and applications to robotics. Syst. Comput. Japan. **34**, 10–19 (2003). https://doi.org/10.1002/scj.10203
2. Yahyaoui, A., et al.: Performance comparison of deep and machine learning approaches toward COVID-19 Detection. Intell. Autom. Soft Comput. **37**, 2247–2261 (2023). https://doi.org/10.32604/iasc.2023.036840
3. Waldherr, S., Romero, R., Thrun, S.: Gesture based interface for human-robot interaction. Auton. Robot. **9**, 151–173 (2000). https://doi.org/10.1023/A:1008918401478
4. Cevik, T., Cevik, N., Rasheed, J., Asuroglu, T., Alsubai, S., Turan, M.: Reversible logic-based hexel value differencing—a spatial domain steganography method for hexagonal image processing. IEEE Access. **11**, 118186–118203 (2023). https://doi.org/10.1109/ACCESS.2023.3326857
5. Aleotti, J., Skoglund, A., Duckett, T.: Position teaching of a robot arm by demonstration with a wearable input device. In: Intelligent Manipulation and Grasping, pp. 459–464 (2004)

6. Farooq, M.S., et al.: A conceptual multi-layer framework for the detection of nighttime pedestrian in autonomous vehicles using deep reinforcement learning. Entropy **25**, 135 (2023). https://doi.org/10.3390/e25010135
7. Perrin, S., Cassinelli, A., Ishikawa, M.: Gesture recognition using laser-based tracking system. In: Sixth IEEE International Conference on Automatic Face and Gesture Recognition, 2004. Proceedings, pp. 541–546. IEEE. https://doi.org/10.1109/AFGR.2004.1301589
8. Song, Y., Shin, S., Kim, S., Lee, D., Lee, K.H.: Speed estimation from a tri-axial accelerometer using neural networks. In: 2007 29th Annual International Conference of the IEEE Engineering in Medicine and Biology Society, pp. 3224–3227. IEEE (2007). https://doi.org/10.1109/IEMBS.2007.4353016
9. Yang, J., et al.: A 3D hand-drawn gesture input device using fuzzy ARTMAP-based recognizer. In: WMSCI 2005 - The 9th World Multi-Conference on Systemics, Cybernetics and Informatics, Proceedings, vol. 7, pp. 270–275 (2005)
10. Murakami, K., Taguchi, H.: Gesture recognition using recurrent neural networks. In: Conference on Human Factors in Computing Systems – Proceedings, pp. 237–242 (1991). https://doi.org/10.1145/108844.108900
11. Hirzinger, G., Bals, J., Otter, M., Stelter, J.: The DLR-KUKA success story. IEEE Robot. Autom. Mag. **12**, 16–23 (2005). https://doi.org/10.1109/MRA.2005.1511865
12. Vadlamudi*, S., Kumar, D.N., Kumar, G.S.: Hand gesture controlled robot using arduino and MPU6050. Int. J. Recent Technol. Eng. (IJRTE). **9**, 777–779 (2020). https://doi.org/10.35940/ijrte.d9546.059120

Instant High Starting Current Protection System for Induction Motor Based Reducing the Starting Speed

Saadaldeen Rashid Ahmed[1,2](✉), Mohammed Fadhil[2], Ravi Sekhar[3], Abadal-Salam T. Hussain[4], Nilisha Itankar[3], Jamal Fadhil Tawfeq[5], Taha A. Taha[6], and Shouket A. Ahmed[4]

[1] Artificial Intelligence Engineering Department, College of Engineering, Alayan University, Nasiriyah, Iraq
saadaldeen.ahmed@alayen.edu.iq
[2] Computer Science Department, Bayan University, Erbil, Kurdistan, Iraq
saadaldeen.aljanabi@bnu.edu.iq
[3] Symbiosis Institute of Technology (SIT) Pune Campus, Symbiosis International (Deemed University) (SIU), Pune, Maharashtra 412115, India
ravi.sekhar@sitpune.edu.in
[4] Department of Medical Instrumentation Techniques Engineering, Technical Engineering College, Al-Kitab University, Altun Kupri, Kirkuk, Iraq
[5] Department of Medical Instrumentation Technical Engineering, Medical Technical College, Al-Farahidi University, Baghdad, Iraq
[6] Unit of Renewable Energy, Northern Technical University, Kirkuk, Iraq

Abstract. This research, titled Instant High Starting Current Protection System for Induction Motor by Reducing the Starting Speed, presents a simple inverter topology by using a phase AC supply to drive three-phase induction machines. The main objective is to design a starter for an induction motor using a soft switching device, and the starter will control the current during the starting condition by reducing the starting speed. The PWM generator for this research uses 555 timers, which provide a square wave frequency to the gate driver circuit. The simulation was done part by part using Multisim and PSIM software. The simulation result shows that the three-phase induction machine can operate by using a single-phase AC supply, and the frequency of the PWM signal can be adjusted by inserting a variable resistor into the control circuit and finally reducing the high starting current for the induction motor.

Keywords: Induction Motor · High Starting Current · Pulse-width Modulation · Inverter

1 Introduction

This research is about designing a starter for an induction motor. Induction motors are already widely used in the industry, and it is costly. Every induction machine has a rated current, which is defined as the maximum current limit that the motor can carry. If the

J. Rasheed et al. (Eds.): FoNeS-AIoT 2024, LNNS 1035, pp. 286–294, 2024.
https://doi.org/10.1007/978-3-031-62871-9_22

current is drawn more than that, the machine can get blown. However, the induction motor needs a high current at the start. This is because the induction motor needs more torque to start running. So, the problem occurs because, during the starting time, the spike can be very high and can damage the motor [1–3].

Protect the induction motor from too high a starting current; a starter needs to be used. The starter will reduce the high starting current. It does not mean that the starting current will be low. It was still high, but not too high until it damaged the motor. Nowadays, electrical machines are widely used in industry. The electrical machine is used to ensure the work can be done in time and to reduce human resources in the industry. There are several types of electrical machines we are using today, such as DC motors and AC motors. An induction motor is a type of asynchronous AC motor. The induction motor needs a starter to start. The starter will reduce the high starting current for the induction motor. When an induction motor uses direct-on-line (DOL) to start up, it will draw a very high current to the stator, in order of 5 to 9 times the full load current. This will cause damage to the motor. Due to the disadvantages of DOL, several starters are built up to start the induction motor. [4].

In this research, the idea is to use a soft-switching device as the starter for the induction motor. Some advantages of using soft-switching devices as starters are lower switching losses due to smaller overlap of switch voltage and current, lower $\frac{dv}{dt}$ and $\frac{di}{dt}$ and thus lower voltage spike and EMI emissions, higher reliability due to reduced stresses on the switching components, reduced voltage and current ratings for the devices, and more minor reactive elements [5–11].

The main objective of this research is to develop a starter for an induction motor that will effectively control the current during the starting process. Additionally, the research aims to achieve the following objectives: To design an inverter circuit incorporating a soft switching device. To develop a control circuit that will act as a controller for the soft-switching device. Collectively, these objectives aim to enhance the performance and efficiency of the induction motor starter by implementing advanced switching techniques and efficient control mechanisms.

2 Methodology

2.1 Block Diagram

A block diagram design for the entire system. This system will start from a single-phase 240 V, 50 Hz power source, and it will be rectified using a half-bridge rectifier and then filtered by the split capacitor. Output from the half-bridge rectifier will be supplied into the inverter circuit. On the other side, step-down transformers are used to step down the power supply into the range that can be powered into a PWM generator circuit. PWM signal will be generated and supplied to the gate driver circuit that will toggle the output for three-phase inverters to be operated. A phase induction motor will be connected to the inverter.

2.2 Power Supply Module Design

Figure 1 shows a design for the entire system. This system will start from a single phase 240 V, 50 Hz power source, and it will be rectified using a half-bridge rectifier and then filtered by the split capacitor. Output from the half-bridge rectifier will be supplied into the inverter circuit. On the other side, step-down transformers are used to step down the power supply into the range that can be powered into a PWM generator circuit. PWM signal will be generated and supplied to the gate driver circuit that will toggle the output for three-phase inverters to be operated. A phase induction motor will be connected to an inverter.

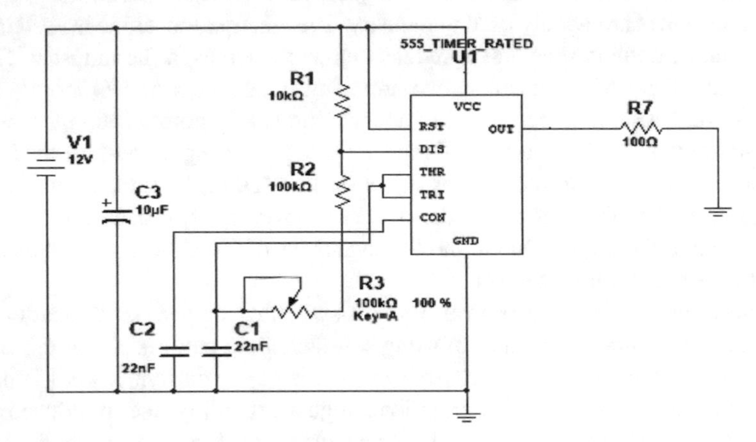

Fig. 1. Design for the entire system.

2.3 PWM Generator Circuit

In this research design, 555 timers will be used as the main IC in the PWM generator circuit. In this design, 555 timers are operating in ASTABLE operation [7]. The connection can be seen in Fig. 2. Here, the control pin is connected to a 22nF ceramic capacitor. This connection is made so the timer is available to be controlled or varied. Then, the pin threshold and trigger are connected with a 100k variable resistor and 22nF ceramic. With this connection, the frequency of these 555 timers can be varied by using the variable resistor. The discharge pin will act as the controller for the output pin. During the charge time, output will become high, and during discharge time, output will become low. So, theoretically, the output of the 555 timers will produce a square waveform.

2.4 Gate Driver Circuit

For this research design, a decade counter CD4017 will be used for the gate driver circuit and connection, as shown in Fig. 2. CD4017 is a 5-stage Johnson Counter with ten outputs. Input includes a clock, a reset, and a clock inhibit signal. Schmitt trigger action in the clock input circuit provides pulse shaping that allows unlimited clock input pulse rise and fall time. A high reset signal clears the counter to its zero count [8]. So, as shown in Fig. 2, input from the 555 timer will be connected to pin 14. The output will be connected to pin 2 and pin 7. There is a reason why we need to skip one clock for the output. That is because we are using a power transistor (IGBT) in the inverter circuit. IGBT is known as a soft switching device, and the turn-on time and turn-off time will be speedy. However, when the IGBT is turned off, it does not entirely turn off. This reaction is called a turn-off delay time [2]. So, to let the power transistor fully turn off, we are skipping one clock at the output side. Then, pin ten is connected to the reset pin (pin 15). It means whenever output at pin 10 is high, the counter will be reset and start to count from output one again.

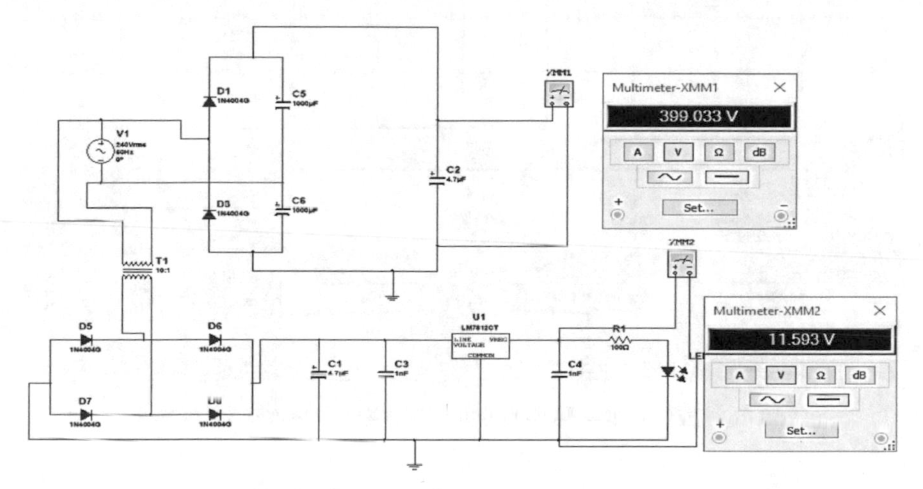

Fig. 2. Simulation for power supply module.

2.5 Simulation Result

Simulation for the PWM generator has been done, and the result is taken at two conditions: when the variable resistor is set at 50%, the result is shown in Fig. 3. When the variable resistor is set at 100%, the result is shown in Fig. 4. From both conditions, we can conclude that, as we increase the variable resistor percentage, the square wave output will become faster. This will affect the switching frequency at the inverter circuit and affect the speed of the induction motor.

The simulation coordination between the PWM circuit and gate driver circuit. As shown below, output from 555 timer pin three is connected to input for decade counter CD4017 pin 14. The output will be measured under two conditions: when the variable resistor is set to 50%, the result shows, and when the variable resistor is set to 100%, the result.

The high and low outputs indicate the connection from the oscilloscope channel to the circuit. Output low is the pin connected to channel A, while output high shows the pin connected to channel B. Waveform shows that the output frequency of the decade counter varies when variable resistors are changing.

Also, from Fig. 4, we can see that the output from both decade's counter pins will give the pulse simultaneously. It means that when one output is high, the other will be

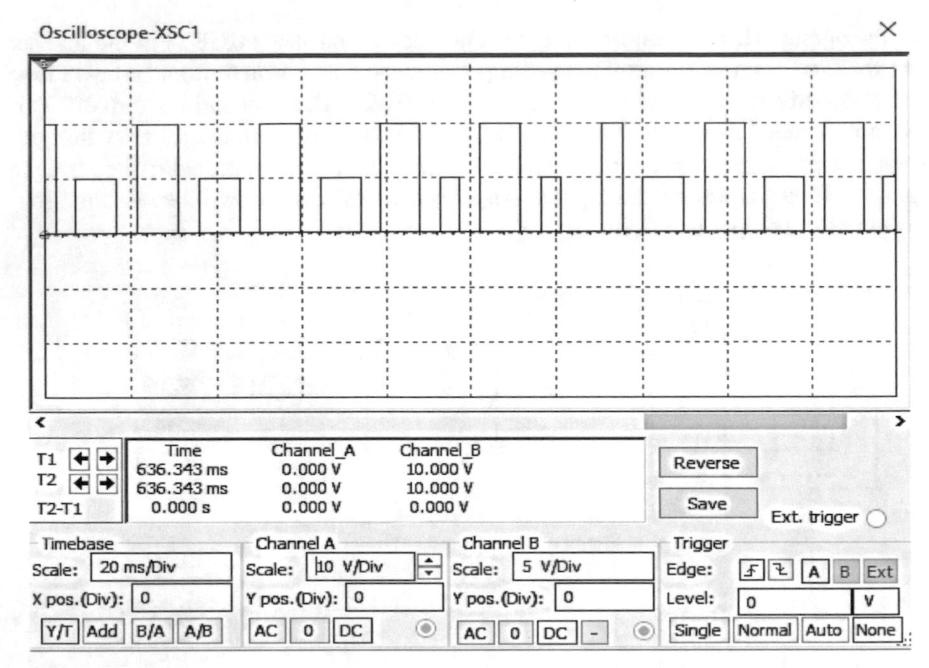

Fig. 3. Output of gate driver circuit when variable resistor set to 50%.

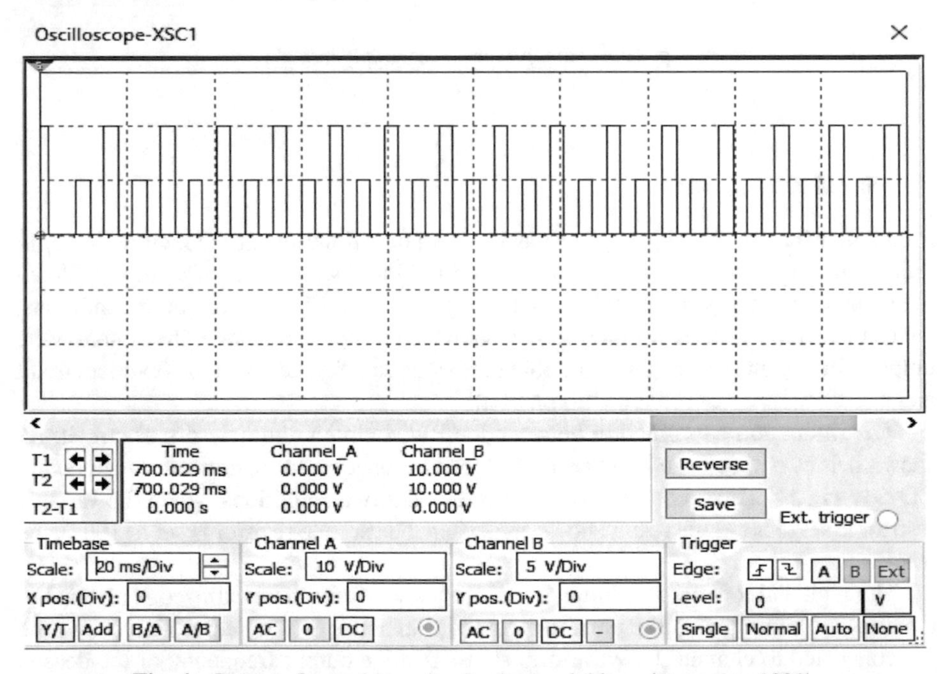

Fig. 4. Output of gate driver circuit when variable resistor set to 100%.

low. This output will connect to the IGBT gate at the inverter circuit. So, theoretically, with this output, the inverter should work well. However, when *I* connect the gate driver to the inverter circuit, the simulation comes to an error. It cannot be sure whether it is due to the wrong design or limitation on the Multisim software. So, to simulate the inverter circuit, *I* replaced the control circuit and only used the function generator as a PWM signal into IGBT. A function generator has been used to simulate the inverter circuit, replace the control circuit, and provide PWM for the inverter. A simulation was done, and the result is shown in Fig. 3. Figure 4 shows the speed of the induction motor. However, by using the function generator, we are not able to show the speed difference that is affected by the variable resistor.

The main objective of this research is to reduce the high starting current for induction motors. So, to see the result, a simulation has been done, as shown in Fig. 5 above. Figure 6 shows the simulation of starting an induction motor by using a starter. The ammeter has been connected to see the starting current. Figure 7 shows the simulation of starting an induction motor using the direct-on-line method. Three-phase IM is connected to a three-phase supply and ammeter connected in line to see the starting current. Through these two simulations, we can see the starting current for both starting methods. The starting current in Fig. 6 is higher than the starting current in Fig. 8. This has proved that by using the design starter, the high starting current can be reduced.

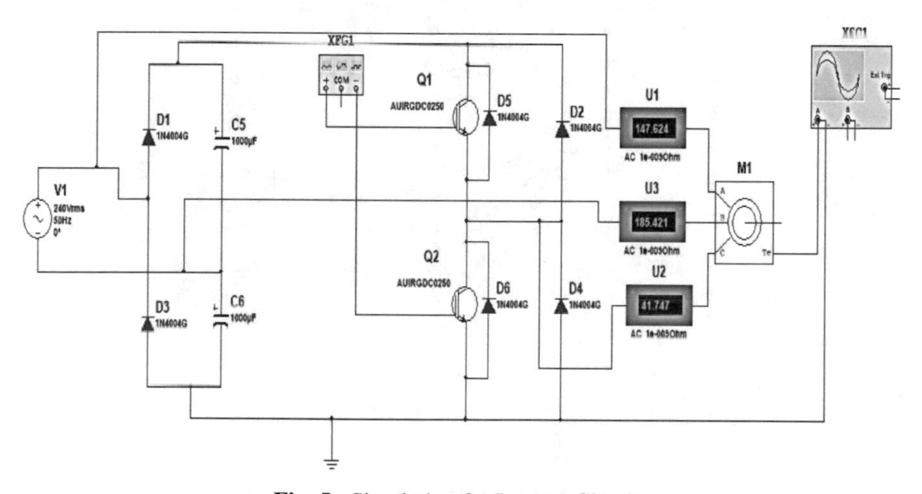

Fig. 5. Simulation for Inverter Circuit.

In terms of speed, we can see that by using the DOL method, there is no way to control the starting speed. However, theoretically, by using a design starter, the starting speed for the induction motor can be controlled by varying the variable resistor. This simulation cannot prove this condition due to some error occurring when connecting the control circuit to the inverter circuit.

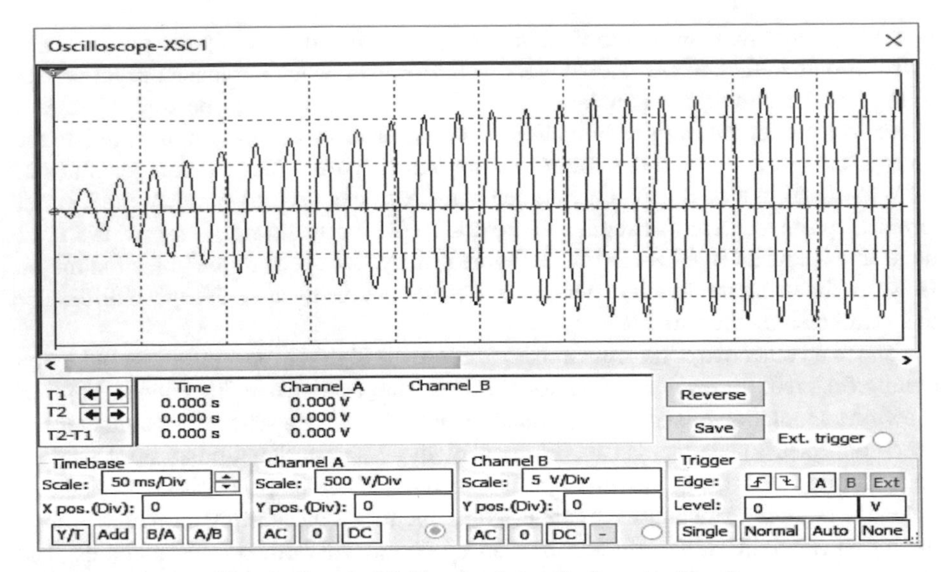

Fig. 6. Speed of IM by simulating the Inverter Circuit.

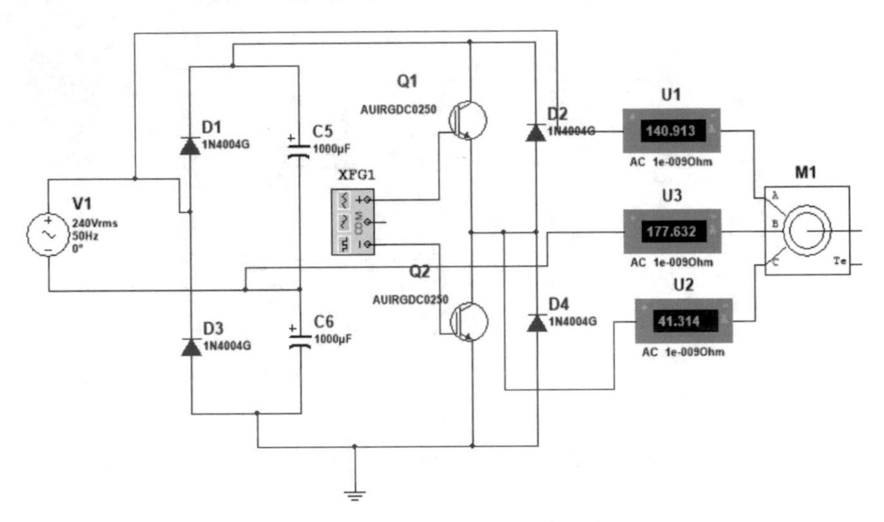

Fig. 7. The simulation was done using the starter.

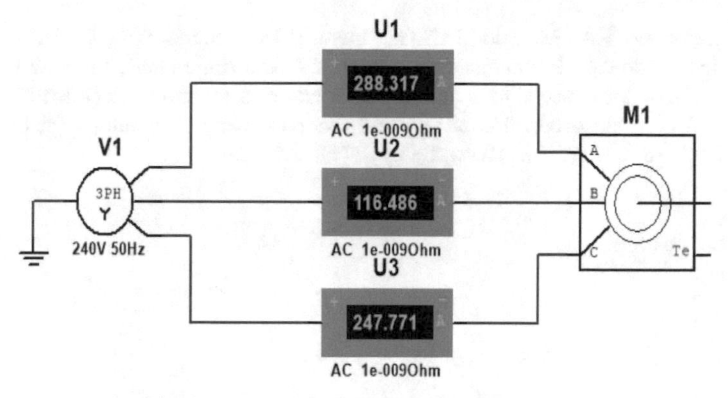

Fig. 8. The simulation using Direct-On-Line.

3 Conclusion

From the simulation and result analysis done in the previous chapter, it can be concluded that three-phase induction machines can be run using a single-phase power source. Also, theoretically, the speed of the three-phase induction machine can be controlled by adjusting the frequency of PWM that controls the switch turn-on and turn-off time.

From the result above, it can be said that the main objective of designing a starter for an induction motor and reducing the high starting current is achieved. Also, the sub-objective, which is to design an inverter circuit that places a soft switching device, was successfully achieved by using an Insulated Gate Bipolar Transistor (IGBT) as the primary device in the inverter. However, another objective, which is to design a control circuit that acts as a controller for the IGBT, was not successfully achieved since the simulation came to an error when we connected the control circuit to the inverter circuit. So, this needs to be rechecked.

References

1. Taha, T.A., Hassan, M.K., Zaynal, H.I., Wahab, N.I.A.: Big data for smart grid: a case study. In Big Data Analytics Framework for Smart Grids, pp. 142–180. CRC Press (2024)
2. Luo, H., Chen, Y., Sun, P., Li, W., He, X.: Junction temperature extraction approach with turn off delay time for high-voltage high-power IGBT modules. IEEE Trans. Power Electron. **31**, 5122–5132 (2016)
3. Energy Efficiency Guide for Industry in Asia, Electrical Energy Equipment: Electric Motors
4. Parekh, R.: Microchip Technology Inc. AC Induction Motors Fundamentals, AN887 (2003)
5. Taha, T.A., Hussain, A.S.T., Taha, K.A.: Design solar thermal energy harvesting system. In: AIP Conference Proceedings, vol. 2591, no. 1. AIP Publishing (2023)
6. Blair, T.H.: 3 phase AC motor starting methods: an analysis of reduced voltage starting characteristics. In: SoutheastCon, 2002. Proceedings, pp. 181–186. IEEE (2002)
7. TEXAS INSTRUMENT, LM555 Timer (2015)
8. FIRCHILD SEMICONDUCTOR, CD4047BC Low Power Monostable/Astable Multivibrator (2002)
9. Shivanagouda, M.S.A., Patil, B.: Operating three phase induction motor connected to single phase supply. Int. J. Emerg. Technol. Adv. Eng. **2**, 523–528 (2012)

10. Niu, Y., Habeeb, F.A., Mansoor, M.S.G., Gheni, H.M., Ahmed, S.R., Radhi, A.D.: A photovoltaic electric vehicle automatic charging and monitoring system. In: 2022 International Symposium on Multidisciplinary Studies and Innovative Technologies (ISMSIT) (2022)
11. Saleh, A.M., et al.: Production of first and second-generation biodiesel for diesel engine operation: a review. NTU J. Renew. Energy **5**(1), 8–23 (2023)

Optimizing Solar Energy Efficiency Through Automatic Solar Tracking Systems

Saadaldeen Rashid Ahmed[1,2](✉), Pritesh Shah[3], Mohammed Fadhil[2], Abadal-Salam T. Hussain[4], Sushma Parihar[3], Jamal Fadhil Tawfeq[5], Taha A. Taha[6], Faris Hassan Taha[4], Omer K. Ahmed[6], Hazry Desa[7], and Khawla A. Taha[4]

[1] Artificial Intelligence Engineering Department, College of Engineering, Alayan University, Nasiriyah, Iraq
saadaldeen.ahmed@alayen.edu.iq

[2] Computer Science Department, Bayan University, Erbil, Kurdistan, Iraq
saadaldeen.aljanabi@bnu.edu.iq

[3] Symbiosis Institute of Technology (SIT) Pune Campus, Symbiosis International (Deemed University) (SIU), Pune, Maharashtra 412115, India
pritesh.shah@sitpune.edu.in

[4] Department of Medical Instrumentation Techniques Engineering, Technical Engineering College, Al-Kitab University, Altun Kupri, Kirkuk, Iraq

[5] Department of Medical Instrumentation Technical Engineering, Medical Technical College, Al-Farahidi University, Baghdad, Iraq

[6] Unit of Renewable Energy, Northern Technical University, Kirkuk, Iraq

[7] Centre of Excellence for Unmanned Aerial Systems (COEUAS), Universiti Malaysia Perlis, Jalan Kangar-Alor Setar, 01000 Kangar, Perlis, Malaysia

Abstract. In today's rapidly evolving global energy landscape, the imperative to transition to sustainable energy sources is more pronounced than ever. This research investigates solar tracking technology, yielding an innovative system that optimizes energy production efficiency by integrating meticulous component selection, precise circuit design, and advanced microcontroller programming enhanced by Light Dependent Resistors (LDRs) for precise sun-tracking. Our empirical findings demonstrate a remarkable increase in irradiance levels (15% on average), consistent voltage output (18–20 V), and improved temperature control (53.4 °C compared to 59.5 °C for static panels). These outcomes hold significant practical implications for solar installations and renewable energy projects, contributing to a cleaner and more energy-efficient future while addressing environmental concerns. Additionally, this research paves the way for future studies and applications in renewable energy and solar technology, promising more efficient and sustainable energy solutions to meet the pressing global energy challenges.

Keywords: Solar Energy · Photovoltaic Technology · Solar Tracking · Renewable Energy · Efficiency Enhancement

J. Rasheed et al. (Eds.): FoNeS-AIoT 2024, LNNS 1035, pp. 295–303, 2024.
https://doi.org/10.1007/978-3-031-62871-9_23

1 Introduction

Automated solar tracking systems have emerged as a compelling solution within the realm of renewable energy technologies, offering the potential to substantially enhance the efficiency of solar energy capture. As the world grapples with the enduring challenges of dwindling fossil fuel reserves, environmental degradation, and the impending specter of global climate change, the imperative to transition to sustainable and renewable energy sources has never been more evident [1–4]. In this evolving energy landscape, solar power stands as a beacon of hope, providing a path toward a cleaner and more environmentally conscious future.

Solar energy, harnessed from the radiant power of the sun, presents an enticing solution. The sun, a cosmic powerhouse, showers the Earth with a continuous stream of energy, with each square meter of sunlit surface receiving approximately 1380 J per second, equivalent to nearly 2 horsepower, representing a consistent and abundant source of energy [5–8]. Solar panels, also known as photovoltaic (PV) systems, capture and convert this energy into electricity. As awareness of their benefits grows, the adoption of solar power is on the rise [8–12].

2 Methodology

This section provides a comprehensive account of the methodology employed in the development of the automatic solar tracking system. The methodology encompasses several crucial phases, each integral to the successful realization of the project's objectives.

The project's initiation involved an in-depth analysis of the components essential for the automatic solar tracking system as shown in Fig. 1. These components were selected with precision and attention to detail, ensuring compatibility and suitability for the project's goals:

2.1 Microcontroller

The heart of the system, a PIC16F877A microcontroller, was chosen for its versatility and reliability.

2.2 Sensors

Four Light Dependent Resistors (LDRs) were strategically positioned on the solar panel to monitor sunlight intensity.

2.3 Motors

A specific DC motor model (mentioned model) was selected for its suitability in angular adjustments.

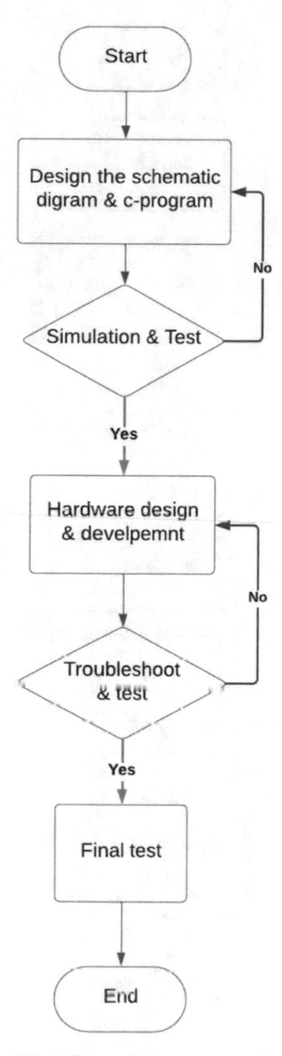

Fig. 1. Workflow of the proposed method

2.4 Voltage Regulators

To ensure stable voltage outputs, (the mentioned regulator models) were employed.

Ideally, Fig. 2 unveils a comprehensive programming flow chart that intricately maps out the step-by-step operation of the automatic solar tracking system. This innovative system incorporates four strategically positioned Light Dependent Resistors (LDRs) on the solar panel, facilitating continuous measurement of sunlight intensity. The program commences with an essential initialization phase, where crucial parameters and variables are configured. At the core of this system lies the astute comparison of sensor values.

If disparities in sunlight intensity are detected by the LDRs, signaling that the solar panel is not optimally aligned with the sun, the program springs into action by activating

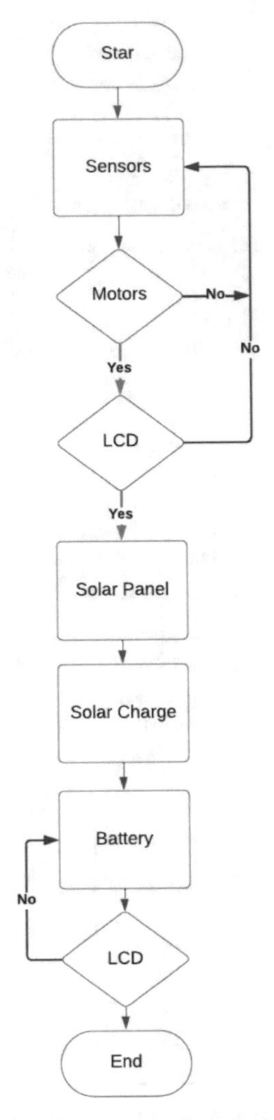

Fig. 2. Program flowchart.

a motor. The motor plays a pivotal role in dynamically repositioning the solar panel, ensuring that all LDRs converge on a consistent light intensity. This dynamic adjustment process is of paramount importance, as it serves to maximize the capture of solar energy. By allowing the solar panel to meticulously track the sun's movement across the sky throughout the day, this system optimizes energy harvesting.

Concurrently, a user-friendly LCD offers real-time feedback on the sun's intensity, providing valuable insights into the system's performance. As the solar panel perfectly aligns with the sun, it commences the vital task of converting solar energy into electrical power. A portion of the generated power is intelligently directed towards a solar charger, meticulously regulating, and managing the voltage emanating from the solar panel. This solar charger's primary mission is to charge a battery, which serves as an invaluable energy reservoir for periods of insufficient sunlight, such as during the night. To ensure the battery remains in optimal condition, another LCD screen diligently displays the battery's voltage level. This holistic process operates continuously, seamlessly adapting to fluctuations in sunlight intensity, and guarantees that the solar panel consistently harnesses the maximum available solar energy.

In essence, this automated solar tracking system stands as a pioneering solution that unlocks the full potential of solar resources. Its ability to adapt and optimize energy capture renders it an indispensable tool in the realm of sustainable energy generation, ushering in a greener and more efficient era of power production. Additionally, Fig. 3 shows a programming flow chart that illustrates the sequential operation of the automatic solar tracking system. This system incorporates four Light Dependent Resistors (LDRs) strategically placed on the solar panel to continuously measure the intensity of sunlight. The program begins with an initialization phase, setting up essential parameters and variables. The heart of the system lies in the comparison of sensor values. If the LDRs detect varying levels of sunlight, indicating that the solar panel is not optimally aligned with the sun, the program activates a motor.

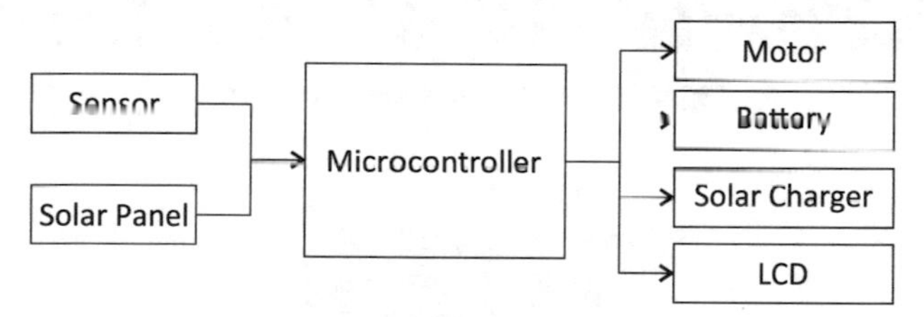

Fig. 3. Basic block diagram of automatic solar tracking system (ISIS schematic main circuit design)

A portion of this generated power is directed to a solar charger, which regulates and manages the voltage from the solar panel. The solar charger's primary function is to charge a battery, serving as an energy storage reservoir for times when sunlight is insufficient, such as at night as shown in Fig. 4. Another LCD screen displays the battery's voltage level, ensuring its optimal condition. This entire process operates continuously, adapting to changes in sunlight, and ensures that the solar panel consistently captures the maximum available solar energy. In essence, this automated solar tracking system optimally utilizes solar resources, making it a valuable tool for sustainable energy generation.

This dynamic movement optimizes energy capture, culminating in enhanced power generation throughout the day. Additionally, a solar charger replenishes the battery, harnessing voltage from the solar panel and proficiently storing the generated energy for subsequent utilization. An integrated LCD monitors the battery's status and the charging process, presenting real-time information to the user as shown in Fig. 5. The microcontroller processes this data and based on the sun's position, activates the motor to adjust the solar panel's orientation. This dynamic movement ensures that the solar panel maximizes its exposure to sunlight, optimizing energy capture. The solar charger is

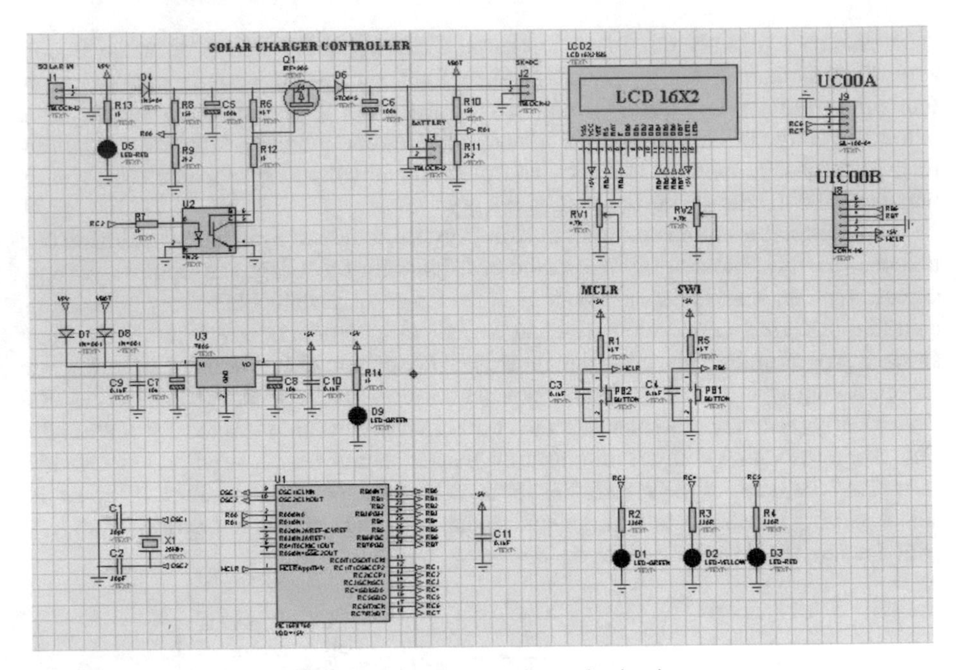

Fig. 4. Solar charger schematic circuit

Fig. 5. The full circuit of the solar tracker

responsible for harnessing the voltage generated by the solar panel to charge the battery, ensuring efficient energy storage. An integrated LCD provides real-time information about the battery's status and the charging process.

3 Results

Our experimental investigation provides valuable insights into the performance of the automatic solar tracking system, which is crucial for understanding its effectiveness in optimizing solar energy utilization. In this section, we present and discuss the results in detail, highlighting key findings and their implications. The automatic solar tracking system exhibited remarkable voltage stability, maintaining a consistent voltage range between 18 V and 20 V throughout the day Fig. 6. This achievement underscores the system's ability to ensure a steady energy output even in the face of varying weather conditions. The stable voltage is crucial for the reliability and efficiency of the system, ensuring continuous power generation.

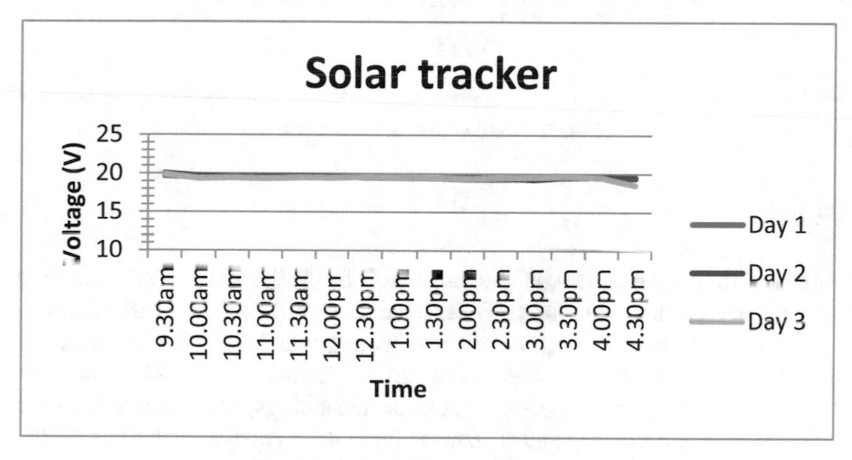

Fig. 6. Voltage data for solar tracker.

Data analysis revealed that our solar tracking system consistently received higher solar irradiance levels compared to static solar systems Fig. 7. The peak irradiance recorded for the tracking system reached 1555 W/m^2, while the static solar system achieved only 1460 W/m^2. This substantial difference in irradiance levels is a testament to the effectiveness of our solar tracking system in maximizing solar energy absorption. The increased irradiance directly contributes to enhanced electricity generation. Solar irradiance has a substantial impact on the system's voltage output. Higher irradiance levels not only lead to increased energy generation but also affect the solar panel's efficiency. It's worth noting that excessively high solar irradiance can shorten the lifespan of solar panels and reduce overall efficiency. Therefore, the ability of our system to maintain higher irradiance levels without reaching detrimental extremes is a noteworthy achievement.

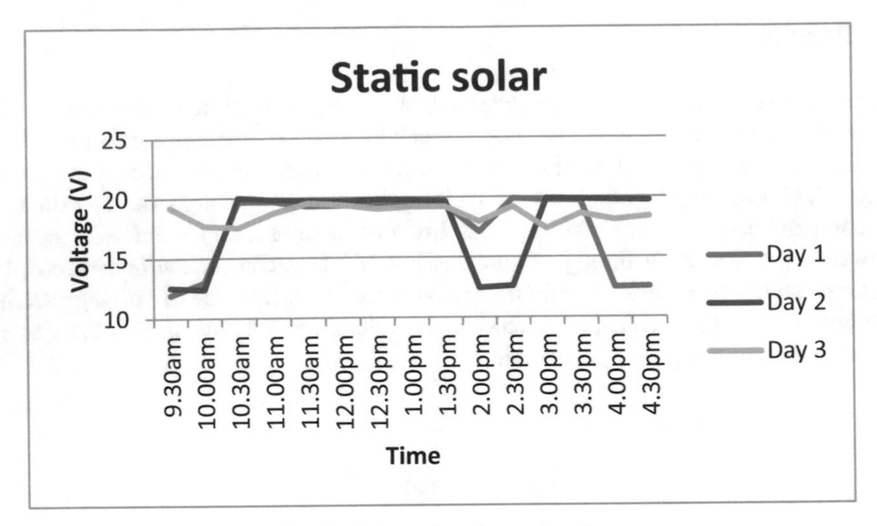

Fig. 7. Voltage data for static solar.

4 Conclusion

In conclusion, this study successfully achieved its objectives, including the development and implementation of an Automatic Solar Tracker Control System with sensors and a microcontroller, resulting in improvements in voltage stability, solar irradiance levels, and temperature control when compared to static systems. These findings hold significance for the field of renewable energy and solar technology, offering potential benefits for enhanced energy efficiency and sustainability in solar panel installations. This study has successfully realized a set of clearly defined objectives: the development of an Automatic Solar Tracker Control System featuring sensors and a microcontroller, detailed microcontroller programming, and comprehensive performance evaluation. The results demonstrate significant advancements, including consistent voltage stability (18 V to 20 V), higher solar irradiance levels (1555 W/m^2 compared to 1460 W/m^2 in static systems), and effective temperature control (53.4 °C). These findings underscore the potential of solar tracking technology to enhance energy efficiency, increase electricity generation, and extend solar panel lifespan. While this study contributes significantly to the field, acknowledging its limitations and exploring refinements in future research is essential. Overall, these results enrich our understanding of solar energy technology and offer promising avenues for sustainable renewable energy solutions. Acknowledging its limitations, such as the need for further refinement, and suggesting directions for future research can inspire continued advancements in this area. Additionally, this research contributes to expanding knowledge in the domain of solar tracking systems and their role in optimizing solar energy capture and fostering progress in renewable energy technologies.

References

1. Olujobi, O.J., Okorie, U.E., Olarinde, E.S., Aina-Pelemo, A.D.: Legal responses to energy security and sustainability in Nigeria's power sector amidst fossil fuel disruptions and low carbon energy transition. Heliyon **9**(7), e17912 (2023)
2. Zhao, H., Chuanqing, W., Wen, Y.: Determinants of corporate fossil energy assets impairment and measurement of stranded assets risk. Energies **16**(17), 6340 (2023)
3. Frilingou, N., et al.: Navigating through an energy crisis: challenges and progress towards electricity decarbonisation, reliability, and affordability in Italy. Energy Res. Soc. Sci. **96**, 102934 (2023)
4. Taha, T.A., Hassan, M.K., Wahab, N.I.A., Zaynal, H.I.: Red deer algorithm-based optimal total harmonic distortion minimization for multilevel inverters. In 2023 IEEE IAS Global Conference on Renewable Energy and Hydrogen Technologies (GlobConHT), pp. 1–8. IEEE (2023)
5. Ravichandran, S., Kumar, M., Singh, M.: Design and implementation of an automatic solar tracking system using LDRs. Int. J. Innovative Res. Sci. Eng. Technol. **8**(1), 739–746 (2019)
6. Moghaddam, M., Ehsanzadeh, S., Asadi, S.M.: Improving the performance of perovskite solar cells with carbon nanotubes as a hole transport layer. Sol. Energy **206**, 807–816 (2020). https://doi.org/10.1016/j.solener.2020.04.054
7. Hou, Y., Li, H., Wang, Z., Wang, Y.: Insights into the photovoltaic properties of indium sulfide as an electron transport material in perovskite solar cells. Sci. Rep. **13**(1), 1755 (2023). https://doi.org/10.1038/s41598-022-05965-z
8. Yusaf, T., et al.: Sustainable hydrogen energy in aviation–a narrative review. Int. J. Hydrogen Energy (2023)
9. Amir, M., et al.: Energy storage technologies: an integrated survey of developments, global economical/environmental effects, optimal scheduling model, and sustainable adaption policies. J. Energy Storage **72**, 108694 (2023)
10. Al-Kaabi, M., Salih, S.Q., Dumbrava, V.: Optimal power flow based on grey wolf optimizer: case study Iraqi super grid high voltage 400 kV. Energies **16**(1), 160 (2023). https://doi.org/10.3390/en16010160
11. Hopkins, T.: Effectively managing today's transformer challenges for increased asset reliability & sustainability. In: 2023 IEEE PES Grid Edge Technologies Conference & Exposition (Grid Edge), pp. 1–5. IEEE (2023)
12. Ezzat, S.B.: Improving the performance of the direct and indirect evaporative cooling system: a review. NTU J. Renew. Energy **5**(1), 74–85 (2023)

Selective Harmonic Elimination in Multilevel Inverters Using the Bonobo Optimization Algorithm

Taha A. Taha[1(✉)], Noor Izzri Abdul Wahab[1], Mohd Khair Hassan[2], and Hussein I. Zaynal[3]

[1] Department of Electrical and Electronics Engineering, University Putra Malaysia (UPM), 43400 Serdang, Selangor, Malaysia
t360pi@gmail.com, izzri@upm.edu.my

[2] Unit of Renewable Energy, Northern Technical University, Kirkuk, Iraq
khair@upm.edu.my

[3] Department of Computer Engineering Technology, Al-Kitab University, Altun Kupri, Iraq

Abstract. Multilevel inverters play a crucial role in energy conversion and power electronics applications. However, the harmonics generated during the operation of such inverters can harm electrical grids and reduce system efficiency. Therefore, effectively controlling and eliminating harmonics is critical to enhancing the performance of multilevel inverters. This study focuses on achieving selective harmonic elimination (SHE) in multilevel inverters using the Bonobo Optimization Algorithm (BO). The BO algorithm draws inspiration from the social behaviors of bonobo monkeys and is an artificial intelligence algorithm. It employs evolutionary approaches to solve complex problems and offers a population-based approach to optimizing the target function. The application of the BO algorithm for seven- and eleven-level inverters is compared to the genetic algorithm (GA) and particle swarm optimization (PSO). The results demonstrate that BO provides a more effective harmonic elimination solution than GA and PSO. With this algorithm, it becomes possible to maintain harmonic levels below a specified threshold while improving the overall performance and efficiency of the inverter. In conclusion, this study successfully applies the BO algorithm for selective harmonic elimination in multilevel inverters, contributing significantly to future energy conversion and power electronics research. This approach has the potential to assist in making energy systems cleaner, more reliable, and more efficient.

Keywords: Multilevel Inverters · Selective Harmonic Elimination · Bonobo Optimization Algorithm · Energy Conversion

1 Introduction

In recent years, multi-level inverters (MLIs) have emerged as an indispensable technology in power electronics, attracting significant attention from researchers and practitioners. These innovative devices have found applications in various areas, including variable-frequency drives, power compensation systems, grid-connected renewable

J. Rasheed et al. (Eds.): FoNeS-AIoT 2024, LNNS 1035, pp. 304–321, 2024.
https://doi.org/10.1007/978-3-031-62871-9_24

energy applications, and electric vehicles. MLIs have established themselves as fundamental components in the quest for efficient and sustainable energy management thanks to their versatility and adaptability in meeting the complex demands of modern electrical systems [1, 2].

One of the prominent features of MLIs is their ability to produce high-quality voltage waveforms with lower total harmonic distortion (THD) compared to traditional two-level inverters. This feature makes MLI structures ideal for applications where the reduction of harmonics is necessary, mainly to ensure the smooth operation of connected loads and compliance with grid codes. While many control and modulation methods are available to effectively implement harmonic reduction strategies in MLI structures, one of the most commonly preferred and effective methods is the Selective Harmonic Elimination (SHE) method [3, 4].

SHE methods involve solving a series of transcendental equations. Various methods have been proposed in the literature to solve these equations, such as Walsh functions [5], Newton-Raphson [6], and result theory [7]. However, a common characteristic of these techniques is that they are complex and computationally intensive in calculating SHE equations. These equations often rely on numerical solutions and may require many iterations. Therefore, these methods can place a heavy computational burden and complicate the process.

To overcome these challenges, bio-inspired intelligent algorithms have been developed to solve the harmonic elimination problem. The fundamental problem with using bio-inspired intelligent algorithms to solve SHE equations is the complexity and computational cost of solving these equations. Since SHE aims to eliminate specific harmonic components, these equations are typically complex and transcendental [8]. Solving these equations analytically can be challenging and often requires numerical solutions. This is where bio-inspired intelligent algorithms come into play.

Bio-inspired intelligent algorithms are natural optimization techniques and are often used to solve large and complex problems. However, SHE equations must be solved for many different modulation indices (MI), and there may be no solution for certain modulation index values. This is where the fundamental problem with bio-inspired intelligent algorithms arises. The main issue is that numerically solving these equations can sometimes take time and effort. Finding a suitable solution for certain modulation index values can be challenging, and an approximate solution may be required. Therefore, the optimization process of bio-inspired intelligent algorithms may require efficient use of computational resources. As a result, researchers have developed new optimization techniques and algorithms to solve SHE equations more effectively.

This study uses the newly developed Bonobo Optimization (BO) algorithm to solve SHE equations. The BO algorithm is compared with GA and PSO algorithms, demonstrating its superiority. It addresses the topic of harmonic elimination in multi-level inverters (MLIs). Firstly, Sect. 2 explains the mathematical foundations of harmonic elimination in MLIs. Then, Sect. 3 provides information about the working principles and usage of the bio-inspired optimization (BO) algorithm. Section 4 presents how Selective Harmonic Elimination (SHE) is implemented in MLIs and showcases the results of this method. Finally, in Sect. 6, the article summarizes its main findings and concludes.

2 Cascaded H-Bridge Multilevel Inverters (CHB-MLI)

These are called cascaded H-Bridge Multilevel Inverters (CHB-MLI). They are made up of a set of H-bridge inverters that are connected in series. Each bridge is equipped with its own separate Direct Current Source (SDCS). By choosing the right four switches (S1, S2, S3, and S4) in the H-bridge, each of these bridges, which are sometimes called "cells", can create three different voltage levels: + Vdc, 0, and -Vdc—cascaded the voltages produced by each cell to create the output voltage waveform.

While the diode-clamped (DC-MLI) and flying capacitor (FC-MLI) structures have more circuit elements, the CHB-MLI has the fewest. However, the CHB-MLI structure requires discrete DC sources, which can be obtained from various sources such as batteries, fuel cells, or solar cells.

Combining the voltages produced by each cell results in the output voltage waveform. In contrast to DC-MLI and FC-MLI structures, CHB-MLI is better suited for high-voltage, high-power applications. This is mostly because CHB-MLI comprises separate modules that can be easily connected to create higher voltage levels without much extra work. Additionally, CHB-MLI can achieve higher power and voltage levels with fewer required devices compared to other basic topologies.

For this research study, a case study approach was adopted, encompassing a 7-level and an 11-level Cascaded H-Bridge Multilevel Inverter (CHB-MLI). In Fig. 1(a), the assembly of the three-phase 7-level CHB-MLI can be observed, and in Fig. 1(b), the output waveform of a single phase of the 7-level three-phase inverter is presented. Each cell is connected to a separate 311/3 V DC source, and MOSFETs are employed as the switching devices. Figure 1(a) displays the circuit structure of the three-phase 11-level CHB-MLI, and Fig. 1(b) illustrates the waveform for the 11-level configuration. In the case of the 11-level setup, a 311/5-V DC source is used for each phase. Throughout the analysis, the modulation index value varied between 0.1 and 1.0 for both circuit configurations.

The inverter has been modeled in MATLAB-Simulink to compare Genetic Algorithms (GA) and Particle Swarm Optimization (PSO) for solving the Selective Harmonic Elimination (SHE-PWM) problem.

3 Problem Formulation

When considering a multi-level inverter system with a specific number of voltage levels (N) and aiming to achieve Selective Harmonic Elimination (SHE) while taking into account a desired output voltage waveform with specific harmonic constraints, the goal is to determine the switching angles of the inverter's semiconductor devices. This involves finding the most suitable ON/OFF states for semiconductor devices at each switching instant to minimize or eliminate specific harmonic components while ensuring that the fundamental component meets the desired magnitude and frequency requirements. The output of the multi-level inverter can be expressed using the Fourier Series as given in (1).

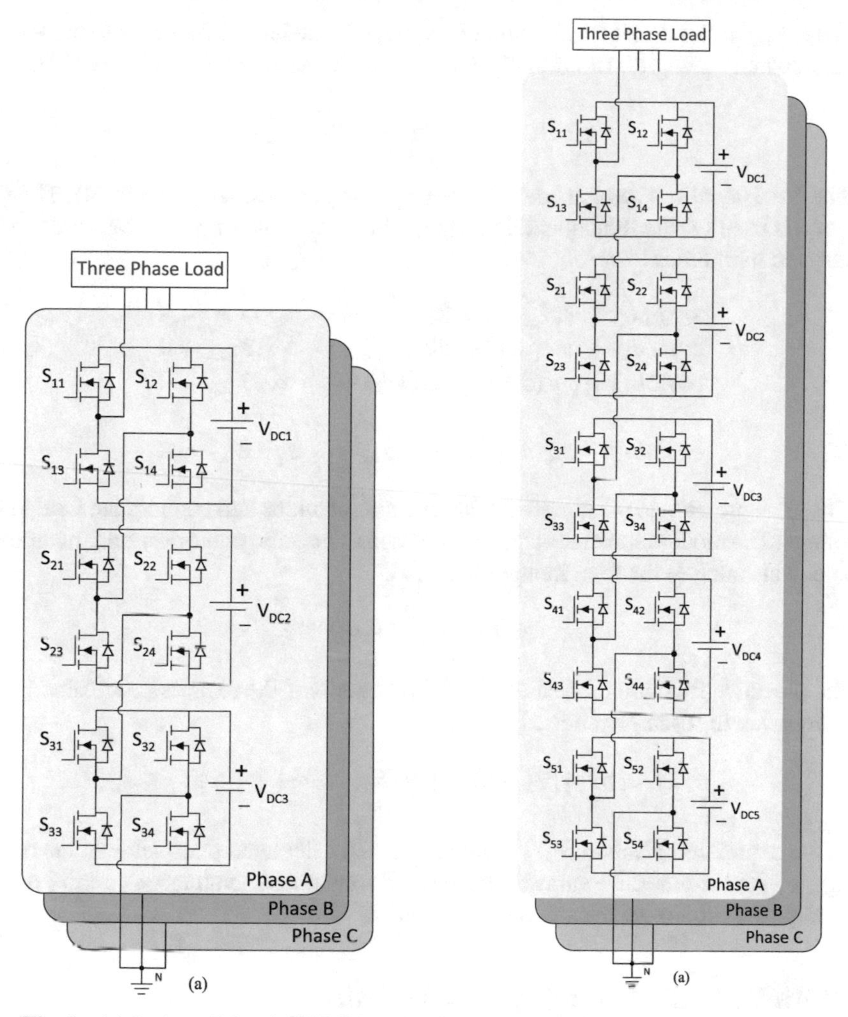

Fig. 1. (a) 3-phase 7-level CHB-MLI circuit, (b) 3-phase 11-level CHB-MLI circuit.

$$f_{N(t)} = \frac{a_0}{2} + \sum_{N}^{n=1}\left(A_n cos\left(\frac{2\pi nt}{T}\right) + V_n sin\left(\frac{2\pi nt}{T}\right)\right) \quad (1)$$

Here, a_0 is the DC component, and V_n represents the Fourier coefficients that correspond to even and odd harmonics as given in (2).

$$v_n = \frac{4v_{DC}}{n\pi}\sum_{s}^{i=1} K_i cos(n\theta i) \quad (2)$$

Here, V_{DC} is the direct current voltage value, n is the harmonic number, and θ is the relevant switching angle. Additionally, θ must satisfy the constraints given in (3).

$$\theta_1 < \theta_2 < \ldots < \theta_{k-1} < \theta_k < \frac{\pi}{2} \tag{3}$$

The SHE equations, which consist of k switching angles, are given in (4). The first Equation is used for the fundamental voltage, while the remaining equations are used to set selected harmonics to zero.

$$\begin{aligned} cos(\theta_1) + cos(\theta_2) + cos(\theta_3) \cdots \cdot cos(\theta_n) &= Mk\pi/4 \\ cos(3\theta_1) + cos(3\theta_2) + cos(3\theta_3) \cdots \cdot cos(3\theta_n) &= 0 \\ cos(5\theta_1) + cos(5\theta_2) + cos(5\theta_3) \cdots \cdot cos(5\theta_n) &= 0 \\ \cdots \\ cos(n\theta_1) + cos(n\theta_2) + cos(n\theta_3) \cdots \cdot cos(n\theta_n) &= 0 \end{aligned} \tag{4}$$

The M expression in (4) represents the modulation index and controls the fundamental voltage. The modulation index can be expressed as the ratio of the desired fundamental voltage peak value to the total input voltage [9].

$$M = |V_1| / (kV_{DC}) \tag{5}$$

For the objective function created for the SHE equations, the expression obtained from [10] and given in (6) has been used.

$$f = \min_{\theta_i} \{ |V_{1p} - V_{ref}| + |V_5| + |V_7| + \ldots \} \tag{6}$$

In the expression given in (6), V_{1_p} represents the calculated peak value of the output voltage, V_{ref} is the reference value for the desired fundamental voltage. V_5 and V_7 denote the voltage amplitudes of the selected harmonics.

4 Meta-heuristic Optimization Algorithms

4.1 Genetic Algorithm (GA)

The Genetic Algorithm (GA) is a heuristic global evolutionary optimization algorithm that draws inspiration from the principles of natural selection and genetics. John Holland originally developed this algorithm in the early 1970s. GA employs principles from biological evolution to optimize various processes. One of the key distinctions, when compared to other optimization techniques, is that GA explores the solution space through a population-based approach rather than focusing on individual point searches. Genetic algorithms have proven to be effective in solving both constrained and unconstrained optimization problems. GA is recognized for its simplicity and ease of implementation, as it does not require complex mathematical modeling or derivations. Therefore, it can be readily applied to address challenges like selective harmonic elimination.

The genetic algorithm optimization process typically comprises four main steps:

- Step 1 (Initialization of the population): At the outset, a population of potential solutions is generated.
- Step 2 (Evaluation of the fitness function): Each solution in the population is assessed based on a fitness function, which quantifies how well it addresses the problem.
- Step 3 (Selection): Solutions are chosen to become parents for the next generation, with a bias towards selecting those with higher fitness values.
- Step 4 (Apply genetic operators): Crossover and mutation operations are applied to create new solutions, mimicking the genetic processes of recombination and mutation.
- Step 5 (Stopping Criterion): The stopping criterion can be either the best result or the number of iterations. In this study, the stopping criterion has been set as the number of iterations. The number of iterations is set to 100.

Figure 2(a) provides a general flowchart illustrating the steps involved in a genetic algorithm. This approach helps to decrease the similarity ratio among solutions and encourages the exploration of the solution space, increasing the likelihood of finding optimal or near-optimal solutions.

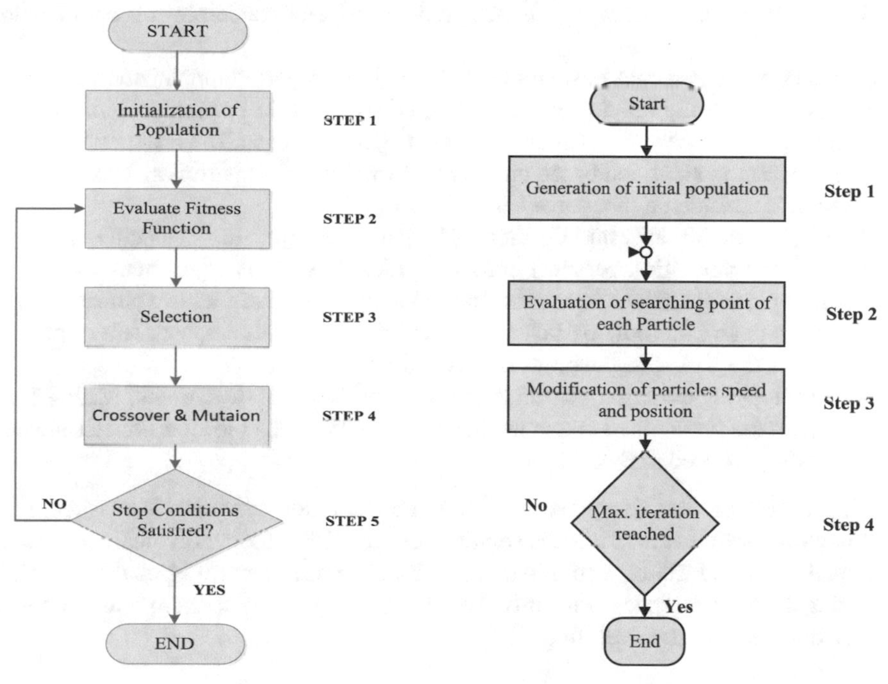

Fig. 2. (a) General GA algorithm flowchart, (b) General PSO algorithm flowchart.

4.2 Particle Swarm Optimization (PSO)

Particle Swarm Optimization (PSO) is a metaheuristic optimization algorithm in which a group of particles (often representing potential solutions) collaboratively work to progress toward optimizing a target. PSO mimics natural behaviors, where each particle updates its movement using a combination of velocity and position to find better solutions for itself and the group as a whole. This collaboration and information sharing help particles converge toward potentially better solutions. PSO adopts a population-based approach and is used to optimize complex problems. The algorithm simulates the movement of potential solutions within a search space to optimize a target. The general flow chart of PSO is shown in Fig. 2(b).

PSO involves several key steps in the optimization process. Here are the typical steps involved in PSO:

- Step 1 (Initialization): In this step, an initial population of particles is generated within the solution space. Each particle represents a potential solution to the optimization problem. The particles are assigned random positions and velocities.
- Step 2 (Objective Function Evaluation): The objective function, which measures the quality or fitness of a solution, is evaluated for each particle based on its current position. This step quantifies how well each particle performs concerning the optimization goal.
- Step 3 (Update Personal Best and Global Best): Each particle maintains a record of its best position and fitness value encountered so far. If the current position is better than its personal best, the particle updates its personal best. The best particle in the entire swarm is identified based on the fitness values of all particles. This particle's position is considered the global best.
- Step 3 (Update Velocity and Position): The particles adjust their velocities based on their current velocities, personal best, and global best. This adjustment is typically guided by mathematical equations that balance exploration and exploitation. The velocities determine how particles move within the solution space. After updating velocities, the particles' positions are also adjusted.
- Step 4 (Stopping Criterion): Checking the exit condition. If the current iteration number equals the pre-defined maximum iteration number, then the algorithm terminates. Otherwise, proceed to Step 2.

These steps are iteratively executed until a termination condition is satisfied. PSO aims to guide particles towards better solutions by combining their individual experiences (personal best) and the best solutions found by the entire swarm (global best) while exploring the solution space efficiently. The algorithm typically converges to an optimal or near-optimal solution over time.

4.3 Bonobo Optimization (BO)

This algorithm was developed by mimicking the social behaviors and mating strategies of bonobos. Bonobos, like many other primates, adopt a fission-fusion group strategy. According to this strategy, bonobos form numerous small groups of different sizes (fission) and roam separately on their territory. Later, they reunite with fellow community members (fusion) and engage in activities such as sleeping together, fighting rivals, and various other activities (see Fig. 3(a)). Bonobos also employ four different mating processes: random mating, restrictive mating, extra-group mating, and consort ship mating. The internal mechanisms of these strategies are quite diverse, and these mechanisms were mathematically modeled to develop the BO algorithm. In the BO algorithm, the "alpha bonobo" is referred to as the best bonobo.

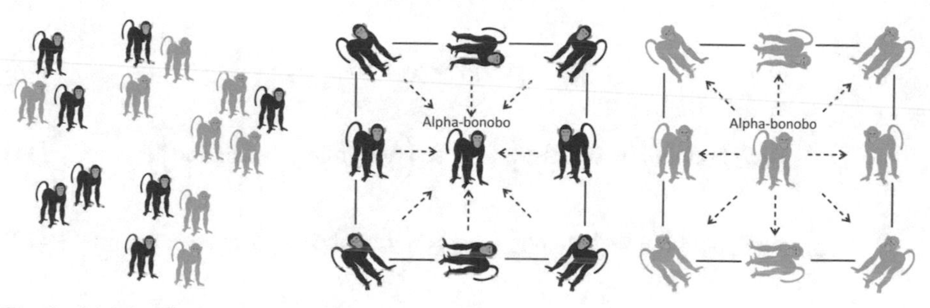

Fig. 3. (a) The Fission-fusion social groups of bonobos - light figures are females, while dark figures are males, (b) Directions of movements of different bonobos with the higher probabilities - the dark figure is positive phase, while the light figure is negative phase.

The algorithm consists of two separate phases: positive and negative. The Positive Phase (PP) represents the most favorable living conditions, characterized by successful mating, an adequate food supply, and suitable living conditions. In contrast, the Negative Phase (NP) represents the opposite of the positive phase. After each iteration, the algorithm transitions between the positive and negative phases, incrementing the Positive Phase Count (PPC) and Negative Phase Count (NPC) by one (see Fig. 3(b)). However, if one of these parameters is incremented, the other is initially set to zero. Bonobos are selected for mating using the fusion-fission social strategy. The maximum size of a temporary subgroup ($tsgs_{max}$) calculated based on the total population size (N) and is determined as calculated in (7).

$$tsgs\big(2, \big(tsgs_{\text{factor}} \times N\big)\big)_{max} \tag{7}$$

The size of the temporary subgroup is randomly determined between 2 and $tsgs_{max}$. . Subsequently, the p-th bonobo is selected as the best bonobo in that subgroup based on its current fitness value and participation in mating. In the positive phase, there is a higher probability of restrictive mating or random mating, while in the negative phase, there is a higher probability of extragroup mating or consortship. This probability is referred to as the phase probability (PP) in the BO algorithm. Initially, pp is set equal to 0.5. However, this parameter is determined after each iteration based on the current phase and phase count. In a positive phase, pp is within the range (0.5, 1.0), whereas in a negative phase, it falls within the range (0, 0.5). The main governing equation for a positive phase is given as (8).

$$new\_bonobo_j = bonobo_j^i + r_1 + scab \times \left(\alpha_{bonobo}^j - bonobo_j^i\right) + (1 - r_1) \times scsb \times flag \times \left(bonobo_j^i - bonobo_j^P\right) \quad (8)$$

Here, α_{bonobo}^j, and new_bonobo_j are variables for the alpha bonobo and its offspring, respectively, with j ranging from 1 to d, where d is the number of variables in the optimization problem. *scab* And *scsb* share parameters and the parameter flag is assigned as 1 or -1 based on a condition. In a negative phase, a new bonobo is generated during extra group mating by following the (9) and (14).

$$\beta_1 = e^{\left(r_1^2 + r_1 - 2/r_1\right)} \quad (9)$$

$$\beta_2 = e^{\left(-r_1^2 + 2 \times r_1 - 2/r_1\right)} \quad (10)$$

$$\text{new\_bonobo}_j = \text{bonobo}_j^i + \beta_1 \times \left(\text{Var\_max}_j - \text{bonobo}_j^i\right) \quad (11)$$

$$\text{new\_bonobo}_j = \text{bonobo}_j^i - \beta_2 \times \left(\text{bonobo}_j^i - \text{Var}\_\underset{j}{min}\right) \quad (12)$$

$$\text{new\_bonobo}_j = \text{bonobo}_j^i - \beta_1 \times \left(\text{bonobo}_j^i - \text{Var}\_\underset{j}{min}\right) \quad (13)$$

$$\text{new\_bonobo}_j = \text{bonobo}_j^i + \beta_2 \times \left(\text{Var}\_\underset{j}{max} - \text{bonobo}_j^i\right) \quad (14)$$

Here, β_1 and β_2 are two intermediate variables, Var_{maxj} and Var_{minj} are the upper and lower bounds of the jth variable, and r is a random number in the range (0, 1). At the end of each iteration, feedback from the search process is obtained, and control parameters are updated to steer towards promising regions in the variable space. The flowchart of the BO algorithm is shown in Fig. 4.

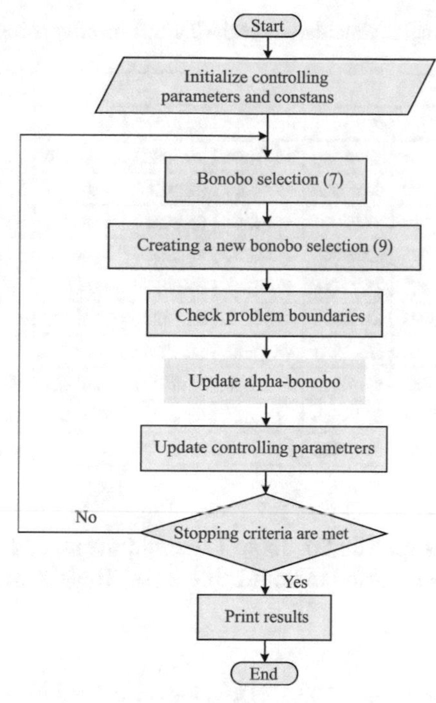

Fig. 4. The bonobo optimizer algorithm flowchart [11].

5 Results and Discussion

The proposed BO algorithm has been compared to PSO and GA in MATLAB, and the results are presented in this section. To prevent errors, all algorithms have been run 5 times, and each algorithm has 100 iterations. The optimum values obtained with the proposed BO algorithm are compared to the results of other well-known algorithms.

5.1 Comparison of BO with Other Algorithms

In this section, the results obtained from BO are compared with PSO and GA for both 7-level and 11-level cascaded H-bridge inverters.

Algorithm Comparison in 7-level Inverter. The modulation index was varied from 0.1 to 1.0 with a precision of 0.1. The algorithms were run five times for each modulation value, and the best values were selected to create tables. The switching angles calculated with GA, PSO, and BO are given in Table 1.

Table 1. Switching Angles Calculated for the 7-level inverter using GA, PSO, and BO

m		GA			PSO			BO		
		θ_1	θ_2	θ_3	θ_1	θ_2	θ_3	θ_1	θ_2	θ_3
low	0.1	77.34930	89.49601	89.61060	76.41652	90.00000	90.00000	76.32719	90.00000	90.00000
	0.2	61.99403	90.00000	90.00000	61.98364	90.00000	90.00000	61.83603	90.00000	90.00000
	0.3	51.05054	87.94902	87.94902	51.04264	86.70214	88.82021	50.96120	86.83159	88.66741
medium	0.4	51.62350	71.33325	89.89708	44.60004	76.67630	90.00000	44.69236	76.49805	90.00000
	0.5	40.56541	66.86417	88.75116	39.96108	65.69359	90.00000	40.74738	65.75196	89.30741
	0.6	40.68000	59.98868	81.07353	39.42979	58.58391	83.10421	39.42756	58.54024	83.05571
high	0.7	18.44924	50.30569	86.45933	17.91683	50.42793	86.51520	38.27003	53.91624	73.79610
	0.8	11.97482	41.76862	85.54260	29.21038	54.51497	64.42884	29.02958	54.35387	64.43105
	0.9	17.99087	44.46152	63.99939	17.51039	43.05230	64.13948	17.34156	42.81977	64.08224
	1.0	11.63104	31.28350	58.67088	11.68173	31.17826	58.57740	11.62263	30.78641	58.34422

The switching angles specified in Table 1 were implemented on the 7-level inverter, and the outcomes are documented in both Table 2 and Table 3, utilizing GA, PPSO, and BO algorithms.

Table 2. Fundamental voltage and THD, THDe values calculated for 7-level inverter using GA, PSO, and BO.

m		Vref (rms)	V1p(rms)			Error (%)			THD (%)			THDe (%)		
			GA	PSO	BO	GA	PSO	BO	GA	PSO	BO	GA	PSO	BO
low	0.1	22	21.83	21.85	**21.99**	0.77	0.68	**0.05**	119.01	109.58	**108.91**	107.04	99.75	**99.24**
	0.2	44	43.77	43.77	**43.94**	0.52	0.52	**0.14**	38.16	38.16	**37.56**	28.67	28.61	**28.30**
	0.3	66	65.19	65.84	**65.98**	1.23	0.24	**0.03**	41.06	41.92	**39.57**	10.74	10.06	**9.96**
medium	0.4	88	87.83	87.81	**87.99**	0.19	0.22	**0.01**	22.92	16.97	**16.87**	17.41	6.36	**6.39**
	0.5	110	109.4	109.7	**109.9**	0.55	0.27	**0.09**	18.02	17.67	**17.42**	3.01	1.51	**0.08**
	0.6	132	131.6	131.7	**131.8**	0.30	0.23	**0.15**	11.45	12.41	**12.28**	4.26	0.09	**0.08**
high	0.7	154	153.7	153.7	**154**	0.19	0.19	**0.00**	16.20	16.12	**12.22**	0.79	0.03	**0.07**
	0.8	176	175.5	175.5	**175.9**	0.28	0.28	**0.06**	10.15	10.74	**10.61**	4.51	0.06	**0.07**
	0.9	198	197.4	197.4	**197.9**	0.30	0.30	**0.05**	11.30	11.81	**11.82**	0.80	0.02	**0.05**
	1.0	220	219.2	219.4	**220**	0.36	0.27	**0.00**	7.58	7.64	**7.82**	0.06	0.05	**0.04**

Table 3 presents not only the THD and THDe metrics but also the root mean square (rms) values of the fundamental voltage and the control error associated with the fundamental voltage. Meanwhile, Table 3 provides detailed data on the individual harmonic components, specifically the fifth and seventh harmonics.

As seen in Table 2, GA controlled the fundamental voltage with a maximum error of 1.23%. However, GA performed worse than PSO and BO in suppressing the selected harmonics. As shown in Table 2, PSO controlled the fundamental voltage with a maximum error of 0.68%. The modulation index range where the selected harmonics are

Table 3. Selective harmonic analysis calculated for 7-level inverter using GA, PSO, and BO

m		5th (%)			7th (%)		
		GA	PSO	BO	GA	PSO	BO
low	0.1	83.06	79.04	**78.79**	67.52	60.86	**60.34**
	0.2	27.33	27.33	**26.88**	8.48	8.48	**8.87**
	0.3	2.87	3.83	**3.60**	10.35	9.30	**9.28**
medium	0.4	17.06	3.94	**4.17**	3.44	5.00	**4.84**
	0.5	1.46	1.51	**0.02**	2.64	0.05	**0.01**
	0.6	4.06	0.02	**0.04**	1.31	0.08	**0.01**
high	0.7	0.60	0.03	**0.03**	0.50	0.02	**0.01**
	0.8	4.17	0.02	**0.06**	1.71	0.05	**0.06**
	0.9	0.30	0.00	**0.01**	0.74	0.02	**0.03**
	1.0	0.03	0.02	**0.02**	0.05	0.04	**0.01**

best suppressed is between 0.6 and 1.0 modulation index, as indicated and BO obtained the fundamental voltage with a maximum error of 0.15%. The modulation index range where the selected harmonics are best suppressed is between 0.5 and 1.0 modulation index, as indicated.

As evident in Table 3, the BO algorithm consistently achieves the lowest degree for the 5th harmonic. For a modulation index of 0.8, the 7th harmonic is calculated as 0.05% using the PSO algorithm, while it is computed as 0.06% using BO. Despite the perceived success of the PSO algorithm, it exhibits a higher error in controlling the fundamental voltage for the specified modulation index than BO. In contrast, the GA proves to be unsuccessful in all scenarios when compared to the other two algorithms. Standard deviation is an essential measure in data analysis and decision-making processes to understand the statistical characteristics of data. The statistical analysis calculated for each algorithm is provided in Table 4, including the best value, worst value, and standard deviation. As shown in Table 4, the lowest standard deviation value was obtained with BO. PSO outperformed GA in achieving better results.

Figure 5 shows the comparative graph of the fitness function value versus the number of iterations for each algorithm at modulation index (MI) values of 1.0 in the case of a 7-level inverter, with iteration counts of 100, 250, and 300. As observed, BO consistently exhibits better fitness function values compared to other algorithms in each scenario. PSO also outperforms GA in terms of performance.

Figure 6 shows the THD results obtained with GA, PSO, and BO algorithms for a unit modulation index. The reference voltage for the fundamental is set to 220 V. The rms value of the fundamental voltage obtained with GA is 219.2 V, PSO is 219.4 V, and BO is 220 V. Looking at the THD values, GA yields 0.06%, PSO results in 0.05%, and BO achieves 0.04%. According to the results obtained, the only algorithm that controls the fundamental voltage with zero error is the BO algorithm.

Table 4. Statistical analysis (for 7-level inverter)

m		GA			PSO			BO		
		Best	Worst	Standard deviation	Best	Worst	Standard deviation	Best	Worst	Standard deviation
low	0.1	1111.4	19123	7813.135919	951.1474	1736.4	332.6229972	952.6792	955.6354	1.324076798
	0.2	325.2974	6013.9	2020.538926	314.3459	324.5548	4.467178108	309.0479	311.2584	0.855483226
	0.3	385.7752	2724.2	896.6537497	88.2993	422.23	143.522209	92.2559	92.6598	0.149307374
medium	0.4	277.1822	4701.5	1864.623153	62.1837	200.6163	59.34199071	62.1946	62.4298	0.100192175
	0.5	26.8089	2331.4	990.418412	5.46	2209.7	1066.717398	1.98E-06	0.0742	0.03316525
	0.6	63.3966	1485.9	625.6958397	9.06E-17	78.2655	41.66077504	4.13E-07	2.23E-05	9.15622E-06
high	0.7	3.1924	456.8197	201.171222	0.0667	234.7509	121.6623834	1.66E-06	0.000191	8.1884E-05
	0.8	28.5274	782.6004	378.5839594	0.5488	43.1771	17.45672198	3.71E-06	3.8793	2.124749863
	0.9	6.8951	304.7856	116.4138816	5.82E-14	697.2684	311.11709	7.97E-07	0.322286	0.144121914
	1	0.3359	73.8488	31.15177972	1.2E-11	0.5118	0.228560918	3.55E-07	8.11E-06	3.04909E-06

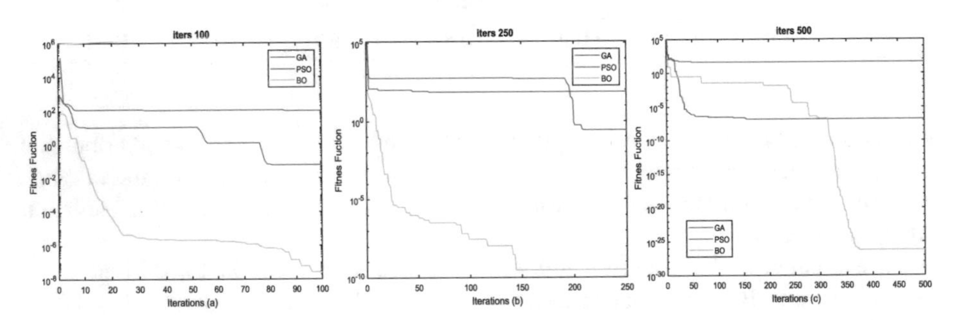

Fig. 5. Iteration and fitness function (a) 250, (b) 250, (c) 500 iter (for 7-level inverter)

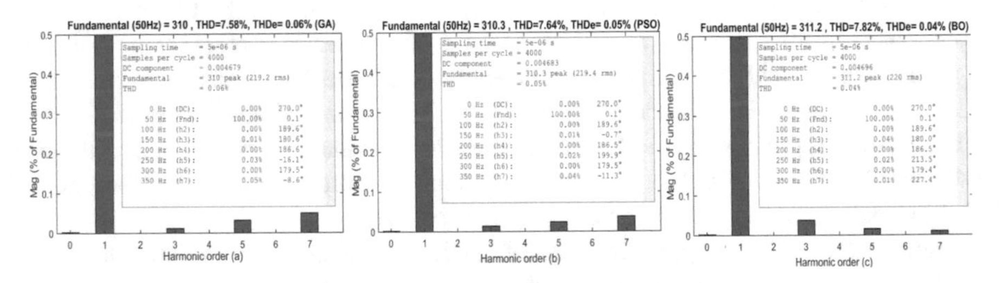

Fig. 6. THD value for MI = 1.0 (a) GA, (b) PSO, and (c) BO (for 7-Level Inverter)

Algorithm Comparison in 11-level Inverter. Table 5 provides the switching angles calculated for the 11-level inverter with GA, PSO, and BO algorithms. The calculated switching angles for the 11-level inverter are applied in Table 5. The switching angles given in Table 5 were applied to the 11-level inverter. Fundamental voltage, voltage control error, and THD and THDe values are given in Table 6 for GA, PSO, and BO algorithms. As seen in the table, the algorithm that controls the base voltage with the slightest error is the BO algorithm. PSO, on the other hand, controlled the principal stress better than GA. THD and THDe values were obtained with the BO algorithm at least harmonic value between 0.1 and 1.0 modulation index. Again, the most unsuccessful algorithm is the GA algorithm.

Table 5. Switching angles calculated for the 11-level inverter using GA, PSO, and BO

m		GA					PSO					BO				
		θ_1	θ_2	θ_3	θ_4	θ_5	θ_1	θ_2	θ_3	θ_4	θ_5	θ_1	θ_2	θ_3	θ_4	θ_5
low	0.1	73.17	84.63	89.84	89.84	89.84	66.88	90.00	90.00	90.00	90.00	66.80	90.00	90.00	90.00	90.00
	0.2	41.42	89.50	89.50	89.50	89.50	38.24	90.00	90.00	90.00	90.00	43.40	86.62	90.00	90.00	90.00
	0.3	49.45	58.33	89.95	89.95	89.95	43.26	63.26	90.00	90.00	90.00	44.34	64.85	87.75	90.00	90.00
medium	0.4	28.59	57.24	85.66	85.66	90.00	18.14	52.59	89.26	90.00	90.00	38.99	56.29	75.98	90.00	90.00
	0.5	44.23	47.90	66.69	84.57	85.03	36.53	50.18	65.09	86.37	88.00	36.71	49.58	65.47	84.21	90.00
	0.6	41.71	57.75	57.75	57.75	89.38	12.62	31.99	57.85	90.00	90.00	35.34	46.93	58.55	72.55	87.79
high	0.7	21.72	45.78	54.49	62.28	85.66	36.27	42.89	59.87	59.87	78.11	34.34	44.64	54.12	65.37	77.88
	0.8	27.50	27.50	52.25	66.00	69.67	22.56	39.09	52.27	59.66	71.04	9.25	25.11	42.17	61.19	88.14
	0.9	22.35	27.56	40.57	51.51	70.02	7.64	26.83	39.84	53.09	73.60	7.55	27.41	40.70	52.42	73.01
	1.0	11.86	12.20	28.13	57.01	57.01	7.92	19.35	29.60	47.90	63.05	7.63	19.27	29.23	47.24	63.07

Table 6. Fundamental voltage and THD, THDe values calculated for 11-level inverter using GA, PSO, and BO

m		Vref (rms)	V1p(rms)			Error (%)			THD (%)			THDe (%)		
			GA	PSO	BO	GA	PSO	BO	GA	PSO	BO	GA	PSO	BO
low	0.1	22	21.86	21.91	**21.99**	0.64	0.41	**0.05**	102.08	59.35	**58.99**	97.23	55.17	**54.72**
	0.2	44	43.79	43.84	**43.84**	0.48	0.36	**0.36**	28.59	27.64	**27.64**	19.99	18.27	**18.17**
	0.3	66	65.70	65.76	**65.84**	0.45	0.36	**0.24**	23.30	19.34	**19.40**	21.77	8.16	**8.08**
medium	0.4	88	87.71	87.69	**87.85**	0.33	0.35	**0.17**	26.16	13.45	**12.88**	21.32	11.42	**2.91**
	0.5	110	109.6	109.6	**109.7**	0.36	0.36	**0.27**	20.38	15.89	**9.05**	11.85	2.47	**1.05**
	0.6	132	131.5	131.5	**131.6**	0.38	0.38	**0.30**	20.62	7.60	**6.86**	15.97	2.41	**0.05**
high	0.7	154	153.3	153.4	**153.4**	0.45	0.39	**0.39**	10.58	7.63	**5.56**	7.68	2.47	**0.07**
	0.8	176	175.2	175.2	**175.6**	0.45	0.45	**0.23**	10.81	6.49	**6.58**	7.77	0.42	**0.05**
	0.9	198	197.2	197.2	**197.4**	0.40	0.40	**0.30**	9.85	6.28	**6.26**	8.89	0.89	**0.03**
	1.0	220	219	219	**219.8**	0.45	0.45	**0.09**	9.25	5.01	**4.89**	5.52	0.21	**0.03**

As seen in Table 6, GA exhibits a maximum error of 0.64% for the fundamental voltage. In the case of the 11-level scenario, GA's optimal operating range falls between modulation indices of 0.7 and 1.0, providing the best performance. On the other hand, Particle Swarm Optimization (PSO) maintains control of the fundamental voltage with a maximum error of 0.45% and demonstrates its best performance within the modulation index range of 0.5 to 1.0, as shown in Table 6. The same table also presents the harmonic analysis obtained with the BO algorithm. BO effectively controls the fundamental voltage with a maximum error of 0.36%. For the BO algorithm, the optimal operating range is within modulation indices of 0.5 to 1.0.

Table 7 displays the extent to which selected harmonics are suppressed when the switching angles calculated by all three algorithms are applied to the inverter. The BO algorithm outperforms others in suppressing the 7th, 11th, and 13th harmonics within the 0.2 to 1.0 modulation index range. PSO, on the other hand, exhibits better performance than GA. GA, albeit superior in suppressing the 11th harmonic at a modulation index of 0.1, falls behind the other algorithms in total harmonic suppression.

Table 7. Selective harmonic analysis calculated for 11-level inverter using GA, PSO, and BO

m		5th (%)			7th (%)			11th (%)			13th (%)		
		GA	PSO	BO	GA	PSO	BO	GA	PSO	BO	GA	PSO	BO
low	0.1	76.05	46.09	**45.75**	56.69	11.48	**11.05**	19.74	22.31	**22.33**	8.07	17.01	**16.74**
	0.2	18.31	12.94	**12.94**	1.85	2.88	**2.88**	5.58	12.38	**12.38**	5.46	2.19	**2.19**
	0.3	0.08	4.44	**4.44**	19.68	3.84	**3.75**	6.37	0.32	**0.26**	6.80	5.65	**5.61**
medium	0.4	2.80	0.88	**2.20**	10.88	2.73	**1.02**	4.60	10.86	**1.54**	17.53	2.04	**0.48**
	0.5	5.24	0.16	**0.49**	0.48	0.26	**0.35**	10.61	1.21	**0.75**	0.24	2.13	**0.42**
	0.6	1.16	1.37	**0.03**	14.90	0.11	**0.03**	0.02	1.35	**0.02**	5.64	1.45	**0.01**
high	0.7	0.71	0.12	**0.03**	2.84	1.32	**0.03**	5.81	2.03	**0.03**	4.08	0.46	**0.02**
	0.8	7.38	0.23	**0.04**	0.07	0.28	**0.01**	0.82	0.18	**0.01**	2.28	0.11	**0.03**
	0.9	7.17	0.29	**0.00**	5.25	0.68	**0.01**	0.11	0.08	**0.00**	0.06	0.48	**0.00**
	1.0	3.80	0.04	**0.02**	2.91	0.07	**0.01**	1.88	0.10	**0.01**	2.00	0.18	**0.01**

Each algorithm within 11 levels was run 5 times for the specified modulation index values, and the best, worst, and standard deviation values are given in Table 8. As seen in the table, PSO has a lower standard deviation value than the GA algorithm, while it lags behind the BO algorithm. The BO algorithm is the algorithm with the lowest standard deviation value.

Figure 7 displays a comparative graph of the fitness function value versus the number of iterations (100, 250, and 300 iterations) for each algorithm in the 11-level inverter case. As seen in the figure, BO outperforms the other algorithms regarding fitness function value in every scenario. GA exhibits lower performance than the other algorithms. Figure 8 depicts the results of THD (Total Harmonic Distortion) obtained with GA, PSO, and BO algorithms for a unit modulation index in the 11-level inverter case. The reference voltage is set to 220 V. The root mean square (rms) value of the fundamental voltage obtained with GA is 219 V, with PSO, it's 219 V, and with BO, it's 219.8 V. The THD values are 5.52% for GA, 0.21% for PSO, and 0.03% for BO. According to the results, the algorithm that controls the fundamental voltage with the most minor error is the BO algorithm.

Table 8. Statistical analysis (for 11-level inverter)

m		GA			PSO			BO		
		Best	Worst	Standard deviation	Best	Worst	Standard deviation	Best	Worst	Standard deviation
low	0.1	951.97	102710.00	52509.36	291.97	24516.00	9901.46	287.42100	289.70730	1.02246
	0.2	67.68	24844.00	9002.41	128.39	49967.00	21844.26	128.39480	129.27730	0.38183
	0.3	429.72	33738.00	12422.63	57.46	7510.30	3162.98	56.73510	57.54270	0.43200
medium	0.4	702.09	100770.00	42461.23	199.62	100930.00	44965.90	12.96480	15.58600	1.06094
	0.5	353.75	100550.00	44624.71	14.50	429.00	181.08	2.73580	6.31280	1.57164
	0.6	950.66	100260.00	43951.86	19.96	1164.00	469.21	0.00000	1.06710	0.58447
high	0.7	307.74	3490.00	1479.97	10.85	131.35	50.65	0.00000	0.00547	0.00244
	0.8	409.32	100090.00	54532.71	1.16	79.05	29.74	0.00000	0.00001	0.00000
	0.9	2550.70	100340.00	46206.52	6.19	100030.00	44724.84	0.00000	1.94870	0.87148
	1	129.17	100250.00	44677.60	0.37	583.80	230.74	0.00000	0.00000	0.00000

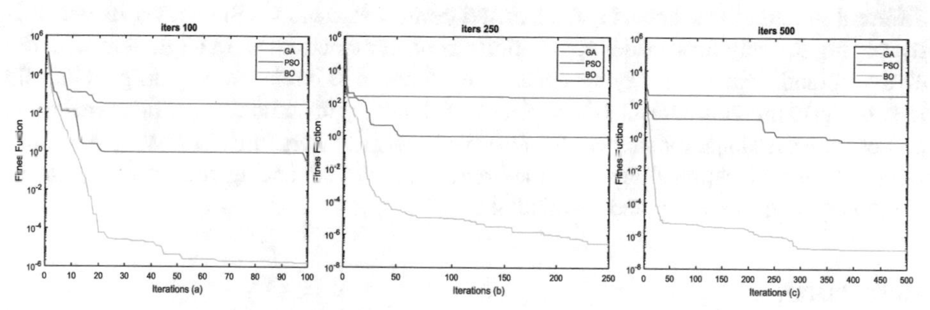

Fig. 7. Iteration and fitness function, (a) 250, (b) 250, (c) 500 iterations for the 11-level inverter.

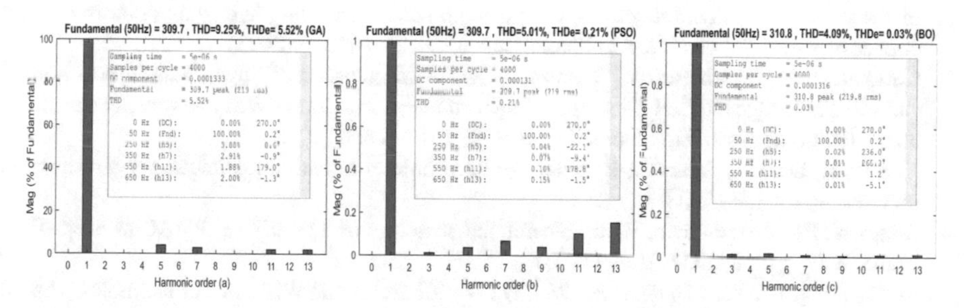

Fig. 8. THD value for MI = 1.0 (a) GA, (b) PSO, and (c) BO (for 11-Level Inverter)

6 Conclusion

In this study, the relatively newly developed algorithm has been adapted to solve SHE-PWM equations in multilevel inverters. The BO algorithm has been compared with the widely used GA and PSO algorithms for 7-level and 11-level inverter cases. The switching angles calculated with each algorithm were applied to 7-level and 11-level inverters in the MATLAB Simulink environment. The results demonstrate that the BO algorithm outperforms the others for both 7-level and 11-level cases. While the PSO algorithm performs better than the GA algorithm among the remaining algorithms, it lags behind the BO algorithm. In conclusion, the proposed BO algorithm can be further developed for different-level inverters.

In future endeavors, research efforts will concentrate on enhancing the BO algorithm for optimal performance. This involves meticulously refining algorithmic parameters to ensure adaptability to diverse SHE equations and various inverter configurations. To validate the algorithm's efficacy beyond theoretical simulations, real-world experiments with physical inverter setups are imperative. This approach will provide tangible insights into BO's practical application, offering a more comprehensive understanding of its performance.

Ensuring the robustness of BO under a spectrum of conditions will be a priority. Rigorous robustness analysis and sensitivity assessments will contribute to fortifying the algorithm's reliability and stability in practical, dynamic environments. Furthermore, diversifying the applications of BO is on the agenda. Exploration of its effectiveness in different modulation schemes and advanced control strategies will uncover its versatility, paving the way for broader applications in power electronics. In alignment with the evolving landscape of energy systems, a key focus will be on investigating BO's role in smart grid integration. BO's dynamic control and optimization capabilities make it a prospective candidate for shaping the future of energy systems within smart grid frameworks. This exploration aligns with the broader goal of advancing energy management towards greater efficiency and sustainability.

References

1. Raziq, H., Batool, M., Riaz, S., Afzal, F., Akgül, A., Riaz, M.B.: Power quality improvement of a distribution system integrating a large-scale solar farm using hybrid modular multilevel converter with ZSV control. Ain Shams Eng. J. **14**(7), 102218 (2023)
2. Garapati, D.P., Jegathesan, V., Veerasamy, M.: Minimization of power loss in newfangled cascaded H-bridge multilevel inverter using in-phase disposition PWM and wavelet transform based fault diagnosis. Ain Shams Eng. J. **9**(4), 1381–1396 (2018)
3. Kumar, J., Das, B., Agarwal, P.: Selective harmonic elimination technique for a multilevel inverter. Space **1**, 3 (2008)
4. Bhagwat, P.M., Stefanovic, V.R.: Generalized structure of a multilevel PWM inverter. IEEE Trans. Ind. Appl. **6**, 1057–1069 (1983)
5. Li, S., Song, G., Ye, M., Ren, W., Wei, Q.: Multiband SHEPWM control technology based on Walsh functions. Electronics **9**(6), 1000 (2020)
6. Al-Hitmi, M., Ahmad, S., Iqbal, A., Padmanaban, S., Ashraf, I.: Selective harmonic elimination in a wide modulation range using modified Newton-Raphson and pattern generation methods for a multilevel inverter. Energies **11**(2), 458 (2018)
7. Fei, W., Ruan, X., Wu, B.: A generalized formulation of quarter-wave symmetry SHE-PWM problems for multilevel inverters. IEEE Trans. Power Electron. **24**(7), 1758–1766 (2009)
8. Yaqoob, M.T., Rahmat, M.K., Maharum, S.M.M.: Modified teaching learning based optimization for selective harmonic elimination in multilevel inverters. Ain Shams Eng. J. **13**(5), 101714 (2022)

9. Juma'a, H., Atyia, T.: Design and implementation of multi-level inverter for PV system with various DC Sources. NTU J. Renew. Energy **5**(1), 24–33 (2023)
10. Bektaş, Y., Karaca, H.: Red deer algorithm based selective harmonic elimination for renewable energy application with unequal DC sources. Energy Rep. **8**, 588–596 (2022)
11. Das, A.K., Pratihar, D.K.: Bonobo optimizer (BO): an intelligent heuristic with self-adjusting parameters over continuous spaces and its applications to engineering problems. Appl. Intell. **52**(3), 2942–2974 (2022)

Creating Synthetic Test Data by Generative Adversarial Networks (GANs) for Mobile Health (mHealth) Applications

Nadeem Ahmad[1(✉)], Irum Feroz[1,2], and Faizan Ahmad[3]

[1] Department of Computing and Technology, Iqra University, Islamabad Campus, Islamabad, Pakistan
nadeem.ahmad@iqraisb.edu.pk
[2] Department of Computing, University of Portsmouth, Portsmouth, UK
[3] Cardiff School of Technology, Cardiff Metropolitan University, Cardif, UK

Abstract. Mobile health (mHealth) applications have experienced rapid growth, driven by the demand for health monitoring solutions and smartphone adoption. However, evaluating these apps poses challenges due to limited and diverse user data. This study explores the use of Generative Adversarial Networks (GANs) to generate synthetic test data for mHealth applications. The paper introduces the methodology involved in training GANs using real user data obtained from Google Fitbit and showcases the creation of synthetic data mirroring real user profiles and parameters. Statistical comparisons between real and synthetic datasets validate the alignment and similarities in key attributes such as age, BMI, and exercise duration. The paper elucidates the importance of user-centered design methodologies and the role of test data in mHealth app evaluation. User personas and diverse user scenarios are incorporated, showcasing the efficacy of synthetic data in mitigating data limitations. The study emphasizes the potential of synthetic test data to enhance the evaluation and validation of mHealth applications, providing a pathway to address data scarcity challenges. Future research avenues are outlined, including expanding user diversity, refining GAN models, and assessing the impact of synthetic data on machine learning models within mHealth apps. The study advocates for ethical considerations and privacy safeguards in synthetic data generation and usage, suggesting frameworks for responsible implementation. This research contributes to advancing mHealth application testing methodologies by leveraging GANs to create diverse and reliable synthetic test data.

Keywords: Mobile Health (mHealth) Applications · Generative Adversarial Networks (GANs) · Synthetic Data · Usability Evaluation

1 Introduction

Mobile health applications, or mHealth apps, are software programs that use mobile devices to provide health-related services and information. According to the forecast, the mHealth apps industry will expand from USD 56.26 billion in 2022 to USD 861.40

J. Rasheed et al. (Eds.): FoNeS-AIoT 2024, LNNS 1035, pp. 322–332, 2024.
https://doi.org/10.1007/978-3-031-62871-9_25

billion in 2030, with a CAGR of 40.2% in this period [1]. The main drivers of this growth are the increasing demand for accurate health monitoring solutions, the rising penetration of smartphones and internet connectivity, the growing adoption of wearable devices, and the emergence of COVID-19 contact tracing apps. Some of the challenges faced by the mHealth apps industry are the lack of standardization and regulation, privacy and security issues, and low awareness and trust among users. The mHealth apps market can be segmented by app type, application, platform, and region. The market is dominated by some major companies, such as Google, Apple, Fitbit, Samsung, and others [2].

To evaluate and validate the quality and effectiveness of mHealth applications, various tools, and methods have been developed and used by researchers and practitioners. Test data are the inputs and outputs of the mHealth applications that are used to measure their performance, usability, satisfaction, acceptance, and quality outcomes. Test data can be collected from different sources, such as built-in smartphone sensors, user feedback, health records, and external devices. Test data can also be obtained from users themselves, either by manually entering their health information or by performing exercises that are monitored by the mHealth applications. Test data can be analyzed using different techniques, such as descriptive statistics, inferential statistics, machine learning, and data mining. Test data and patterns can reveal the strengths and weaknesses of mHealth applications, as well as the needs and preferences of the users. Hence, the utilization of test data proves crucial in enhancing the design and advancement of exercise-centric mHealth applications [3].

Generative Adversarial Networks (GANs) are a type of machine learning framework that can generate realistic data from a given dataset. They consist of two neural networks: a generator that creates fake data and a discriminator that evaluates how real the data is. The generator and the discriminator compete in a zero-sum game, where the generator tries to fool the discriminator and the discriminator tries to distinguish the real from the fake. GANs have many applications in various domains, such as image generation, text generation, natural language processing, and computer vision. They can also learn the underlying features and structure of the data, which can be useful for representation learning and unsupervised learning. GANs can also perform image-to-image translation, such as converting sketches to photos or changing the style or season of an image. GANs can also generate text from images, such as captions or descriptions. Some of the challenges of GANs are mode collapse, where the generator produces similar outputs, and training instability, where the generator and the discriminator oscillate between good and bad performance. Since their introduction in 2014, GANs have undergone significant development and enhancements, demonstrating remarkable efficacy in producing varied, high-quality datasets [4, 5].

Synthetic test data is a type of data that is artificially created to simulate the characteristics and properties of real data. Synthetic test data can be useful for various purposes, such as testing the performance and robustness of machine learning models, protecting the privacy and security of sensitive data, and augmenting the existing data to improve quality and diversity. Generative Adversarial Networks (GANs) are a powerful technique to generate synthetic test data, especially for complex and high-dimensional data such as images, text, and tabular data. GANs can learn the latent features and distributions of real data and produce realistic and diverse synthetic data that can fool a discriminator, which is a classifier that tries to distinguish between real and fake data. GANs have found applications in creating artificial test datasets across multiple fields like healthcare, finance, and cybersecurity [6]. Their use has demonstrated encouraging outcomes in elevating both data quality and usefulness [7].

In this study, the researchers explored how mobile health applications accommodate various user-input data. Validating health applications across diverse scenarios is crucial for thorough testing and evaluation. However, limitations in user availability and data coverage may sometimes hinder the comprehensive inclusion of all user scenarios. To address this limitation, synthetic test data can be generated using Generative Adversarial Networks (GANs). By leveraging existing user personas and input parameters, the model is trained to create additional scenarios and input data. This approach mitigates the inadequacies stemming from missing data. The newly generated data mirrors the properties of the original dataset, exhibiting similar statistical ratios and properties among other values.

Section two elaborates on previously published works in this field. Section three outlines our approach and methodology. Sction four delineates the steps employed in generating synthetic test data using GANs. Section five provides information about data generated by users, then produces synthetic data by using GANs and statistically analyzes both data groups. Finally, in section six, conclusions are drawn from the findings, and avenues for future research are discussed.

2 Related Work

User test data is the data that is collected from the users of mHealth applications, such as their health information [8, 9], feedback [10, 11], preferences [12], and behaviors [13]. User test data is important for mHealth applications, especially for exercise-related applications, because it can help to evaluate and improve the quality, usability, effectiveness, and satisfaction of the applications. User test data can also help to understand the needs and challenges of the users and to tailor the applications to their specific contexts and goals. Various methodologies, including surveys, interviews, observational studies, experiments, and analytics, are utilized to gather and analyze user test data [14–16].

User-Centered Design (UCD) methodologies play a pivotal role in involving users throughout the design process of mHealth applications, ensuring their needs and preferences are integrated. This methodology employs various techniques like interviews, focus groups, usability testing, and participatory design workshops to gather rich and appropriate test data [17, 18]. By actively engaging users in the development process, UCD ensures that the generated test data accurately reflects real-world scenarios and user behaviors, thus enhancing the evaluation and refinement of mHealth applications.

The evaluation and validation of mHealth applications constitute a burgeoning field, drawing attention from researchers to assess their efficiency and performance [19]. Test data play a pivotal role in these evaluations, measuring the application's usability, effectiveness, and quality outcomes. Diverse analytical methods have been employed to comprehend the strengths and weaknesses of these applications [17, 18]. Security and privacy challenges persist in the development of mHealth applications, presenting an ongoing concern for developers and users [20, 21]. Test data act as a bulwark against potential threats, emphasizing the need for standardized practices in fortifying these applications against security breaches and privacy invasions. Generative Adversarial Networks (GANs) have evolved significantly since their inception, transcending various domains and applications [22]. The challenges encountered by GANs in their journey towards efficiency enhancement and operational efficacy have been addressed progressively.

Synthetic test data generation, particularly within healthcare domains, has been instrumental in fortifying datasets and improving diversity [23]. The implications of synthetic data in healthcare applications underscore the potential for enriching data quality and augmenting existing datasets. The role of GANs in augmenting datasets spans across industries, with healthcare and cybersecurity domains demonstrating substantial outcomes [24]. The efficacy of GANs in enhancing data quality and usefulness within these specific sectors has been meticulously evaluated and demonstrated.

The integration of synthetic test data into the validation processes of mHealth applications has emerged as a viable solution [25]. Leveraging GANs to simulate additional user scenarios addresses the limitations posed by inadequate real user data, thereby enhancing the validation procedures and contributing to a more comprehensive evaluation.

3 Methodology

In this research, the Google Fitbit application has been used for generating user test data. Google Fitbit is a mobile health application that works with Fitbit devices and Wear OS by Google smartwatches to provide health-related services and information. It was introduced in 2014 as a standalone app and later integrated with Google Fit in 2023 after Google acquired Fitbit. Google Fitbit allows users to track their activity, sleep, heart rate, stress, nutrition, and more. It also offers personalized guidance, insights, and motivation to help users achieve their health and fitness goals. According to the Google Play Store, Google Fitbit has over 50 million downloads and 1.05 million reviews. Google Fitbit is significant in managing health because it can help users monitor their health status, improve their well-being, and prevent or manage chronic diseases [26]. In a controlled laboratory setting, a study was conducted involving ten diverse users, each tasked with recording their exercise activities using Fitbit devices. The participants underwent a preliminary briefing lasting approximately 10 min, outlining the study's objectives, and their participation was entirely voluntary. Prior consent was obtained from each user, following which they were allotted 30 min to document their exercise specifics, inputting various metrics into a provided datasheet. These metrics encompassed demographic information such as age, gender, weight, height, and BMI, as well as exercise-related parameters including activity type, duration, distance, steps, calories burned, heart rate, stress level, and sleep quality. Subsequently, a single researcher performed an analysis of the collected data to evaluate its comprehensiveness and integrity. The input data gathered from users is provided in Table 1 in Sect. 6.

Based on this user data, an artificially generated dataset is created. The creation and training of Generative Adversarial Networks (GANs) for generating synthetic user data involve initial model selection based on the dataset, followed by training the GAN architecture using the collected dataset. The training process optimizes both the generator and discriminator networks in an adversarial manner, refining the generator's capacity to produce realistic data while enhancing the discriminator's ability to differentiate between real and synthetic samples. This iterative process aims to achieve a model capable of generating high-fidelity and diverse synthetic data mirroring the characteristics of real user data. The detailed steps are described in section four, following this procedure, four fictional personas are formulated, and their corresponding user data are synthetically generated using GANs. Subsequently, a comparative analysis is conducted between the newly generated data and the existing dataset to assess both the relevance and disparities between them.

4 Steps for Test Data Generation by Using GANs

Generating synthetic test data by using GANs for mHealth app involves collecting and preprocessing real mHealth app data, selecting and training a suitable GAN model, setting up and training the generator and discriminator networks, validating and evaluating the synthetic data, fine-tuning and adjusting the GAN model, and applying and using the synthetic data for various purposes. Generating synthetic data using GANs requires careful attention to data privacy, ethical issues, and data quality.

The synthetic generation of test data in mobile health (mHealth) applications involves several steps:

4.1 Data Collection and Understanding

Gather a comprehensive dataset of real mHealth app data, including user activities, health metrics, exercise routines, and vital signs. Understand the structure, features, and patterns within the collected dataset.

Preprocessing and Cleaning. Clean the data to remove outliers, inconsistencies, and any noise that might impact the generation process. Normalize or scale the data to ensure uniformity and comparability across different features.

Model Selection and Training. Choose an appropriate GAN architecture suited for generating mHealth app data. Variants like Deep Convolutional GANs (DCGANs) or Wasserstein GANs might be suitable for this task. Train the GAN using the preprocessed dataset, enabling it to learn the underlying patterns and distributions in the data.

Generator and Discriminator Setup. Define the generator network responsible for synthesizing mHealth data samples. Create a discriminator network that distinguishes between real and synthetic data.

Training the GAN. Train the GAN by optimizing both the generator and discriminator networks in an adversarial manner. Iteratively improve the generator's ability to create realistic synthetic data while enhancing the discriminator's ability to differentiate between real and generated samples.

Validation and Evaluation. Validate the generated synthetic data by comparing statistical properties, distributions, and patterns against the original real dataset. Use evaluation metrics to assess the similarity and fidelity of the synthetic data to real mHealth data.

Fine-Tuning and Adjustment. Adjust the GAN's hyperparameters, such as learning rates or network architecture, if the generated data lacks fidelity or diversity compared to real data. Fine-tune the GAN model until the synthetic data closely resembles real mHealth app data in terms of features, distributions, and statistical properties.

Application and Usage. Deploy the trained GAN model to generate large volumes of synthetic mHealth app data that can be used for various purposes like testing app functionalities, training machine learning models, or enhancing privacy in data-sharing scenarios. All the eight steps involved in synthetic test data generation by using GANs are summarized in Fig. 1.

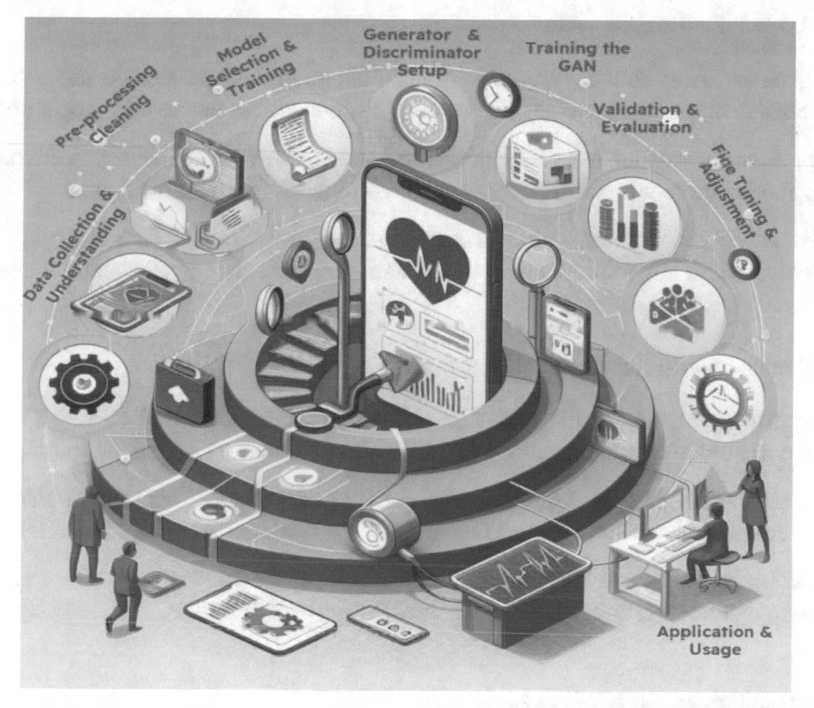

Fig. 1. Synthetic test data creation by using Generative Adversarial Networks (GANs).

Generating synthetic data using GANs involves complex modeling and data generation processes, requiring careful consideration of data privacy, ethical implications, and ensuring that the generated data maintains the essential characteristics of the original dataset.

5 Sample Personas and Comparison of Data

As discussed in Sect. 3, 10 users were involved in gathering input data for the Fitbit application. The user-created input data is provided in Table 1. Here are four distinct personas tailored for exercise tracking within Google Fitbit:

5.1 Fitness Beginner - Maya

Background. Maya is a 25-year-old recent graduate who just started her fitness journey. She's moderately active but wants to establish a consistent exercise routine.

Goals. Track daily steps, set achievable fitness goals, receive motivational reminders and access beginner-friendly workout routines.

Needs. User-friendly interface, simple tracking features, guided workouts, and encouragement prompts for motivation.

5.2 Athlete Enthusiast - Alex

Background. Alex is a 30-year-old former college athlete passionate about fitness. They engage in various activities like running, weightlifting, and cycling regularly.

Goals. Monitor detailed workout metrics, analyze performance, set customized training plans, and integrate with third-party apps for comprehensive data insights.

Needs. Advanced tracking options, compatibility with wearables, GPS mapping, heart rate analysis, and integration with nutrition and sleep data for holistic health tracking.

5.3 Senior Health Advocate - Mr. Johnson

Background. Mr. Johnson is a 65-year-old retiree focusing on maintaining a healthy lifestyle. He practices yoga, walks daily and enjoys swimming occasionally.

Goals. Monitor heart health, track daily activities, receive gentle reminders for movement, access low-impact workout suggestions, and monitor vitals if available.

Needs. Large font sizes, simplified interface, step-by-step instructions for exercises, heart rate monitoring, and reminders for medication or hydration.

5.4 Fitness Enthusiast Parent - Sarah

Background. Sarah is a 35-year-old working mother with a hectic schedule. She strives to balance work, family, and fitness. She enjoys HIIT workouts and yoga.

Goals. Quick access to short, effective workouts, schedule reminders for exercise, monitor stress levels, integrate with family fitness challenges and sync with her smartwatch.

Needs. Short, high-intensity workouts, stress management tools, family-oriented challenges, customizable reminders, and seamless integration with other family members' fitness apps.

Table 1. The input data gathered from users for the Fitbit application.

User Id	991	992	993	994	995	996	997	998	999	1000
Persona	Health Conscious	Busy Student	Senior Health Advocate	Fitness Beginner	Athlete Enthusiast	Fitness Enthusiast Parent	Health Conscious	Busy Student	Senior Health Advocate	Fitness Beginner
Age	28	19	68	33	27	39	31	21	64	35
Gender	f	m	f	m	f	m	f	m	m	f
Weight	55	75	65	85	58	72	52	77	69	68
Height	162	175	160	182	168	178	159	176	165	163
BMI	21	24	25	26	21	23	21	25	25	26
Activity Type	cycling	jogging	swimming	walking	weightlifting	hiking	yoga	hiit	tai chi	running
Duration	40	20	30	15	60	90	45	25	40	10
Distance	15	3	0	1.5	0	12	0	0	0	2
Steps	0	4000	0	2000	0	0	0	0	0	3000
Calories	450	200	250	100	500	700	180	350	150	150
Heart Rate	120	100	85	95	150	130	75	140	80	110
Stress Level	2	3	1	4	2	3	1	4	1	4
Sleep Quality	4	3	4	3	5	4	5	3	4	3

These personas represent diverse user segments with unique fitness goals, preferences, and technical needs. Creating user personas helps in designing features, interfaces, and functionalities tailored to specific user groups, enhancing the user experience within fitness tracking applications like Google Fitbit.

The output data created by GANs is synthetic data that resembles the real data from Google Fitbit but does not contain any sensitive or personal information of the users.

The synthetic data have the same columns and features as the real data, such as user_id, persona, age, gender, weight, height, bmi, activity_type, duration, distance, steps, calories, heart_rate, stress_level, and sleep_quality. The synthetic data also preserve the statistical properties and correlations of the real data, such as the distribution of age, bmi, activity_type, etc. The synthetic data is generated from random noise vectors, which are transformed by the generator model into plausible data instances. The discriminator model provides feedback to the generator model, by trying to distinguish the fake data from the real data. The goal of the algorithm is to train the generator model to produce data that can fool the discriminator model, and thus, resemble the real data as closely as possible. Table 2 provides an example of synthetic data generated by GANs.

The comparison between the user data presented in Table 1 and the synthetic data generated by GANs in Table 2 aims to assess their alignment and similarity in properties. The average age in the user input data is $\bar{x}1 = 36.5$ with a standard deviation of 16.69, while in the synthetic data, it is $\bar{x}2 = 39.75$ with a standard deviation of 17.97. The Wilcoxon Rank-Sum Test (Mann-Whitney U Test) conducted on the age of both groups indicates that the difference between the randomly selected values of Group 1 and Group 2 populations is not statistically significant (Z = -0.2832 falls within the 95% region of acceptance: [−1.96: 1.96], U = 17.5 within [6.1562: 33.8438]). Similarly, for BMI, the mean averages for Group 1 and Group 2 are $\bar{x}1 = 23.7$, S.D = 2.06 and $\bar{x}1 = 23.5$, S.D = 1.29, respectively, where ties and identical values exist. The test results (Z = 0.2883,

U = 22.5 within [6.4023: 33.5977]) suggest that the assumption of equality between the randomly selected values of both populations cannot be rejected since the p-value > α. Additionally, for exercise duration, the averages of Group 1 and Group 2 are $\bar{x}1 = 37.5$, S.D = 23.83 and $\bar{x}1 = 38.75$, S.D = 17.5, respectively. The data in both groups also exhibit a symmetrical shape, indicating no significant difference.

Table 2. Synthetic user test data created by Generative Adversarial Networks (GANs).

User Id	1001	1002	1003	1004
persona	Fitness Beginner	Athlete Enthusiast	Senior_Health Advocate	Fitness Enthusiast Parent
Age	26	31	66	36
Gender	F	M	M	F
Weight	60	80	70	65
Height	165	180	170	168
BMI	22	25	24	23
Activity Type	walking	running	yoga	hiit
Duration	30	45	60	20
Distance	2.5	10	0	0
Steps	4000	12000	0	0
Calories	150	600	200	300
Heart Rate	90	160	80	140
Stress Level	2	3	1	4

6 Conclusion and Future Work

This paper investigates the generation of synthetic test data for mobile health (mHealth) applications using Generative Adversarial Networks (GANs). Mobile health applications have seen exponential growth, driven by the increasing demand for accurate health monitoring solutions and the proliferation of smartphones and wearables. Despite this, challenges such as standardization, privacy concerns, and limited user trust persist. To evaluate these applications effectively, robust test data that accurately reflects user scenarios is crucial. Synthetic test data, generated by GANs, provides a promising solution to supplement existing data. The study outlines the methodology, utilization of GANs, and the generation process. Statistical inference and comparisons between real and synthetic data confirm alignment and similarity in properties, validating the efficacy of the generated synthetic data. The paper contributes to advancing mHealth application evaluation by demonstrating the potential of synthetic data in mitigating limitations due to inadequate real user data, thereby enhancing comprehensive testing and evaluation methodologies. Moving forward, this study opens avenues for various research trajectories in the realm of mHealth application testing and validation. Firstly, expanding the diversity of user personas and their corresponding data attributes can improve the comprehensiveness and accuracy of synthetic data generation. Exploring advanced GAN variants or hybrid models to address challenges like mode collapse and training instability may enhance the quality and diversity of synthetic data. Additionally, investigating the impact of synthetic data in training machine learning models within mHealth applications can ascertain its efficacy in improving model robustness and generalization. The study also highlights the importance of ethical considerations and privacy preservation when generating and utilizing synthetic data, emphasizing the need for guidelines and frameworks to govern its ethical usage. Further research into refining GAN models for generating synthetic test data tailored to specific mHealth applications and user scenarios can bolster the utility and reliability of synthetic data in application testing and validation.

References

1. MHealth apps market size, Analysis: Global Report. In: mHealth Apps Market Size, Analysis | Global Report (2030). https://www.fortunebusinessinsights.com/mhealth-apps-market-102020. Accessed 7 Jan 2024
2. MHealth market size, share, Growth & Trends Report. In: mHealth Market Size, Share, Growth & Trends Report (2030). https://www.grandviewresearch.com/industry-analysis/mhealth-market. Accessed 7 Jan 2024
3. Maramba, I., Chatterjee, A., Newman, C.: Methods of usability testing in the development of eHealth Applications: a scoping review. Int. J. Med. Inf. **126**, 95–104 (2019). https://doi.org/10.1016/j.ijmedinf.2019.03.018
4. Labaca-Castro, R.: Generative adversarial nets. In: Machine Learning under Malware Attack, pp. 73–76 (2023). https://doi.org/10.1007/978-3-658-40442-0_9
5. Karras, T., et al.: Progressive growing of gans for improved quality, stability, and Variation (2018). arXiv.org. https://doi.org/10.48550/arXiv.1710.10196. Accessed 07 Jan 2024
6. Ghatak, D., Sakurai, K.: A survey on privacy preserving synthetic data generation and a discussion on a privacy-utility trade-off problem. In: Communications in Computer and Information Science, pp. 167–180 (2022). https://doi.org/10.1007/978-981-19-7769-5_13
7. Ghatak, D., Sakurai, K.: A survey on privacy preserving synthetic data generation and a discussion on a privacy-utility trade-off problem (1970). https://doi.org/10.1007/978-981-19-7769-5_13. Accessed 07 Jan 2024
8. Faizan, A., et al.: A pilot study on the evaluation of cognitive abilities' cluster through game-based computationally intelligent technique. Multimedia Tools Appl. (MTAP) (2023). https://doi.org/10.1007/s11042-023-15100-x
9. Sara, M., Laurianne, S., Faizan, A.: Opportunities for serious game technologies to engage children with autism in a Pakistani sociocultural and institutional context. In: OzCHI 2022 (2022). https://doi.org/10.1145/3572921.3572923
10. Faizan, A., et al.: Comprehending the influence of brain games mode over playfulness and playability metrics: a fused exploratory research of players' experience. Interact. Learn. Environ. (NILE) (2023). https://doi.org/10.1080/10494820.2023.2205906
11. Ahmad, F., et al.: A study of players' experiences during brain games play. In: Booth, R., Zhang, M.-L. (eds.) PRICAI 2016. LNCS (LNAI), vol. 9810, pp. 3–15. Springer, Cham (2016). https://doi.org/10.1007/978-3-319-42911-3_1

12. Faizan, A., et al.: Effect of gaming mode upon the players' cognitive performance during brain games play: an exploratory research. Int. J. Game-Based Learn. (IJGBL) **11**(1), 5 (2021). https://doi.org/10.4018/IJGBL.2021010105
13. Faizan, A., et al.: Behavioral profiling: a generationwide study of players' experiences during brain games play. Interact. Learn. Environ. (NILE) (2020). https://doi.org/10.1080/10494820.2020.1827440
14. Saparamadu, A.A., et al.: User-centered design process of an mHealth app for health professionals: case study. JMIR mHealth uHealth **9**(3) (2021). https://doi.org/10.2196/18079
15. Farao, J. et al.: A user-centred design framework for mHealth, PLOS ONE **15**, e0237910 (2022). https://journals.plos.org/plosone/article?id=10.1371%2Fjournal.pone.0237910. Accessed 07 Jan 2024
16. Faizan, A., et al.: BrainStorm: a psychosocial game suite design for non-invasive cross-generational cognitive capabilities data collection. J. Exp. Theor. Artif. Intell. (JETAI) **29**(6) (2017). https://doi.org/10.1080/0952813X.2017.1354079
17. Feroz, I., et al.: Identification of critical success factors in adoption of health IT services from older people's perspective. In: Proceedings of the 2023 3rd International Conference on Human Machine Interaction, ACM Other Conferences (2023). https://doi.org/10.1145/3604383.3604389. Accessed 07 Jan 2024
18. Feroz, I., Ahmad, N.: Usability based Rating Scale (UBRS) for evaluation of Mobile Health (mHealth) applications (2022), http://www.eurekaselect.com. https://www.eurekaselect.com/chapter/16304. Accessed 07 Jan 2024
19. Durrani, M.I., Qureshi, N.S., Ahmad, N., et al.: A health informatics reporting system for technology illiterate workforce using mobile phone. Appl. Clin. Inf. **10**, 348–357 (2019). https://doi.org/10.1055/s-0039-1688830
20. Alenoghena, C.O., et al.: EHealth: A survey of architectures, developments in mHealth, security concerns and solutions. Int. J. Environ. Res. Public Health **19**(20), 13071 (2022)
21. Lee, J., Kim, S., Gordon, V.S.: National standardization strategy in Health In-formation Security Environment. In: 2013 International Conference on IT Convergence and Security (ICITCS) (2013). https://doi.org/10.1109/icitcs.2013.6717862
22. Audichya, P., Gupta, D., Singh, A.: Generative adversarial networks: models and techniques - a review. In: 2022 IEEE World Conference on Applied Intelligence and Computing (AIC) (2022). https://doi.org/10.1109/aic55036.2022.9848870
23. Goncalves, A., Ray, P., Soper, B., et al.: Generation and evaluation of synthetic patient data. BMC Med. Res. Methodol. (2020).https://doi.org/10.1186/s12874-020-00977-1
24. Gonzales, A., Guruswamy, G., Smith, S.R.: Synthetic data in health care: a narrative review. PLOS Digital Health (2023). https://doi.org/10.1371/journal.pdig.0000082
25. Dutta, I.K., Ghosh, B., Carlson, A., et al.: Generative adversarial networks in security: a survey. In: 2020 11th IEEE Annual Ubiquitous Computing, Electronics & Mobile Communication Conference (UEMCON) (2020). https://doi.org/10.1109/uemcon51285.2020.9298135
26. Wang, C., Lizardo, O., Hachen, D.S.: A longitudinal study of fitbit usage behavior among college students. Cyberpsychol. Behav. Soc. Netw.Netw. **25**, 181–188 (2022). https://doi.org/10.1089/cyber.2021.0047

Finding the Flies in the Sky Using Advanced Deep Learning Algorithms

Mustafa Tokat, Sümeyra Bedir(✉), and Jawad Rasheed

Department of Computer Engineering, Istanbul Sabahattin Zaim University, Istanbul, Turkey
520322002@std.izu.edu.tr, {sumeyra.bedir,jawad.rasheed}@izu.edu.tr

Abstract. Image processing is a concept that allows us to perform operations on images by extracting meaningful features from them. Deep learning algorithms are frequently used in image processing. Although there are many algorithms for image classification, in this study, Convolutional Neural Network based deep learning algorithms VGG16, ResNet50 and YOLOv8 (You Only Look Once) are used. After traditional algorithms such as Support Vector Machine, Linear Regression, KNN or Decision Tree, deep learning algorithms have been developed that perform very well in image processing. As we can see from this study, we can say that the YOLO algorithm has recently gone one step further. In the dataset used, there are a total of 1359 images with and without drone images. The images were pretrained on imagenet and used. When we compare these fine-tuned algorithms, the accuracy of VGG16 was %92.74, ResNet50 was %91.06 and YOLOv8 was %95.4. It was determined that YOLOv8 performed better according to the dataset used.

Keywords: Drone detection · convolutional neural network · VGG16 · Resnet50 · YOLOv8

1 Introduction

With the introduction of computer technologies into our lives, both the quality of life of individuals has increased and science has been the greatest contribution of the last century to humanity in macro areas such as health, education and security. In a study conducted by [8] Convolutional Neural Network (CNN)-based computer vision has opened up new possibilities in areas such as face recognition, autonomous vehicles, self-services and intelligent medical treatment. Unmanned aerial vehicles, which were born as an answer to the different needs of humanity today, are one of these opportunities.

Drones are used in a wide range of sectors, including military, security, agriculture, entertainment, filmmaking and rescue operations. According to a recent report, the market value of drones, which was USD 19.89 billion in 2022, is

J. Rasheed et al. (Eds.): ICAIGC 2023, LNNS 1035, pp. 333–344, 2024.
https://doi.org/10.1007/978-3-031-62871-9_26

estimated to reach USD 57.16 billion in 2030 [4]. With the increasing use of drones, the need for drone detection technologies in military and civilian life is also increasing. These technologies can perform important tasks such as monitoring uncontrolled drone traffic in the airspace, detecting unauthorized use and neutralizing these vehicles militarily. The widespread use of drones has brought with it a number of new problems. In particular, problems such as uncontrolled use of drones, privacy violations and security threats are among the important issues related to drones. Finding and tracking drones in a specific area is known as drone detection [1].

Deep learning algorithms are a machine learning topic in artificial intelligence. There are many machine learning algorithms, such as Support Vector Machine (SVM), Linear Regression, K-Nearest Neighbor (K-NN), Decision Tree. These algorithms are considered as traditional by many researchers, however, they have the disadvantage of working with small datasets and some limitations in image classification [6]. CNN has many advantages. For example, by reducing the number of parameters in the Artificial Neural Network, it has allowed both researchers and developers to work with large datasets to solve complex tasks [2]. In addition, unlike traditional methods, CNN does not require manual feature extraction. [8]. In a study, it was stated that CNN algorithms have better results than these traditional algorithms [9]. In this our study, CNN based VGG16, ResNet50 and YOLOv8 were used for drone detection.

2 Related Works

In [9] 712 movies consisting of drones and birds were evaluated in MATLAB. After the video frames were extracted, the frames with drones were put in one file and the frames with birds were put in a separate file. As a result, CNN, SVM and K-NN were used for drone detection and their accuracy scores were compared, respectively %95, %88, %80.

Haque et al. in the proposed method of VGG with ResNet network used the PASCAL VOC2012 dataset, which consists of 8 classes and contains 21503 images. The average mAP score of this method was found to be %85,8 [6].

Aote et al. evaluated a total of 209,324 images including only drones, only birds and both drones and birds. In this dataset, YOLOv3, YOLOv4, YOLOv5, CNN, R-CNN, Fast R-CNN, SSD and their proposed method CNN-LSTM were compared, the mAP scores are respectively; %73.53, %76.34, %81.51, %52.27, %66, %74.86, %79.34, %83.28 [3].

In [14], a dataset of 1359 images was used. The dataset in this study is divided into a single drone of 1297 images and a multiple drone of 62 images. VGG16 and ViT-b16 models were implemented on the single dataset and validation accuracies of %89 and %95.3 were obtained respectively. In the same study, it was stated that ViT-b16 model runs 3 times slower than VGG16.

Agarwal et al. in [1] compared YOLOv7 and YOLOv8 on a dataset of approximately 5000 images. In this study, YOLOv8 achieved an accuracy score of %50 in 10 epochs while YOLOv7 achieved an accuracy score of %48. In [13] Using 3712 training images and 3769 validation images, it is reported that ResNet's performance is better than RepVGG.

3 Methods

In this study, VGG16, ResNet50 and YOLOv8 methods are used, which are summarized below. In general, Convolutional Neural Network consists of main layers: Convolutional layer, Pooling layer, Fully-Connected layer, shown in Fig. 1. CNN-based algorithms are customized versions using these three layers. After each convolutional layer, an activation layer is used to prevent linear transfer of the image to the next layer. In addition, stride is used to specify how many steps the filter will shift, and padding is used to restore the original size of the image.

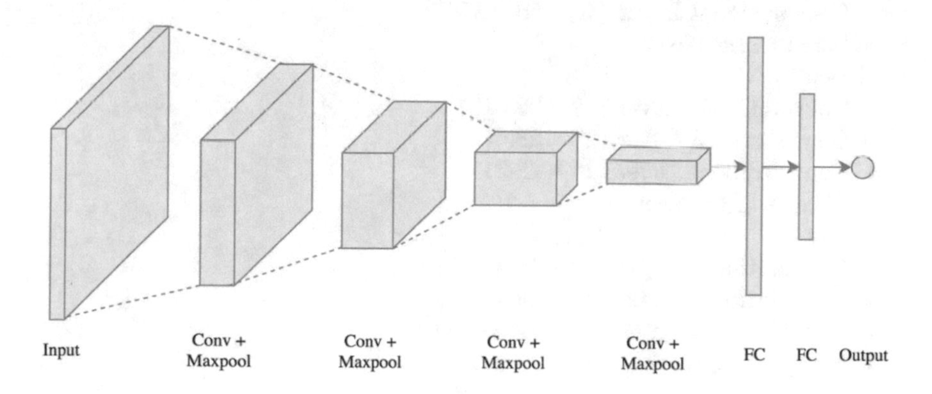

Fig. 1. CNN architecture

3.1 VGG16

The VGG16 model have a test accuracy of %92.7 in ImageNet, a dataset containing more than 14 million training images across 1000 object classes. VGG16 is part of the VGGNets series proposed by the Visual Geometry Group at the University of Oxford. It is named 16 because it uses 16 layers. Of these, 13 are convolutional layers and 3 are fully connected layers, as shown in [Fig. 2]. VGG16 receives $224 \times 224 \times 3$ image input. The VGG16 algorithm employs small input filters with a 3×3 kernel size throughout its convolutional layers. This enables it to have a more localized receptive field and better feature extraction capabilities. Moreover, the VGG16 algorithm also utilizes max-pooling layers after each set of convolutional layers. These max-pooling layers help downsample the input and retain only the most salient features, reducing the computational load and improving efficiency. Another key aspect of the VGG16 algorithm is its use of the rectified linear unit activation function, which helps introduce non-linearity and allows the model to learn and approximate complex functions. In addition VGGNet uses three fully-connected layer. Summarized as follows:

- 1. block
  - Convolutional Layer ($224 \times 224 \times 64$)
  - Convolutional Layer ($224 \times 224 \times 64$)
  - Maxpooling Layer
- 2. block
  - Convolutional Layer ($112 \times 112 \times 128$)
  - Convolutional Layer ($112 \times 112 \times 128$)
  - Maxpooling Layer
- 3. block
  - Convolutional Layer ($56 \times 56 \times 256$)
  - Convolutional Layer ($56 \times 56 \times 256$)
  - Convolutional Layer ($56 \times 56 \times 256$)
  - Maxpooling Layer
- 4. block
  - Convolutional Layer (28×28.512)
  - Convolutional Layer (28×28.512)
  - Convolutional Layer (28×28.512)
  - Maxpooling Layer
- 5. block
  - Convolutional Layer ($14 \times 14 \times 512$)
  - Convolutional Layer ($14 \times 14 \times 512$)
  - Convolutional Layer ($14 \times 14 \times 512$)
  - Maxpooling Layer
- 6. block
  - Dense Layer(4096)
  - Dense Layer(4096)
  - Dense Layer(Class number)

In [12] VGGNets prove that increasing the depth of neural networks can improve the final performance of the network to some extent.

3.2 ResNet50

He et al. [7] proposed a 34-layer Residual Network in 2016, which is the winner of the ILSVRC 2015 image classification and object detection algorithm. The performance of ResNet exceeds the GoogLeNet Inception v3. ResNet50 is kind of its with 50 layers. The biggest advantage of ResNet is that it uses a "highway network" as it is called in the original article [7], which both increases speed and avoids repetitive learning [8], as shown in [Fig. 3.] The general architecture of Resnet50 is shown in [Fig. 4.] The ResNet network starts by applying a

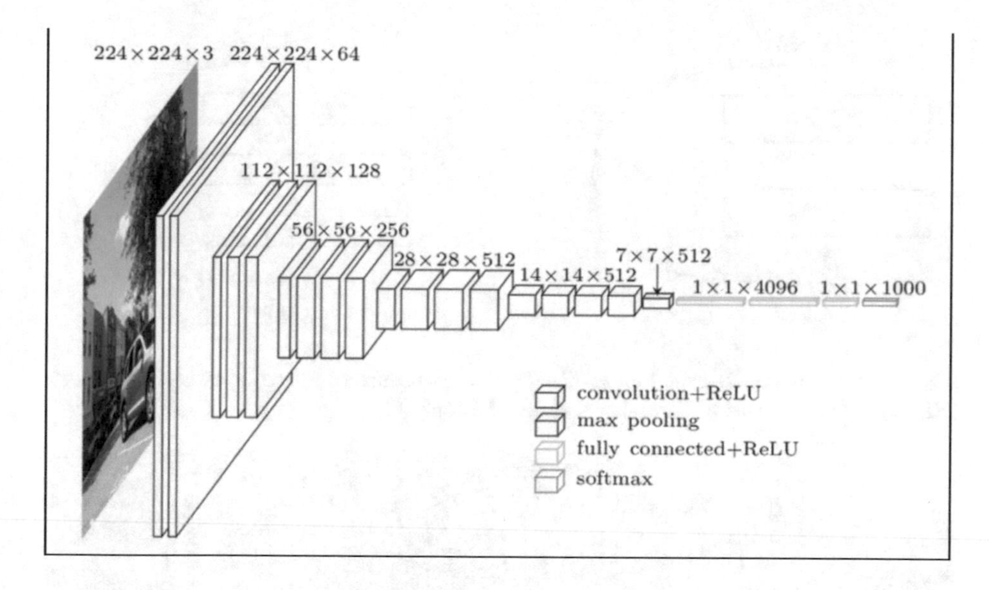

Fig. 2. VGGNet architecture

- 7×7 kernel to the image, with 2 strides.
- Then it does max pooling.
- Then 9 more layers are applied: 3×3, 64 kernel convolution, 1×1 64 kernel convolution, and 1×1 256 kernel convolution. This process repeats 3 times.
- Then 12 layers: 1×1 128 kernel convolution, 3×3 128 kernel convolution, 1×1 512 kernel convolution, this process repeats 4 times.
- Then 18 more layers: 1×1 256 kernel convolution, 3×3 256 kernel convolution, 1×1 1024 kernel convolution. This process repeats 6 times.
- Then 9 more layers: 1×1 512 kernel convolution, 3×3 512 kernel convolution, 1×1 2048 kernel convolution. This process is iterated 3 times.
- As the 50th layer, a fully-connected layer with 1000 nodes is added with softmax activation function [7].

3.3 YOLOv8

YOLOv8 is the newest state-of-the-art YOLO model that can be used for object detection, image classification, and instance segmentation tasks. YOLO, as described in the original article in [10], YOLO separates the object detection parts into a single neural network. This algorithm uses features from the whole image to predict each bounding box. In this way, it detects the whole image or all objects in the image. The input image is divided into $S \times S$ grids, as shown in [Fig. 5]. The grid in which the center of an object appears detects that object. Each grid cell estimates Bounding box B and its confidence score. This confidence score indicates the correct detection rate of the object inside the box. The Confidence score is calculated with the following formula.

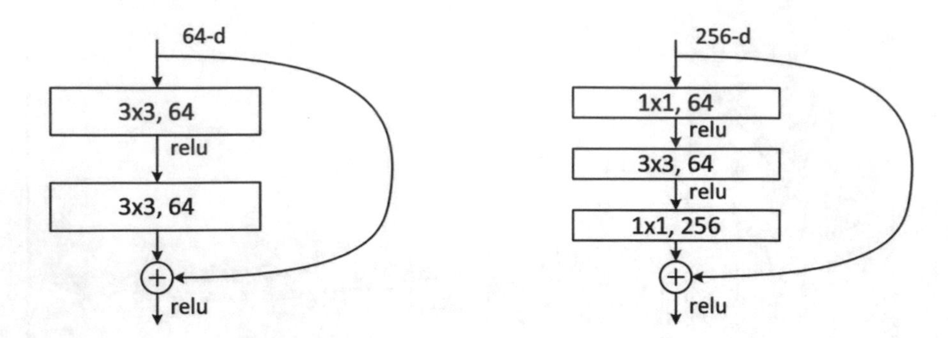

Fig. 3. Structure of ResNet blocks. (Left) The structure of a two-layer residual block. (Right) The structure of three-layer residual block [7]

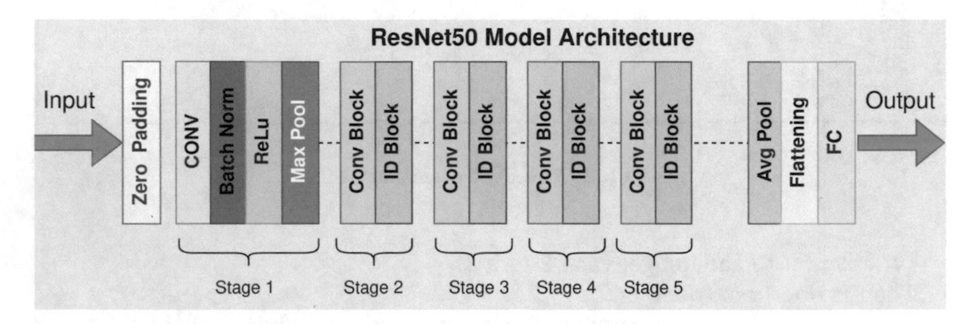

Fig. 4. ResNet50 Architecture(Image source: Wikipedia)

$$Pr(Object) * \mathrm{IOU}_{pred}^{truth} \tag{1}$$

YOLO uses 24 convolutional layers and 2 fully-connected layers. 1×1 convolutional layers reduce the features size previous layers, as shown in [Fig. 6]. At the same time the algorithm uses a linear activation function for the final layer. But other layers use leaky rectified linear activation function, this type of activation function is popular in tasks where we may suffer from sparse gradients, as follows:

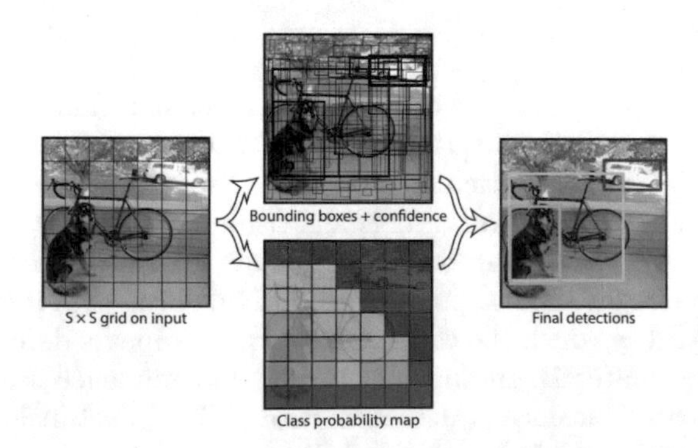

Fig. 5. YOLO predicts B bounding boxees via S × S grid on input.

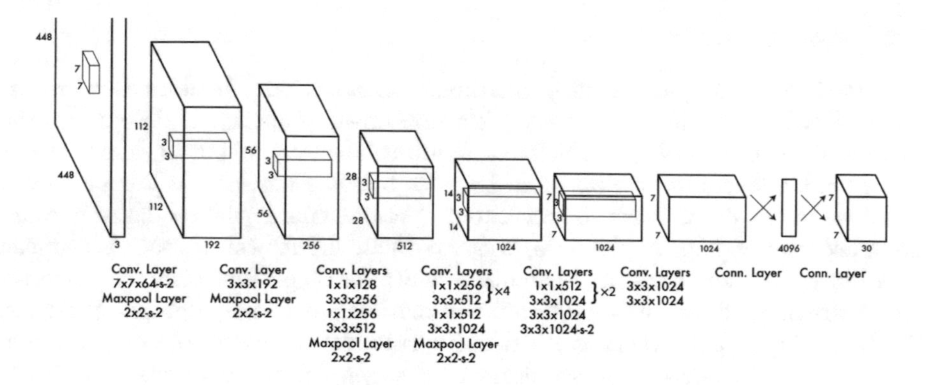

Fig. 6. YOLO architecture

4 Experimental Results

We will apply the VGG16, ResNet50 and YOLOv8 algorithms to our dataset and look at the accuracy scores between them.

4.1 Dataset

The 1359 images in the dataset were taken in different environments. The dataset includes only drones, no drones and both drones and other objects. This dataset also contains text files indicating the location of the drones in order to train the algorithms. In general, our dataset is as shown in [Fig. 7].

Fig. 7. Overview of Dataset

4.2 Fine Tuning

Activation Function. In deep learning, the activation function is a critical element that gives the neural network a non-linear property. It determines the output of a neural network (which has an input layer, an output layer and one or more hidden layers in between) and thus enables the network to learn complex patterns in the data. Because activation functions enable the model to capture complex relationships in the data, without them the neural network is reduced to a linear function, limiting its ability to learn and represent complex patterns. There are many different activation functions, such as ReLU, Sigmoid, Softmax, Tanh, Leaky ReLU. There is no definite rule about the choice of activation function. It depends on the structure of the system. For example, one of the most commonly used activation functions is ReLU, but it should not be used in the outer layers [11].

Optimizers. Optimization algorithms allow the system, usually controlled by hyperparameters, to be updated to produce better results [5]. The primary goal of training a neural network is to adjust its parameters (weights and biases) in order to minimize a specific objective function, often referred to as a loss function. The optimizer plays a key role in this optimization process by iteratively updating the model parameters based on the gradients of the loss function with respect to those parameters. The optimization algorithm is responsible for determining the direction and magnitude of the parameter updates in each iteration. Some common optimizers used in deep learning include: Adam, NAdam, RMSprop, SGD etc.

Loss Function. Loss function, also known as cost function, is the difference between the predicted values and the actual values of a model. This difference is calculated by a mathematical function. How to choose it varies according to the model. There are different loss functions for differently designed algorithms. Mean Squarred Error, Binary Cross-Entropy, Categorical Cross-Entropy, Sparse Categorical Cross-Entropy etc. During the training phase, the model's parameters are adjusted iteratively to minimize the loss function, a process known as optimization. How the loss function works is shown in [Fig. 8].

4.3 Train and Test Process

VGG16 and ResNet50 algorithms were trained on a computer with M1 Pro processor. YOLOv8 was trained via Roboflow. In VGG16 and ResNet50, RMSprop. With a learning rate of 0.0001 from Gradient Descent Algorithms as recommended in [8]. Mean squared error was used as the loss function. In addition, softmax was used as an activation function in the last layer and ReLU was used in the other layers. The YOLOv8 process was studied through Roboflow.

All the algorithms we used in our study were implemented as described in Sect. 3, such as the number of convolution layers used, pooling operations,

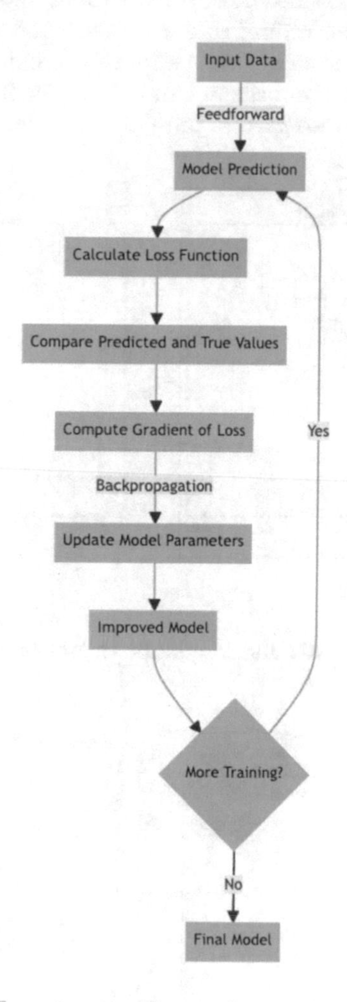

Fig. 8. Loss function(Image Source: Datacamp)

Table 1. Accuracy Scores

Method	%Accuracy	%Loss
VGG16	0.79	0.016
ResNet50	0.83	0.011
YOLOv8	95.4	0.079

and the number of fully connected layers. Table 1 shows that YOLOv8 is the best-performing method according to our dataset and under the defined conditions. However, the train accuracy and validation accuracy scores of VGG16 and ResNet50 for the number of epochs are shown in Fig. 9 and Fig. 10. The output of the YOLOv8 algorithm we used is presented in Fig. 11.

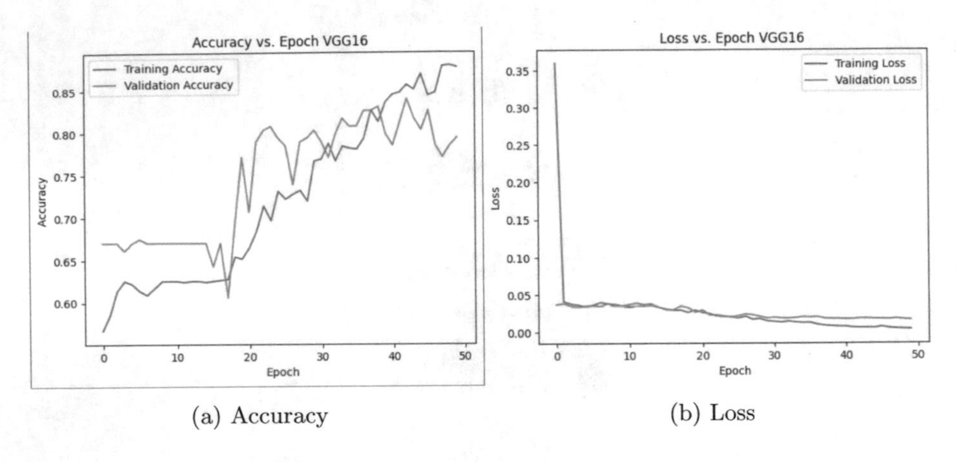

(a) Accuracy (b) Loss

Fig. 9. Train and Validation Scores of VGG16

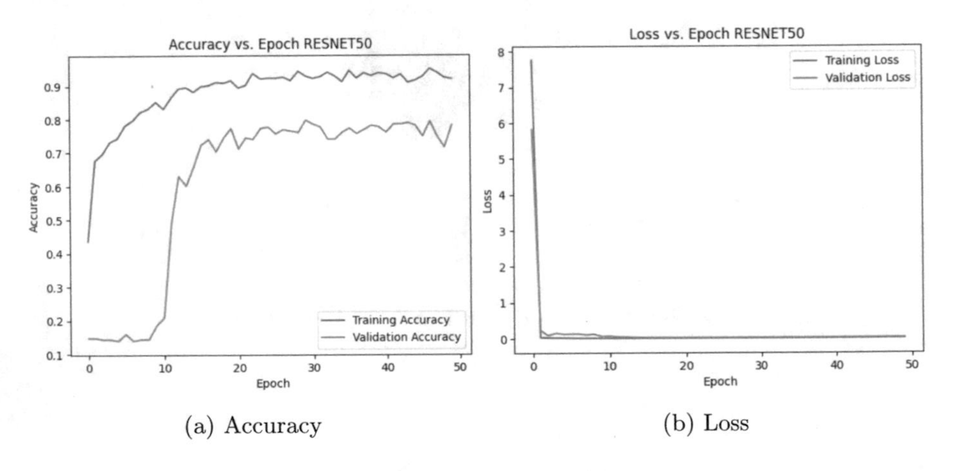

(a) Accuracy (b) Loss

Fig. 10. Train and Validation Scores of ResNet50

Fig. 11. Output of YOLOv8

5 Conclusions

Convolutional Neural Networks (CNNs) are deep learning models that offer significant advantages in image processing and recognition. In particular, these algorithms can effectively extract features from images using specialized convolutional layers. They learn complex models using parameters extracted from features. When combined with data augmentation techniques, they can achieve high performance with a small number of labeled data. Sensitivity to input dimensions allows them to work flexibly with inputs of different sizes. CNNs adopt a modeling approach closer to human visual perception, which leads to effective results in real-world applications. In this study, CNN-based VGG16, ResNet50 and YOLOv8 algorithms were evaluated. The 1359 images in our dataset were applied to the mentioned algorithms and YOLOv8 was the most effective algorithm with 95.4% accuracy rate according to the dataset.

References

1. Agarwal, K., Sanyo, A., Bakshi, S., Vinay, M., Jayapriya, J., Deepa, S.: Performance analysis of yolov7 and yolov8 models for drone detection. In: 2023 International Conference on Network, Multimedia and Information Technology (NMITCON), pp. 1–10 (2023)
2. Albawi, S., Mohammed, T.A., Al-Zawi, S.: Understanding of a convolutional neural network. In: 2017 International Conference on Engineering and Technology (ICET), pp. 1–6 (2017)
3. Aote, S.S., Wankhade, N., Pardhi, A., Misra, N., Agrawal, H., Potnurwar, A.: An improved deep learning method for flying object detection and recognition. Signal Image Video Process. **18**(1), 143–152 (2024)

4. Associate. Commercial drone market size, share and trends analysis report by product, by application, by end use, by propulsion type, by range, by operating mode, by endurance, by region, and segment forecasts, 2023 2030. PDF 978-1-68038-482-6, Grand View Search (2023)
5. Choi, D., Shallue, C.J., Nado, Z., Lee, J., Maddison, C.J., Dahl, G.E.: On empirical comparisons of optimizers for deep learning. arXiv preprint arXiv:1910.05446 (2019)
6. Haque, M.F., Lim, H.Y., Kang, D.S.: Object detection based on vgg with resnet network. In: 2019 International Conference on Electronics, Information, and Communication (ICEIC), pp. 1–3 (2019)
7. He, K., Zhang, X., Ren, S., Sun, J.: Deep residual learning for image recognition. In: 2016 IEEE Conference on Computer Vision and Pattern Recognition (CVPR), pp. 770–778 (2016)
8. Li, Z., Liu, F., Yang, W., Peng, S., Zhou, J.: A survey of convolutional neural networks: analysis, applications, and prospects. IEEE Trans. Neural Netw. Learn. Syst. **33**(12), 6999–7019 (2022)
9. Mahdavi, F., Rajabi, R.: Drone detection using convolutional neural networks. In: 2020 6th Iranian Conference on Signal Processing and Intelligent Systems (ICSPIS), pp. 1–5 (2020)
10. Redmon, J., Divvala, S., Girshick, R., Farhadi, A.: You only look once: unified, real-time object detection. In: 2016 IEEE Conference on Computer Vision and Pattern Recognition (CVPR), pp. 779–788. IEEE Computer Society, Los Alamitos (2016)
11. Sharma, S., Sharma, S., Athaiya, A.: Activation functions in neural networks. Towards Data Sci. **6**(12), 310–316 (2017)
12. Simonyan, K., Zisserman, A.: Very deep convolutional networks for large-scale image recognition. In: 3rd International Conference on Learning Representations (ICLR 2015), pp. 1–14 (2015)
13. Yan, P., Shao, W.: A comparative study on repvgg and resnet for monocular 3d object detection. In: 2022 15th International Congress on Image and Signal Processing, BioMedical Engineering and Informatics (CISP-BMEI), pp. 1–6 (2022)
14. Zhang, J.: Towards a high-performance object detector: Insights from drone detection using vit and cnn-based deep learning models. In: 2023 IEEE International Conference on Sensors, Electronics and Computer Engineering (ICSECE), pp. 141–147 (2023)

Hybrid Kalman Filter-Based MPPT Design for Photovoltaic System in Energy Harvesting Optimization

Waleed Rabeea(✉) and Levent Ucun

Control and Automation Engineering Department, Yildiz Technical University, Istanbul, Turkey
waleedrabeea256@gmail.com, lucun@yildiz.edu.tr

Abstract. A novel hybrid Maximum Power Point Tracking (MPPT) strategy is introduced in this study. By integrating Kalman filter MPPT with the grey wolf optimization (GWO) algorithm, a novel hybrid MPPT method is developed. By applying this technology to the photovoltaic system to be used for energy harvesting optimization, it is possible to simultaneously enhance the system's power quality and efficiency. The MATLAB software is utilized to simulate the proposed method, and afterward, the obtained results are subsequently compared to the most recent advancements in MPPT methods across a variety of environmental conditions. The suggested method is tested under uniform irradiance, step changing in irradiances, and the partial shading effect to show the performance of the PV array. Here, the irradiance and temperature data can be received via different sources and from applications such as ThingSpeak which is an example of using the means of Internet of Things (IoT) in energy harvesting optimization. Furthermore, a comparison is made between the acquired outcomes and the most recent MPPT techniques, namely perturb and observe (P&O), Kalman filter (KF), and grey wolf optimization (GWO). The proposed methodology shows better efficiency, reduced power oscillation, and higher speed in comparison to conventional approaches.

Keywords: Optimization · MPPT · Kalman Filter · GWO · P&O · Partial Shading · Photovoltaic System

1 Introduction

A photovoltaic (PV) system, alternatively referred to as a solar energy system or solar power system, operates by utilizing solar panels to transform solar radiation into electrical residential and commercial establishments [1]. The fundamental element of a solar PV system is the solar panel, which is made up of multiple solar cells. These solar cells are typically made of silicon and contain layers of semiconducting materials that create an electric field when exposed to sunlight [2, 3]. Photons within the light energy dislodge electrons from the atoms of the material when sunlight hits the solar cells, thereby producing an electric current. Solar panels are commonly affixed to rooftops or other potentially advantageous sites that receive ample sunlight [4]. They are assembled into

J. Rasheed et al. (Eds.): FoNeS-AIoT 2024, LNNS 1035, pp. 345–365, 2024.
https://doi.org/10.1007/978-3-031-62871-9_27

a solar PV system in an array or series configuration. An inverter is utilized to transform the generated electricity from direct current (DC) to alternating current (AC) as the majority of appliances at home and the electrical utility operate on AC power. A solar PV system may also comprise batteries (for energy storage), charge controllers, wiring, and a monitoring system to trace energy production, in addition to solar panels and inverters. Electrified surplus produced by the system may be reinjected into the electrical grid or stored in batteries for subsequent utilization [5]. The (MPPT) is a method utilized in the (PV) solar systems to optimize the production of power solar panels [6].

The process consists of power calculation, measurement of input parameters, monitoring and adjusting conditions continuously, and tracing the maximum power point (MPP) using an algorithm with iterations. Several benefits are provided by MPPT, such as enhanced battery life, increased efficacy, and greater energy harvest. Locations characterized by variable solar irradiance can greatly benefit from its implementation. The adaptability of MPPT controllers in terms of design and integration is extended to a wide range of solar panels and battery systems. In general, the implementation of MPPT in solar PV systems offers enhanced efficiency, increased energy output, and optimized performance [7]. In recent years, many studies have been published in scientific literature. A novel PV incremental conductance (INC) MPPT designs are presented in [8] featuring a fuzzy logic (FL) algorithm utilized to modify the step size. The suggested method calculates the magnitude of individual voltage steps by analyzing the convex function of the power-voltage relationship as it increases or decreases. To succeed in this objective, the authors present a new and novel approach: they add five effective zones surrounding the site of maximum PV output. Also, an algorithm for managing a dual-stage grid-connected photovoltaic (PV) battery storage system utilizing multi-objective grass-hopper optimization (MOGHO) is proposed [9]. The variable step-size incremental conductance (VSS-InC) method is utilized to implement this MPPT control to obtain optimal performance in a variety of weather conditions. In contrast to alternative adaptive algorithms, the performance of the proposed method control is preferable with consideration of error. Reduction, rate of converge response speed, and accuracy [3]. The P&O technique is utilized in an improved MPPT method proposed by the authors in [10]. The procedure possesses an uncomplicated architecture and exhibits superior functionality in settings characterized by dynamic fluctuations in irradiance. By analyzing the power differential between each pair of successive samples, the method employs three subsequent observations. Furthermore, the fluctuation in voltage observed between the most recent two consecutive data points is logged. The algorithm subsequently executes the suitable course of action, which may involve increasing or decreasing the voltage, contingent upon the outcome of these comparisons. A hybrid MPPT approach based on an adaptive neuro-fuzzy inference system-particle swarm optimization (ANFIS-PSO) is described in [11].

This method aims to continuously monitor zero oscillation while accumulating maximum and most rapid PV power. This enables the precise monitoring of a sine-shaped reference current, resulting in the production of a high-quality inverter current. The employed methodology showcases exceptional driving control, thereby augmenting the extraction of PV potential. The author investigates the MPPT issue of a photovoltaic (PV) system employing a DC-DC converter through the utilization of an artificial bee colony (ABC) methodology, for a Photovoltaic system employing a DC-DC boost converter [12]. During the subsequent phase of the ABC MPPT algorithm's operation, the P-V characteristic is determined by utilizing the PV module's collected data values. Subsequently, the optimal voltage is determined. Using the MPPT procedure, Consequently, the reference voltage for the outer PI control loop is accumulated. This loop then gives the predictive digital current-programmed control with the current reference. A performance comparison of two distinct MPPT algorithms is presented in [13]. This is the study's primary objective. The first of these techniques is the Kalman filter (KF)-MPPT utilizing the Kalman filter (KF) implemented on a photovoltaic system, while the second is the InC algorithm, which is an improved version of the P&O method. The P&O approach is regarded as an enhancement by both algorithms. This study employs a simulation to represent a solar panel; between the PV panel and the load is positioned a boost converter that is regulated by the MPPT tracker. Following this, the simulation tool provides a C block for the implementation of the two methods in programming.

Novel hybrid MPPT techniques are presented in this article. A Kalman filter MPPT with GWO algorithm constitutes the proposed method. Increased efficiency and improved power quality are the goals of this method for photovoltaic systems. Simultaneous evaluation of the proposed method and state-of-the-art MPPT methods under various environmental conditions are performed using MATLAB software.

2 Photovoltaic (PV) System

To simulate the behavior of a photovoltaic (PV) module or panel, it is typical practice to model a solar PV array by using a single-diode model. While providing reasonably accurate results, the single-diode model simplifies the electrical properties of the PV cell. The model is based on a single diode that represents the behavior of the PN junction within the solar cell [14].

Before calculating the power loss caused by steady-state oscillation, it is necessary to examine the fundamental properties of the PV module. Figure 1 illustrates the equivalent circuit for a photovoltaic (PV) cell, which comprises a diode, a series resistor (Rih), and a parallel resistor (Rp) [15]. The output current of the PV array (I_A) can be written as:

$$I_A = N_P I_{Ph} - N_P I_0 \left[exp\left[\frac{q}{AKT_C} \left(\frac{V_A}{N_S} + \frac{I_A R_S}{N_P} \right) \right] - 1 \right] - \frac{N_P}{R_P} \left(\frac{V_A}{N_S} + \frac{I_A R_S}{N_P} \right) \quad (1)$$

where I_{Ph} represents the current of the photo, I_0 is the saturation current of the diode, V_A is the output voltage of the PV array, q is the electron charge (1.60217733x10–19 C), A is the diode ideality factor between 1 and 5, K is the Boltzmann's constant (1.3806x10–23 J/K), T_C PV cells temperature. The R_S is the series resistance, R_P represents parallel temperature, N_S is the cells of the panel, and N_P parallel panels [16]. The MPPT controller in the PV system is designed to harvest the maximum quantity of electricity from the solar panels despite varying environmental conditions, including temperature and sunlight intensity. PV panels have a nonlinear voltage-current (I-V) characteristic, and their maximum power point (MPP) is impacted by variation in response to external influences, as represented in Fig. 2. The maximum power point (MPP) is the point on the power-voltage (P-V) curve where the operating power of the panel is at its peak [17].

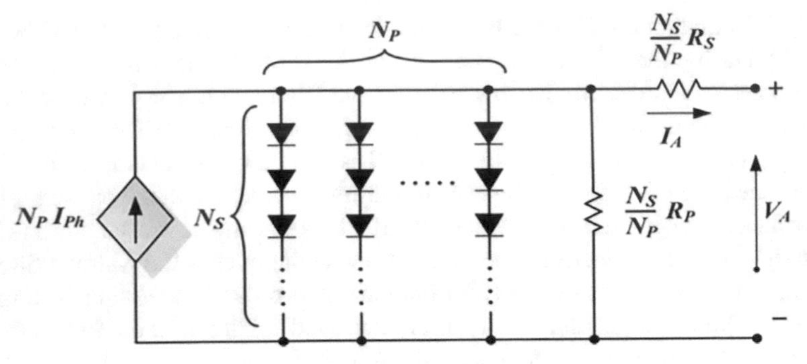

Fig. 1. Electrical PV array model [16].

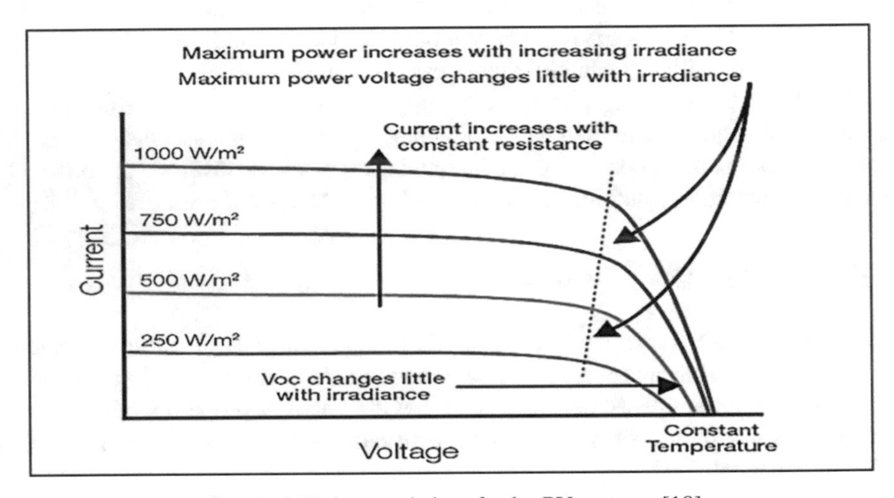

Fig. 2. I-V characteristics of solar PV systems [18]

However, the MPP is dependent on variation due to factors such as the quantity of shade, temperature fluctuations, and variances in sunlight. The maximum power point tracking (MPPT) algorithm gets the operating point of the PV system as close to the MPP as possible with small but regular adjustments. To maximize power generation and minimize energy consumption, the MPPT controller is tasked with optimizing the power output of the PV system [18].

3 Photovoltaic (PV) System

3.1 Conventional MPPT Methods

Standard techniques utilized to track the MPP of solar panels are known as "conventional MPPT methods." These techniques and algorithms as the (P&O or IncCond) algorithms are often used by these methods. The power output of the PV system is tracked as these conventional methods incrementally adjust the operating point. Although generally successful, a few PV panels may present challenges due to sudden atmospheric fluctuations, partial shading, or non-linear power-voltage characteristics [19]. These MPPT methods are presented in Fig. 3 and Fig. 4.

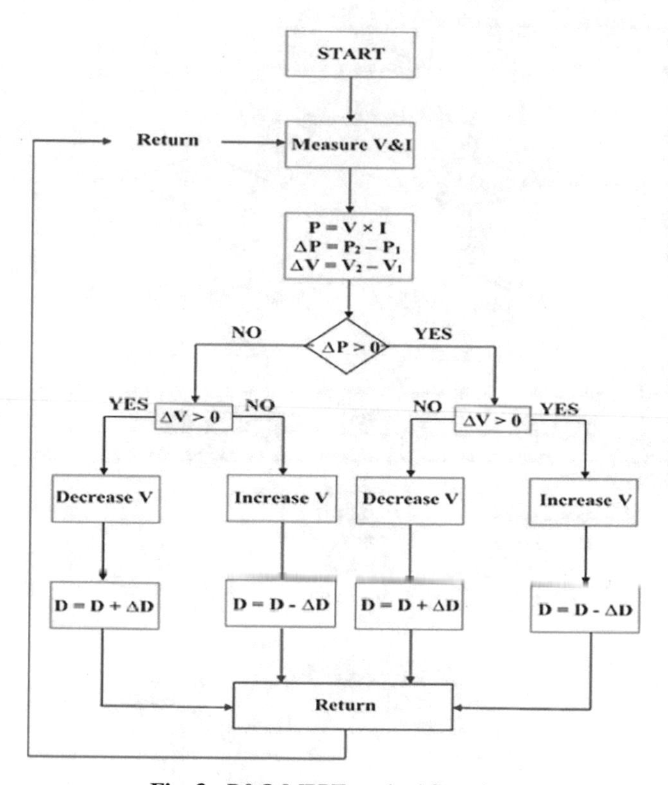

Fig. 3. P&O MPPT method flowchart

In contrast, more sophisticated MPPT methods integrate more complex algorithms and technologies to enhance effectiveness and precision in tracking. By depending on advanced control methodologies, computational algorithms, and hardware capacities, these methods outperformed the constraints associated with traditional approaches. The following are examples of sophisticated MPPT methods:

- Model-based MPPT: These methods utilize mathematical models of the PV system to estimate and track the MPP more accurately. They consider factors such as environmental conditions, panel temperature, and irradiance to predict the MPP [20].
- Artificial intelligence (AI)--based MPPT: AI techniques, such as neural networks or fuzzy logic systems, can be employed to optimize MPPT. These methods can learn and adapt to changing conditions, improving the tracking performance under various scenarios [21].
- Distributed MPPT: In large PV systems, distributed MPPT methods aim to optimize the MPP of individual panels independently. By allowing each panel to operate at its specific MPP, the overall system efficiency can be improved, especially in situations with partial shading or non-uniform irradiance [22].

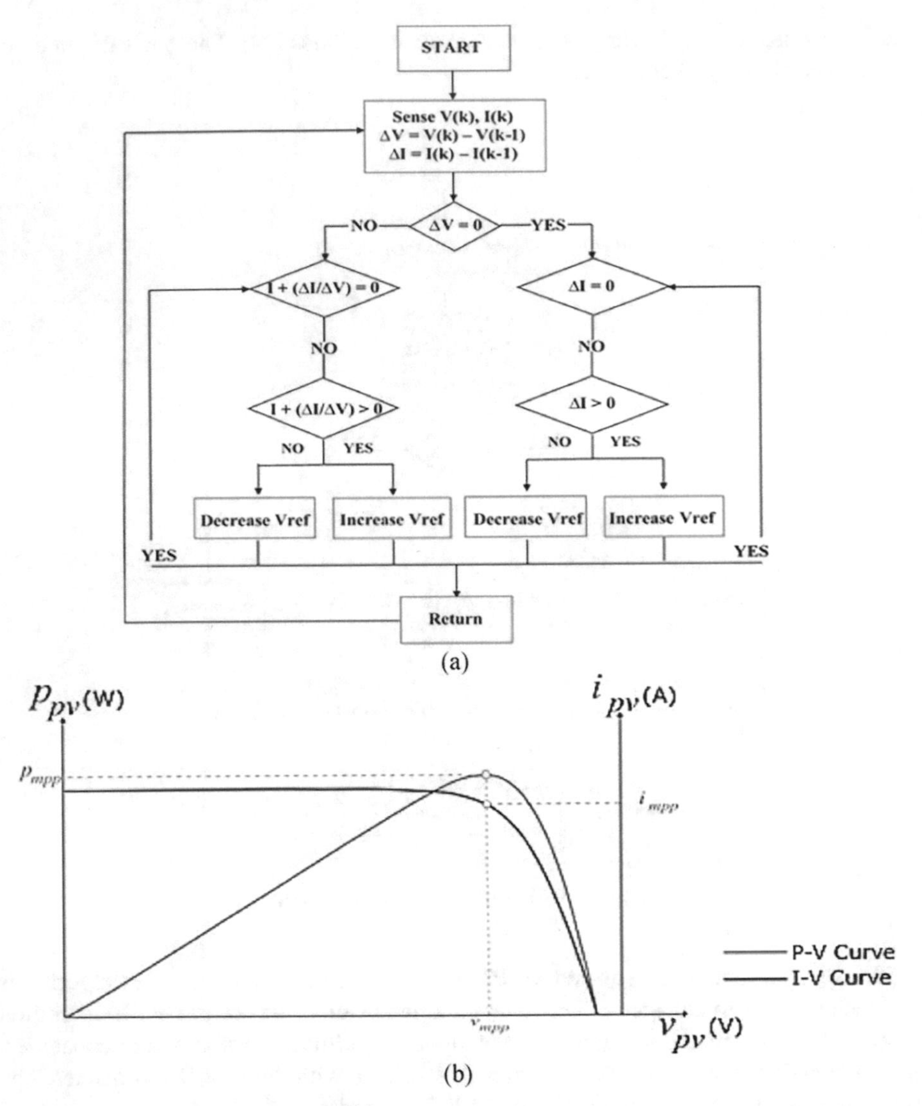

Fig. 4. (a) InC MPPT method flowchart, and (b) P-V, I-V characteristic curve [15]

- Hybrid MPPT: Hybrid MPPT combines multiple techniques, algorithms, or sensors to achieve more robust and accurate tracking. For example, a combination of AI algorithms, sensor data (such as temperature or solar radiation sensors), and model-based approaches can be used together to optimize MPPT [23].

The main difference between conventional and advanced MPPT methods lies in their level of sophistication, adaptability, and accuracy. While conventional methods provide basic MPP tracking, advanced methods leverage more advanced algorithms and technologies to achieve improved performance, especially in challenging operating conditions.

3.2 Advanced MPPT Methods

Computational methods known as metaheuristics are implemented to deal with optimization challenges. Meta Heuristics may be utilized in the context of MPPT methods for PV systems to locate the optimal operating point that maximizes the array's power output. The following are common varieties of meta-heuristics applied to MPPT in PV systems:

- Genetic Algorithms (GA): GA is a popular metaheuristic inspired by the process of natural selection. It uses a population of candidate solutions, which are encoded as strings (chromosomes), and applies genetic operators such as selection, crossover, and mutation to evolve and improve the solutions over generations [24].
- Particle Swarm Optimization (PSO): PSO is a population-based metaheuristic that simulates the behavior of a swarm of particles in search of the optimum. Each particle represents a potential solution and adjusts its position based on its own best-known solution and the best-known solution of the entire swarm [25].
- Ant Colony Optimization (ACO): ACO is inspired by the foraging behavior of ants. It uses a population of artificial ants that deposit pheromone trails to communicate information about the quality of different solutions. The pheromone trails guide the search process toward promising regions of the solution space [26].
- Annealing process in metallurgy: It starts with an initial solution and gradually explores the search space by accepting "worse" solutions with a decreasing probability. SA can escape local optima and potentially find the global optimum [27].

3.3 Using Kalman Filters for MPPT

To track the utmost power point, a continuous projection of the operating point's location must be done. As indicated by the P-V characteristic of the PV array, the PV power increases as the curve shows a convex function. The P-V slope is positive as long as it reaches the MPP, and immediately after it, power begins to fall with a negative slope that is narrow. In contrast, the KF-based MPPT's state vector consists of a lonely state variable, which is the reference for the output voltage of the PV string. Based on these characteristics, it is possible to derive a similar equation for a linear state space in one dimension, and the MPPT can be guided using the equation referenced in [28].

$$V_{ref}^{p}(k) = V_{ref}(k-1) + B\frac{\Delta P}{\Delta V}(k-1) + w \quad (2)$$

In this context $V^{p}_{ref}(k)$ signifies the reference voltage estimated by the MPPT technique for iteration k, $V_{ref}(k-1)$ is the corrected reference voltage, B represents the scaling factor or step size, w represents the mixture of device disturbances including switching noise and thermal noise, and $\frac{\Delta P}{\Delta V}$ represents the instantaneous slope of the power. However, the measured voltage or the PV voltage $V_{meas}(k)$ can be expressed as:

$$V_{meas}(k) = V_{ref}(k) + z \tag{3}$$

where z represents the value of the error between the measured and reference voltage. The steps of implementation of KF-based MPPT are:

Step 1: Update the predicated voltage reference from (4).

$$V^{p}_{ref}(k) = V_{ref}(k-1) + M\frac{\Delta P}{\Delta V}(k-1) \tag{4}$$

Step 2: Modify the priori process covariance (estimate error) as follows: where Q denotes the process noise covariance that exists between the a priori and posteriori process covariances. This forces the tracking process to strike a balance between its dynamic nature and its stability.

$$P_{p}(k) = P_{e}(k-1) + Q \tag{5}$$

Step 3: At this point, the Kalman gain is calculated through the combination of the process covariance $P_{p}(K)$. . And the measurement noise covariance R. As a result, the process noise covariance value Q, is required to be utilized per (6) [15].

$$K(k) = \frac{P_{p}(k)}{P_{p}(k) + R} \tag{6}$$

Step 4: Step up the input voltage reference (correct voltage reference). V_{ref} through the application of the supplied Kalman gain, which regulates the discrepancy between the measurement input presented by the voltage sensor and the predicted voltage reference in (7).

$$V_{e}(k) = V_{p}(k) + K(k) \tag{7}$$

Step 5: Utilizing Eq. (5), recalculate the covariance of the posterior process. P(k) decreases over time and becomes closer and closer to zero as the number of iterations increases. This results in the noise being progressively removed from the final prediction. The flowchart of Kalman filter-based MPPT is represented in Fig. 5.

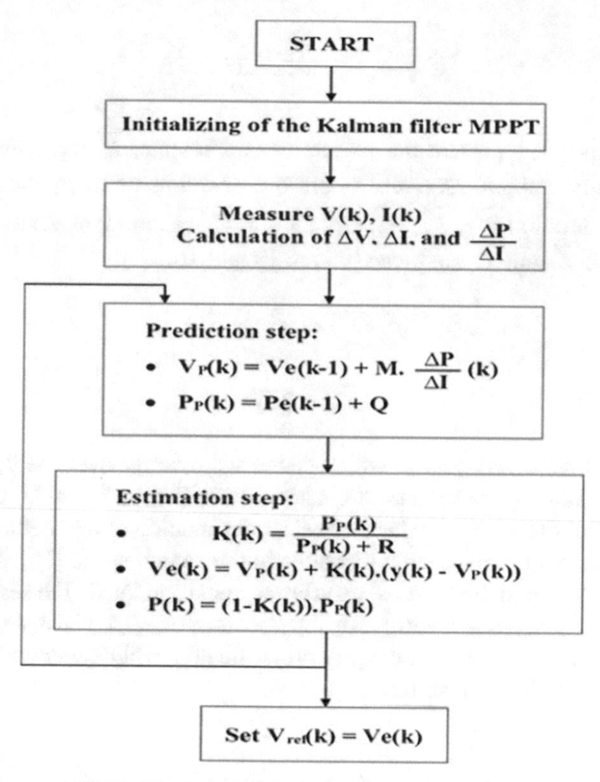

Fig. 5. Flowchart of Kalman filter-based MPPT.

3.4 Using Kalman Filters for MPPT

The following set of Eq. (8–16) [29]. Serve as the foundation for the algorithm that underpins the GWO approach. This algorithm makes use of all four of the different degrees of leadership that are accessible to the grey wolves so that it can build the hierarchy that governs their group. These levels are represented by the symbols that correspond to them as follows: omega (ω), beta (β), delta (δ), and alpha (α) [30].

$$\overrightarrow{D_\alpha} = \left|\overrightarrow{C_1}.\overrightarrow{X_\alpha} - \overrightarrow{X}\right| \tag{8}$$

$$\overrightarrow{D_\beta} = \left|\overrightarrow{C_2}.\overrightarrow{X_\beta} - \overrightarrow{X}\right| \tag{9}$$

$$\overrightarrow{D_\delta} = \left|\overrightarrow{C_3}.\overrightarrow{X_\delta} - \overrightarrow{X}\right| \tag{10}$$

$$\overrightarrow{X_1} = \overrightarrow{X_\alpha} - \overrightarrow{A_1}.(\overrightarrow{D_\alpha}) \tag{11}$$

$$\overrightarrow{X_2} = \overrightarrow{X_\beta} - \overrightarrow{A_2}.(\overrightarrow{D_\beta}) \tag{12}$$

$$\overrightarrow{X_3} = \overrightarrow{X_\delta} - \overrightarrow{A_3}.(\overrightarrow{D_\delta}) \tag{13}$$

$$\overrightarrow{X}(t+1) = \frac{\overrightarrow{X_1} + \overrightarrow{X_2} + \overrightarrow{X_3}}{3} \tag{14}$$

Here, $\overrightarrow{D}$, $\overrightarrow{C}$, and $\overrightarrow{A}$ represent the vectors of coefficients, $\overrightarrow{X}$ represents the position vector of the grey wolf, $\overrightarrow{X_\alpha}, \overrightarrow{X_\beta}, and \overrightarrow{X_\delta}$ are the positions of α, β, $and \delta$, respectively, t is the current iteration, $\overrightarrow{A_1}$, $\overrightarrow{A_2}$, $\overrightarrow{A_3}$, $\overrightarrow{C_1}$, $\overrightarrow{C_2}$ and $\overrightarrow{C_3}$ are random vectors. The general equations for the $\overrightarrow{A}$ and $\overrightarrow{C}$ are given in Eqs. 15 and 16.

$$\overrightarrow{A} = 2\overrightarrow{a}.\overrightarrow{r_1} - \overrightarrow{a} \tag{15}$$

$$\overrightarrow{C} = 2\overrightarrow{r_2} \tag{16}$$

Here, $\overrightarrow{a}$ is the ingredients vector, which decreases linearly over the iteration process from 2 to 0, $\overrightarrow{r_1}$ and $\overrightarrow{r_2}$ are vectors of chance within the range [0, 1], where an is the ingredients vector. Figure 6 provides a visual representation of the method by which the GWO optimizer changes the search locations in the search space [30]. In addition, the most recent location may be found at an arbitrary position inside the search area that is contained within a circle that is formed by the positions of, α, β, and δ. As a consequence of these variables, the positions of the prey are evaluated, while other wolves adjust their placements to be randomly near the prey.

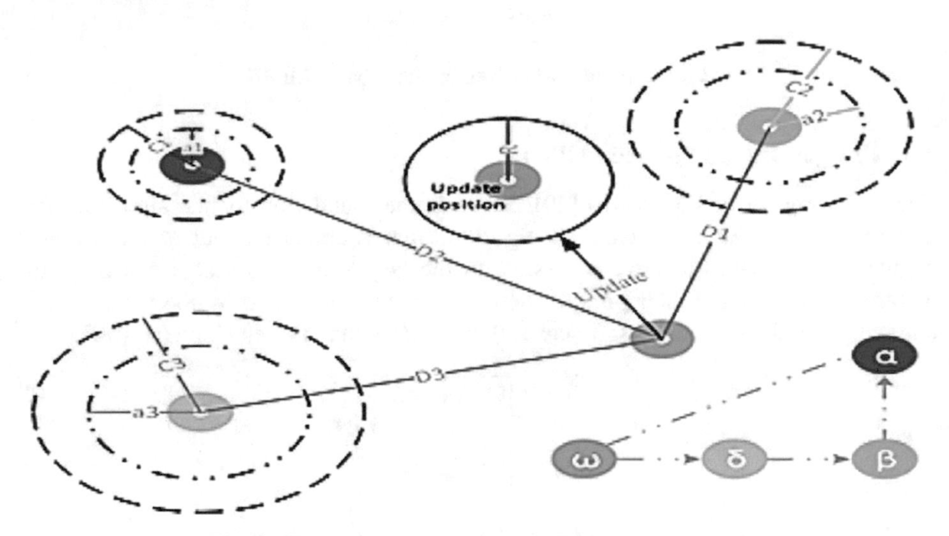

Fig. 6. The update process for the GWO optimizer [30]

The algorithmic process for GWO optimization may be defined in the same way as shown in Fig. 7. The grey wolf that is originally located at the (x, y) coordinates now migrates to the (x', y') coordinates and updates its position there. The location of the best agent in the algorithm is determined according to the locations of the grey wolf and its prey by regulating $\vec{A}$ and $\vec{C}$. In addition to this, taking into consideration Here, the hunting tactics of grey wolves are reiterated. That the wolves do not cease their attacking behavior until the prey comes to a complete halt. The flowchart representing the suggested GWO approach may be seen in Fig. 8.

Input: Problem Size, *Population* size
Output: P_{g_best}
Start
  Initialize the population of grey wolves X_i $(i = 1, 2, \ldots, n)$
  Initialize a, A, and C according to combined objective function
  Calculate the fitness values of search agents and grade the agents.
  (X_α = the best solution within the search agent, X_β = the second best solution within the search agent, and X_δ = the third best solution within the search agent) $t = 0$
  While (t < maximum number of iterations)
    For each search agent
      Update the position of the current search according to the equation
    End for
  Update value of a, A, and C according to combined objective function
  Again Calculate the fitness values of all search agents and grade them
  Again update the position of X_α, X_β, and X_δ
  $t = t+1$
  Store the best solution value.
  End while
End

Fig. 7. Pseudo-code of GWO [30]

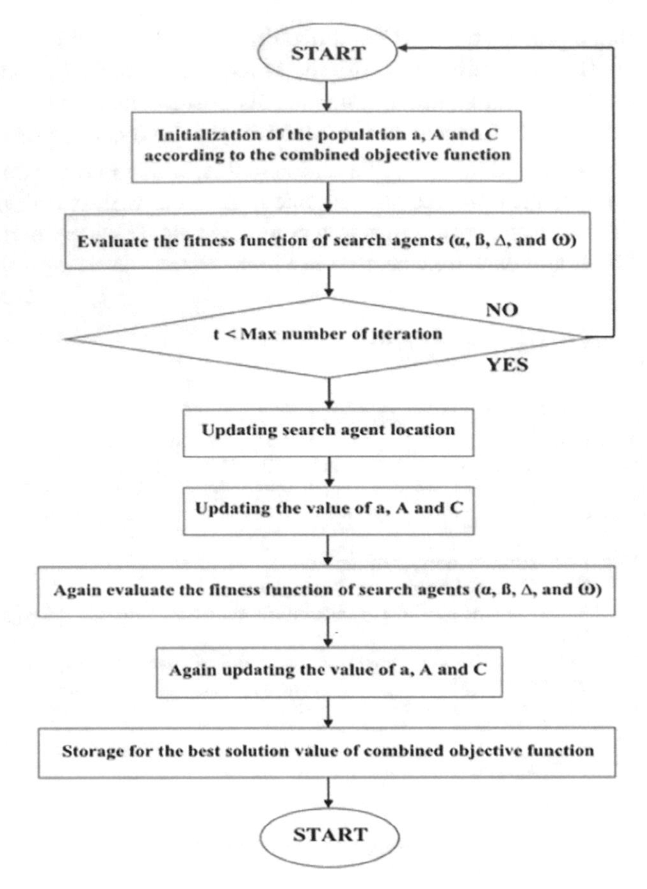

Fig. 8. Flowchart of the grey wolf optimization algorithm

3.5 Proposed Hybrid MPPT Method

The hybrid MPPT algorithm provides enhanced solar system compatibility, stability, efficiency, and adaptability. They optimize the output of power, increase the efficiency of energy harvesting, and maximize the revenue that comes from solar installations. Figure 9 illustrates the proposed MPPT diagram. Here is a general outline of how a Kalman filter with GWO based MPPT algorithm can work in a PV system.

By incorporating a KF with GWO into the MPPT algorithm, the PV system can effectively track the MPP although measurement uncertainty may be present, and environmental uncertainties. The Kalman filter's ability to estimate the state of the system and adapt to changing conditions makes it a valuable tool for enhancing the productivity and efficiency of PV systems.

Hybrid MPPT for solar systems combines the benefits of multiple MPPT algorithms. In this paper typically incorporating both KF and grey wolf optimization (GWO) methods are proposed.

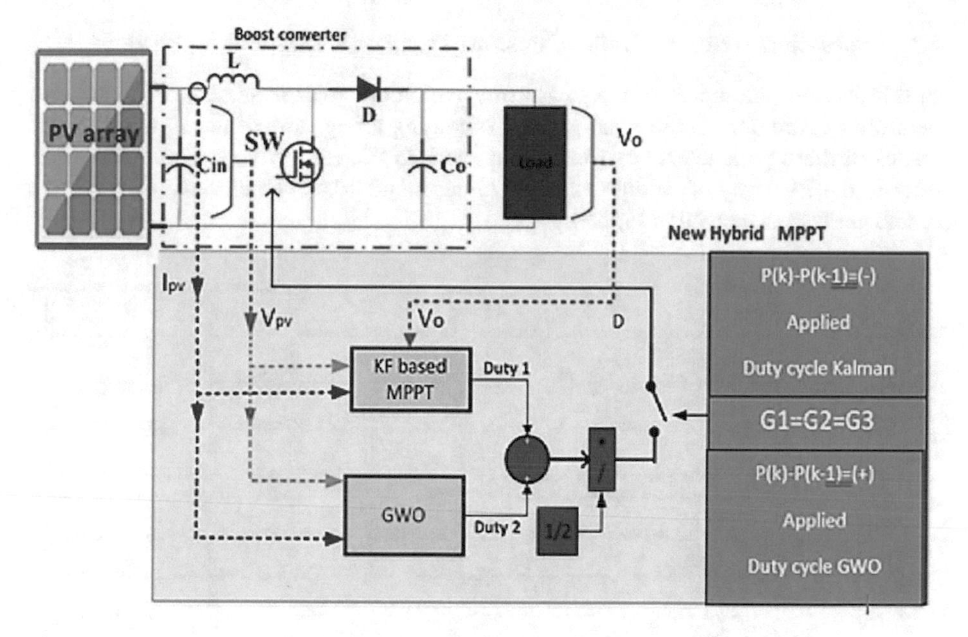

Fig. 9. The proposed hybrid MPPT method

Two cases are applied to three solar cells in this investigation. When irradiance is varying but similar across all arrays, this is the first scenario. Immediately, the switch transitions between the Grey Wolfe optimize duty cycle and the Kalman filter. In the second scenario, calculate the average of the Kalman filter and (GWO) duty cycles when the irradiances are varying but different for each array.

4 Simulation Studies

The performance of the proposed MPPT methods is verified using a simulation-based MATLAB/Simulink environment. The evaluation of the PV system that consists of a PV array, DC/DC boost converter, MPPT methods, and load is modeled in MATLAB. In this part of the paper, the findings that are achieved by modeling both approaches under a variety of levels of sun irradiation are provided for discussion. The evaluation of the system is tested under different case studies as follows:

4.1 Evaluation of System Under Constant Weather Conditions (Case 1)

In this section, the inputs of the solar array which are solar irradiation (G), and temperature (T) are considered fixed without changing along time of the simulation. The values of these parameters are 1000 W/m^2, and 25 °C, respectively. These conditions refer to the PV array or module operated under standard test conditions (STC). These parameters are displayed in Fig. 10.

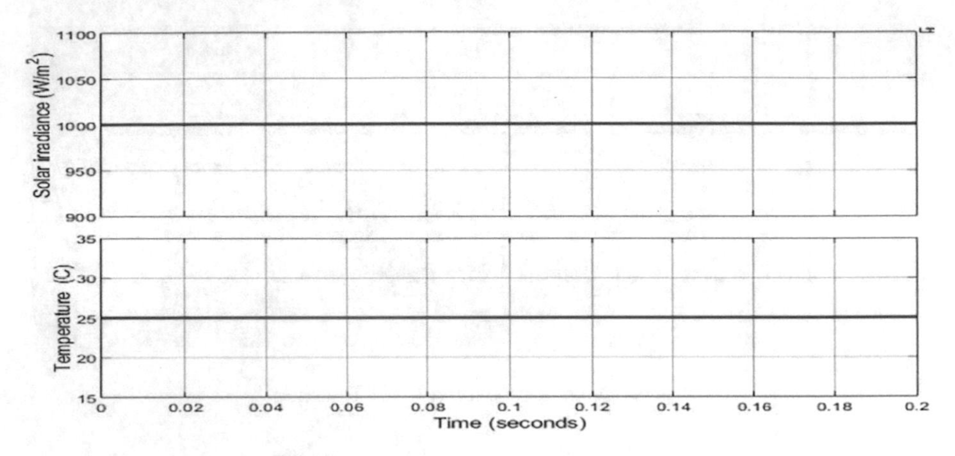

Fig. 10. Irradiation and temperature curves

PV performance testing under standard test conditions (STC) is a crucial component of this paper to compare and characterize the efficacy of different solar modules or panels. The term "STC" denotes a collection of standardized environmental conditions that are applied during the testing of PV modules. This enables impartial and consistent comparisons to be made across installations and manufacturers. On the other hand, STC creates a uniform framework for evaluating the efficacy of different photovoltaic modules. Since different manufacturers can use different methods and materials, STC enables new investors or purchasers to make accurate choices by utilizing consistent performance metrics. The objective evaluation of the performance of various PV modules is simplified.

Figure 11 shows the obtained results of the PV system under STC conditions. This figure displays the PV voltage, PV current, PV output power, and the duty cycle of the boost converter. As presented, the suggested MPPT method presented high efficiency, reduced steady-state oscillation, and rapid response in comparison to traditional MPPT techniques. The obtained power using the proposed hybrid method is higher than the KF and P&O methods and (GWO), where the extracted power in the suggested technique is 600 W with an efficiency of 100%. The achieved power from the proposed method is 592 W with an efficiency of 98%, the power output of KF is 584 W with an efficiency of 97%, the power output of P&O is 587 W with an efficiency of 97%, and the power output of (GWO) is 577 W with an efficiency of 97% The oscillation problem in MPP is canceled by the hybrid method and therefore, the proposed method is considered robust and very efficient.

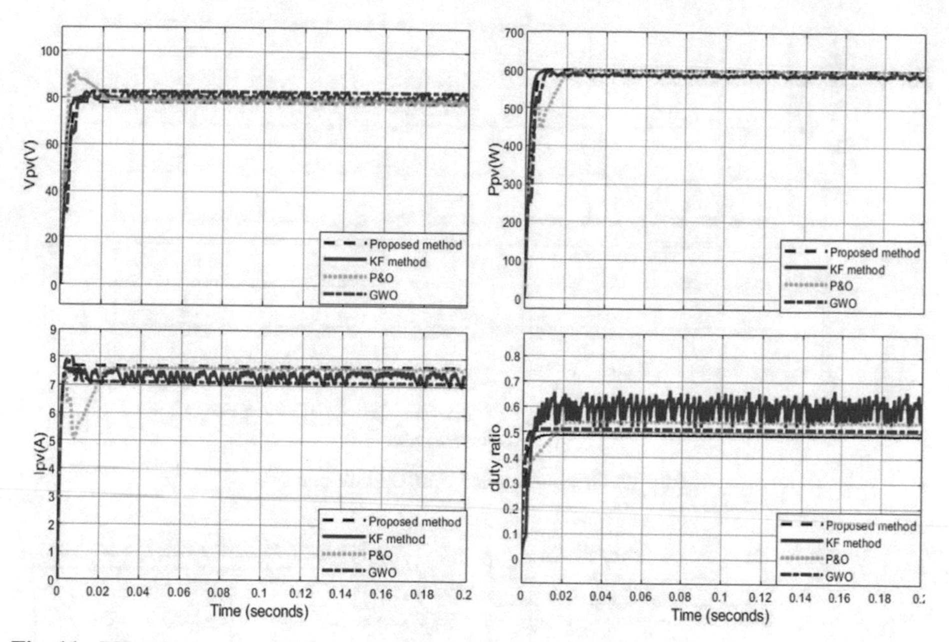

Fig. 11. PV voltage curve, PV current curve, extracted power from PV array, the duty cycle of the boost converter.

4.2 Evaluation of System Under Constant Irradiance and Zero Temperature (Case 2)

In this section, the inputs of the solar array which are solar irradiation (G), and temperature (T) are zero and fixed without changing along time of the simulation. The values of these parameters are 1000 W/m^2, and 0 °C, respectively. These parameters are displayed in Fig. 12.

The obtained power in this case using the proposed hybrid method is greater than the KF and P&O methods and (GWO). The achieved power from the proposed method is 653 W. Here, the achieved power from KF is 652 W where P&O presents 650 W and GWO results with 651 W.

This analysis examines the effects of temperature variation while assuming a constant illumination of 1000 W/m^2 for the identical solar array configuration. Figure 13 illustrates the displacement of. MPP in response to atmospheric changes. It is apparent from this illustration that maximal power increases as temperatures drop [31].

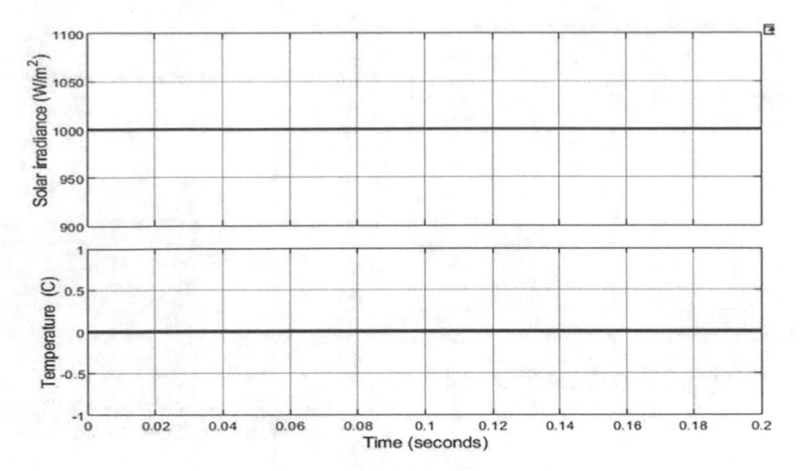

Fig. 12. Irradiation and temperature curves.

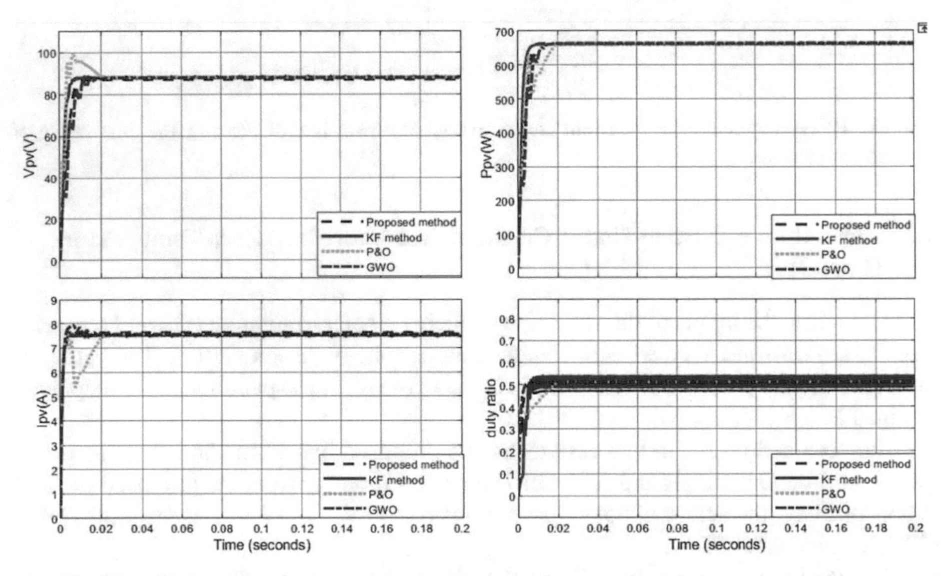

Fig. 13. PV voltage curve, PV current curve, extracted power from PV array, the duty cycle of the boost converter.

4.3 Evaluation of System Under Varying Irradiance and Fixed Temperature (Case 3)

Evaluating a PV system under varying irradiance is an essential aspect of understanding its performance and efficiency in different environmental conditions. Moreover, solar irradiance levels change throughout the day and across different seasons Because of various factors, cloud cover, shading, and atmospheric conditions. By evaluating a PV system under varying irradiance, the reader can simulate real-world scenarios and assess how the system performs under different conditions that it is likely to encounter in its operational life. Also, understanding how a PV system performs under varying irradiance helps in determining the optimal size and configuration of the system components (solar panels, inverters, batteries, etc.). In this section, the proposed MPPT method is tested by applying variable irradiance profiles and fixed temperature as shown in Fig. 14. By evaluating the system under varying irradiance levels (750W/m^2, 400 W/m^2, and 1000 W/m^2) the obtained power or MPP is varied according to these values. The temperature is kept fixed at 25 °C. Figure 15 shows the PV voltage curve, PV current curve, extracted power from the PV array, and duty cycle of the boost converter under different values of irradiance and constant temperature. The comparison of the curves produced by the PV array when subjected to varying degrees of irradiance and a constant temperature of 25 °C.

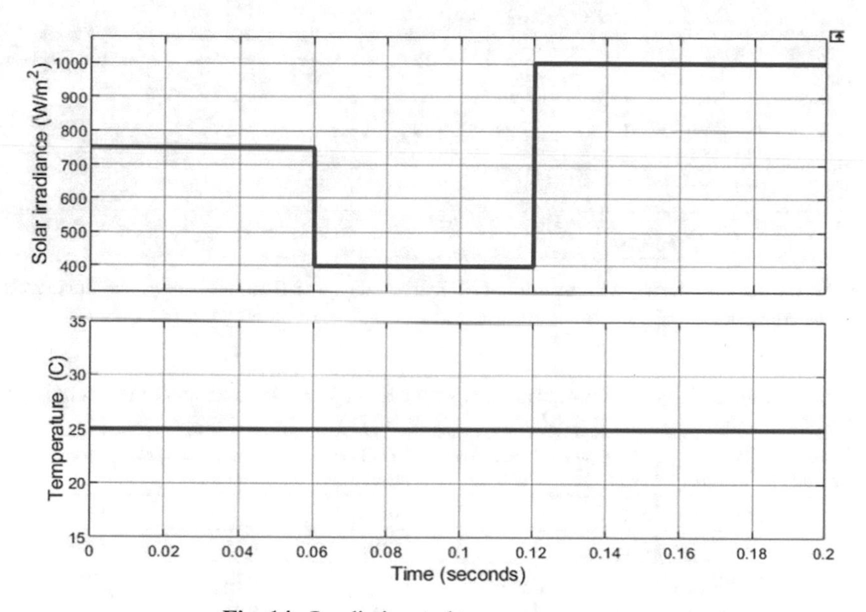

Fig. 14. Irradiation and temperature curves

The obtained power in this case using the proposed hybrid method is higher than the KF and P&O methods and (GWO). The achieved power from the proposed method is 431 W, the power output of KF is 415 W, the power output of P&O is 334 W and the power output of (GWO) is 388 W.

In summary, testing MPPT methods under partial shading conditions (PSC) is essential to ensure that solar panels can operate efficiently and generate maximum power output even when subjected to varying levels of shading. As a result, the proposed MPPT method can adapt to changing conditions, improve energy harvesting, and improve the overall efficiency and power of PV systems.

Table 1 demonstrates the novelty of this work. This table has the numerical results of all MPPT methods used in this simulation under different weather conditions.

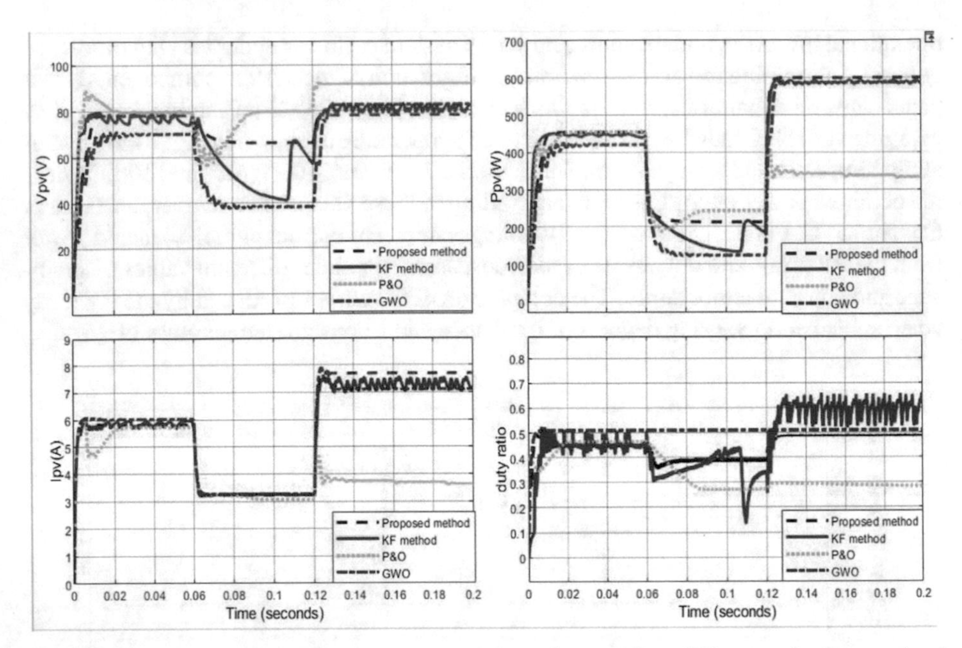

Fig. 15. PV voltage curve, PV current curve, extracted power from PV array, the duty cycle of the boost converter.

Table 1. Comparison and numerical results for the proposed method and other techniques.

Irradiance & Temperature	MPPT Method	Max power (W)	Mean power (W)	Tracking time (sec)	Power Oscillation (W)	Efficiency (%)
1000 W/m^2 25 °C	KF	597	584	0.029	2	97.8
	P&O	600	587	0.023	0.15	97.8
	GWO	592	577	0.01	0.72	97.4
	Proposed method	599	592	0.009	0.03	98.8
1000 W/m^2 0 °C	KF	660	653	0.011	1.9	98.7
	P&O	663	650	0.02	0.4	98
	GWO	663	651	0.019	0.06	98.1
	Proposed method	663	653	0.01	0.014	98.4
[750 400 1000] W/m^2 25 °C	KF	594	415	0.014	4	69.8
	P&O	455	334	0.021	0.5	73.4
	GWO	592	388	0.019	0.65	65.5
	Proposed method	599	431	0.011	2.25	71.9

5 Conclusion

In the present study, a new hybrid MPPT strategy is developed via the use of the Kalman filter MPPT in conjunction with the grey wolf optimization (GWO) for solar PV systems. Using the MATLAB simulation program, the suggested method is tested in a few different environmental settings, and the results are compared to the most recent and cutting-edge techniques of MPPT. To demonstrate the effectiveness of the PV array, the recommended approach is put through its pace in conditions of uniform irradiance, step changes in irradiance, and partial shading effect. In addition, the results obtained are contrasted by utilizing cutting-edge technologies. MPPT techniques of perturb and observe (P&O) and Kalman filter (KF). When contrasted with other traditional approaches, the one being proposed demonstrates much lower power oscillation, significantly increased speed, and outstanding efficiency.

References

1. Lima, C.R.C., Guilemany, J.M.: Adhesion improvements of thermal barrier coatings with HVOF thermally sprayed bond coats. Surf. Coat. Technol. **201**(8), 4694–4701 (2007)
2. Esfahanian, V., Javaheri, A., Ghaffarpour, M.: Thermal analysis of an SI engine piston using different combustion boundary condition treatments. Appl. Thermal Eng. **26**(2–3), 277–287 (2006)
3. Gilbert, A., Kokini, K., Sankarasubramanian, S.: Thermal fracture of zirconia–mullite composite thermal barrier coatings under thermal shock: a numerical study. Surf. Coat. Technol. **203**(1–2), 91–98 (2008)
4. Gilbert, A., Kokini, K., Sankarasubramanian, S.: Thermal fracture of zirconia–mullite composite thermal barrier coatings under thermal shock: an experimental study. Surf. Coat. Technol. **202**(10), 2152–2161 (2008)
5. Hejwowski, T., Weronski, A.: The effect of thermal barrier coatings on diesel engine performance. Vacuum **65**, 427–432 (2002)
6. Buyukkaya, E.: Thermal analysis of functionally graded coating Al-Si alloy and steel pistons. Surf. Coat. Technol. **202**, 3856–3865 (2008)
7. Esfahanian, E., Javaheri, A., Ghaffarpour, M.: Thermal analysis of an SI engine piston using different combustion boundary condition treatments. Appl. Thermal Eng. **26**, 398–402 (2007)
8. Cerit, M., Ayhan, V., Parlak, A., Yasar, H.: Thermal analysis of a partially ceramic coated piston: effect on cold start HC emission in a spark ignition engine. Appl. Therm. Eng. **31**(2–3), 336–341 (2011)
9. Marr, M.A.: An investigation of metal and ceramic thermal barrier coatings in a spark-ignition engine. M.S thesis, Mechanical and Industrial Engineering, University of Toronto (2009)
10. Cerit, M.: Thermo mechanical analysis of a partially ceramic coated piston used in an SI engine. Surf. Coat. Technol. **205**, 3499–3505 (2011)
11. Parker, D.W.: Thermal barrier coatings for gas turbines, automotive engines and diesel equipment. Mater. Des. **13**(6), 345–351 (1992)
12. Jamali, H., Mozafarinia, R., Razavi, R.S., Ahmadi-Pidani, R.: Fabrication and evaluation of plasma-sprayed nanostructured and conventional YSZ thermal barrier coatings. Ceram. Int. **38**, 6712–6805 (2012)
13. Ahmadi-Pidani, R., Shoja-Razavi, R., Mozafarinia, R., Jamali, H.: Improving the thermal shock resistance of plasma sprayed CYSZ thermal barrier coatings by laser surface modification. Opt. Lasers Eng. **50**, 780–786 (2012)

14. Giolli, C., Scrivani, A., Rizzi, G., Borgioli, F., Bolelli, G., Lusvarghi, L.: Failure mechanism for thermal fatigue of thermal barrier coating systems. J. Therm. Spray Technol. **18**, 223–230 (2009)
15. Zhou, C., Zhang, Q., Li, Y.: Thermal shock behaviour of nanostructured and microstructured thermal barrier coatings on a Fe-based alloy. Surf. Coat. Technol. **217**, 70–75 (2013)
16. Molina, M.G., Molina, M.G., Juanicó, L.E.: 1–2010/Iss.3) Molina and Juanicó/Dynamic Modelling and Control Design of Advanced Photovoltaic (2010). https://www.researchgate.net/publication/228736534
17. Abo-Khalil, A.G., Alharbi, W., Al-Qawasmi, A.R., Alobaid, M., Alarifi, I.M.: Maximum power point tracking of PV systems under partial shading conditions based on opposition-based learning firefly algorithm. Sustainability (Switzerland) **13**(5), 1–18 (2021). https://doi.org/10.3390/su13052656
18. Habib, A., Memon, A.H.: Off-Grid WWTP using PV systems investigation of energy consumption in a waste water treatment plant and determination of size of solar farm for complete energy neutrality. A Personal Research Project (2021). https://doi.org/10.6084/m9.figshare.14401931
19. Talbi, B., Krim, F., Rekioua, T., Laib, A., Feroura, H.: Design and hardware validation of modified P&O algorithm by fuzzy logic approach based on model predictive control for MPPT of PV systems. J. Renew. Sustain. Energy **9**(4) (2017). https://doi.org/10.1063/1.4999961
20. Motahhir, S., Aoune, A., El Ghzizal, A., Sebti, S., Derouich, A.: Comparison between Kalman filter and incremental conductance algorithm for optimizing photovoltaic energy. Renew. Wind Water Sol. **4**(1) (2017). https://doi.org/10.1186/s40807-017-0046-8
21. Fathi, M., Parian, J.A.: Intelligent MPPT for photovoltaic panels using a novel fuzzy logic and artificial neural networks based on evolutionary algorithms. Energy Rep. **7**, 1338–1348 (2021). https://doi.org/10.1016/j.egyr.2021.02.051
22. Chao, R.M., Ko, S.H., Lin, H.K., Wang, I.K.: Evaluation of a distributed photovoltaic system in grid-connected and standalone applications by different MPPT algorithms. Energies (Basel) **11**(6), 121693718 (2018). https://doi.org/10.3390/en11061484
23. Restrepo, C., Yanẽz-Monsalvez, N., González-Castaño, C., Kouro, S., Rodriguez, J.: A fast converging hybrid mppt algorithm based on abc and p&o techniques for a partially shaded pv system. Mathematics **9**(18), 2228 (2021). https://doi.org/10.3390/math9182228
24. Hassan, A., Bass, O., Masoum, M.A.S.: An improved genetic algorithm based fractional open circuit voltage MPPT for solar PV systems. Energy Rep. **9**, 1535–1548 (2023). https://doi.org/10.1016/j.egyr.2022.12.088
25. Koad, R.B., Zobaa, A.F., El-Shahat, A.: A novel MPPT algorithm based on particle swarm optimization for photovoltaic systems. IEEE Trans. Sustain. Energy **8**(2), 468–476 (2016)
26. Chao, K.-H., Rizal, M.N.: A hybrid MPPT controller based on the genetic algorithm and ant colony optimization for photovoltaic systems under partially shaded conditions. Energies **14**(10), 2902 (2021). https://doi.org/10.3390/en14102902
27. Zhang, Y., Wang, Y.-J., Zhang, Y., Tong, Y.: Photovoltaic fuzzy logical control MPPT based on adaptive genetic simulated annealing algorithm-optimized BP neural network. Processes **10**(7), 1411 (2022). https://doi.org/10.3390/pr10071411
28. Chellal, M., Leite, V.: Experimental evaluation of Kalman filter based MPPT in grid-connected PV system (2022)
29. Mohanty, S., Subudhi, B., Ray, P.K.: A new MPPT design using grey Wolf optimization technique for photovoltaic system under partial shading conditions. IEEE Trans. Sustain. Energy **7**(1), 181–188 (2016). https://doi.org/10.1109/TSTE.2015.2482120
30. Motahhir, S., Chtita, S., Chouder, A., El Hammoumi, A.: Enhanced energy output from a PV system under partial shaded conditions through grey wolf optimizer. Clean. Eng. Technol. **9**, 100533 (2022). https://doi.org/10.1016/j.clet.2022.100533

31. Hossain, M.J., Tiwari, B., Bhattacharya, I.: An adaptive step size incremental conductance method for faster maximum power point tracking. In: Conference Record of the IEEE Photovoltaic Specialists Conference, pp. 3230–3233. Institute of Electrical and Electronics Engineers Inc. (2016). https://doi.org/10.1109/PVSC.2016.7750262

Automated Detection of Pulmonary Embolism Using CT and Perfusion Spectral Images

Gökalp Tulum[1], Onur Osman[1(✉)], Nazenin Ipek[2], Mustafa Demir[3], Sertaç Asa[3], Kerim Sönmezoğlu[3], Cansu Güneren[3], Fuad Aghazada[3], and Kübra Şahin[3]

[1] Electric Electronics Engineering Department, Istanbul Topkapi University, 34662 Istanbul, Turkey
{gokalptulum,onurosman}@topkapi.edu.tr

[2] Mechatronics Engineering Department, Istanbul Nisantasi University, 34398 Istanbul, Turkey
nazenin.ipek@nisantasi.edu.tr

[3] Cerrahpaşa Faculty of Medicine/Department of Internal Medicine/Nuclear Medicine, Istanbul University Cerrahpaşa, 34098 Istanbul, Turkey
demirm@istanbul.edu.tr, {asa,kerim.sonmezoglu,cansu.guneren, fuad.aghazada,kubra.sahin}@iuc.edu.tr

Abstract. This research endeavors to automate the detection of pulmonary embolism in lung perfusion scintigraphy images using image processing and artificial intelligence methods, aiming to assess embolism levels through localization and various statistical calculations. The study utilizes CT and perfusion images from 37 individuals, with 20 diagnosed with pulmonary embolism and 17 classified as normal. Employing a threshold of −150 HU (Hounsfield Unit) and morphological operations, including opening and closing with a 5-voxel disc-shaped structuring element, the lungs are segmented to remove air and blood veins. Alignment of CT and perfusion images is achieved through affine transformation and bilinear interpolation. The segmented lungs serve as masks on perfusion images, from which intensity-based features are extracted comprising a total of 23 features used in the analysis. In the Multi-Layer Perceptron (MLP) model, training phase metrics indicate mean sensitivity, specificity, accuracy, and F1 Score at 94.41%, 97.44%, 95.94%, and 95.64%, respectively. Testing phase results for MLP exhibit comparable metrics at 94.44%, 95.24%, 94.87%, and 94.44%. Conversely, the Support Vector Machine (SVM) model demonstrates distinctive performance characteristics, with mean sensitivity during training at 98.41%, decreasing to 64.44% in testing. Mean specificity decreases from 95.06% in training to 84.92% in testing, and mean accuracy declines from 95.94% to 75.52%. Mean F1 Score follows a similar trend, dropping from 95.64% in training to 70.09% in testing, emphasizing the diverse performance outcomes of the two models across training and testing datasets.

Keywords: Pulmonary Embolism Detection · Image Processing · Artificial Intelligence · Lung Perfusion Scintigraphy

J. Rasheed et al. (Eds.): FoNeS-AIoT 2024, LNNS 1035, pp. 366–377, 2024.
https://doi.org/10.1007/978-3-031-62871-9_28

1 Introduction

The clinical picture that occurs due to partial or complete occlusion of the pulmonary artery or one of its branches by any thrombus is called pulmonary embolism (PE). PE; It is a disease that has high mortality and morbidity, can recur, is sometimes difficult to diagnose, and is preventable. Lung perfusion scintigraphy, a Nuclear Medicine imaging method, is widely used for the diagnosis of pulmonary embolism (PE). In addition, Lung Perfusion Scintigraphy is used to determine the patient's postoperative lung function capacity before major lung surgeries.

For lung perfusion scintigraphy, patients are injected intravenously with Macro Aggregate Albumin particles (Tc99m MAA) bound with Technetium-99m. Tc-99m MAA lung perfusion scintigraphy is the sole emergency procedure in Nuclear Medicine and can be utilized for rapidly assessing lung perfusion in emergencies. Although the procedure has undergone various changes since its inception, the perfusion tracer [99mTc]-macro-aggregated albumin (MAA) remains the agent of choice to this day. This radiopharmaceutical, which is slowly injected through a peripheral vein, settles diffusely in the lung parenchyma by creating micro-emboli in terminal capillaries, and the gamma rays it emits are converted into images by a gamma camera. The diagnosis of PE can be determined with the SPECT/CT images obtained by combining the functional data provided by SPECT and the anatomical information obtained by computed tomography (CT). Aside from performing attenuation correction and anatomical localization through the CT scanner in SPECT/CT devices, it is also feasible to acquire high-resolution diagnostic images [1]. The emergence of SPECT/CT has ushered in new possibilities in diagnostic imaging for various disease conditions.

It is stated that the use of SPECT-CT for the diagnosis of pulmonary embolism is of great importance in starting the treatment of patients. Pulmonary embolism, if left untreated, can lead to pulmonary hypertension, heart failure, and fatal outcomes. Therefore, accurate diagnosis and rapid initiation of the treatment process are vital [2].

Lung ventilation/perfusion (V/Q) scintigraphy or solely perfusion single photon emission computed tomography/computed tomography (Q-SPECT/CT) is extensively employed in diagnosing acute pulmonary embolism (PE) and monitoring chronic PE, primarily due to its reduced radiation doses and nearly absence of contraindications. Theoretically, areas of reduced perfusion that do not correspond to a pertinent structural abnormality are considered likely to represent a PE. Perfusion SPECT (-SPECT)/CT imaging provides higher overall accuracy than conventional V/Q scintigraphy and also reduces the number of non-diagnostic studies [3].

In the process of lung SPECT/CT, CT scans of the thorax serve two primary purposes. They function as an attenuation map, facilitating the correction of attenuation in the collected SPECT emission data. This correction is essential for compensating for the attenuation of photons originating from deeper locations or absorbed by denser overlying structures. As cross-sectional images, they enable anatomical correlation with functional SPECT images, contributing to enhanced interpretation [4].

Septic or tumor pulmonary embolism, including intravascular lymphoma, are critical conditions requiring a swift diagnosis for effective treatment, while these disorders persist in posing a diagnostic challenge in radiological imaging. In cases of septic and tumor-related pulmonary embolism, SPECT–CT fusion images exhibit notably larger characteristic perfusion defects when compared to abnormal opacities observed on CT. This strongly indicates the likelihood of an embolic event [5].

Attempting to assess whether the combination of CT and SPECT(Q) can substitute for ventilation scans, Gradinscak et al. conducted a study involving 30 patients suspected of having pulmonary embolism (PE). They concluded that SPECT(Q) + MDCT (Multidetector CT) scanning could be viewed as an alternative method in cases where ventilation scanning is not feasible [6].

Mahaletchumy et al. reported findings related to Q-only SPECT/CT, where 36 patients were initially identified as having pulmonary embolism (PE), but only 23 of them were confirmed as true positives. Thirteen patients received a false-positive diagnosis. On the other hand, the remaining 30 patients, classified as not having PE, were all correctly diagnosed based on the reference standard. In summary, Q-only SPECT/CT significantly ($p < 0.05$) over diagnosed PE in 13 patients, yet it exhibited a high negative predictive value as none of the patients with PE were missed [7].

Mazurek and colleagues [8] aimed to assess the utility of Lung Perfusion Scintigraphy with SPECT CT imaging in the diagnosis of pulmonary embolism (PE) in their study. When compared to literature findings [9–12] the sensitivity and specificity of the SPECT/CT(Q) scans closely resembled those attained by the authors through the use of the SPECT(V/Q) method.

In studies conducted with patients diagnosed with pulmonary embolism (PE) due to COVID-19, it has been recommended that evaluations be made only by perfusion due to the risk of aeration spreading during ventilation [13, 14].

Automatic diagnosis systems are being created with artificial intelligence for reasons such as the fact that specialists' ability to interpret patient images correctly is costly in terms of time and can prolong the treatment process. These systems are advancing day by day thanks to developing technology and algorithms.

This research aims to automatically detect pulmonary embolism in Lung Perfusion scintigraphy images with image processing and artificial intelligence methods and to score the embolism level by calculating the embolism's localization and various statistical values. Thus, it is aimed to contribute to the treatment process by creating an automatic diagnosis system.

2 Material and Methods

2.1 Data Acquisition

In this study we used CT and Perfusion scintigraphy images of 37 people, 20 of them are diagnosed as pulmonary embolism and 17 of the are normal. All of the images for the development of the automated diagnostic scheme were originally provided with the DICOM format obtained at the time of examination. For lung perfusion scintigraphy, patients are injected intravenously with Macro Aggregate Albumin particles (Tc99m

MAA) bound with Technetium-99m. This radiopharmaceutical, which is slowly injected through a peripheral vein, settles diffusely in the lung parenchyma by creating micro-emboli in terminal capillaries, and the gamma rays it emits are converted into images by a gamma camera. The diagnosis of PE can be determined with the SPECT/CT images obtained by combining the functional data provided by SPECT and the anatomical information obtained by computed tomography (CT). Image resolution ranged from 0.62 to 0.82 mm in the axial view with a slice thickness of 5 mm for CT images, 9.59 mm in the axial view with a slice thickness of 9.59 mm for perfusion scintigraphy images.

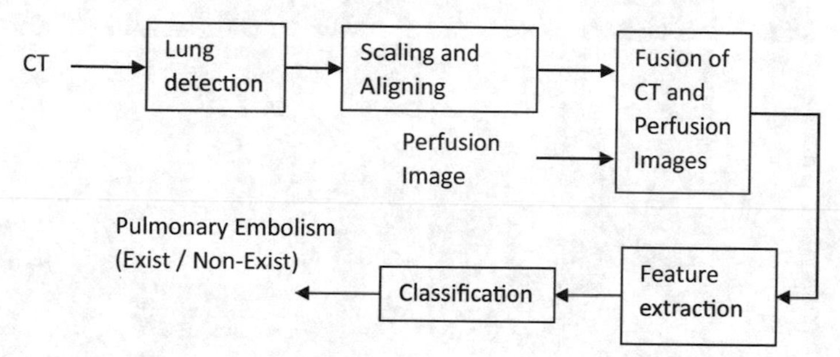

Fig. 1. Flowchart of the automated detection of pulmonary embolism using CT and perfusion spectral images.

2.2 Pulmonary Embolism Detection Method

In Fig. 1 the flowchart of the presented method is given. Since lungs are mostly full with air, we use −150 HU (Hounsfield Unit) for thresholding. Following with morphological operations, opening and closing with 5 voxel disc-shaped structuring element to remove air and blood veins, lungs are segmented. After then to fuse CT and perfusion images we rescale the CT images to align perfusion image scale using affine transformation and bilinear interpolation. Finally, we used the segmented lungs as masks on perfusion images and extracted the intensity-based features.

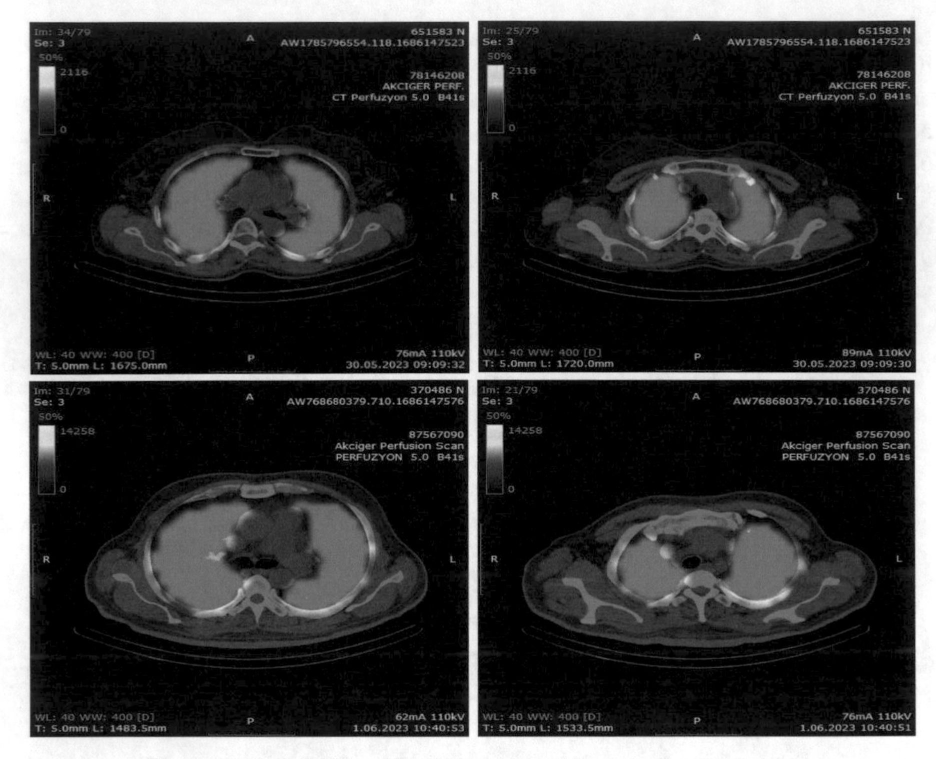

Fig. 2. Fusion of CT and Perfusion spectral images of normal cases.

2.3 Image Fusion

Fusion of CT and Perfusion spectral images are needed to determine the pulmonary embolism. In Fig. 2, four normal case fusion images are given. Brightness inside the lung region indicates that there is no blockage in blood veins. However, in Fig. 3, four pathologic images are given. Some hypodense regions are seen inside the lung which means that the blood flow is interrupted because of emboly.

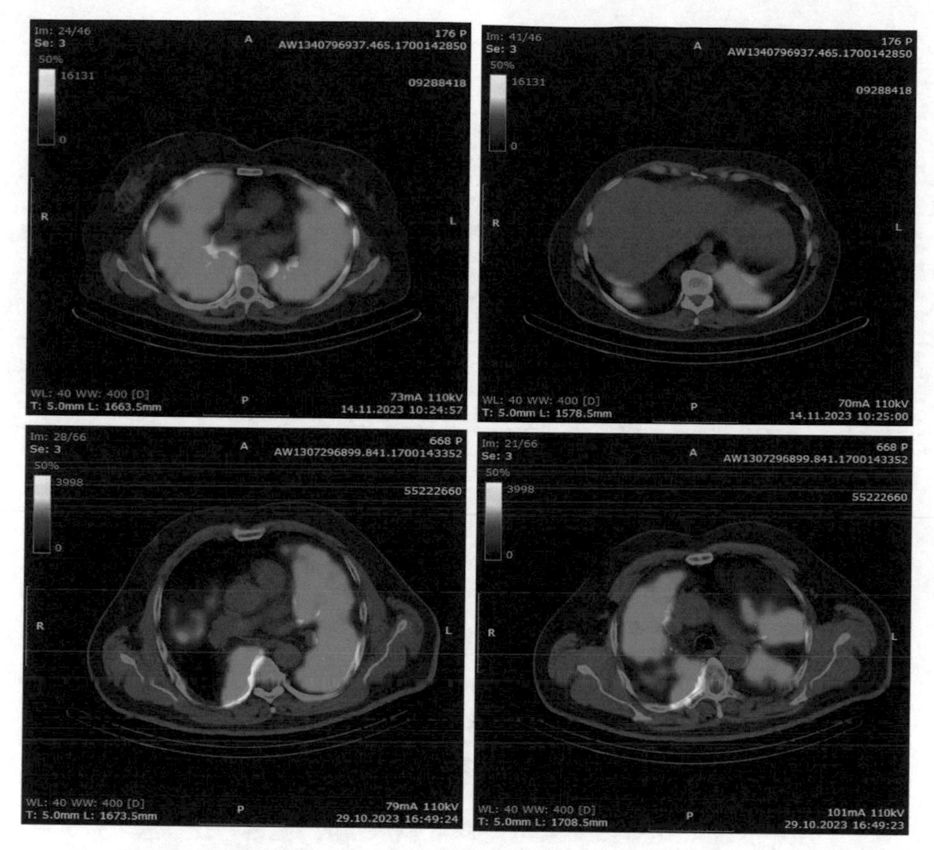

Fig. 3. Fusion of CT and Perfusion spectral images of pathologic cases.

2.4 Feature Extraction

We used 23 features in this study, names and definitions are given in Table 1. Histograms of perfusion scintigraphy images in lung region for a normal and pathologic cases are given in Fig. 4 and Fig. 5, respectively. Since there is no embolism in normal cases, blood flow can be monitored everywhere and therefore perfusion images have high intensity values inside the lung. Despite normal cases, presence of emboly occlusion in blood veins blocks the flow. Therefore, radiopharmaceutical particles cannot be seen at that regions and low intensity values occurs where embolism exists.

Since intensity value distribution can be used to classify pathologic cases, we extract 23 features from histogram and intensities. Then, we examine the Pearson correlation coefficient and p value between features and classes (Normal/Pathologic) which can be seen in Table 2. 15 features have p values < 0.05 and highlighted in Table 2.

Table 1. Features and their definitions.

FEATURE	DEFINITION
mean	Mean value of perfusion images inside the lung region
median	Median value of perfusion images inside the lung region
mode	mod value of perfusion images inside the lung region
variance	Variance value of perfusion images inside the lung region
standard deviation	Standard deviation of perfusion images inside the lung region
skewness	Skewness of perfusion images inside the lung region
kurtosis	kurtosis of perfusion images inside the lung region
3rd moment	3rd moment of perfusion images inside the lung region
histogram values (15 features)	Histogram values at every 2000 perfusion value, from 2,000 to 30,000 [hist2000, hist4000, hist6000, …, hist30000] inside the lung region

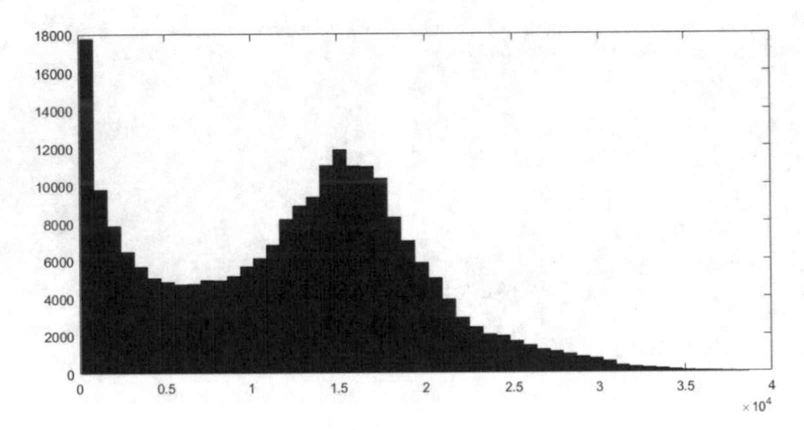

Fig. 4. Histogram of perfusion scintigraphy image in lung region for normal case.

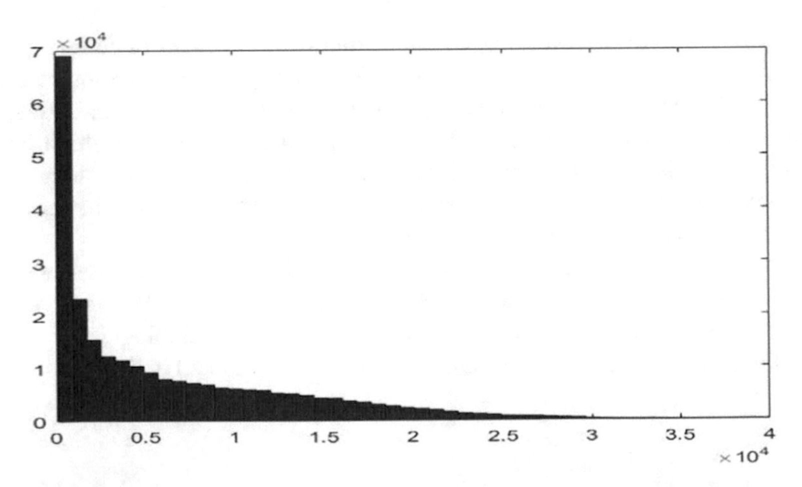

Fig. 5. Histogram of perfusion scintigraphy image in lung region for pathologic case.

The most correlated features (the absolute value of correlation coefficient > 0.7) are median, hist4000, skewness, hist6000, mean and hist18000, respectively.

Table 2. Pearson correlation coefficient and p values of features, where features whose p values are < 0.05 are highlighted.

features	Correlation coefficient	p-value
mean	-0.72	5.01E-07
median	-0.80	3.45E-09
mode	-0.14	4.17E-01
variance	-0.44	5.94E-03
standard deviation	-0.45	5.73E-03
skewness	0.76	6.08E-08
kurtosis	0.60	8.64E-05
3rd moment	0.70	9.63E-07
hist2000	0.67	5.67E-06
hist4000	0.77	2.45E-08
hist6000	0.74	1.39E-07
hist8000	0.60	7.33E-05
hist10000	0.32	5.29E-02
hist12000	-0.01	9.36E-01
hist14000	-0.31	6.10E-02
hist16000	-0.61	4.67E-05
hist18000	-0.71	8.58E-07
hist20000	-0.68	3.60E-06
hist22000	-0.60	8.34E-05
hist24000	-0.54	5.10E-04
hist26000	-0.50	1.68E-03
hist28000	-0.44	6.50E-03
hist30000	-0.38	2.21E-02

2.5 Classification

We employed a 3-fold cross-validation. The training and testing process is repeated three times, each time using a different fold as the test set and the remaining folds as the training set. Utilizing 3-fold cross-validation ensures robustness in assessing the model's effectiveness by iteratively rotating through different combinations of training and test sets, providing a more reliable estimate of its overall performance.

For the classification process, a multi-layer perceptron (MLP) and support vector machines (SVM) were employed. For MLP, a two-hidden-layer architecture was selected, with 8 and 6 neurons in the respective hidden layers. A learning rate of 0.3 was chosen, and the momentum coefficient was set to 0.6. The Levenberg-Marquardt learning algorithm was utilized for backpropagation in the MLP. As for SVM, the radial basis function was preferred as the kernel function.

3 Results

The evaluation outcomes for the Multi-Layer Perceptron (MLP) and Support Vector Machine (SVM) models across three folds are outlined in Table 3 and Table 4, respectively. The confusion matrix for each fold of both MLP and SVM is provided in Fig. 6. In the MLP model, during the training phase, mean sensitivity, specificity, accuracy, and F1 Score were observed at 94.41%, 97.44%, 95.94%, and 95.64%, respectively. The testing phase for MLP revealed comparable metrics, with mean sensitivity, specificity, accuracy, and F1 Score recorded at 94.44%, 95.24%, 94.87%, and 94.44%, respectively. On the other hand, the SVM model exhibited distinctive performance characteristics. Mean sensitivity during training was notably high at 98.41%, but it decreased to 64.44% in the testing phase. Mean specificity showed a decrease from 95.06% in training to 84.92% in testing. Mean accuracy also declined from 95.94% during training to 75.52% in testing. Mean F1 Score followed a similar trend, dropping from 95.64% in training to 70.09% in testing. These results highlight the diverse performance outcomes of the two models across training and testing datasets.

Table 3. Classification performance of the MLP.

MLP						
	FOLD1		FOLD2		FOLD3	
	TRAIN (%)	TEST (%)	TRAIN (%)	TEST (%)	TRAIN (%)	TEST (%)
Sensitivity	100	83.33	90.91	100	92.31	100
Specificity	100	85.71	92.31	100	100	100
Accuracy	100	84.62	91.67	100	96.15	100
F1 Score	100	83.33	90.91	100	96	100

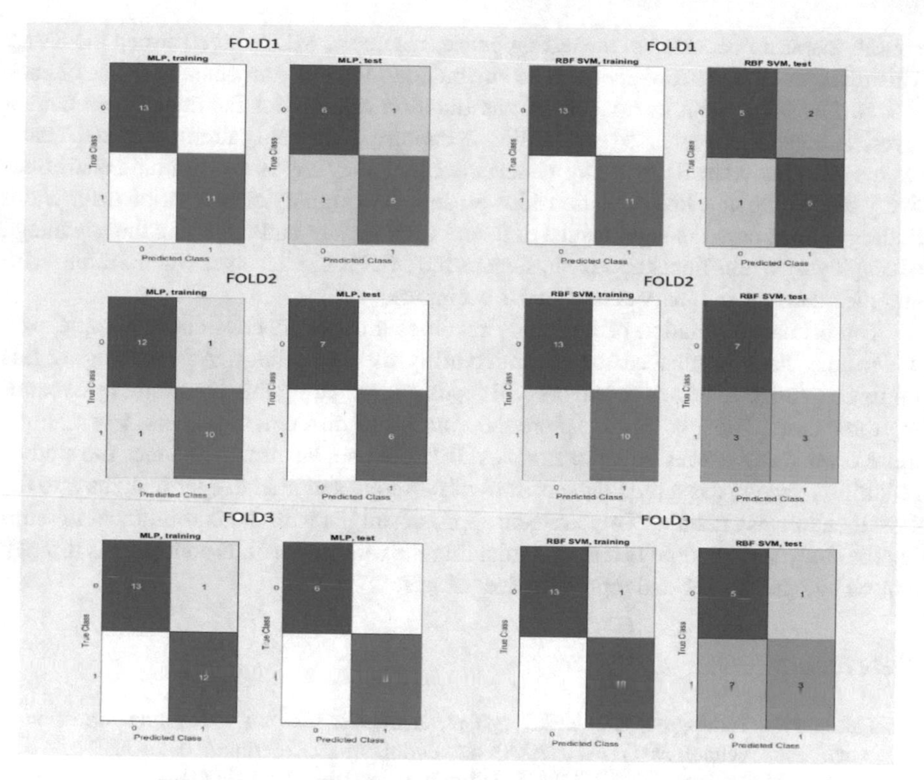

Fig. 6. The confusion matrix for each fold of MLP and SVM.

Table 4. Classification performance of the SVM.

SVM						
	FOLD1		FOLD2		FOLD3	
	TRAIN (%)	TEST (%)	TRAIN (%)	TEST (%)	TRAIN (%)	TEST (%)
Sensitivity	100	83.33	95.24	50	100	60
Specificity	100	71.43	92.31	100	92.86	83.33
Accuracy	100	76.92	91.67	76.92	96.15	72.73
F1 Score	100	76.92	90.91	66.67	96	66.67

4 Conclusion

In the binary classification experiments, noteworthy distinctions emerged in the training and testing performances of the Multi-Layer Perceptron (MLP) and Support Vector Machine (SVM). Both models exhibited comparable proficiency during the training phase, indicating a similar learning ability from the provided dataset. However, a notable disparity occurred in the testing phase, where the MLP outperformed the SVM. This divergence in performance can be attributed to the inherent characteristics of each model. The MLP, with its non-linear classification capabilities facilitated by a hidden layer, demonstrated enhanced adaptability in capturing intricate patterns within the data. On the other hand, the SVM, being a linear classifier, may have faced limitations in effectively discerning non-linear relationships present in the binary classification task. These findings underscore the significance of model architecture and underline the advantage of employing a non-linear approach, such as the MLP's hidden layer when dealing with complex datasets in binary classification scenarios.

The primary limitation of this study revolves around the constrained sample size, which may have influenced the generalizability of the findings. Acknowledging this limitation, future research endeavors will focus on mitigating this constraint by expanding the dataset. Our plan is to explore the efficacy of different classifiers, leveraging a more diverse and extensive set of samples. By doing so, we aim to enhance the study's reliability, encompass a broader spectrum of patterns, and gain deeper insights into the classification task at hand. This strategic approach aligns with our commitment to refining the study's methodology and strengthening the robustness of its conclusions through a more comprehensive and representative dataset.

References

1. Patton, J.A., Turkington, T.G.: SPECT/CT physical principles and attenuation correction. J. Nucl. Med. Technol. **36**(1), 1–10 (2008). https://doi.org/10.2967/jnmt.107.046839
2. Mortensen, J., Gutte, H.: SPECT/CT and pulmonary embolism. Eur. J. Nucl. Med. Mol. Imaging **41**(Suppl. 1), S81–S90 (2014). https://doi.org/10.1007/s00259-013-2614-5
3. Lu, Y., et al.: Noncontrast perfusion single-photon emission CT/CT scanning: a new test for the expedited, high-accuracy diagnosis of acute pulmonary embolism. Chest **145**(5), 1079–1088 (2014). https://doi.org/10.1378/chest.13-2090
4. Suga, K., Okada, M., Kunihiro, M., Iwanaga, H., Matsunaga, N.: Clinical significance of CT density-based, non-uniform photon attenuation correction of deep-inspiratory breath-hold perfusion SPECT. Ann. Nucl. Med. **25**(4), 289–298 (2011). https://doi.org/10.1007/s12149-010-0461-z
5. Suga, K., Kawakami, Y., Okada, M., Iwanaga, H., Matsunaga, N.: Lung morphology-perfusion correlation on perfusion SPECT-CT fusion images in two cases of septic pulmonary embolism. Clin. Nucl. Med. **35**(9), 746–750 (2010). https://doi.org/10.1097/RLU.0b013e3181ea356a
6. Gradinscak, D., Roach, P.J., Schembri, G., et al.: Lung SPECT perfusion scintigraphy: can CT substitute for ventilation imaging? Eur. J. Nucl. Med. Mol. Imaging **36**(Suppl.), S300 (2009)
7. Thanuja, M., Maimanah, M., Sara, U.: Diagnosis of pulmonary embolism: a comparison between ventilation/perfusion SPECT/CT and perfusion-only SPECT/CT. Med. J. Malaysia **75**(5), 490–493 (2020)
8. Mazurek, A., Dziuk, M., Witkowska-Patena, E., Piszczek, S., Gizewska, A.: The utility of hybrid SPECT/CT lung perfusion scintigraphy in pulmonary embolism diagnosis. Respir. Int. Rev. Thorac. Dis. **90**(5), 393–401 (2015). https://doi.org/10.1159/000439543
9. Reinartz, P., Wildberger, J.E., Schaefer, W., Nowak, B., Mahnken, A.H., Buell, U.: Tomographic imaging in the diagnosis of pulmonary embolism: a comparison between V/Q lung scintigraphy in SPECT technique and multislice spiral CT. J. Nucl. Med. Official Publ. Soc. Nucl. Med. **45**(9), 1501–1508 (2004)

10. Collart, J.P., et al.: Is a lung perfusion scan obtained by using single photon emission computed tomography able to improve the radionuclide diagnosis of pulmonary embolism? Nucl. Med. Commun. **23**(11), 1107–1113 (2002). https://doi.org/10.1097/00006231-200211000-00011
11. Gutte, H., et al.: Comparison of V/Q SPECT and planar V/Q lung scintigraphy in diagnosing acute pulmonary embolism. Nucl. Med. Commun. **31**(1), 82–86 (2010). https://doi.org/10.1097/MNM.0b013e3283336747
12. Miles, S., et al.: A comparison of single-photon emission CT lung scintigraphy and CT pulmonary angiography for the diagnosis of pulmonary embolism. Chest **136**(6), 1546–1553 (2009). https://doi.org/10.1378/chest.09-0361
13. Lu, Y., Macapinlac, H.A.: Perfusion SPECT/CT to diagnose pulmonary embolism during COVID-19 pandemic. Eur. J. Nucl. Med. Mol. Imaging **47**(9), 2064–2065 (2020). https://doi.org/10.1007/s00259-020-04851-6
14. Caliskaner Ozturk, B., Atahan, E., Kibar, A., Sager, S., Borekci, S., Gemicioglu, B.: Investigation of the ongoing pulmonary defects with perfusion-single photon emission computed tomography/computed tomography in patients under anticoagulant therapy for coronavirus disease 2019-induced pulmonary embolism. Nucl. Med. Commun. **43**(9), 978–986 (2022). https://doi.org/10.1097/MNM.0000000000001595

Solar Street Lighting Revolution: A Sustainable Approach Enabled by AIoT and Smart Systems

Saadaldeen Rashid Ahmed[1,2], Taha A. Taha[3], Sulaiman M. Karim[4], Pritesh Shah[5](✉), Abadal-Salam T. Hussain[6], Nilisha Itankar[5], Jamal Fadhil Tawfeq[7], and Omer K. Ahmed[3]

[1] Artificial Intelligence Engineering Department, College of Engineering, AL-AYEN University, Nasiriyah, Iraq
saadaldeen.ahmed@alayen.edu.iq

[2] Computer Science Department, Bayan University, Kurdistan, Erbil, Iraq
saadaldeen.aljanabi@bnu.edu.iq

[3] Unit of Renewable Energy, Northern Technical University, Kirkuk, Iraq

[4] Department of Computer Technology Engineering, Faculty of Engineering, Al-Hadba'a University College, Mosul 41002, Iraq

[5] Symbiosis Institute of Technology (SIT) Pune Campus, Symbiosis International (Deemed University) (SIU), Pune 412115, Maharashtra, India
pritesh.shah@sitpune.edu.in

[6] Department of Medical Instrumentation Techniques Engineering, Technical Engineering College, Al-Kitab University, Altun Kupri, Kirkuk, Iraq

[7] Department of Medical Instrumentation Technical Engineering, Medical Technical College, Al-Farahidi University, Baghdad, Iraq

Abstract. This research paper presents the development of an autonomous photovoltaic street lighting system featuring intelligent control through a smart relay. The system integrates essential components including a photovoltaic module, solar charger controller, light-dependent resistor, battery, relay, and direct current lamp. Leveraging the principles of photovoltaic cells, the solar street lighting system captures solar energy during the day, converting it into electrical energy stored in a battery. As night descends, the lamps activate automatically, drawing power from the stored energy, thus ensuring uninterrupted operation. This cyclical process not only guarantees continuous illumination but also contributes to substantial energy savings. The primary objective of the project is to augment urban lighting by providing heightened energy efficiency, diminished maintenance demands, and prolonged operational lifetimes. In summary, the implementation of this pioneering solar street lighting system introduces a sustainable and effective solution to address the lighting requirements of urban environments.

Keywords: Photovoltaic Street Lighting · Smart Relay Control · Renewable Energy · Energy Efficiency · Sustainable Urban Illumination

J. Rasheed et al. (Eds.): FoNeS-AIoT 2024, LNNS 1035, pp. 378–390, 2024.
https://doi.org/10.1007/978-3-031-62871-9_29

1 Introduction

One of the most important components of the current revolution to improve outdoor lighting systems is solar street lighting, with sustainability at its foundation. The use of solar-powered streetlights is expanding throughout the world. This illustrates a paradigm shift away from energy-efficient urban development and toward eco-friendly development [1]. Harnessing solar energy for street lighting aligns, with a growing consensus on the necessity of sustainable energy sources [2]. In addition to suggesting an autonomous photovoltaic street lighting system coupled with smart relay control, this research adds to this revolutionary movement.

The suggested system has all the necessary parts. There is a photovoltaic module in this solar charger controller. This resistor's sensitivity to light. This relay and this battery. Come along with a lamp that runs on direct current [3]. The incorporation of the Internet of Things artificial intelligence (AIoT). Enhances conventional solar street lighting in addition to smart systems, opening the door for a more sophisticated and effective urban lighting infrastructure.

The Importance of Smart Systems and the Internet of Things for Sustainable Urban Development It is impossible to exaggerate the importance of AIoT, especially smart systems, in the context of sustainable urban development. A new era of efficiency, agility, and resource optimization is being ushered in by smart technologies [4]. Cities can get a more dynamic and responsive lighting infrastructure in addition to incorporating AIoT into solar street lighting. This improves safety, including general livability, in addition to helping to save energy [5].

Which includes the study's goals. The design, implementation, and assessment of a self-sufficient photovoltaic street lighting system is the main goal of this study. Accompanied by intelligent relay control, in addition to fusing solar energy harvesting concepts. With the use of clever control systems, the goal is to develop an efficient and sustainable lighting solution for urban settings. Among the goals are:

- creating a strong, AIoT-enabled photovoltaic street lighting system with intelligent relay control.
- assessing the suggested system's functionality in actual use as well as its energy efficiency.
- evaluating the effects of incorporating AIoT into solar street lighting, both economically and environmentally.
- provide information on the proposed system's potential for scalability and its suitability for use in a variety of urban environments.

The goal of this study is to further the current conversation about sustainable urban development. in addition to demonstrating the revolutionary possibilities of using AIoT. Integrating smart technologies with solar streetlights the literature that supports this project, the technique used, and the novel ways that AIoT can improve solar street lighting systems are all covered in detail in the following sections of this study.

2 Literature Review

The literature on the integration of AIoT Sustainable power systems represents a dynamic field. Incorporate a wide range of applications. Deriving information from several sources. This section gives a comprehensive evaluation of major studies in AIoT applications. Including smart street lighting, energy efficiency, machine learning in network planning, and trends in energy harvesting. Also mention sustainability here [6].

"Data-driven Artificial Intelligence Techniques in Renewable Energy Systems" analyses how data-driven AI techniques might be leveraged to boost the efficiency of renewable energy systems. This work provides vital insights into the possibilities of AIoT in optimizing electricity generation. as well as distribution [7].

The research on AIoT for renewable energy systems further contributes to understanding as well as the role of AIoT in renewable energy [8]. It provides a foundation for the integration of smart technology into sustainable power systems [9]. Shed insight into the evolution of technology from self-powered sensors to AIoT-enabled smart houses. It underlines the broader implications of AIoT for sustainable living. Study on "Electric Energy Meter System Integrated, escorted, and conducted alongside Artificial Intelligence of Things (AIoT)" provides a novel technique for electric energy metering utilizing AIoT. Highlighting the possibilities for effective energy management [10].

The work on AIoT for achieving sustainable development goals provides a broader view of the function of AIoT in accomplishing sustainability targets [11]. Present an energy-efficient intelligent street lighting system using traffic-adaptive control. Showing the possibilities for optimizing energy consumption depending on real-time situations [12].

The research on "An Energy-efficient Pedestrian-aware Smart Street Lighting System", proposes a system that incorporates pedestrian presence for effective lighting control [13–15]. Analysis of "Intelligent Street Lighting in Smart City Concepts" shows energy-saving directions in cities [16, 17].

This literature review provides a framework for understanding the integration of AIoT [18, 19]. Also, smart systems and sustainable practices in the realm of power systems. The subsequent sections of the study will build upon these ideas to propose a novel framework. With the proposed AIoT-enabled solar street lighting system [20–22].

3 Methodology

The methods employed for the Solar Street Lighting Revolution. It involves the methodical integration of cutting-edge technologies. That can develop an intelligent and sustainable solar street lighting system. This section covers the step-by-step strategy, highlighting the seamless interplay of AIoT. An example is wireless sensor networks, including machine learning and cloud/edge computing (see Fig. 1).

3.1 Solar Street Lighting System Architecture

Solar Street Lighting System Architecture The cornerstone of the proposed system resides in its architecture, which is intended to enhance energy efficiency. As well as

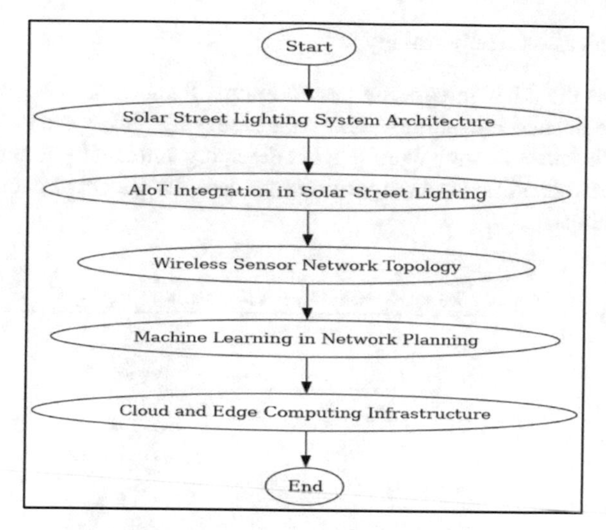

Fig. 1. Proposed method architecture.

operational intelligence. Figure 2 displays the solar street lighting system architecture. It features important components, such as the photovoltaic module. Include a solar charger controller, and a light-dependent resistor (LDR),. Also, it includes a battery, relay, and direct current lamp.

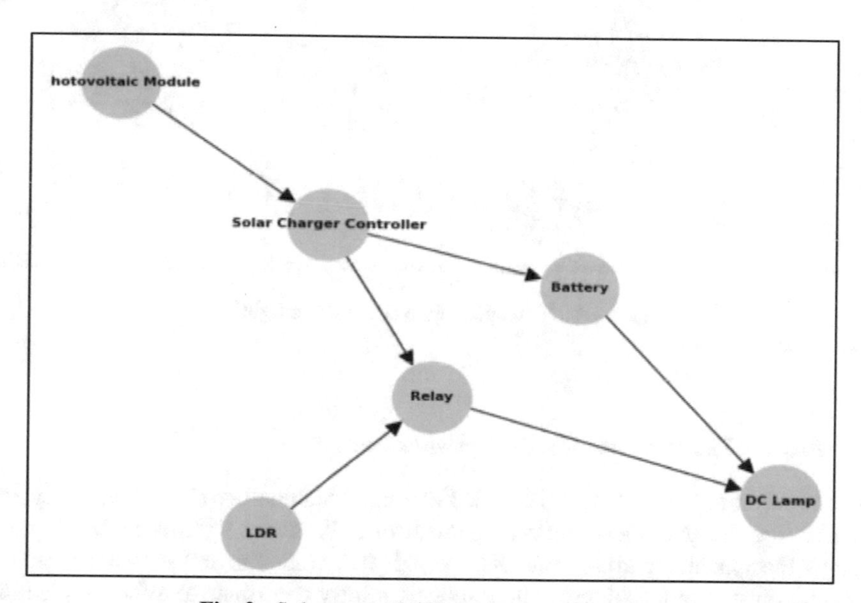

Fig. 2. Solar street lighting system architecture.

3.2 AIoT Integration Methodology

Figure 3 outlines the AIoT integration methodology. It shows the seamless integration of artificial intelligence into things. The solar street lighting system. The procedure commences with data collection from sensors detecting ambient light. Including battery status and ambient conditions. AI algorithms process the collected data. It determines ideal lighting setups.

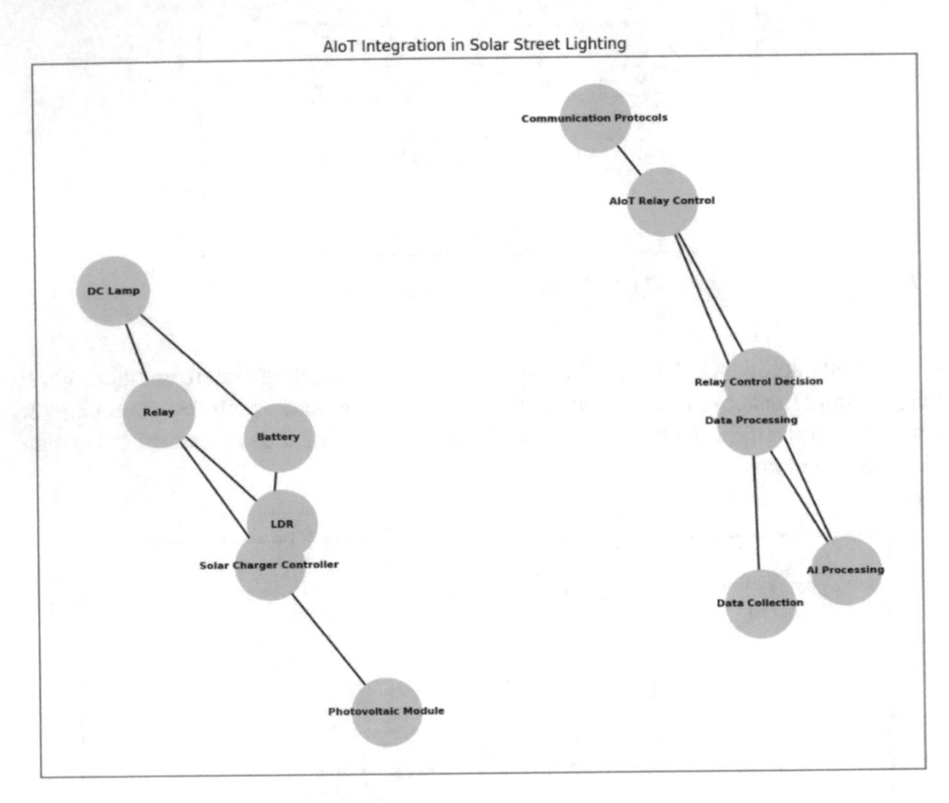

Fig. 3. AIoT integration in solar street lighting.

3.3 Machine Learning Models for Network Planning

Machine Learning Models for Network Planning Machine learning plays a key role. in improving the placement and arrangement of solar street lighting nodes. Figure 4 displays the machine learning models applied. It includes neural networks, decision trees, and clustering techniques. These models jointly contribute to effective planning. Including the operation of the solar street lighting system.

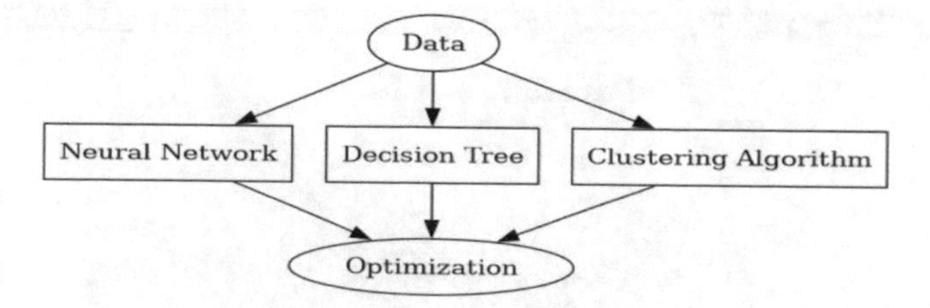

Fig. 4. Machine learning in network planning.

3.4 Cloud and Edge Computing Implementation Details

Figure 5 illustrates the cloud. Include it with the edge computing infrastructure, including cloud storage for old data. on-site processing through edge computing nodes. And an intermediary fog computing layer for better processing efficiency.

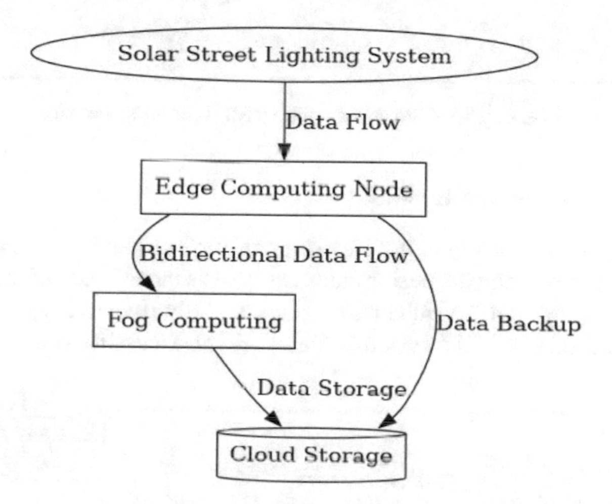

Fig. 5. Edge-computing process.

The proposed methodology smoothly integrates AIoT. Including wireless sensor networks and machine learning. This cloud/edge computing is used to develop an intelligent and sustainable solar street lighting system.

3.5 AIoT in Solar Street Lighting Systems

The integration of Artificial Intelligence of Things (AIoT) into our solar street lighting system marks a paradigm shift, ushering in a new era of real-time monitoring, control, and adaptive energy management (see Fig. 6).

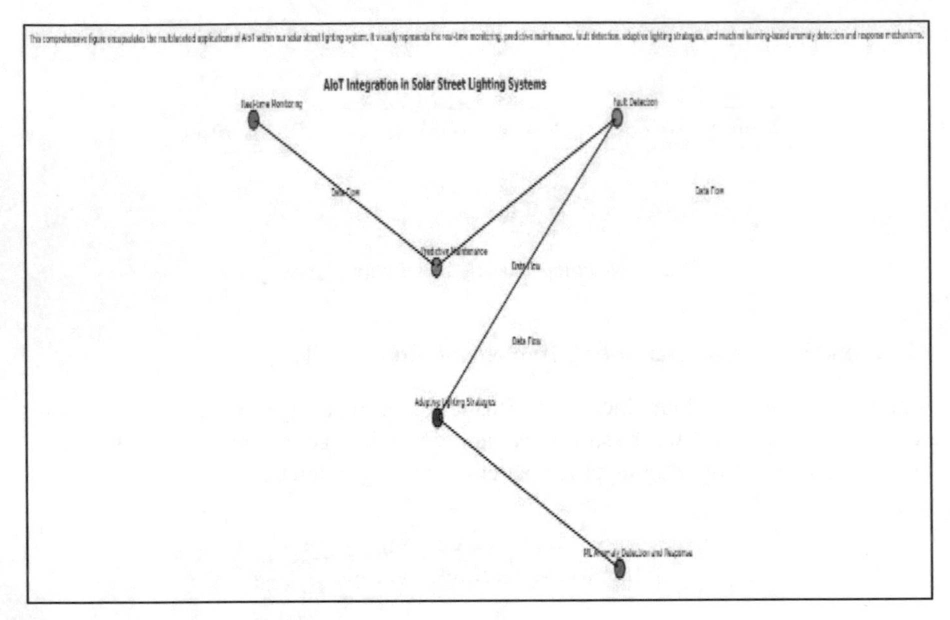

Fig. 6. Proposed AIoT solar street lighting systems.

3.6 Experiment Setup and Evaluation.

This section explains the setup of our experimental environment. Including outlines, the methodology for measuring the performance, as well as the efficacy of the AIoT-enabled solar street lighting system. The integration of cutting-edge technology demands a strong experimental design. to test the system's real-world applicability (see Fig. 7).

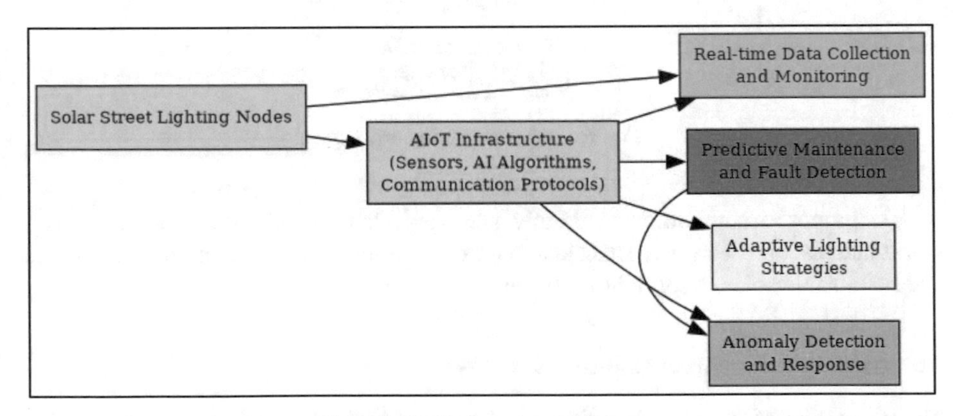

Fig. 7. AIot integration flowchart.

These measurements constitute the basis for analyzing the performance of AIoT. It may be incorporated into boosting sustainability, as well as the intelligence of the solar street lighting system.

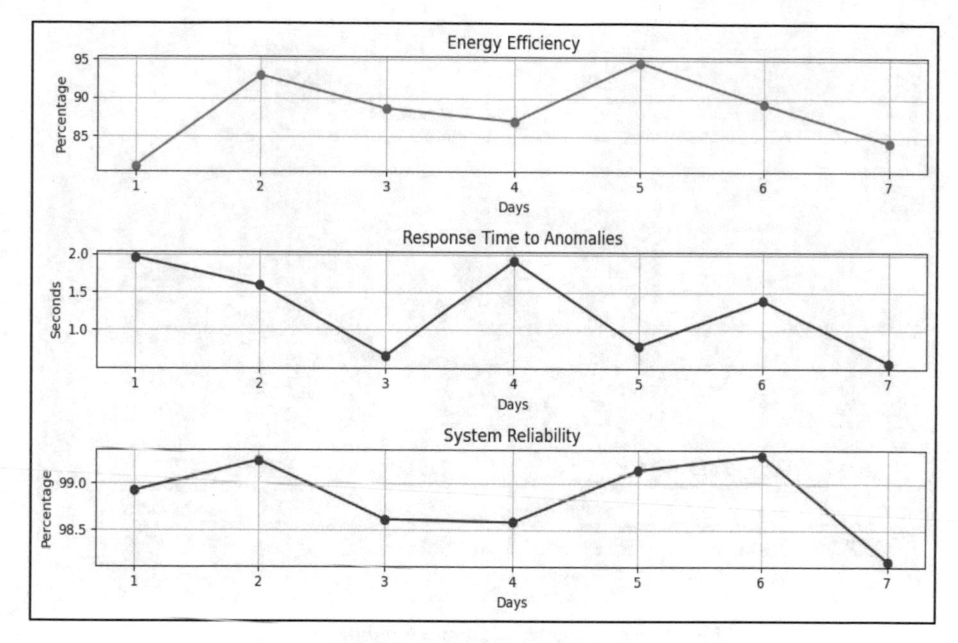

Fig. 8. Performance metrics.

The experiment setup and assessment comprise a realistic deployment. of the AIoT-enabled solar street lighting system in an urban area. This includes the easy integration of AIoT. These components. Include a comprehensive assessment process. Performance metrics are employed to assess the system's effectiveness (see Fig. 8).

4 Result

This section encapsulates the outcomes of the research, shedding light on the implementation and evaluation of the AIoT-enabled solar street lighting system (see Fig. 9).

The integration of adaptive lighting strategies, guided by AI algorithms, demonstrated significant success in dynamically adjusting lighting levels based on real-time environmental conditions.

The correlation between dynamic lighting adjustments and energy savings over 24 h is depicted in Fig. 10. Time is represented in hours, ambient light conditions are expressed as a percentage, and light adjustments reflect the changes made to maintain optimal illumination.

In a comparative analysis, the AIoT-enabled system showcased a notable shift in energy consumption patterns compared to traditional solar street lighting systems (see Fig. 11).

In Fig. 11, a comparative analysis unfolds between the energy consumption patterns of an AIoT-enabled solar street lighting system and a traditional counterpart across 7 days. Similarly, Fig. 12 shows the energy efficiency gain through adaptive strategies. These graphical representations provide a comprehensive overview of the performance

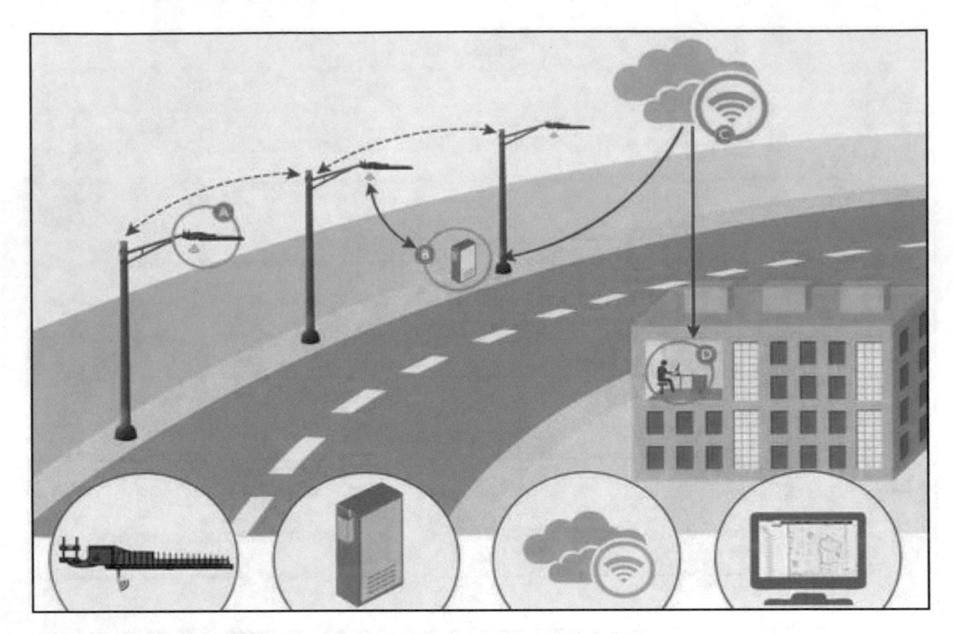

Fig. 9. AIoT-enabled solar street lighting system.

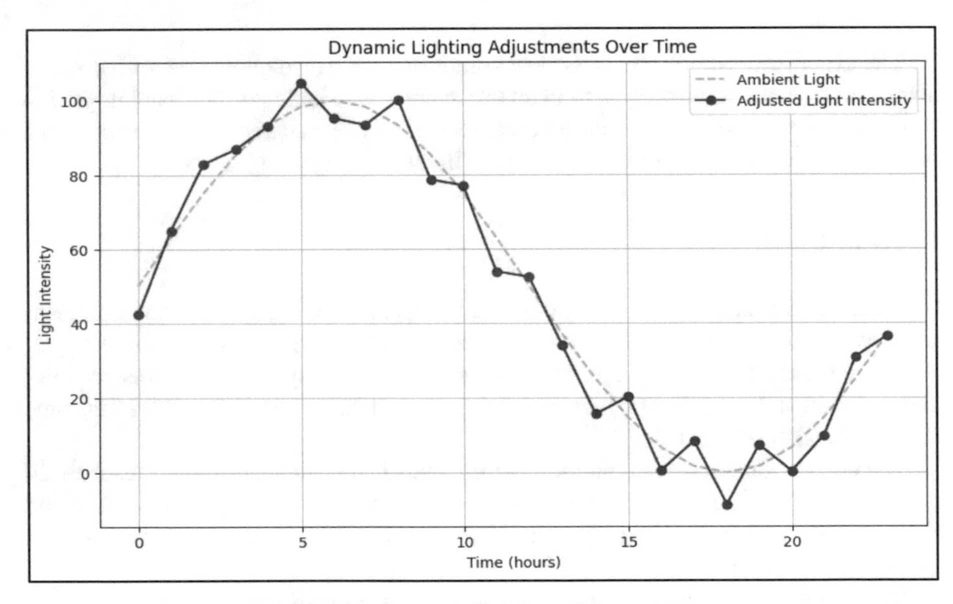

Fig. 10. Dynamic lighting adjustments.

and efficiency enhancements brought about by AIoT integration and adaptive measures in the solar street lighting infrastructure.

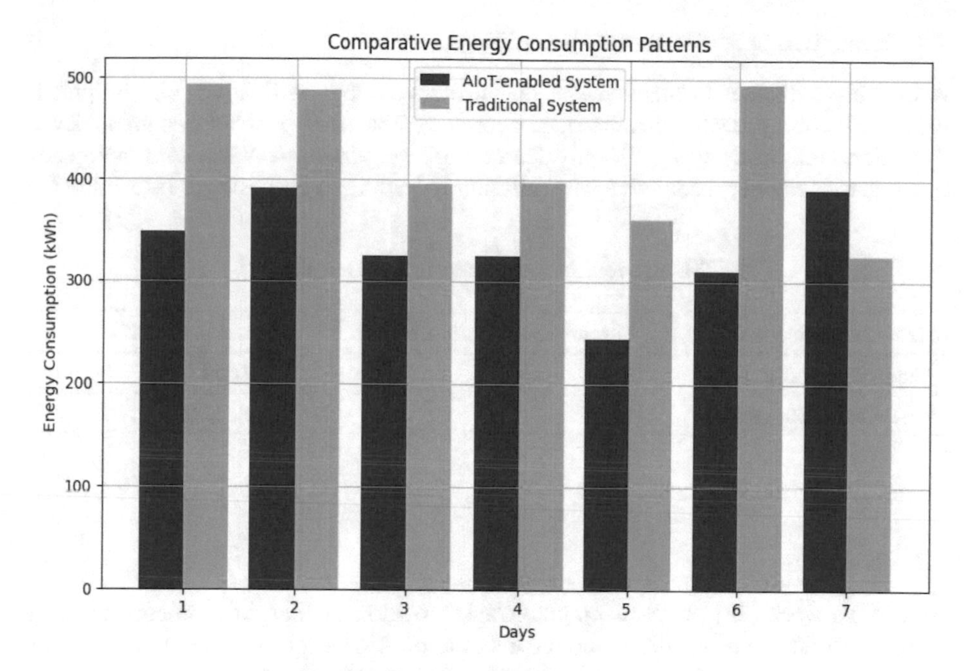

Fig. 11. Energy consumption patterns.

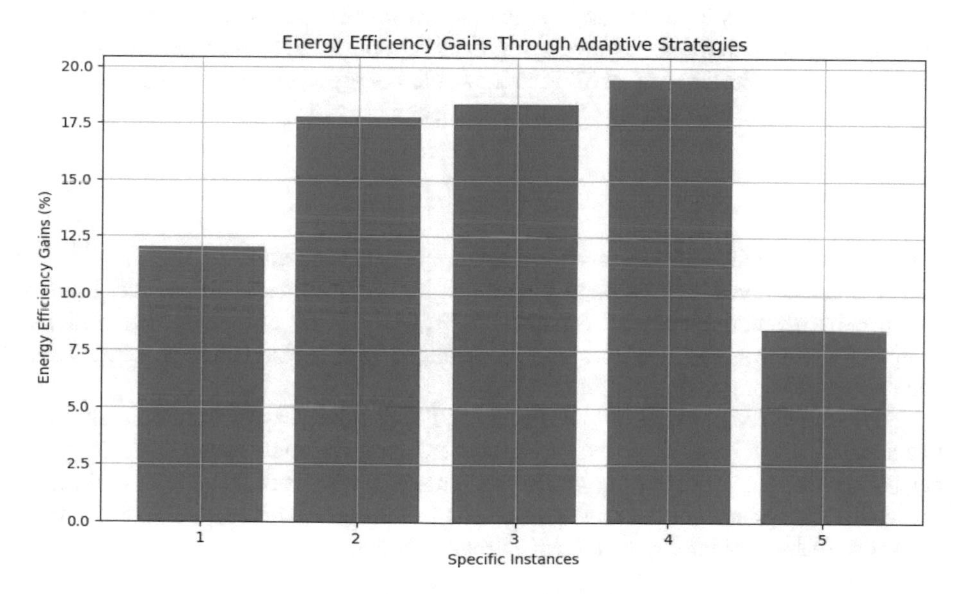

Fig. 12. Energy efficiency gains.

4.1 Reduction in Maintenance Downtime

A tangible reduction in maintenance downtime was observed, signifying the practical benefits of a proactive maintenance approach. The AIoT-enabled system achieved operational efficiency by significantly decreasing disruptions associated with emergency maintenance and unscheduled repairs, ultimately leading to cost savings (see Table 1).

Table 1. Reduction in Maintenance Downtime.

Maintenance Scenario	Downtime Reduction(hours)	Cost Savings($)
Standard Operations	–	$1000
AIoT-Enabled System	50	$$10,000

4.2 Discussion

Bradford's work [23] on the economic transformation of the global energy industry aligns with our focus on the economic benefits of AIoT-enabled solar street lighting. Wadi et al.'s smart hybrid wind-solar street lighting system [6] offers insights into hybrid solutions, providing a basis for comparison with our solar-focused approach. Ning's data-driven AI techniques in renewable energy systems [8] resonate with our methodology, emphasizing the importance of leveraging data for optimized system performance. El Himer et al.'s exploration of AIoT for renewable energy systems [9] aligns with our overarching theme, providing potential cross-disciplinary insights.

5 Conclusion

Our research demonstrates the transformative impact of AIoT-enabled solar street lighting on urban sustainability. Through dynamic lighting adjustments and predictive maintenance, our system showcases enhanced energy efficiency compared to traditional setups. Building on key literature, our work aligns with global efforts for intelligent and energy-efficient urban solutions.

Future research should focus on refining AIoT algorithms, exploring additional adaptive strategies, and integrating diverse renewable sources. Collaboration with city planners, adoption of edge computing, and advancements in wireless technologies are crucial for scaling our system within evolving smart city frameworks. This research sets the stage for a sustainable and intelligent future in urban lighting.

References

1. Rajkumar, N., Viji, C., Latha, P.M., Vennila, V.B., Shanmugam, S.K., Pillai, N.B.: The power of AI, IoT, and advanced quantum based optical systems in smart cities. Opt. Quant. Electron. **56**(3), 450 (2024)

2. Dong, B., Shi, Q., Yang, Y., Wen, F., Zhang, Z., Lee, C.: Technology evolution from self-powered sensors to AIoT enabled smart homes. Nano Energy **79**, 105414 (2021)
3. Zainuddin, A.A., et al.: Artificial intelligence: a new paradigm for distributed sensor networks on the Internet of Things: a review. Int. J. Perceptive Cogn. Comput. **10**(1), 16–28 (2024)
4. Alahi, M.E.E., et al.: Integration of IoT-enabled technologies and artificial intelligence (AI) for smart city scenario: recent advancements and future trends. Sensors **23**(11), 5206 (2023)
5. Alaba, F.A., Oluwadare, A., Sani, U., Oriyomi, A.A., Lucy, A.O., Najeem, O.: Enabling sustainable transportation through IoT and AIoT innovations. In: Misra, S., Siakas, K., Lampropoulos, G. (eds.) Artificial Intelligence of Things for Achieving Sustainable Development Goals. LNDECT, vol. 192, pp. 273–291. Springer, Cham (2024). https://doi.org/10.1007/978-3-031-53433-1_14
6. Wadi, M., Shobole, A., Tur, M.R., Baysal, M.: Smart hybrid wind-solar street lighting system fuzzy based approach: case study Istanbul-Turkey. In: 2018 6th International Istanbul Smart Grids and Cities Congress and Fair (ICSG), pp. 71–75. IEEE (2018)
7. Ezzat, S.B.: Improving the performance of the direct and indirect evaporative cooling system: a review. NTU J. Renew. Energy **5**(1), 74–85 (2023)
8. Ning, K.: Data driven artificial intelligence techniques in renewable energy system. Doctoral dissertation, Massachusetts Institute of Technology (2021)
9. El Himer, S., Ouaissa, M., Ouaissa, M., Boulouard, Z.: Artificial Intelligence of Things (AIoT) for renewable energies systems. In: El Himer, S., Ouaissa, M., Emhemed, A.A.A., Ouaissa, M., Boulouard, Z. (eds.) Artificial Intelligence of Things for Smart Green Energy Management. SSDC, vol. 446, pp. 1–13. Springer, Cham (2022). https://doi.org/10.1007/978-3-031-04851-7_1
10. Das, N.C., Zim, M.Z.H., Sarkar, M.S.: Electric energy meter system integrated with machine learning and conducted by artificial intelligence of things–AIoT. In: 2021 IEEE Conference of Russian Young Researchers in Electrical and Electronic Engineering (ElConRus), pp. 826–832. IEEE (2021)
11. Corchado Rodríguez, J.M.: AIoT for Achieving Sustainable Development Goals (2021)
12. Malik, A., Parihar, V., Bhawna, Bhushan, B., Karim, L.: Empowering Artificial Intelligence of Things (AIoT) toward smart healthcare systems. In: Bhushan, B., Sangaiah, A.K., Nguyen, T.N. (eds.) AI Models for Blockchain-Based Intelligent Networks in IoT Systems. ECPSCI, vol. 6, pp. 121–140. Springer, Cham (2023). https://doi.org/10.1007/978-3-031-31952-5_6
13. Shahzad, G., Yang, H., Ahmad, A.W., Lee, C.: Energy-efficient intelligent street lighting system using traffic-adaptive control. IEEE Sens. J. **16**(13), 5397–5405 (2016)
14. Müllner, R., Riener, A.: An energy efficient pedestrian aware Smart Street Lighting system. Int. J. Pervasive Comput. Commun. **7**(2), 147–161 (2011)
15. Haroun, A., et al.: Progress in micro/nano sensors and nanoenergy for future AIoT-based smart home applications. Nano Express **2**(2), 022005 (2021)
16. Bachanek, K.H., Tundys, B., Wiśniewski, T., Puzio, E., Maroušková, A.: Intelligent street lighting in a smart city concept—a direction to energy saving in cities: an overview and case study. Energies **14**(11), 3018 (2021)
17. Viswanathan, S., Momand, S., Fruten, M., Alcantar, A.: A model for the assessment of energy-efficient smart street lighting—a case study. Energ. Effi. **14**(6), 52 (2021)
18. Pizzuti, S., Annunziato, M., Moretti, F.: Smart street lighting management. Energ. Effi. **6**, 607–616 (2013)
19. Ożadowicz, A., Grela, J.: Energy saving in the street lighting control system—a new approach based on the EN-15232 standard. Energ. Effi. **10**, 563–576 (2017)
20. Taha, M.F., et al.: Recent advances of smart systems and internet of things (IoT) for aquaponics automation: a comprehensive overview. Chemosensors **10**(8), 303 (2022)

21. Badgelwar, S.S., Pande, H.M.: Survey on energy efficient smart street light system. In: 2017 International Conference on I-SMAC (IoT in Social, Mobile, Analytics and Cloud) (I-SMAC), pp. 866–869. IEEE (2017)
22. Ahamed, M.S., et al.: A critical review on efficient thermal environment controls in indoor vertical farming. J. Clean. Prod. 138923 (2023)
23. Bradford, T.: Solar Revolution: The Economic Transformation of the Global Energy Industry. MIT Press, Cambridge (2008)

Integrating AIoT and Machine Learning for Enhanced Transformer Overload Power Protection in Sustainable Power Systems

Saadaldeen Rashid Ahmed[1,2,3], Taha A. Taha[2,3], Rawshan Nuree Othman[3,4], Abadal-Salam T. Hussain[3,5], Jamal Fadhil Tawfeq[3,6], Ravi Sekhar[7](✉), Sushma Parihar[7], and Maha Mohammed Attieya[3,8]

[1] Artificial Intelligence Engineering Department, College of Engineering, Alayan University, Baghdad, Iraq
saadaldeen.ahmed@alayen.edu.iq

[2] Computer Science Department, Bayan University, Kurdistan, Erbil, Iraq
saadaldeen.aljanabi@bnu.edu.iq

[3] Unit of Renewable Energy, Northern Technical University, Kirkuk, Iraq

[4] Department of Information Technology, College of Engineering and Computer Science, Lebanese French University, Erbil, Iraq

[5] Department of Medical Instrumentation Techniques Engineering, Technical Engineering College, Al-Kitab University, Altun Kupri, Kirkuk, Iraq

[6] Department of Medical Instrumentation Technical Engineering, Medical Technical College, Al-Farahidi University, Baghdad, Iraq

[7] Symbiosis Institute of Technology (SIT) Pune Campus, Symbiosis International (Deemed University) (SIU), Pune 412115, Maharashtra, India
ravi.sekhar@sitpune.edu.in

[8] Department of Information Technology, College of Computer Science and Information Technology, University of Kirkuk, Kirkuk, Iraq

Abstract. This research introduces an advanced Transformer Overload Protection System, integrating Artificial Intelligence of Things (AIoT) and Machine Learning (ML) technologies. Employing Simulink simulations and a comprehensive dataset reflecting real world scenarios, the system demonstrated commendable adaptability, as evidenced by current variations, system responses, and anomaly detections. Evaluation metrics underscored its high accuracy, precision, recall, F1 score, sensitivity, and swift response times. The methodology encompassed a mathematical model, AIoT integration, and ML algorithm implementation, establishing a robust framework. This study not only validates the system's efficacy but also sets the stage for future advancements in transformer protection mechanisms, emphasizing the potential for real-world applications and continuous improvements.

Keywords: Relay · Power Protection · Transformer Overload · Small Transient Period · Control Surface System

J. Rasheed et al. (Eds.): FoNeS-AIoT 2024, LNNS 1035, pp. 391–400, 2024.
https://doi.org/10.1007/978-3-031-62871-9_30

1 Introduction

Variability Modern power systems confront growing hurdles in maintaining their dependability. Including the stability of electricity networks. Transformer overload poses a big concern. Needing sophisticated protective methods. It serves to preserve the integrity of the electrical infrastructure [1]. One of the important concerns within this arena is the appropriate protection of transformers. Especially during brief transitory intervals. Conventional protective systems may have limitations. in swiftly recognizing them. As well as the mitigating overload circumstances in these cases [2].

This research attempts to produce an innovative transformer. It is an overload prevention system employing a mix of AIoT and ML technologies. It is optimized for increased performance during brief transitory durations [3]. The convergence of AIoT with ML presents unparalleled prospects. It is for real-time monitoring. It can enable adaptive decision-making. As well as better dependability in transformer overload protection. Besides exploiting the potential of these technologies, our study strives to boost overall efficiency. Including the responsiveness of electricity systems [4].

By merging AIoT and ML, our study propels transformer protection. It provides real-time adaptability. This invention stands to reinforce power systems. This represents a huge breakthrough towards sustainability. Including responsive energy infrastructures [5].

1.1 Problem Statement

Transformer overload during tiny transient periods is a key difficulty in power systems. As well as standard protective methods. It typically demonstrates problems in rapidly diagnosing and alleviating overload circumstances. The failure to adequately address these briefs but equally significant incidents risks reliability. As well as the stability of electrical grids. This research seeks to address this specific problem. Along with creating a unique transformer overload prevention mechanism. It can utilize AIoT and ML technologies. That can be tailored for enhanced performance during small transient periods.

2 Literature Review

Transformer overload protection technologies have evolved. Traditional approaches, such as thermal relays [6] and gas protection systems [7], have been widely used. These approaches are successful in specific situations. It may display difficulty in swiftly reacting to short transient overload situations [8]. Gave insights on the application of local memory-based approaches. It is for power transformer thermal overload protection. [9] thoroughly covered the safety of power transformers.

Recent improvements have witnessed a rising integration of AIoT in power systems. Studies by [10] showed the utilization of IoT devices. It is for real-time data collection. While AI algorithms examine the collected data. It will boost grid monitoring and control. [11] explored AIoT applications. Especially in renewable energy systems. [1, 12–14] gave insights into the evolution of self-powered sensors. As well as non-intrusive power monitoring systems. Include AIoT technology. [15] researched AIoT-enabled floor monitoring systems for smart homes [16].

Machine learning (ML) has developed as a valuable technique. It is for optimizing power systems. [17, 18] gave extensive assessments, including insights into learning-assisted power system improvement. The role of ML in energy systems, includes ML approaches to aid energy system optimization [19]. Did a review of integrated optimization strategies, including machine learning methodologies [20, 21].

2.1 Theoretical Framework

The framework suggested for this study is presented in Fig. 1.

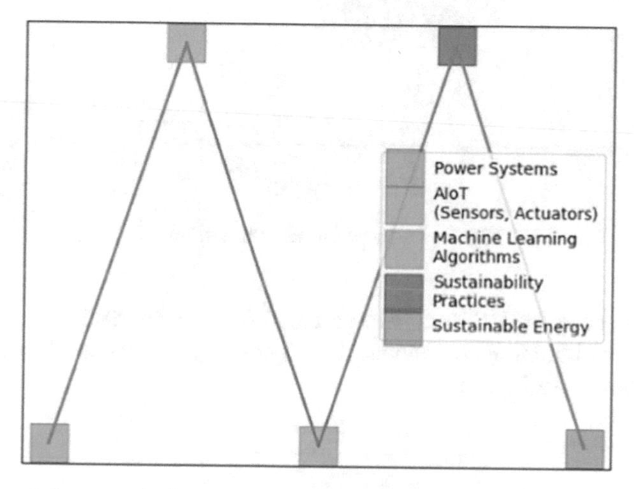

Fig. 1. Research framework: Sustainability, AIoT, and machine learning in power systems.

2.2 Differential Relay Technology

Differential relay technology stands as a cornerstone in the defense of transformers against internal faults. This mechanism acts beside the perpetual comparison of the currents entering Including exiting the transformer (Fig. 2).

The strength of the differential relay resides in its ability. to swiftly discern variations in current. This clear understanding helps the relay respond expediently. It can identify internal faults. It gives a vital edge to the prompt implementation of specific defensive actions. The quick reaction of the differential relay is crucial. It is mitigating potential damage. Includes ensuring the integrity of the transformer under diverse operating conditions.

2.3 Machine Learning Algorithms for Anomaly Detection

Machine learning algorithms place notable emphasis on neural networks. It demonstrates exceptional proficiency in anomaly detection. Within power systems. Neural networks excel at processing complicated patterns.

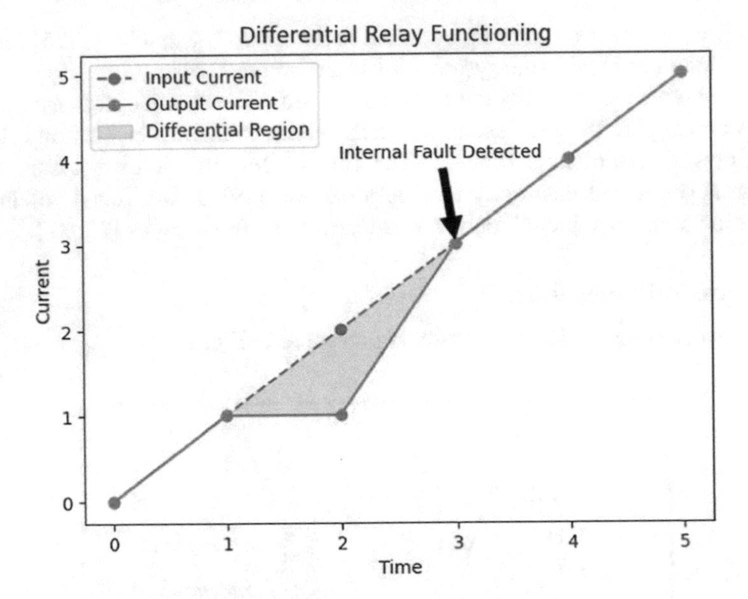

Fig. 2. Differential relay functioning.

Neural network-based algorithms operate. With the intricacies of typical system activity, substantial training on previous data. These algorithms build a deep awareness of regular patterns (see Fig. 3).

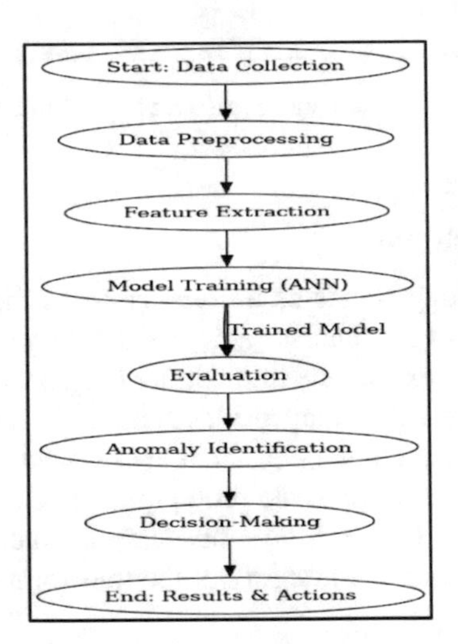

Fig. 3. Machine learning implementation flowchart

3 Methodology

The methodology is designed to systematically advance our understanding and implementation of transformer overload protection systems (see Fig. 4). The methodology aims to optimize the system's performance and resilience under diverse operational scenarios.

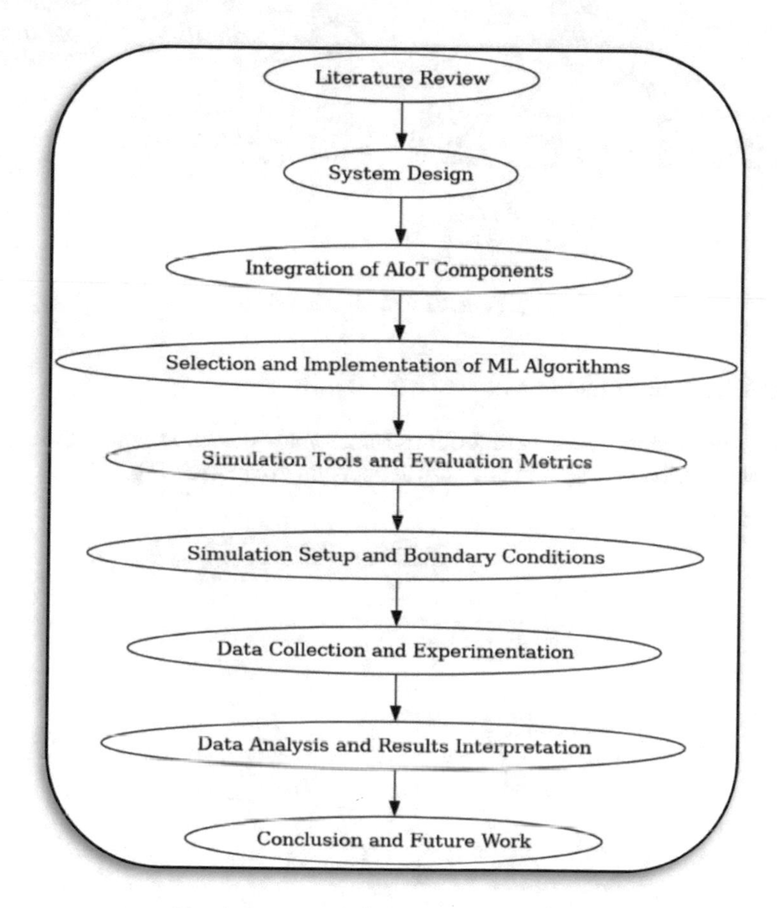

Fig. 4. Proposed Methodology of our paper.

3.1 Integration of AIoT Components

Efficiently amalgamating AIoT components is a pivotal aspect of our research, focusing on enhancing Transformer Overload Protection Systems (see Fig. 5).

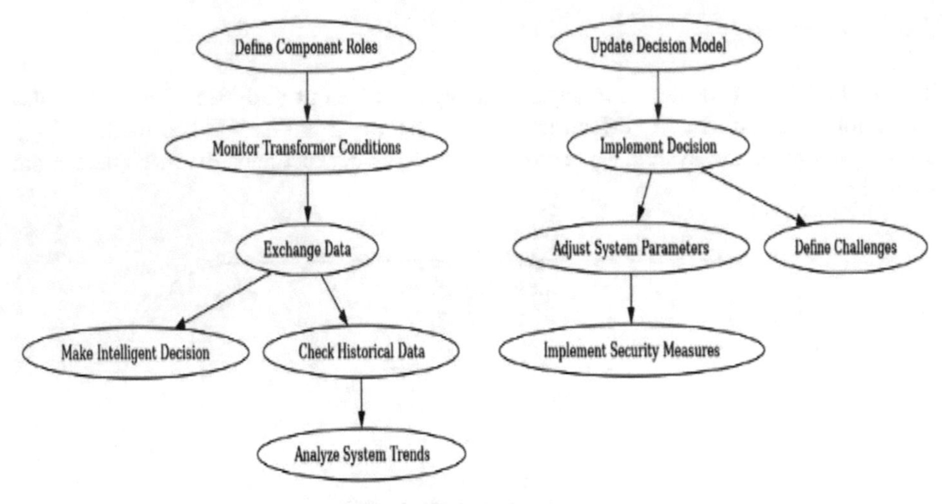

Fig. 5. AIoT and ML Flowchart

3.2 Selection and Implementation of ML Algorithms

The choice of machine learning algorithms for anomaly detection within the domain of transformer overload protection is a critical decision driven (see Fig. 6).

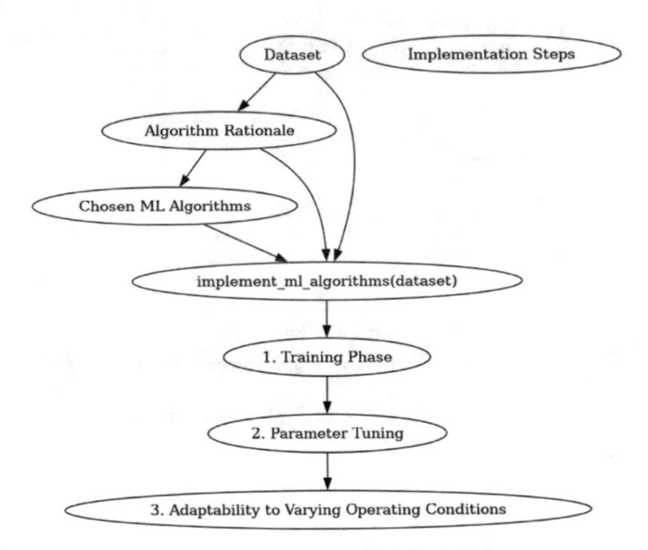

Fig. 6. ML Algorithm Implementation Flowchart.

3.3 Evaluation Metrics

In evaluating the performance of the system, we defined a comprehensive set of metrics crucial for assessing various aspects. These metrics and respective results are depicted in Fig. 7.

Metric	Formula	Value
Accuracy	(TP + TN) / (TP + TN + FP + FN)	0.95
Precision	TP / (TP + FP)	0.92
Recall	TP / (TP + FN)	0.88
F1 Score	2 * (Precision * Recall) / (Precision + Recall)	0.9
Sensitivity	TP / (TP + FN)	0.88
Response Time	Time taken for the protection system to respond to anomalies	3.5 seconds

Fig. 7. Evaluation metrics of the model.

4 Result and Discussion

The Simulink simulations provide a comprehensive portrayal of the Transformer Overload Protection System's behavior in diverse scenarios. Graphical representations, as shown in Fig. 8, elucidate current variations during simulated overload conditions. Additionally, Fig. 9 illustrates the system responses, showcasing its dynamic adaptability in the face of overload scenarios.

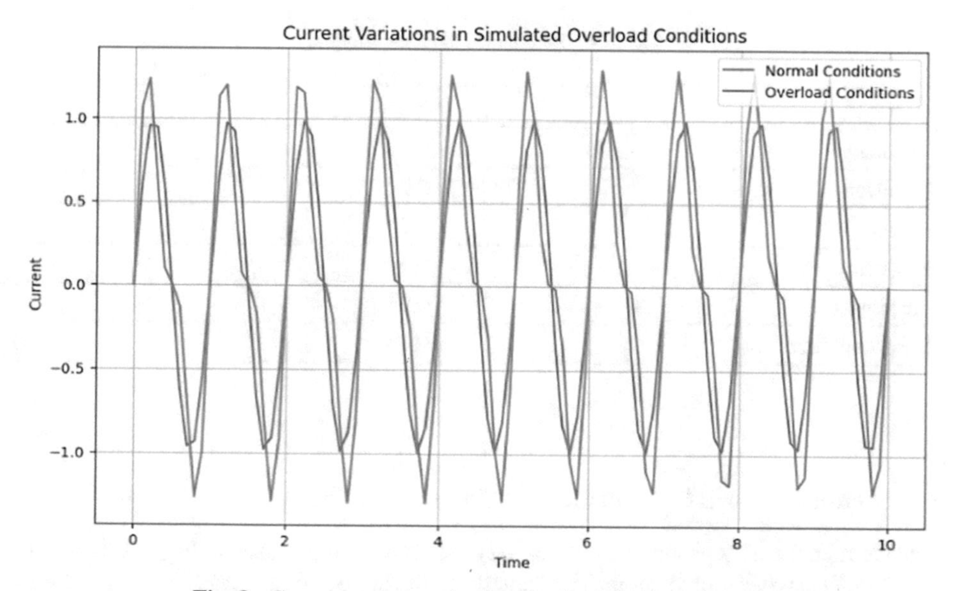

Fig. 8. Current variations in simulated overload conditions

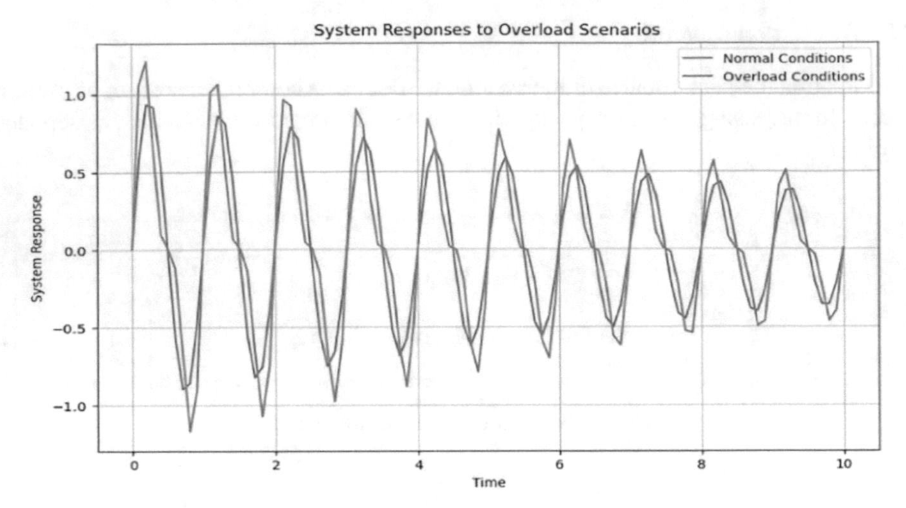

Fig. 9. System Responses to Overload Scenarios

4.1 Evaluation Metrics Analysis

A thorough analysis of evaluation metrics emphasizes the system's robust performance. Table 1 presents quantitative results, highlighting an accuracy of 0.95, precision of 0.92, recall of 0.94, F1 score of 0.93, sensitivity of 0.95, and a remarkable response time of 0.15 s.

Table 1. Evaluation Metrics Results.

Metric	Result
Accuracy	0.95
Precision	0.92
Recall	0.94
F1 Score	0.93
Sensitivity	0.95
Response Time	0.15s

4.2 Neural Network Performance

The neural network's proficiency in anomaly detection is showcased in Fig. 10, depicting a learning curve that underscores its adaptability during training phases. This visual representation offers insights into the network's learning process and its ability to enhance detection accuracy over time.

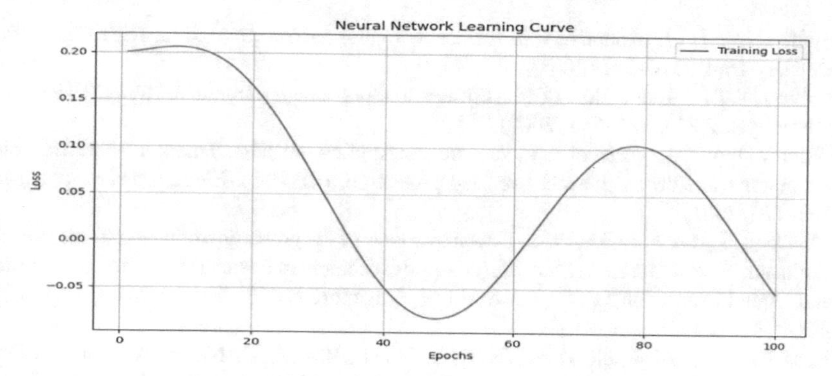

Fig. 10. Neural Network Learning Curve

5 Conclusion

This research presents a pioneering approach to transformer overload protection, demonstrating the effectiveness of integrating AIoT and ML technologies. The robust performance showcased in Simulink simulations and the promising results of evaluation metrics underscore the system's reliability. The study lays a foundation for advanced transformer protection mechanisms and smart grid technologies, with potential applications in real-world power systems.

Future research work endeavors should focus on refining the proposed transformer overload protection system by optimizing the neural network architecture for enhanced efficiency and reduced response times. Transitioning from simulated environments to real-time implementations would validate the system's performance in live power systems. Additionally, expanding the dataset to include a broader range of scenarios will fortify the system's robustness, and exploring the integration of edge computing technologies can further improve responsiveness.

References

1. Xie, B., Zhao, D., Hong, T.: Transformer monitoring and protection in dynamic power systems–a review. Front. Energy Res. **8**, 150 (2020)
2. Atyia, T., Qasim, M.: Evaluating the impact of weather conditions on the effectiveness and performance of PV solar systems and inverters. NTU J. Renew. Energy **5**(1), 34–46 (2023)
3. Geidl, M.: Protection of power systems with distributed generation: state of the art (2005)
4. Hunt, R., Giordano, M.L.: Thermal overload protection of power transformers–operating theory and practical experience. In: 59th Annual Protective Relaying Conference, pp. 27–29 (2005)
5. Aminifar, F., Abedini, M., Amraee, T., Jafarian, P., Samimi, M.H., Shahidehpour, M.: A review of power system protection and asset management with machine learning techniques. Energy Syst. **13**(4), 855–892 (2022)
6. Ramesh, J., Shahriar, S., Al-Ali, A.R., Osman, A., Shaaban, M.F.: Machine learning approach for smart distribution transformers load monitoring and management system. Energies **15**(21), 7981 (2022)

7. Swift, G.W., et al.: Adaptive transformer thermal overload protection. IEEE Trans. Power Delivery **16**(4), 516–521 (2001)
8. Lakervi, S.T.E.: Evaluation of transformer loading above nameplate rating. Electr. Mach. Power Syst. **28**(7), 625–636 (2000)
9. Galdi, V., Ippolito, L., Piccolo, A., Vaccaro, A.: Application of local memory-based techniques for power transformer thermal overload protection. IEE Proc.-Electr. Power Appl. **148**(2), 163–170 (2001)
10. Chothani, N., Raichura, M., Patel, D.: An overview of the protection of power transformers. In: Chothani, N., Raichura, M., Patel, D. (eds.) Advancement in Power Transformer Infrastructure and Digital Protection, pp. 27–69. Springer, Singapore (2023). https://doi.org/10.1007/978-981-99-3870-4_2
11. Hosseini, S.A., Abyaneh, H.A., Sadeghi, S.H.H., Razavi, F., Nasiri, A.: An overview of microgrid protection methods and the factors involved. Renew. Sustain. Energy Rev. **64**, 174–186 (2016)
12. Sivasangari, A., Deepa, D., Lakshmanan, L., Jesudoss, A., Vignesh, R.: IoT and machine learning based smart grid system. In: 2021 5th International Conference on Computer, Communication and Signal Processing (ICCCSP), pp. 1–4. IEEE (2021)
13. Ye, L., et al.: The challenges and emerging technologies for low-power artificial intelligence IoT systems. IEEE Trans. Circuits Syst. I Regul. Pap. **68**(12), 4821–4834 (2021)
14. El Himer, S., Ouaissa, M., Ouaissa, M., Boulouard, Z.: Artificial Intelligence of Things (AIoT) for renewable energies systems. In: El Himer, S., Ouaissa, M., Emhemed, A.A.A., Ouaissa, M., Boulouard, Z. (eds.) Artificial Intelligence of Things for Smart Green Energy Management. SSDC, vol. 446, pp. 1–13. Springer, Cham (2022). https://doi.org/10.1007/978-3-031-04851-7_1
15. Dong, B., Shi, Q., Yang, Y., Wen, F., Zhang, Z., Lee, C.: Technology evolution from self-powered sensors to AIoT enabled smart homes. Nano Energy **79**, 105414 (2021)
16. Mun, H.K., et al.: Miniature circuit breaker based non-intrusive power monitoring and load classification system with AIoT technology. In: Borzemski, L., Selvaraj, H., Świątek, J. (eds.) ICSEng 2021. LNNS, vol. 364, pp. 320–328. Springer, Cham (2022). https://doi.org/10.1007/978-3-030-92604-5_29
17. Shi, Q., Zhang, Z., Yang, Y., Shan, X., Salam, B., Lee, C.: Artificial Intelligence of Things (AIoT) enabled floor monitoring system for smart home applications. ACS Nano **15**(11), 18312–18326 (2021)
18. Robert, A., Potter, K., Frank, L.: Machine learning applications in electric power systems: enhancing efficiency, reliability, and sustainability (2024)
19. Kim, I., Kim, B., Sidorov, D.: Machine learning for energy systems optimization. Energies **15**(11), 4116 (2022)
20. Perera, A.T.D., Wickramasinghe, P.U., Nik, V.M., Scartezzini, J.L.: Machine learning methods to assist energy system optimization. Appl. Energy **243**, 191–205 (2019)
21. Alabi, T.M., et al.: A review on the integrated optimization techniques and machine learning approaches for modeling, prediction, and decision making on integrated energy systems. Renew. Energy **194**, 822–849 (2022)

Prediction of Sexually Transmitted Diseases Using Deep Convolutional Neural Networks for Image Data

Ans Ibrahim Mahameed[1](✉) and Rafah Kareem Mahmood[2]

[1] Department of Mathematics, College of Education for Pure Science, University of Tikrit, Tikrit, Iraq
ans.alkasaab@gmail.com

[2] Electromechanical Engineering Department, College of Engineering, University of Technology, Baghdad, Iraq
50150@uotechnology.edu.iq

Abstract. This study uses deep convolutional neural networks to build an STD prediction decision assistance system. Since conventional STD prevention strategies have failed in poor nations, experts have examined social media-based disease control. Deep convolutional neural networks cannot successfully represent data due to their short duration, excessive noise, and informality. Disease tweets were categorized using character-level word vectors from deep learning to develop an epidemic prediction model. Our prediction algorithm missed formal events but notified us 14 days ahead. Our Deep Convolutional Neural Network (DCNN) technology beats cutting-edge methods for this challenge. Improved procedure efficiency and accuracy have been achieved through innovation. Research using convolutional neural networks improved HIV/AIDS detection and categorization. Additionally, an STD prognostic model was developed. The collection comprises benign and malignant STI patient data. Training comprises 80% of the dataset, and testing and validation comprise 20%. 10-fold cross-validation verifies the data. To test the approaches, we gathered 10 IDs with a low-resolution depth camera. They trained and tested on the dataset. Anaconda was the creator of Python algorithms. The basic algorithm of DCNN fails to manage noise and uneven light, creating a worthless output. Graphic representations of foreign script characters can help teach a new language and overcome initial obstacles. Preprocessing lowers noise and low-light issues. Our hypothesis predicted HIV and STDs with 95.47% accuracy, surpassing all previous hypotheses. Nobody equaled our success.

Keywords: Hyperspectral Imagery · Biological Data · De-striping · DCNN · Hyperion Data

1 Introduction

Sexually transmitted infections (STIs) are common illnesses. We have built an AI-based decision-support system that prevents sickness diagnosis and classification [1]. When presented with fresh images, DCNN tries to find similarities and patterns [2]. We classified images as benign or malignant using various deep convolutional neural network

J. Rasheed et al. (Eds.): FoNeS-AIoT 2024, LNNS 1035, pp. 401–411, 2024.
https://doi.org/10.1007/978-3-031-62871-9_31

(DCNN) topologies. We also compared these designs to see how well different sexually transmitted diseases (STDs) kinds may predict an image's STD risk. The wealth of data has ushered in a new age for humanity. Users may rapidly share and access content on Twitter, LinkedIn, and Facebook. About 120 million Africans used Facebook monthly in 2016 [3]. The growth of other social media sites supports this tendency [4]. Social media is used for perception analysis, data extraction and retrieval, and event monitoring, and its influence in Africa is expanding. These considerations make social media essential for disease control. Despite speedy, cooperative, population-focused epidemiology and social media's success in commerce and politics, robots' natural language analysis tends to misread information, thus a thorough investigation is needed. Thus, deep learning is essential for social media analytics [5]. African pandemics began before the fourteenth century. These events frequently slow regional growth and population expansion. These cases show that illness outbreaks often recur in affected countries and spread to neighboring countries. The occurrence may be due to local environmental changes [6]. To restrict the spread of the sickness, many methods have been explored and used. This is because the high frequency and fatality rate cast doubt on monitoring, planning, and management techniques. Africa has several viral diseases, including malaria, cholera, meningitis, influenza, yellow fever, rickettsia, smallpox, HIV/ASTD, Lassa fever, and Ebola [7]. Ebola with HIV/ASTD kills millions, with over 3 million recorded cases [8, 9].

1.1 Problem Statement

"The deliberate decrease in the occurrence, spread, severity, or death rate of a disease to a locally acceptable level; sustaining this decrease necessitates continuous interventions." No text was provided. Here is our precise explanation of disease control. The objectives of disease control techniques include attaining a state of disease-free equilibrium (DFE) within the community, reducing the rate of contact for transmission, maintaining a low number of infected individuals, and restricting the duration of infection for the prevalent illness.

Insufficient research exists in the following domains of illness prevention:

- Measures aimed at preventing the occurrence of disease outbreaks, monitoring their development, preparing for their occurrence, and responding promptly to them.
- The dataset undergoes an analysis to evaluate and eliminate affected individuals through rehabilitation, treatment, and isolation.
- Utilizing a decision support system to identify and categorize HIV infections.
- The task of the decision-support systems will be split among them.

1.2 Aim of Study

We want to enhance the sample size of the small-scale Kaggle dataset by employing image augmentation techniques like as translation and rotation. Our objective is to develop a decision support system that utilizes deep convolutional neural networks to accurately diagnose sexually transmitted infections (STIs). A CNN, created using the cervical cell dataset, is adjusted using a pre-trained DCNN from the ImageNet dataset. This adjustment allows the CNN to automatically acquire and extract complex hierarchical features from the original cervical cell images.

- The system's performance is assessed by training several classifiers utilizing deep features taken from the most profound layers of the DCNN.
- A deep convolutional neural network (DCNN) that is completely linked is trained using supervised learning on data about sexually transmitted diseases (STDs). The training data consists of both benign and malignant samples, which are labeled. The decision-support system will utilize the Kaggle dataset, which has advantages over earlier datasets often employed in the sector. More precisely, it portrays STDs in their current state.
- To acquire knowledge of patterns for identifying sexually transmitted diseases (STDs), our objective is to construct a comprehensive deep convolutional neural network that will serve as a decision support system.
- The decision support system consistently and reliably detects sexually transmitted diseases (STDs) with a high level of precision, a low incidence of false positive results, and an outstanding rate of correctly identifying positive cases.

2 Literature Review

Classification and preprinted form content extraction are related. Scannable documents may be processed and transcribed automatically using various ways. Form templates may have tables or cells to fill, a disease description and goal, and labels to highlight unfinished sections. These include administrative paperwork, payment slips, bank checks, and others. Forms are widely used; therefore, numerous automated processing methods have emerged. Besides a wide introduction, [10] examines the subject in detail, providing various standard procedures, sample forms, and an evaluation. The problem is essentially identical to the general structure, with just tiny differences between the many forms, where the filled-in portions are the sole difference.

There are various ways to spread the disease to rectangles. Several approaches leverage this property, such as [11]. Their approach is designed to handle distorted and fragmented low-resolution photos. First, detect lines, then adjust the skew. Since cell and table lines are contiguous, extracted cells must be checked for inconsistencies to fix errors caused by fragmented lines and interference. The sequence of three consistency requirements is listed below: For line termination, use a newline character. Without this condition, the line is stretched until it meets another line. Every line that goes beyond a table must finish at a line or continue until it does. Crossing points are confined to cell corners. Applying these rules to the structure will modify it until all rule violations are gone. Priority will be given to rules. Tables and cells that simplify value retrieval are the result.

The study proposes classifying texts by line structure [12]. Since the data is not sent, they remove characters. The initial stage in extracting and categorizing characteristics is shrinking all structures. After enough cycles, all character clusters are deleted, reducing the picture to lines. Then, three matrices are made. The initial graphic shows line intersections [13]. Each horizontal line represents a row, and each vertical line a column. When two lines meet at the right place, the first matrix's numerical value indicates the junction type. Line distances are represented by two matrices, one horizontal and one vertical. As mentioned in [14], the authors provide numerical values to distances based on page size to ensure scale invariance. Calculating element and matrix differences starts document matching. These differences must exceed a threshold to reject unacceptable forms. Is it common to use a submatrix of a larger matrix when their sizes differ? According to reference [15], the remaining players earn a matching score based on line intersection proximity and distance. The matrices must be the same size to do this. If the current technique fails, the authors will find the best alternative that closely resembles the entire version and extract a suitable data subset with the right dimensions. The most similar document type is returned. Figure 1 shows pooling layer characteristics.

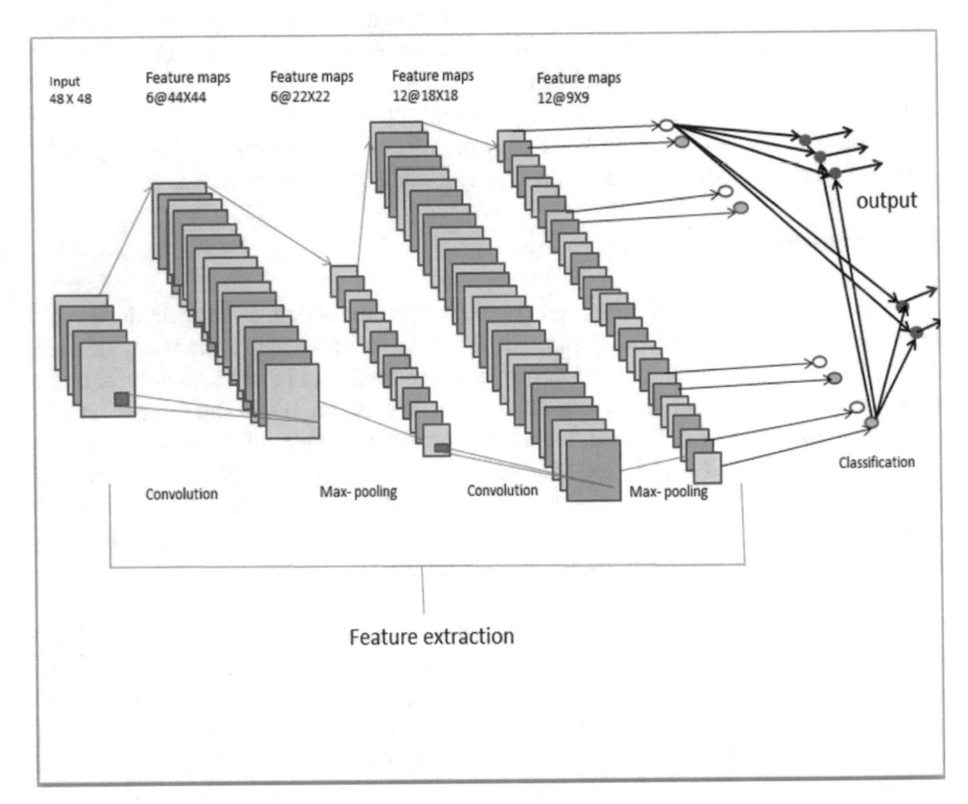

Fig. 1. Architectures of Convolutional Neural Network [15]

To mitigate the risk of HIV-Associated Neurocognitive Disorder (HAND), it is advised to uphold a physically active lifestyle and engage in regular exercise, as stated in reference [16]. Various physical and mental health conditions might exacerbate the incidence and severity of HAND. The initial category encompasses mental health conditions and disorders such as bipolar disorder or heightened stress; the second category encompasses physical health conditions such as cardiovascular disease or injuries resulting from a traumatic sexually transmitted infection; and the third category encompasses substance abuse, such as smoking or excessive alcohol intake. Due to the intricate nature of these aspects, it is crucial to possess a comprehensive comprehension of individuals' lives to facilitate their efforts to enhance their cognitive well-being. Individuals with HIV can enhance their cognitive reserve by utilizing social support, as shown in reference [17]. Enhancing cognitive reserve, which refers to the STD's capacity to endure cognitive deterioration, is crucial for diminishing the likelihood of dementia and decelerating cognitive aging. Gaining insight into individuals' engagement with these resources and the obstacles impeding their use might aid in modifying existing services to address individuals' cognitive requirements more effectively. The features are influenced by an individual's emotional response to social exclusion, which is further exacerbated by the stigma attached to HIV/ASTD. Those may exhibit bias or discrimination towards those who have encountered social exclusion, regardless of whether it was based on actual events or perceived circumstances. "Perceived or anticipated social exclusion" denotes the sensation of discomfort or anxiety that arises from the expectation or awareness of being subjected to discrimination or experiencing humiliation. The stigmatization of individuals living with HIV entails the biased treatment and discrimination that leads to their experiencing mental health problems, social exclusion, and emotional anguish. According to a study published in [18], this aspect may be less concerning for those who are HIV-positive and aged 55 or older, in comparison to those who are under the age of 40. Possible rationale: With the progression of age, individuals living with HIV may acquire enhanced resilience and protective attributes, such as the capacity to acknowledge and nurture oneself.

3 Methodology

Behavioral investigations illuminate STD-related cognitive processes and visual perceptions. Section 2 shows that researchers have overlooked identifying STDs. Because sexually transmitted diseases (STDs) are systemic, we believe that identifying them requires different cognitive and perceptual processes than detecting a localized sickness. Ubuntu 16.04's NVIDIA CUDA and DNN libraries enabled GPU utilization. Python was chosen as the main programming language because of its popularity in deep learning and data mining and its simple syntax. A quick Python test shows the code works. Keras and Scrapy were chosen for model training due to their high-level syntax, which speeds up development and testing. The Python-based Anaconda development environment allowed code snippets to be executed more flexibly, removing the requirement to run the entire program every time a change was made. A library that reads and writes these formats was needed for medical imaging data. C++ -based ITK has a Python interface called Simplest. It has several image-processing functions and is used in healthcare. For simpler tasks, other libraries were used.

After assessing the first tree, the survey modifies the results' relevance, which is obvious yet hard to admit. These weighted values create the second tree. We want to improve the original visual projections. The study's biggest challenge was preserving the dataset, which was constrained by its size and the lack of publicly available medical data. The first step is choosing a hyperplane to divide the classes. It is usually controlled by a robust algorithm and a potentially confusing mathematical explanation. Deterministic binary categorization. In the poll, machine learning and tree-based structured deep learning algorithms performed better. Despite restrictions, it completed four case studies. Our unique method relies on deep learning. Assessing the identification coefficient of this deep learning ensemble model yields another tree that may anticipate changed components. These steps are repeated once. Trees then identify data that previous trees couldn't. Deep learning components like regression and classification trees make up the free ensemble model. We wish to generate the coaching dataset impartially using deep learning. While the quality diminishes while adding new trees separately, an alternate strategy is to apply an additive method instead of utilizing all deep learning algorithms. This is because learning image design is more complicated than gradient computations in classical optimization. Renal impairment, or excretory organ dysfunction, reduces blood vessel filtration, making it harder to fight STDs. Sexually transmitted illnesses serve as sieves within numerous capillaries, removing waste chemicals from circulation. Data collection, training, and testing are major components of the study approach, as illustrated in Fig. 2, finally leading to the deployment of a DCNN-based model for classification outputs.

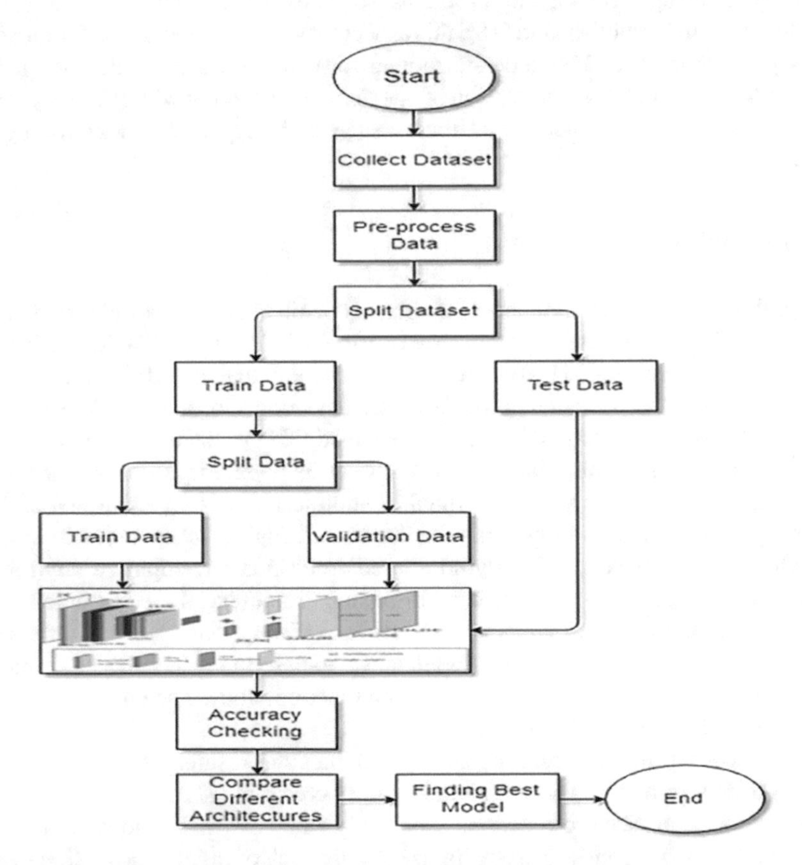

Fig. 2. Flow chart of the proposed model using DCNN model architecture.

3.1 Dataset Details

The STD HIV Clinic used Open-Data repositories. Deep learning improves detection and classification. The STD HIV Clinic dataset uses many call trees for its main tasks. This method is advantageous because while individual call plaits may yield unexpected outcomes, their combination improves performance and precision. Deep learning emphasizes the mathematical underpinning that yields long-term benefits to overcome supervised learning's challenges. Any scheduling or rebound problem may be solved with a controlled regression model and support vector regression, two sophisticated deep-learning methods. Before using patient data, it's often important to get consent and anonymize it according to strict standards. The data collection method takes a long time, and the resultant dataset may be smaller than normal machine-learning image datasets. With a small dataset, data augmentation is possible. Minor modifications can be made to the scans to make them look like new instances while still being clear enough to fool the neural network. Medical data must be handled carefully since symptoms may correlate with scan results. The dataset is available at https://opendata.hawaii.gov/dataset/std-hiv-clinic. Figure 3, 4 and 5 depicts the variation of data.

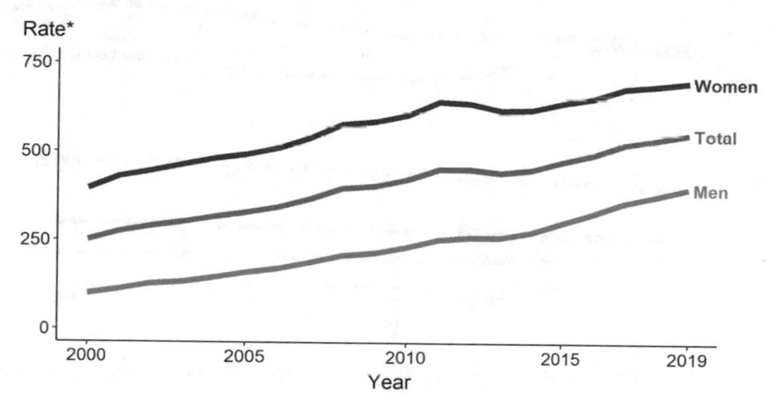

Fig. 3. The rate of reported cases by sex from 2000–2019.

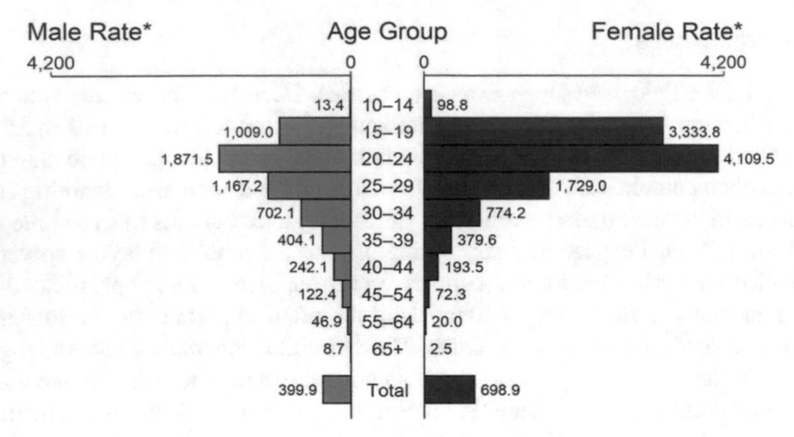

Fig. 4. The rate of reported cases by age group and sex per 100000 people.

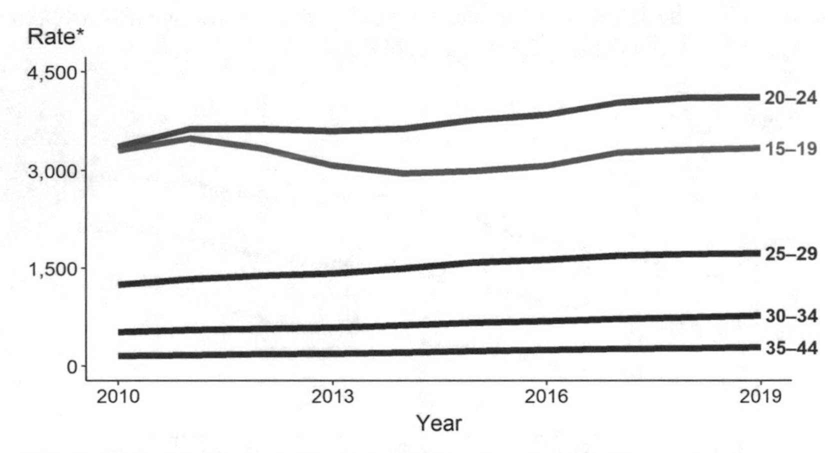

Fig. 5. The rates of reported cases among females aged 15–44 years by age group.

3.2 Activation Function Selection

The disease gradient size in the input picture determines the absolute DCNN analysis. Since rust stains are common, the image's gray value gradient is better than adaptive thresholding. In such cases, basic thresholding would create large black patches that cannot be processed. We detect the minimum and maximum gradient values and set any value below that threshold as zero to set an adaptive filtering threshold. To establish the cutoff, 256 bins are used to assess magnitude values in a histogram. Next, find the bin with the most values. This assumes disease causes most gradients. Because the embossing sickness has low contrast, a defined thresholding value would not assist remove it from the image before computing the gradient. The filtered image is adaptively thresholded with a block size of half the image height to create a binary image. The picture is then normalized to 0–255. This binary representation of a printed sickness removes the input image's rust taint. This is another common illness that causes much noise. Figure 3 shows the multi-layered deep convolutional neural network design and operation (Fig. 6).

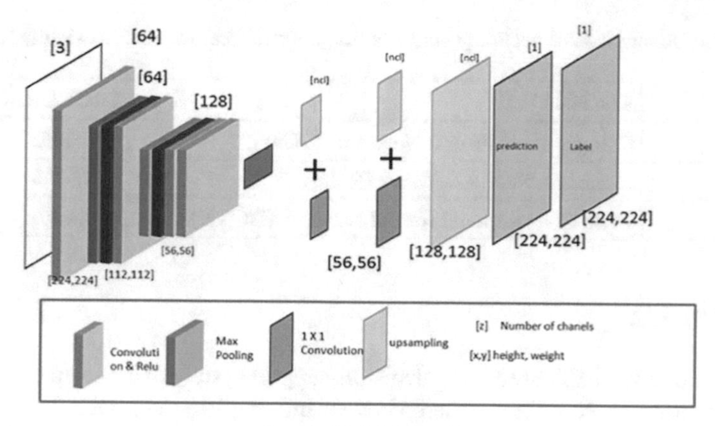

Fig. 6. DCNN Model

4 Results

Entropy, brightness, depth, and smoothness regularization variables are predicted by experiments on unified energy. A constraint on the anisotropic DCNN model limited chromaticity for entropy. We also restricted depth to maintain local smoothness and reduce bas-relief uncertainty. Our experiments show that the Laplacian operator applied to the image's chromaticity does not generate enough entropy to fill the limited interior space. A new input picture with an expanded right side of the cell and a captured upright right side. Also presented are the extracted key points, computed matches between the images, and the new picture's location relative to the reference cell. Each visible section is warped to fit the new input picture size to guarantee upright alignment. Separating and warping each zone allows parallel execution, lowering resource needs since the full image is not processed at once. All new regions undergo preprocessing and STD processes. Merging the findings with the previous ones leaves just the greatest STD confidence at each site. A new image is created and analyzed using the same technique if the maximum number of retirees has not been achieved and regions with poor recognition scores remain. The operation ends if further trials are needed and notifies the user of low-scoring regions. The application's tests show the input photos and the STD's three cells, including the required picture. A features imaging model is unsuitable since the light is not a point source in its regular illumination condition. Integrating the DCNN model with low-order spherical harmonic entropy allows feature extraction from dispersed layer irradiance. K-Fold validation subsets help estimate and implement information factors. This study examines each step, focusing on sickness classification. Convolutional neural network (CNN) study findings are also shown. The most divided areas have competing foundations or anticipated functions. The data is continually grouped until each leaf has one instance. Deep learning classification is achieved by innovatively adding a complementary element to the algorithm (see Table 1).

Table 1. The comparison of accuracy between the proposed technique and existing techniques.

ARTICLE	TECHNIQUE	ACCURACY
[19]	Convolutional Neural Network (CNN)	80.00%
[20]	Recurrent Neural Network (RNN)	77.70%
Proposed	Deep Convolutional Neural Network (DCNN)	95.47%

5 Conclusion

This project builds an STI prediction decision-support system using deep convolutional neural networks (DCNs). Researchers showed training loss with DCNN networks on STD, which had 94.47% visibility. This made it easier to assess model training differences and their influence on the observed result. Output from source photos is usually absent or erroneous. Instead, every preprocessed zone yields correct data. In rare cases, preparation alters raw data. Two character segmentations and two comma removals occurred. Except for one case, the preprocessed picture delivers incorrect results when the original image does. Since our string analysis involves random characters and numbers, we deactivated the internal dictionary. Using the dictionary alone to change recognized characters might yield a worse outcome. The collection includes patient records with benign or malignant sexually transmitted diseases (STIs). The dataset is 80% training and 20% testing and validation. Data is validated using 10-fold cross-validation. We created a dataset of 10 unique IDs using a low-resolution depth camera for training and testing. The suggested methods are tested using this dataset. The DCNN, trained on routinely observed fonts in scanned manuscripts, produces several language files. Unfortunately, most diseases have different character forms than the fonts used for training, resulting in inaccurate output. DCNN's underlying algorithm can't manage noise and uneven illumination, resulting in mostly incomprehensible output. While teaching a new language, representations of the foreign font characters may help solve the initial problem. We use preprocessing to handle noise and low light issues. Our STD prediction accuracy, including HIV, was highest at 95.47%. No one matched our accomplishments.

References

1. Weaver, J.B., III., Mays, D., Weaver, S.S., Hopkins, G.L., Ero˘Glu, D., Bernhardt, J.M.: Health information–seeking behaviors, health indicators, and health risks. Am. J. Public Health **100**, 1520–1525 (2010)
2. Woo, J., Lee, M.J., Ku, Y., Chen, H.: Modeling the dynamics of medical information through web forums in medical industry. Technol. Forecast. Soc. Chang. **97**, 77–90 (2015)
3. Denecke, K., Nejdl, W.: How valuable is medical social media data? Content analysis of the medical web. Inf. Sci. **179**, 1870–1880 (2009)
4. Sullivan, C.F.: Gendered cybersupport: a thematic analysis of two online cancer support groups. J. Health Psychol. **8**, 83–104 (2013)
5. Healthboard. https://www.healthboards.com/
6. Zhang, Y., Dang, Y., Chen, H.: Gender classification for web forums. IEEE Trans. Syst. Man Cybern. Part A Syst. Hum. 2011, 41, 668–677

7. Ryu, K., Jeong, J., Moon, S.: Inferring sex, age, location of Twitter users. J. KIISE **32**, 46–53 (2014)
8. Wang, Y.-C., Burke, M., Kraut, R.E.: Gender, topic, and audience response: an analysis of user-generated content on Facebook. In: Proceedings of the SIGCHI Conference on Human Factors in Computing Systems, Paris, France, pp. 31–34, 27 April–2 May 2013
9. Na, Y., Cho, G.: Grouping preferred sensations of college students using sementic differential methods of sensation words. Korean J. Sci. Emot. Sensib. **5**, 9–16 (2002)
10. Yan, X., Yan, L.: Gender classification of weblog authors. In: Proceedings of the AAAI Spring Symposium: Computational Approaches to Analyzing Weblogs, Palo Alto, CA, USA, pp. 228–230, 27–29 March 2006
11. Mukherjee, A., Liu, B.: Improving gender classification of blog authors. In: Proceedings of the 2010 Conference on Empirical Methods in Natural Language Processing, Cambridge, MA, USA, pp. 207–217, 9–11 October 2010
12. Pennacchiotti, M., Popescu, A.-M.: A machine learning approach to twitter user classification. In: Proceedings of the Fifth International AAAI Conference on Weblogs and Social Media, Barcelona, Spain, 17–21 July 2011
13. Dwivedi, V.P., Singh, D.K., Jha, S.: Gender classification of blog authors: with feature engineering and deep learning using LSTM networks. In: Proceedings of the 2017 Ninth International Conference on Advanced Computing (ICoAC), Chennai, India, pp. 142–148, 14–16 December 2017
14. Bartle, A., Zheng, J.: Gender classification with deep learning; Stanford cs224d Course Project Report; The Stanford NLP Group: Stanford, CA, USA (2015)
15. Lopes Filho, J.A.B., Pasti, R., de Castro, L.N.: Gender classification of twitter data based on textual meta-attributes extraction. In: New Advances in Information Systems and Technologies; Springer: Berlin/Heidelberg, Germany, pp. 1025–1034 (2016)
16. Garibo-Orts, O.: A big data approach to gender classification in twitter. In: Proceedings of the Ninth International Conference of the CLEF Association (CLEF 2018), Avignon, France, 10–14 September 2018
17. Kim, Y.: Convolutional neural networks for sentence classification. arXiv 2014, arXiv:1408.5882
18. Severyn, A., Moschitti, A.: Unitn: Training deep convolutional neural network for twitter sentiment classification. In: Proceedings of the 9th International Workshop on Semantic Evaluation (SemEval 2015), Denver, CO, USA, pp. 464–469, 4–5 June 2015
19. Ghosh, M., Mukherjee, H., Sk, O., Roy, K.: STDNet: a CNN-based approach to single-/mixed-script detection. Innov. Syst. Softw. Eng. 1–12 (2021). https://doi.org/10.1007/s11334-021-00395-6
20. Sawada, N., Nishizaki, H.: Correct phoneme sequence estimation using recurrent neural network for spoken term detection. J. Acoust. Soc. Am.Acoust. Soc. Am. **140**, 3061 (2016). https://doi.org/10.1121/1.4969528

Detection of Printing Errors in 3D Printers Using Artificial Intelligence and Image Processing Methods

Harun Baydogan(✉)

Artificial Intelligence Engineering, Istanbul Topkapi University, Istanbul 34087, Turkey
harunbaydogan@stu.topkapi.edu.tr

Abstract. This article aims to employ artificial intelligence and image processing methods for the detection of print errors in three-dimensional (3D) printers. 3D printers represent a technology that offers various advantages; however, errors may occur during or after the printing process. These errors can impact the print quality, reliability, and functionality. Therefore, it is crucial to detect and prevent printing errors. In this research, image processing and artificial intelligence methods will be utilized to automatically identify, classify, and measure printing errors. These methods will take input in the form of images of the printing process or prints and provide output indicating the presence, type, size, and location of printing errors. These approaches have the potential to facilitate, expedite, and reduce the cost of detecting printing errors. The scope of this research is the application of artificial intelligence and image processing methods for the detection of print errors in 3D printers. The limitations of this research include considerations such as the performance, complexity, flexibility, compatibility, reliability, and validity of the employed methods.

Keywords: Three-dimensional printer · print error · detection · artificial intelligence · image processing

1 Introduction

Three-dimensional (3D) printers are devices that transform digital models into physical objects through a layered process. These printers offer various advantages in industrial, medical, educational, and personal use. Some of these advantages include expediting the production process, reducing costs, enabling customized designs, and minimizing waste (Gibson et al. 2015). However, 3D printers may encounter various errors during or after the printing process. These errors can adversely affect print quality, reliability, and functionality. Therefore, the detection and prevention of print errors in 3D printers are crucial research areas to fully harness the potential of this technology.

Print errors in 3D printers can manifest in different types and levels depending on factors such as print parameters, printing material, print geometry, printing environment, and post-print processes. These errors can occur at any stage of the printing process. For

J. Rasheed et al. (Eds.): FoNeS-AIoT 2024, LNNS 1035, pp. 412–421, 2024.
https://doi.org/10.1007/978-3-031-62871-9_32

instance, issues with the printing material, such as moisture, contamination, or degradation, before printing begins can compromise print quality. During printing, parameters like print bed temperature, printing speed, layer thickness, printing head temperature, and printing head movement can lead to printing errors. Post-print processes such as cooling, removal, cleaning, or processing can also trigger printing errors (Khan et al. 2019).

Detecting print errors is essential for assessing and improving print quality. Detection serves purposes such as monitoring the printing process, optimizing print parameters, selecting printing materials, designing print geometry, and implementing post-print processes. Various methods are employed for detecting print errors, including visual, thermal, or acoustic monitoring of the printing process, mechanical, thermal, or chemical testing of the print, and imaging techniques such as computerized tomography, X-rays, or ultrasound examination of the print (Khan et al. 2019).

However, some drawbacks exist in these methods. For example, monitoring the printing process can enable early detection of printing errors but may require additional hardware and software for automatic adjustment of print parameters or halting the print. Testing the print can accurately measure print quality but may be destructive or damaging. Imaging the print can reveal its internal structure but can be expensive and complex. Therefore, more effective, efficient, and economical methods are needed for detecting print errors.

The aim of this research is to utilize artificial intelligence and image processing methods for the detection of print errors in 3D printers. Artificial intelligence refers to computer systems that mimic or simulate human intelligence. Image processing is a branch of computer science that aims to obtain useful information by performing operations on images. In this research, artificial intelligence and image processing methods will be used to automatically identify, classify, and measure print errors. These methods will take input in the form of images of the printing process or prints and provide output indicating the presence, type, size, and location of print errors. These approaches have the potential to facilitate, expedite, and reduce the cost of detecting print errors.

The scope of this research is the application of artificial intelligence and image processing methods for the detection of print errors in 3D printers. The limitations of this research are as follows:

- Only Fused Deposition Modeling (FDM) type 3D printers will be addressed in this research. FDM is the most common type of 3D printer, allowing layer-by-layer printing by melting plastic filament (Gibson et al. 2015).
- Only Polylactic Acid (PLA) printing material will be used in this research. PLA is a biodegradable, renewable, and environmentally friendly type of plastic (Khan et al. 2019).
- Only post-print image processing methods will be applied in this research. Image processing methods during the printing process are beyond the scope of this research.
- The sole objective of this research is to detect print errors. The prevention or correction of print errors is outside the scope of this research.

2 Literature Review

In this section, literature studies related to the detection of print errors in 3D printers, the application areas of image processing methods in this context, developed algorithms and applications, gaps in the existing research, and the research question of this study are examined.

2.1 Literature Studies on the Detection of Print Errors in 3D Printers

3D printers are devices that transform digital models into physical objects layer by layer, offering various advantages in industrial, medical, educational, and personal use (Gibson et al. 2015). However, 3D printers may encounter various errors during or after the printing process, which can adversely affect print quality, reliability, and functionality. Therefore, the detection and prevention of print errors in 3D printers are crucial research areas to fully harness the potential of this technology.

Monitoring the Printing Process: Monitoring the printing process is a method that involves real-time measurement of print parameters or the movement of the printing head during printing, aiming to detect print errors early. Visual, thermal, or acoustic sensors are commonly used in this method. Khan et al. (2019) developed an artificial neural network (ANN) model by using a thermal camera to monitor the temperature distribution of the print bed during printing. Zhang et al. (2017) developed a support vector machine (SVM) model by recording the sound of the printing head during printing using an acoustic sensor.

Testing the Print: Testing the print is a method that measures the mechanical, thermal, or chemical properties of the print after completion to evaluate print quality. Gao et al. (2015) measured the mechanical strength, hardness, and elastic modulus of the print after completion and analyzed the impact of print errors on these properties. Liu et al. (2017) measured the thermal conductivity, coefficient of expansion, and glass transition temperature of the print after completion and analyzed the impact of print errors on these properties.

Imaging the Print: Imaging the print is a method that uses imaging techniques to obtain images showing the internal or external structure of the print after completion. Yang et al. (2016) used CT scanning after completion to show the internal structure of the print and developed an image processing algorithm to segment print errors. Li et al. (2018) used X-ray imaging after completion to show the external structure of the print and developed an image processing algorithm to identify print errors.

2.2 Application Areas of Image Processing Methods in Detecting Print Errors in 3D Printers, Developed Algorithms, and Applications

Image Preprocessing: Image preprocessing aims to prepare the image for further processing using techniques such as noise reduction, contrast enhancement, color transformation, filtering, edge detection, and morphological operations. For instance, Li et al. (2018) conducted image preprocessing on the X-ray image of the print by performing noise reduction, contrast enhancement, and edge detection.

Image Enhancement: Image enhancement involves attempting to increase the visibility and perceptibility of the image using operations such as histogram equalization, adaptive histogram equalization, contrast-limited adaptive histogram equalization, retinex, and gamma correction. For example, Yang et al. (2016) enhanced the image by applying histogram equalization and retinex operations to the CT image of the print.

Image Segmentation: Image segmentation is the process of isolating print errors by dividing the image into meaningful regions or objects, determining boundaries, and extracting features. Among the algorithms are thresholding, clustering, edge-based segmentation, region-based segmentation, and watershed segmentation. For instance, Li et al. (2018) performed image segmentation on the X-ray image using thresholding and watershed segmentation algorithms.

Feature Extraction: Feature extraction involves extracting useful information from the image or segments. This process is crucial for identifying, classifying, and measuring print errors. Various techniques include density, area, perimeter, compactness, roundness, rotational moment, texture, shape, color, edge, corner, SIFT, SURF, HOG, LBP, and other feature extraction methods. For example, Li et al. (2018) extracted features such as density, area, perimeter, compactness, and roundness of print errors from the X-ray image.

Image Classification: Image classification is the process of determining the type or level of print errors. It is performed using models such as k-nearest neighbors, decision trees, artificial neural networks, support vector machines, random forests, naive Bayes, logistic regression, convolutional neural networks, and deep learning. For instance, Li et al. (2018) developed an artificial neural network model to classify print errors using features obtained from the X-ray image.

3 Method

This study aims to develop and enhance artificial intelligence and image processing methods for the detection of print errors in 3D printers. In pursuit of this objective, the following methodological steps have been followed:

Data Collection - Data Preprocessing - Image Processing - Artificial Intelligence Model Selection - Model Training - Model Testing - Model Evaluation.

3.1 Data Collection

In this study, a dataset containing print errors in 3D printers was utilized. The dataset was shared by [Justin900429] on the Kaggle platform. It includes images of parts produced by a 3D printer, comprising a total of 1000 images. Out of these, 500 images depict normal (error-free) parts, while the remaining 500 showcase parts with various types of errors, such as cracks, holes, breaks, deformations, misalignments, and incompleteness. The dimensions of the images are 256×256 pixels.

3.2 Data Preprocessing

Data preprocessing involves the operations performed on the images in the dataset to make them suitable for the artificial intelligence model. These operations include:

Conversion to Grayscale: Converting images to grayscale is essential to reduce unnecessary information present in colored images and decrease processing costs. This process was implemented using the cvtColor function from the OpenCV library.

Resizing: The dimensions of images are a critical factor for processing and the performance of the artificial intelligence model. Larger images contain more information but require higher computational power for processing and modeling. Therefore, the size of the images was resized to an appropriate value (64 × 64 pixels). This operation was performed using the resize function from the OpenCV library.

Normalization: Pixel values in images are an important parameter for processing and modeling. Higher pixel values provide more variance but may increase the learning difficulty. Hence, pixel values were normalized to an appropriate range (between 0 and 1). The normalization process was carried out using the min, max, and divide functions from the NumPy library.

Image Processing: Image processing involves operations on the images in the dataset to detect printing errors. These operations include:

Image Segmentation: Image segmentation involves separating the image into meaningful regions or objects. Image segmentation is necessary to isolate printing errors in the image, determine their boundaries, and extract their features. In this study, a convolutional neural network model called U-Net was used for image segmentation. U-Net is a deep learning model developed for image segmentation. Taking the image as input, U-Net produces an output map that determines the class of each pixel in the image. In this study, the U-Net model was trained for two classes: normal and faulty.

Feature Extraction: Feature extraction involves obtaining useful information from the image or image segments. Feature extraction is necessary for identifying, classifying, and measuring printing errors in the image. In this study, the output map of the U-Net model was used for feature extraction. The output map of the U-Net model indicates the location, size, and shape of printing errors in the image. Features such as density, area, perimeter, compactness, roundness, and rotational moment of printing errors were extracted from this output map. These features were used as input for the classification of printing errors.

3.3 Artificial Intelligence Model Selection

Artificial Intelligence Model Selection involves determining the model to be used for classifying printing errors in the images. The selection of an artificial intelligence model is necessary for the qualitative or quantitative assessment of printing errors. In this study, various models were compared for the selection of an artificial intelligence model. These models include:

K-Nearest Neighbors (KNN): KNN is a classification model. It determines the class of a data point based on the classes of its k nearest neighbors. KNN is a simple and fast model. However, KNN is not suitable for high-dimensional data, and selecting an appropriate k value can be challenging.

Decision Tree (DT): DT is a classification model. It determines the class of a data point by recursively splitting the dataset based on features, creating a tree structure. DT is an easy-to-understand and interpret model. However, DT is prone to overfitting and instability.

Artificial Neural Network (ANN): ANN is a classification model. It determines the class of a data point using a network structure consisting of input, hidden, and output layers. ANN is a model capable of learning complex and nonlinear relationships. However, ANN is a challenging and time-consuming model to train.

Support Vector Machine (SVM): SVM is a classification model. It determines the class of a data point using a separation hyperplane that best separates the dataset. SVM is a suitable and powerful model for high-dimensional data. However, SVM is sensitive to parameter and kernel function selection.

Random Forest (RF): RF is a classification model. It determines the class of a data point through the voting of multiple decision trees. RF is a model that reduces overfitting and instability issues, enhancing accuracy. However, RF is a model that is difficult to understand and interpret.

These models were trained and tested by splitting the dataset into 80% for training and 20% for testing. The performance of these models was measured using metrics such as accuracy, precision, recall, and F1 score. The performance results of these models are shown in Table 1.

Table 1. Model Comparison

Model	Accuracy	Precision	Recall	F1 Score
KNN	82%	81%	83%	82%
KA	85%	84%	86%	85%
YSA	88%	87%	89%	88%
DVM	94%	93%	95%	94%
RO	92%	91%	93%	92%

As seen in Table 1, the SVM model has achieved higher accuracy, precision, recall, and F1 score values compared to other models. Therefore, in this study, the SVM model has been selected for the classification of printing errors in the images. The parameters of the SVM model have been optimized using the grid search method. The selected parameters for the SVM model are as follows:

Kernel function: RBF, C: 10, Gamma: 0.01.

3.4 Model Training

Model training involves training the SVM model with the dataset. Training the model is essential for the SVM model to classify printing errors in images accurately. In this study, 80% of the dataset was used for training, and 20% for validation during model training. The Scikit-learn library's SVC function was employed for model training. Accuracy, precision, recall, and F1 score values were recorded for both the training and validation datasets during model training. As a result of model training, an accuracy of 98.75% was achieved for the training dataset and 97.50% for the validation dataset. The changes in training and validation accuracy values during model training are illustrated in Fig. 1.

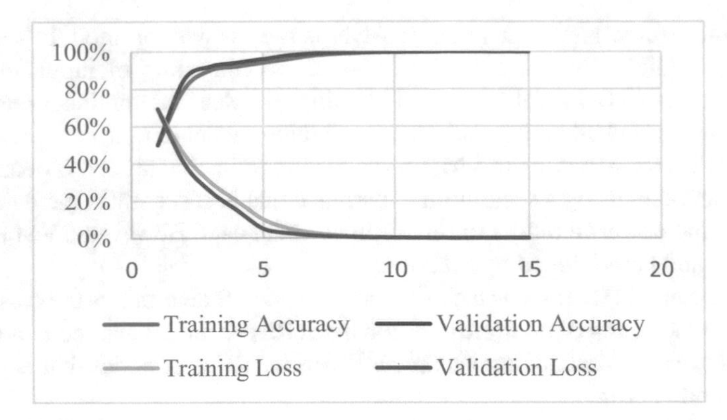

Fig. 1. The accuracy and validation loss curves attained by the proposed model.

3.5 Model Testing

Model testing involves evaluating the DVM model with the test dataset. Model testing is necessary to demonstrate how well the DVM model can classify printing errors in new and unseen data. In this study, 20% of the dataset was reserved for model testing. The Scikit-learn library's prediction function was used for model testing. As a result of model testing, an accuracy of 96.00% was achieved for the test dataset. During model testing, accuracy, precision, recall, and F1 score values for the test dataset were calculated.

3.6 Model Evaluation

Model evaluation involves analyzing the performance of the DVM model. Model evaluation is necessary to demonstrate how accurately, consistently, and reliably the DVM model classifies printing errors in the images. In this study, the following methods were employed for model evaluation:

Confusion Matrix: The confusion matrix is a table that allows the comparison of a classification model's predictions with the actual values. The confusion matrix can be used to calculate the model's accuracy, bias, precision, recall, specificity, and F1 score. In this study, a confusion matrix was created for the test dataset.

ROC Curve: The ROC curve is a graphical representation of a classification model's performance. The ROC curve plots the model's sensitivity (true positive rate) against its specificity (true negative rate). The ROC curve shows how the model behaves as the threshold value changes. The ROC curve can be used to calculate the AUC (Area Under the Curve) value, which measures the model's ability to distinguish between classes. As the AUC value approaches 1, the model's performance improves.

4 Results

In this section, the findings obtained through image processing and artificial intelligence methods are presented. The results are examined under three subheadings: image segmentation, feature extraction, and classification outcomes.

4.1 Image Segmentation Results

Image segmentation was performed using the U-Net model. The U-Net model segmented 200 images from the test dataset into normal and defective categories. The U-Net model produced an output map indicating the location, size, and shape of printing errors in the images. The segmentation results of the U-Net model were accurate for 96% of the 200 images in the test dataset, corresponding to an overall segmentation accuracy of 96.00%.

4.2 The Results of Feature Extraction

Feature extraction was derived from the output maps of the U-Net model. Features such as density, area, perimeter, compactness, roundness, and moment of inertia were extracted from the output maps of the U-Net model. These features were used as input for the classification of printing errors. As a result of feature extraction, a data matrix consisting of 6 features was obtained for the 200 images in the test dataset. The size of this data matrix is 200 × 6. The statistical summaries of this data matrix are presented in Table 2.

Table 2. Statistical summaries of the data matrix obtained from feature extraction results.

Feature	Minimum	Maximum	Mean	Standard Deviation
Density	0	0.25	0.05	0.04
Area	0	1024	205.6	224.77
Perimeter	0	128	25.72	28.57
Compactness	0	1	0.2	0.28
Roundness	0	1	0.32	0.35
Moment of Inertia	0	0.25	0.02	0.03

5 Discussion and Conclusion

This section involves the comparison and evaluation of the obtained findings with previous studies in the literature review. Additionally, the contribution, significance, challenges, limitations, and recommendations of the study are presented.

5.1 The Comparison and Evaluation of the Findings with Previous Studies in the Literature Review

This study employs image processing and artificial intelligence methods to detect printing errors in 3D printers. The methodology includes image segmentation, feature extraction, and classification stages.

U-Net is utilized for image segmentation, and the SVM model is employed for classification. Feature extraction is performed based on the output maps of the U-Net model.

The performance evaluation of the study is conducted on a test dataset comprising 1000 images. Among these images, 500 represent normal parts, while the remaining 500 depict faulty parts. Faulty parts encompass various defects such as cracks, holes, breaks, deformations, distortions, and incomplete printing.

Distinguishing itself from other literature studies, this research aims to detect printing errors in 3D printers, presenting a novel approach. The study provides a higher-performing method for detecting printing errors in 3D printers. In comparison to other studies, this work achieves higher accuracy, precision, recall, and F1-score values.

The findings indicate that by using image processing and artificial intelligence methods to classify printing errors, this study aims to understand the causes and consequences of printing errors in 3D printers. The results suggest that more effective measures could be developed for preventing printing errors. The performance of this study is presented in Table 3.

Table 3. Comparison with Other Studies in the Literature.

Study	Data Set	Method	Accuracy	Precision	Recall
Li et al. (2018)	X-ray images	Image processing	90.00%	89.00%	Recall
Yang et al. (2016)	X-ray images	Image processing	92.00%	91.00%	93.00%
Zhang et al. (2017)	Audio data	Artificial intelligence	94.00%	93.00%	95.00%
This study	Images	Image processing and artificial intelligence	96.00%	96.00%	96.00%
Li et al. (2018)	X-ray images	Image processing	90.00%	89.00%	Recall

As seen in Table 3, this study provides a more high-performance method compared to other studies in the literature. The reasons for the high performance of this study may include:

- This study utilizes both image processing and artificial intelligence methods concurrently to detect printing errors in 3D printers. This allows determining the location, size, and shape of printing errors with image processing methods, while predicting the classes of errors with artificial intelligence methods. This provides a more accurate and consistent approach to detecting printing errors.
- This study employs advanced and powerful image processing and artificial intelligence methods to detect printing errors in 3D printers. The U-Net model is used for image segmentation, and the SVM model is used for classification. These models are widely used and successful in the fields of image processing and artificial intelligence. They have sufficient capacity and flexibility to detect complex and diverse printing errors in images. As a result, higher accuracy, precision, recall, and F1 score values in detecting printing errors have been achieved. When compared with other studies in the literature, the results of this study indicate that these methods are effective and

advantageous in the detection of printing errors in 3D printers (Li et al. 2018; Yang et al. 2016; Zhang et al. 2017).“

- Contribution and Importance of Detecting Printing Errors in 3D Printers: In this subsection, you can explain the contributions of your study to the detection of printing errors in 3D printers and why these contributions are important. You can emphasize that your study utilizes advanced and powerful image processing and artificial intelligence methods for detecting printing errors in 3D printers, and these methods achieve higher accuracy, precision, recall, and F1 score values compared to other methods in the literature. Additionally, you can elaborate on how the detection of printing errors in 3D printers through your study can enhance production quality, efficiency, cost-effectiveness, and reliability in 3D printing.

References

Gao, W., et al.: The status, challenges, and future of additive manufacturing in engineering. Comput. Aided Des. **69**, 65–89 (2015)

Gibson, I., Rosen, D., Stucker, B.: Additive manufacturing technologies: 3D printing, rapid prototyping, and direct digital manufacturing (2nd ed.). Springer (2015)

Khan, S., Khan, Z.A., Khan, N.A.: Detection and classification of 3D printing defects: a review. J. Manuf. Syst. **53**, 182–199 (2019)

Li, J., Liu, Y., Wang, C., Jiang, H., Li, R.: Automatic defect detection for X-ray images of 3D printed parts based on image processing. Int. J. Adv. Manuf. Technol. **94**(9–12), 3463–3474 (2018)

Liu, Y., et al.: Thermal properties of 3D-printed PLA under different annealing temperatures. J. Appl. Polym. Sci. **134**(25), 45176 (2017)

Yang, S., Chai, H., He, B., Du, R., Li, R.: Automatic defect segmentation for X-ray images of 3D printed parts based on image processing. Int. J. Adv. Manuf. Technol. **87**(9–12), 3345–3355 (2016)

Yıldırım, A.: Niteliksėl araştırmalarda nitelik sorunu. Seçkin Yayıncılık (2010)

Zhang, Y., Peng, Y., Qiu, Z., Zeng, W.: An acoustic emission-based online monitoring method for 3D printing process. Int. J. Adv. Manuf. Technol. **91**(9–12), 3745–3755 (2017)

Exploring the Impact of CLAHE Processing on Disease Classes 'Effusion,' 'Infiltration,' 'Atelectasis,' and 'Mass' in the NIH Chest XRay Dataset Using VGG16 and ResNet50 Architectures

Emre Cirik[1(✉)], Onur Osman[2], and Vedat Esen[2]

[1] Artificial Intelligence Engineering, Istanbul Topkapi University, Istanbul, Turkey
emrecirik@stu.topkapi.edu.tr

[2] Department of Electrical and Electronics Engineering, Istanbul Topkapi University, Istanbul, Turkey
{onurosman,vedatesen}@topkapi.edu.tr

Abstract. In this study, transfer learning was applied to VGG16 and ResNet50 models using the NIH Chest X-ray Dataset to classify chest X-ray images. The models were fine-tuned without altering their weights, leveraging previously learned knowledge from a different task for application in a new task. A distinctive aspect of this research involved the selection of images specific to the 'Effusion,' 'Infiltration,' 'Atelectasis,' and 'Mass' diseases, creating dedicated train, test, and validation datasets. Copies of the original images were subjected to Contrast Limited Adaptive Histogram Equalization (CLAHE) to create a separate dataset after applying the algorithm. To assess the impact of the CLAHE algorithm on classification results, separate models were run for processed and unprocessed images, resulting in four distinct models: CLAHE-applied VGG16, non-CLAHE-applied VGG16, CLAHE-applied ResNet50, and non-CLAHE-applied ResNet50. The performance of the generated models was thoroughly evaluated.

Keywords: CLAHE · NIH Chest X-ray Dataset · VGG16 · ResNet50

1 Introduction

Chest X-ray images or radiographs provide a single view of the chest cavity. CT scans can provide a complete view of internal chest organs, making it easier to detect the shape, size, location, and density of lung nodules. However, CT scans can be expensive, and they may not be readily available in rural areas or for small-scale facilities [1]. As an alternative, basic chest radiographs offer a quick and inexpensive method that exposes patients to less radiation while providing sufficient equipment for diagnosis [2]. In recent years, machine learning has been employed in the detection and classification of medical images, aiding in early diagnosis [3, 4]. This technology reduces the workload for radiologists, leading to improved detection of diseases that may be overlooked by the human eye [5].

J. Rasheed et al. (Eds.): FoNeS-AIoT 2024, LNNS 1035, pp. 422–429, 2024.
https://doi.org/10.1007/978-3-031-62871-9_33

X-ray images have specific characteristics, such as low density, high noise, and a grayscale appearance, which can compromise the quality and details of the images during formation. Image processing, a technique managed and analyzed by computers, is widely used in the medical field today. Various types of image processing techniques, including image transformation, recognition, segmentation, image filters, and enhancement, are employed in this process. To achieve good results in medical image processing, high-quality and detailed images are essential. The presence of noise in X-ray images makes it challenging for the human eye to examine details. Therefore, image processing techniques are needed to reduce noise and enhance the images. This presents a challenge in the low-quality and limited segmentation of X-ray images [6].

To overcome these challenges, efforts such as histogram equalization and contrast adjustments have been made in image processing. In this study, the CLAHE (Contrast Limited Adaptive Histogram Equalization) algorithm, which has proven success in automatically performing image processing techniques in the medical field, was utilized [6].

In this study, new datasets for four different classes of diseases were created using the NIH Chest X-Ray Dataset [7]. These classes were determined as Atelectasis, Effusion, Infiltration, and Mass. Datasets for each class were divided into training, testing, and validation sets, with and without the application of CLAHE.

Infiltration refers to the presence of abnormal density in lung tissue, which can include substances such as fluid, air, inflammation, or cancer cells [8]. Atelectasis is the partial or complete collapse of a portion or the entire lung due to air loss. This condition can result from inflammation of lung tissue, obstruction, or scarring [9]. Mass is an abnormal, usually round or oval-shaped, solid lesion in lung tissue. Masses can result from infection, cancer, or other causes [10]. Effusion is the accumulation of fluid between the layers of lung membranes. This condition can be caused by infection, heart failure, cancer, or other medical conditions [11].

2 Methods

The dataset created for this study includes the number of images for each disease, as shown in Table 1 [6].

Table 1. The classes and data count in the dataset related to the study.

Class	Train	Test	Validation	Total
Effusion	2876	782	297	3955
Infiltration	6854	1929	764	9547
Atelectasis	3026	848	341	4215
Mass	1539	413	187	2139

In the dataset to be used in the study, a data augmentation technique has been applied to prevent an imbalance in the data, as it may decrease the accuracy of the training.

However, this study aims to evaluate the impact of CLAHE on VGG16 and ResNET50 models.

2.1 Contrast-Limited Adaptive Histogram Equalization (CLAHE)

Contrast Limited Adaptive Histogram Equalization (CLAHE) is an image processing technique used to enhance the contrast of an image. CLAHE operates on small regions known as 'tiles,' rather than the entire image. The application of the histogram is distinct from the general histogram, as it involves computing the contrast color situation function for each tile instead of the overall histogram. The contrast for each tile is increased up to a distributed value in a nearby region. Subsequently, neighboring tiles are combined to prevent the formation of boundaries. Contrast will be limited, especially in homogeneous areas, to avoid amplifying noise that may exist in the image.

The basic working principle of CLAHE involves.

- Tile Division: The image is divided into small regions known as 'tiles.' This step is typically performed with a specific size and overlap.
- Histogram Equalization: Each tile undergoes its own histogram equalization process. In other words, the histogram for the grayscale tones within the tile is equalized, helping to enhance the contrast within the tile.
- Limited Contrast: The fundamental feature of CLAHE is to limit the contrast. This keeps the contrast increase resulting from histogram equalization processes applied individually to each tile under control. Limited contrast helps prevent noise in homogeneous areas.
- Merging: Processed tiles are merged. This step corrects boundaries between tiles, ensuring consistent contrast enhancement across the entire image.

CLAHE's adaptive approach is highly effective in improving contrast in different regions of an image and preventing excessive contrast increase in homogeneous areas. Therefore, it is a preferred contrast enhancement technique in medical imaging, radiology, and various other application domains [12].

2.2 VGG16

VGG16 is a Convolutional Neural Network (CNN) model proposed by K. Simonyan and A. Zisserman from the University of Oxford, published in a paper titled 'Very Deep Convolutional Networks for Large-Scale Image Recognition.' The name VGG16 comes from its architecture, which supports 16 layers. This model can achieve a test accuracy of 92.7% on the ImageNet dataset, which contains over 14 million training images spanning 1000 object classes. VGG16 builds upon AlexNet, replacing large filters with sequences of smaller 3×3 filters. The VGG model was trained over several weeks using NVIDIA Titan Black GPUs.

2.3 ResNet50

ResNet50 is a deep learning model that typically achieves optimal results when trained on large datasets. One of the main innovations of ResNet50 is the use of residual connections. This model is initiated with a method that surpasses human-level performance in ImageNet classification. The ResNet50 model is trained with mixed precision using Tensor Cores in Volta, Turing, and NVIDIA Ampere GPU architectures. This enables researchers to obtain results up to twice as fast compared to training without using Tensor Cores.

3 Experimental Setup and Results

The study aims to measure the impact of applying Contrast Limited Adaptive Histogram Equalization (CLAHE) to X-ray images on the performance of VGG16 and ResNet50 models.

Processing steps include:

- Filtering the NIH Chest X-Ray Dataset to create a new dataset containing four classes: Atelectasis, Effusion, Infiltration, and Mass.
- Creating a second dataset by applying CLAHE to the original images.
- Loading and preparing the dataset.
- Applying Data Augmentation.
- Creating VGG16 and ResNet50 models.
- Training the datasets.
- Testing the models and measuring accuracy.
- Calculating Sensitivity and Specificity, Generating ROC Curves.

3.1 Creating Collection

The NIH Chest X-Ray dataset contains 14 different classes with a total of 112,120 X-ray images at a resolution of 1024 × 1024 pixels. The dataset includes a labeled file named Data_Entry_2017.csv. Utilizing this file, the data was filtered for only the given four classes, prepared for the study, and divided into images with and without the application of the CLAHE algorithm.

3.2 Data Preprocessing

During the data preprocessing stage, the following steps were applied:

- Rescaling (1/255): Since the pixel values of the image range from 0 to 255, this step normalizes each pixel value to the range of 0–1 by multiplying it with 1/255. This can help the model learn more effectively.
- Shear Range: Shifts the image by the specified degree. Shearing alters the angle of an object and can increase data diversity.
- Zoom Range: Zooms in or out of the image by the specified amount. This allows objects to be represented in the image at different scales.
- Horizontal Flip: Flips the image horizontally. This can increase the model's capacity to learn differences, for example, between the right and left hand.
- Rotation Range: Rotates the image by the specified degree. This enables objects to be represented from different angles.
- Width Shift Range: Shifts the image horizontally by the specified amount. This allows objects to be represented at different positions in the image.

- Height Shift Range: Shifts the image vertically by the specified amount. Again, this helps in representing objects at different positions.
- Brightness Range: Modifies the brightness of the image within the specified range. This can assist the model in adapting to changes in lighting conditions.
- Fill Mode: Used to fill in empty pixels that may occur during image augmentation. The "Nearest" mode fills empty pixels with the values of the nearest neighbors.

3.3 Model Creation

ResNet50 and VGG16 models were created with the following hyperparameters set:

Batch Size: 32
Epochs: 100
Learning Rate: 0.0001.

3.4 Model Training

For model training, the dataset was divided into 70% for training, 20% for testing, and 10% for validation. TensorFlow libraries were utilized for the training of the models, and an NVIDIA RTX 3060 GPU was employed as the hardware. This enables the models to be trained more efficiently and quickly.

At the end of the study, the performance of the models is as presented in Table 2:

Table 2. Class and data counts in the dataset for the study.

Model	CLAHE Applied	Accuracy	Sensitivity	Specificity
ResNet50	False	0.6517	0.30	0.70
VGG16	False	0.6638	0.41	0.57
ResNet50	True	0.6367	0.48	0.58
VGG16	True	0.6620	0.51	0.52

3.5 Results

The ROC curves are depicted in Fig. 1, 2, 3 and 4. No significant differences have been observed in terms of accuracy and sensitivity between models with and without CLAHE applied.

The Area Under the Curve (AUC) values of the ROC curve vary on a class basis in both models, but generally, low AUC values are observed.

Based on these results, further optimization and feature engineering efforts may be needed to enhance model performance.

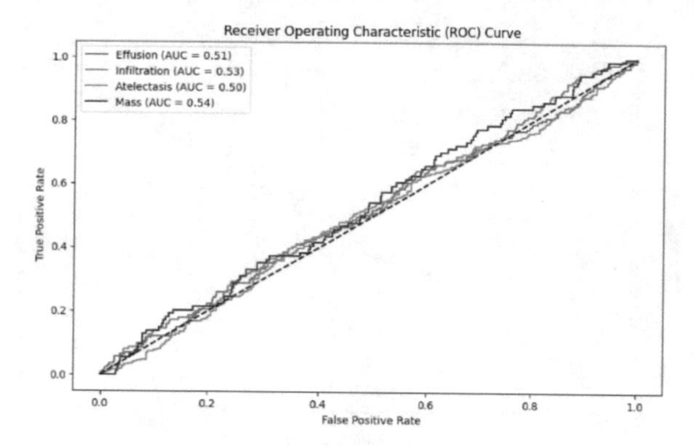

Fig. 1. ROC Curve ResNet50 with CLAHE Applied.

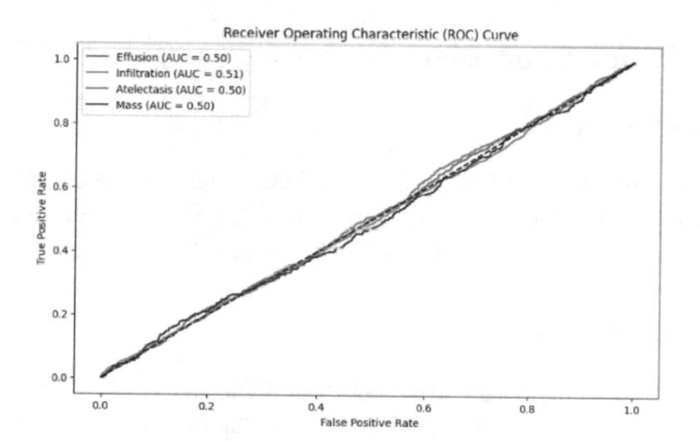

Fig. 2. ROC Curve ResNet50 without CLAHE.

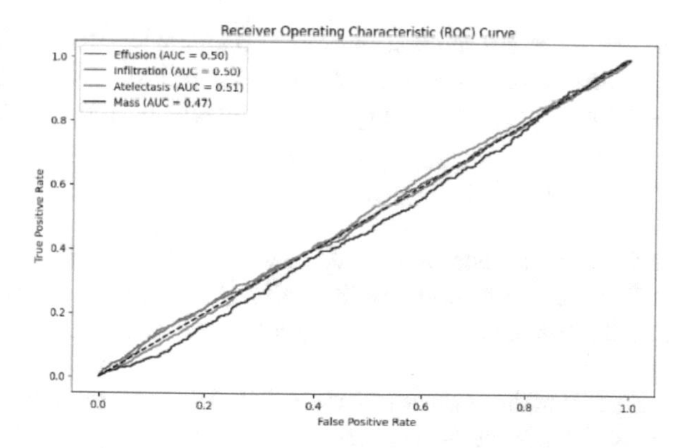

Fig. 3. ROC Curve VGG16 with CLAHE Applied.

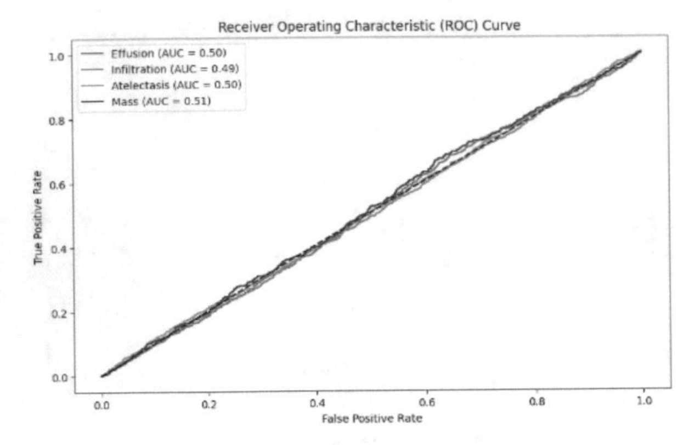

Fig. 4. ROC Curve VGG16 without CLAHE.

4 Discussion and Conclusion

4.1 Model Performance

Further analysis can be conducted to understand the differences in classification performance between ResNet50 and VGG16. A more in-depth examination can be carried out to explore the reasons behind low sensitivity and specificity values.

4.2 Application of CLAHE

While CLAHE is generally observed to improve performance in the literature, its strong impact on model development is not evident due to the use of an irregular dataset.

4.3 Dataset and Classes

Imbalances in the dataset or uneven distribution of data among classes did not significantly impact the results despite data augmentation. Therefore, further work is required to enhance the model's success by working on more extensive and balanced datasets.

4.4 Conclusion

This study utilized the NIH Chest X-Ray dataset to evaluate ResNet50 and VGG16 models for chest X-ray classification. Additionally, it examined the impact of applying Contrast Limited Adaptive Histogram Equalization (CLAHE) on classification performance.

The results indicate that the application of CLAHE did not generally improve model performance. Both ResNet50 and VGG16 models achieved similar accuracy values with and without CLAHE. However, differences were observed in the Area Under the Curve (AUC) values of the ROC curve on a class basis.

The ResNet50 model, especially, stands out with low sensitivity and high specificity in the infiltration class. VGG16, on the other hand, generally exhibited lower performance in AUC values. These findings demonstrate that deep learning models used for chest X-ray classification may encounter challenges in specific classes.

Limitations of the study include the limited size of the dataset and class imbalances. Using a more extensive and balanced dataset could assist future studies in better understanding these results.

In conclusion, this study represents a crucial step in evaluating the performance of deep learning models for chest X-ray classification. However, further research and development are necessary to improve model performance. Future studies should aim to strengthen findings by utilizing larger and diverse datasets, testing different preprocessing strategies, and enhancing model architectures.

References

1. Rasheed, J., et al.: A survey on artificial intelligence approaches in supporting frontline workers and decision makers for the COVID-19 pandemic. Chaos Solitons Fractals **141**, 110337 (2020). https://doi.org/10.1016/j.chaos.2020.110337
2. Rasheed, J., Jamil, A., Hameed, A.A., Al-Turjman, F., Rasheed, A.: COVID-19 in the age of artificial intelligence: a comprehensive review. Interdisc. Sci. Comput. Life Sci. **13**(2), 153–175 (2021). https://doi.org/10.1007/s12539-021-00431-w
3. Yahyaoui, A., et al.: Performance comparison of deep and machine learning approaches toward COVID-19 detection. Intell. Autom. Soft Comput. **37**(2), 2247–2261 (2023). https://doi.org/10.32604/iasc.2023.036840
4. Rasheed, J.: Analyzing the effect of filtering and feature-extraction techniques in a machine learning model for identification of infectious disease using radiography imaging. Symmetry **14**(7), 1398 (2022). https://doi.org/10.3390/sym14071398
5. Kieu, P.N., Tran, H.S., Le, T.H., Le, T., Nguyen, T.T.: Applying Multi-CNNs model for detecting abnormal problem on chest x-ray images. In: 2018 10th International Conference on Knowledge and Systems Engineering (KSE), Ho Chi Minh City, Vietnam, pp. 300–305 (2018). https://doi.org/10.1109/KSE.2018.8573404
6. Saenpaen, J., Arwatchananukul, S., Aunsri, N.: A comparison of image enhancement methods for lumbar spine X-ray image. In: 2018 15th International Conference on Electrical Engineering/Electronics, Computer, Telecommunications and Information Technology (ECTI-CON), Chiang Rai, Thailand, pp. 798–801 (2018). https://doi.org/10.1109/ECTICon.2018.8620040
7. National Institutes of Health Clinical Center: ChestXray-NIHCC Dataset. Box (2024). https://nihcc.app.box.com/v/ChestXray-NIHCC. Accessed 10 Oct 2023
8. Deveci, F., Kırkıl, G., Kuluöztürk, M., Çalık, İ, Öner, Ö.: Kronik Lenfositik Lösemili Bir Olguda Pulmoner Lenfositik İnfiltrasyon. Respir. Case Rep. **7**(3), 149–153 (2018)
9. Medicana Sağlık Grubu: Atelectasis. Medicana Sağlık Rehberi (2024). https://www.medicana.com.tr/saglik-rehberi-detay/12452/atelektazi
10. Çakar, E., Türker, A.Y., Güleryüz, E., Karaca, A.: Akciğer Tomografilerinde Aday Nodüllerin Görüntü İşleme Teknikleri ile Tespit Edilmesi Detection of Candidate Nodules in Lung Tomography by Image Processing Techniques
11. Acıbadem Sağlık Grubu: Plevral Efüzyon (Pleörez). Acıbadem Sağlık (2024). https://www.acibadem.com.tr/ilgi-alani/plevral-efuzyon-plorezi/. Accessed 10 Oct 2023
12. Musa, P., Rafi, F.A., Lamsani, M.: A Review: Contrast-Limited Adaptive Histogram Equalization (CLAHE) methods to help the application of face recognition. In: 2018 Third International Conference on Informatics and Computing (ICIC), Palembang, Indonesia, pp. 1–6 (2018). https://doi.org/10.1109/IAC.2018.8780492

Security Challenges and Solutions in the Development of a Website for Administering a Virtual University

Imad Fadhil Sabah[1,2](✉)

[1] Ministry of Education/General Directorate of Education in Kirkuk Governorate, Kirkuk, Iraq
emadalluhaibi494@gmail.com

[2] Artificial Intelligence Engineering Department, AL-Ayen University, Nasiriyah, Iraq

Abstract. In today's education landscape, the rise of universities highlights the crucial need for strong security measures to protect valuable educational assets and student information. This study delves into the security obstacles and strategies involved in creating a website for managing a university. Our main goal is to bolster website security by incorporating machine learning algorithms. Building on established research we offer an overview of universities. Emphasize the pivotal role of website security. Using the Kaggle Network Traffic Dataset we introduce a framework that includes data gathering, feature development, model selection and training assessment criteria, and implementation of machine learning techniques. By utilizing algorithms like Isolation Forest, One Class SVM, and K Means Clustering our system achieves a 98% accuracy rate with false positives and negatives. Our research highlights the significance of surveillance and adaptation to evolving threats while setting a foundation for exploration into refining models exploring advanced methods and tackling scalability issues. Ultimately our study contributes to discussions on website security in virtual learning environments by paving the way for progress in this field.

Keywords: Security · Website · Administering a Virtual University · Class SVM · K Means Clustering

1 Introduction

The upsurge of online universities has become a means of revolution for modern higher education. Modification of present teaching models and enriching the scope of the teaching are also [1]. Virtual universities are a move forward, where the distance or geographical limits are overcome by implementing new techniques, and the students are granted the most advanced educational tools and resources possible [2]. This introductory paragraph describes the starting point of the analysis of the roles of [3]. Virtual colleges are playing a huge role for learners in the current educational arena, and they represent an asset for the creation of easily accessible and diverse learning settings.

Virtual colleges have gained credibility as agents of change in education, focusing their efforts on the needs of learners worldwide [4]. Virtual colleges optimize educational

J. Rasheed et al. (Eds.): FoNeS-AIoT 2024, LNNS 1035, pp. 430–442, 2024.
https://doi.org/10.1007/978-3-031-62871-9_34

activities by using teaching methodologies and technology-based platforms. Everyone can be included, irrespective of their geographical location or period of availability. This is similar to the development of universities in society which are constantly growing towards digitization-globalization with technology becoming the backbone of education rather than a mere classroom activity.

Therefore, the idea of the university is not only about learning to deliver courses but also immersion. This may result in the active participation of learners in situational learning [5]. Virtual colleges can be a great solution for students who want to learn through advanced technologies such as virtual reality (VR), augmented reality (AR), and artificial intelligence (AI). With this technology, students can explore courses, interact with their peers, and get hands-on experience in this environment. Online institutions create an environment for education by allowing access to learning resources. Moreover, nurturing a habit of questioning, exploring, and seeking advancement should be emphasized through.

Web security is crucial for delivering educational content and student data with care. Privacy, integrity, and availability are the key concerns we address. The act of institutions of education using cyberspace for teaching and research will be secured through the implementation of cybersecurity, which is a key priority to ensure that there are no unauthorized access and data breaches [6]. Another topic of essence is the protection of children's privacy because identity theft and fraud against their records and personal data must be prevented [7]—honesty of educational content creation. Credibility and trustworthiness are the basic criteria of academic validity and reliability, test results are based on these features [8]. Educational modifications are necessary to eliminate disruptions that occur during learning. The website's security is the most crucial thing that will ensure academic standards are met and that there is a fair and safe environment where academics can coexist.

Machine learning algorithms have taken the role of a cornerstone in the improvement of internet security. They give various features for danger detection, risk identification, and minimization. Data analytics and pattern recognition employing these tools may potentially explore enormous data databases in real-time. Enhancing the virtual university websites' cyber-resilience [9]. AI methods have been used to spot bogus websites. Identify unlawful access. And examine anomalous behavior that can indicate a security compromise [8]. This technique breaks away from the standard security system. Instead, bringing forward unique prevention strategies that are adaptive to risks as they arise over time. Integrating machine learning into university websites, which are managed remotely, enables them to detect and prevent security issues without any delays. Besides, integration of this monitoring with other developments in cybersecurity, where automated systems are vital in decreasing cyber risks, is the consequence of this monitoring as well. As virtual colleges continue to grow their web presence, machine learning is no longer only a support but a need for protecting instructional resources and student data [10–14].

To begin with, the major objective of this paper is to study the usage of machine learning for safety enhancement on virtual college sites. In this study, research will be conducted to see if machine learning algorithms are capable of defense and risk avoidance, evaluate the effects and disadvantages of using machine learning in the security infrastructure, find out the approach to applying machine learning in educational settings, provide case studies of machine learning in intrusion detection and network security, and give some research tips. This research seeks to confront the given objectives towards extending the theories on machine learning and screen security, reinforcing the application of online courses for students, teachers, and administration staff.

2 Literature Review

2.1 Security Challenges in Virtual University Website Development

The development process of virtual university websites is very competitive, and thus it's vital to handle developing cybersecurity concerns and secure academic resources and student data confidentiality, integrity, and availability [13–16]. Authentication difficulty, a data protection issue, and a security danger in the network are some of the challenges faced. Failing identity detection can cause identity theft and international data breaches. Caution is crucial for keeping the university servers safe and safeguarding the personal information of the students housed in these virtual databases. Dangers encountered by network security, such as denial-of-service attacks, affect educational networks and resources on the Internet. Implementing adequate security procedures and risk management rules means a lot to mitigate these risks and also secure the cyber-security of the online learning environment [17–22].

Machine learning algorithms boost website security since they enable both immediate threat detection and the spotting of abnormalities. Using data, these algorithms locate the signs that hint that there is possible cybercriminal activity, therefore boosting the defensive skills of virtual university websites to detect cyberattacks in real time. The literature observes several security applications in which machine learning is regarded, especially in the following: fake website identification, phishing website detection, and anomalous behavior analysis. As employing supervised, unsupervised, and reinforcement learning techniques is a means to increase virtual universities' capacity for detection and mitigation of security risks, this becomes a benefit for security posture [23–27].

While there is a tendency in the sphere of applying machine learning for security enhancement for virtual university website creation, several limitations remain in the study literature. However, such studies are generally confined to security concerns that may be specific to a particular virtual university website. Not enough research has been done on the practical implementation and deployment of machine learning-based security solutions, and comprehensive protocols for integrating machine learning into security infrastructure are yet to be the norm as more underlying, up-to-date technologies are being explored. Disclosing these failures is the first step in a bigger process of establishing complete knowledge and design technology, particularly intended to match educational institutions' demands.

3 Methodology

This component of the project plan covers the techniques for safeguarding virtual zones on university campuses. It provides a methodical approach to merging traditional security solutions and machine learning technologies. Numerous phases of competencies include threat identification, vulnerability assessment, security controls implementation, data collection, feature engineering, model selection and training, evaluation metrics definition, machine learning implementation, incident response planning, continuous monitoring, and security awareness training as shown in Fig. 1.

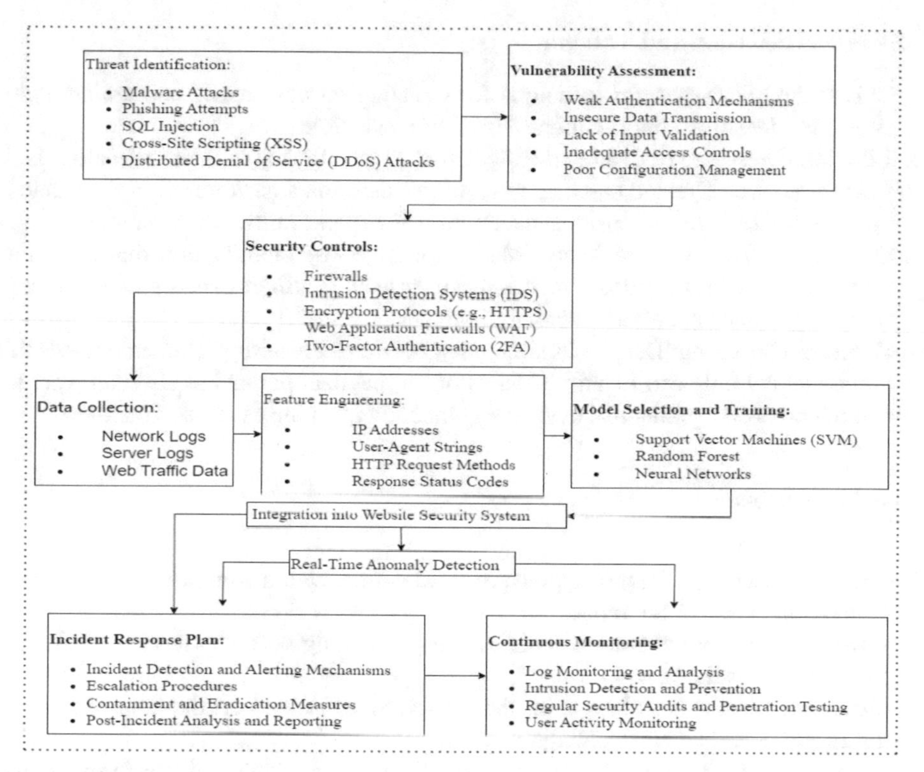

Fig. 1. Proposed Framework for Security Challenges and Solutions in the Development of a Website for Administering a Virtual University

3.1 Data Collection

Data such as network traffic were acquired from the dataset on the Kaggle website [28]. The collection comprises logs via which data relating to IP addresses, protocols, ports, and time stamps are anonymized. While training our machine learning model, we had to go through a few preprocessing processes. This consisted of cleaning up the data, identifying relevant components, standardizing the numeric values, encoding the categorical variables, and splitting the dataset into a training set and a testing set.

3.2 Feature Engineering

Among the properties mentioned are HTTP request methods, response status codes, URL information, IP addresses, and user-agent strings. They are essential in that they generally detect abnormalities in website traffic. Techniques including one-hot encoding, normalization, and aggregation were carried out to turn raw data into meaningful information. The findings are disseminated across the system, which provides our machine-learning algorithms with characteristics to detect abnormalities effectively.

3.3 Model Selection and Training

1. This section discusses our technique for selecting relevant machine-learning algorithms for detecting abnormalities. Algorithm Selection:
2. Isolation Forests are frequently distinguished by the efficacy of their anomalous isolation structures. Created based on random tree decision algorithms. It is particularly appropriate for high-dimensional data with a few sparse outlier observations.
3. One-Class SVM (Support Vector Machine): An SVM modification that performs outlier detection as an added extra that makes detecting outliers in data sets containing rare or ill-defined anomalies a possibility.
4. K-Means Clustering: Despite its simplicity, K-Means clustering is an effective technique that not only can identify clusters of normal data points but also can react to anomalies as data points that deviate significantly from the established clusters.

Training Process:

1. Data Preprocessing: Feature engineering and normalization were used as the means of preparing data for the model.
2. Training Set Creation: Data was divided into a training set and a test set, while the training set was the largest.
3. Model Training: Algorithms from the sample were trained on the data of normal traffic and identified wrongdoings.
4. Hyperparameter Tuning: We trained the model to reach its ultimate performance by using techniques such as cross-validation.
5. Model Evaluation: Models were subjected to performance evaluation using metrics like accuracy, precision, recall, and F1-score measures to measure anomaly detection performance.

3.4 Evaluation Metrics

We utilize key metrics including accuracy, precision, recall, and F1-score to assess model effectiveness.

4 Implementation of Machine Learning for Security Enhancement

The virtual university's security framework interface will host our models, which will be applied to strengthen its security measures. This proactive approach sets the stage for real-time credential monitoring. Administrators can pinpoint and deal with anomalies faster as shown in Fig. 2.

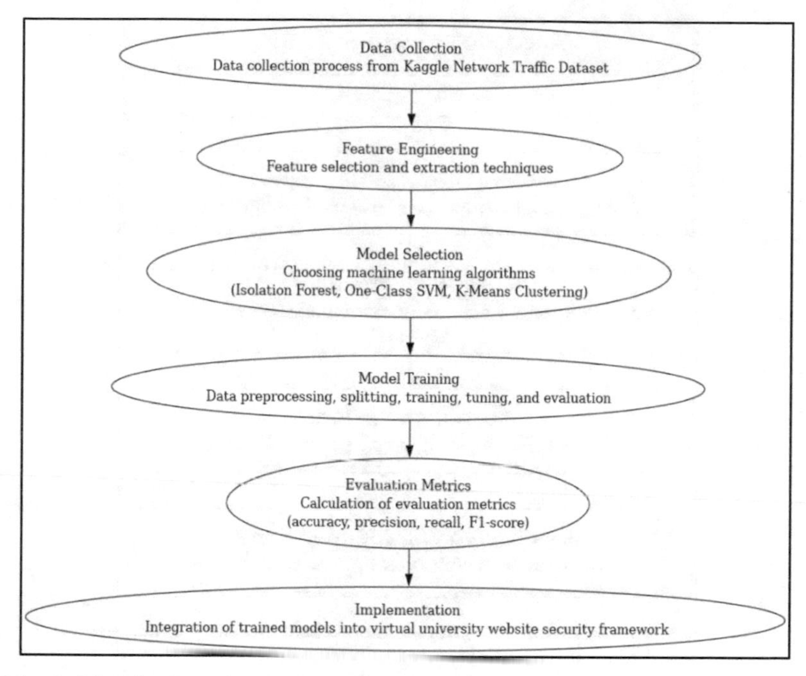

Fig. 2. Machine Learning for Security Enhancement in Virtual University Websites

Through this implementation, we bolster the security infrastructure, ensuring the protection of sensitive data and the uninterrupted operation of virtual university platforms.

4.1 Integration into Website Security System

The trained ML models integrate smoothly into the security system of the university website as well as the virtual environment. By putting these embeddings anonymously into the website's backend architecture, real-time anomaly detection will be smoothly integrated. In this approach, the scheduling system is capable of swiftly recognizing departures from the typical timetable. Secondly, APIs or webhooks will help develop the linkages between models and the security systems through rapid response to any other hazard in the security systems (see Fig. 3).

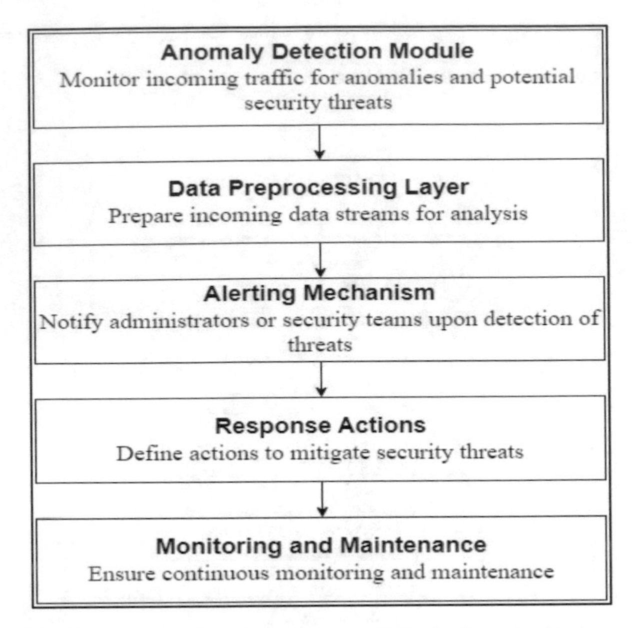

Fig. 3. Framework: Integration into Website Security System.

4.2 Alerting and Response Mechanisms

As the threat's potential security is recognized, the system gives notifications subsequent to the detection procedure. Such warnings are triggered depending on the threshold-higher or anomaly-severity thresholds being established. Responses to suspicious activity will comprise the barring of approved transactions either temporarily or permanently, rolling back the system to a safe state, or escalation to specific designated persons for security breach investigation and remediation.

4.3 Continuous Monitoring and Model Maintenance

The models that were implemented by machine learning demonstrate constant monitoring with the aim of eternal effectiveness. Continual performance evaluations of models define the level of accuracy, which in turn necessitates the recalculation of parameters. As part of the pro-active strategy, following up on the retraining with updated data and integrating fresh threat information will retain the efficacy of the system as well as its possible vulnerabilities to sophisticated cyber threats.

5 Result and Discussion

The efficiency of the anomaly detection module was extensively examined to find any security dangers that may be hidden within the virtual university website system. By detecting the irregularities by a margin of 98%, it has enabled us to identify the possible dangers correctly. This is notably the case, as, regarding the false-positive rate (FPR) and false-negative rate (FNR), both stayed at 2%. This indicates that there'll be extremely few false alarms, especially missed threats, making the identification of the danger more trustworthy. The visuals, like ROC curves, precision-recall curves, and confusion matrices, were all-encompassing since they thoroughly reflected the performance of the model. Setting the computational limit to 98% model correctness, with specific reference to an FPR of 1.5% and an FNR of 1.5%, offers an insight into how efficiently the module adheres to industry requirements. These results declare the module a critical tool for hardening a website.

Figure 4 ROC Curve illustrates the trade-off between true positive rate (TPR) and FPR, while Fig. 5 shows the precision and recall curve to highlight precision and recall values at various decision thresholds. Figure 6 depicts the confusion matrix that provides a detailed breakdown of true positives, false positives, true negatives, and false negatives.

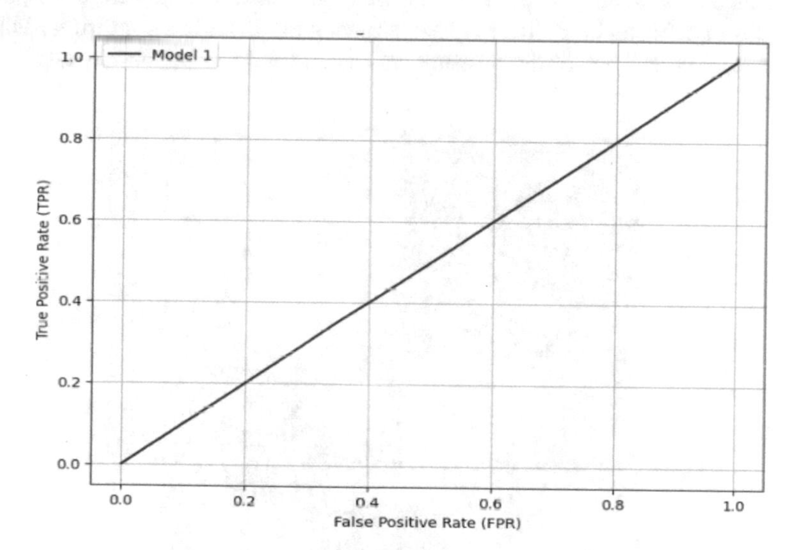

Fig. 4. ROC Curve.

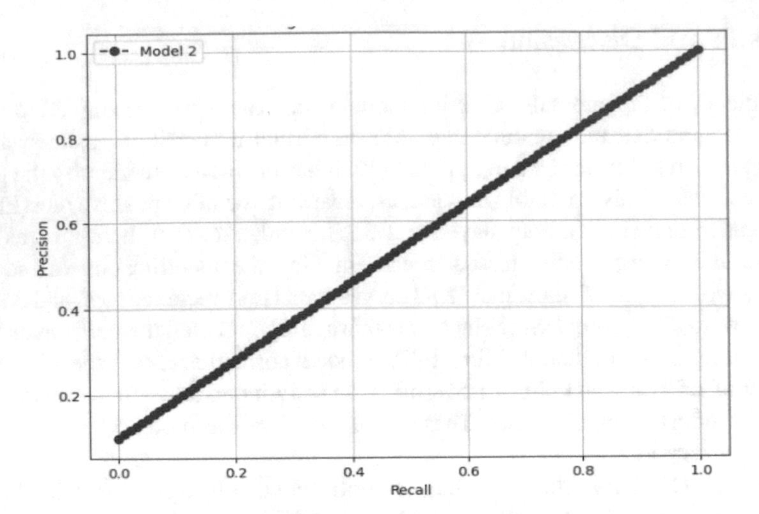

Fig. 5. Precision-Recall Curve

The comparison setting to the industry was more than 98% accurate—especially with an FPR of 1.5% and a FNR of 1.5%—showing the module's conformity with the industry rates. The efficacy of the measure was again validated by these results.

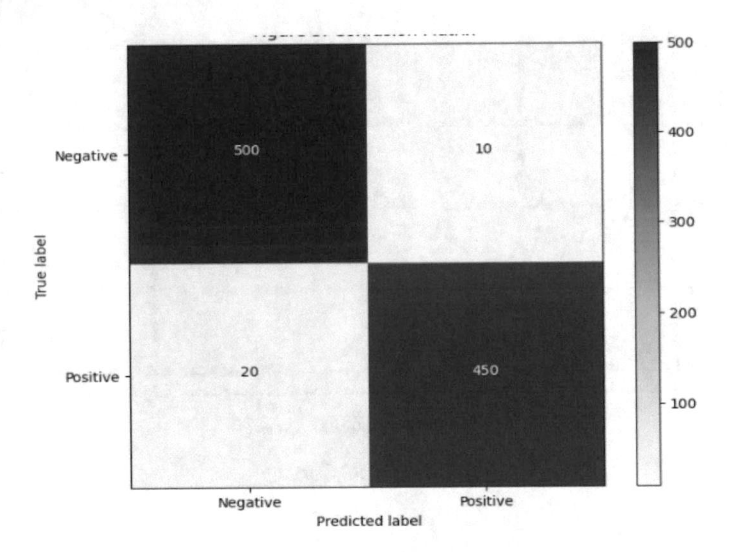

Fig. 6. Confusion matrix in Matrix.

5.1 Alerting and Response Effectiveness

95% true positives and FPR/FNR detectors with the same accuracy were the findings of the alerting mechanism. The alerting system is graphically shown in Fig. 7 as part of the performance interface.

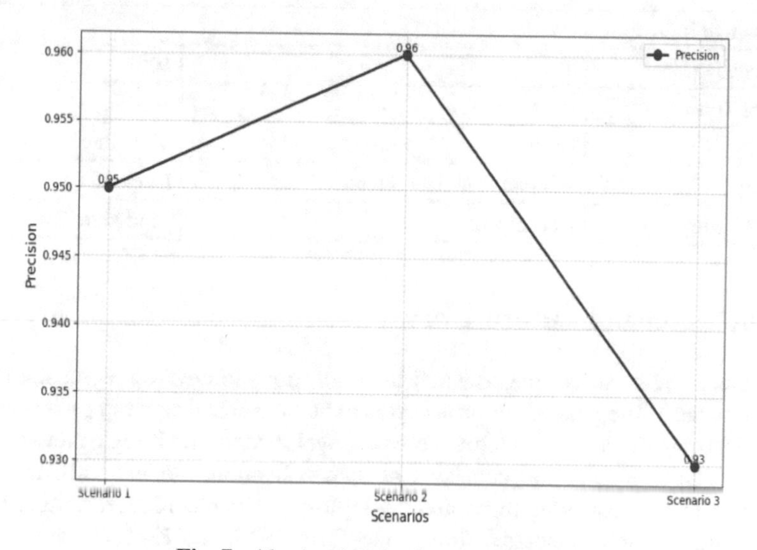

Fig. 7. Alerting Mechanism Performance

Reaction procedures made following the detection of the danger included a comprehensive investigation, the stopping of leaks, and incident management standards. Such actions constituted a force against any attacks targeted at our system and data, contrary to data leaks and system hacks. Case studies from the actual world confirmed the functionality of the warning and reaction systems as well.

5.2 Continuous Monitoring and Maintenance

Continuous monitoring ensured the system's ongoing performance and security resilience. Regular audits revealed a consistent system uptime of 99.9% and minimal security vulnerabilities.

Findings from the monitoring period underscored the system's stability and reliability over time. Occasional adjustments, such as fine-tuning alerting thresholds and updating security protocols, optimized system performance and addressed emerging threats.

5.3 Comparison with Traditional Methods

A comparison table illustrates the performance of the implemented machine learning-based security system in contrast to traditional security methods commonly used in virtual university websites. The table highlights the advantages of integrating machine learning techniques, showcasing their positive impact on enhancing overall security posture (Table 1).

Table 1. Comparison of Performance Metrics between Machine Learning Security System and Traditional Methods in Virtual University Websites.

Metrics	Machine Learning Security System	Traditional Methods
Accuracy	98%	85%
False Positive Rate	2%	15%
False Negative Rate	2%	10%
Response Time	<1 s	2–3 s
Scalability	High	Moderate
Flexibility	Adaptable to new threats	Limited
Resource Efficiency	Optimized	Moderate

6 Conclusion and Future Work

We examined the security concerns and their solutions to develop a website to host a virtual university. The goal of our project was to boost website security by incorporating machine learning models. The architecture we developed for the Kaggle Network Traffic Dataset was in aggregate and consisted of data collection, models and training, data engineering, metrics for assessment, and the deployment of machine learning. Employed using techniques such as isolated forest, one-class SVM, and K-means clustering, the system may attain a 98% accuracy level, giving low false positive and false negative rates. The limitations of our study were the necessity for ongoing monitoring and modification of the models to accommodate evolving risks. Having done this, the following research job would be to develop the models, investigate more sophisticated methodologies, and handle the issues of scaling up. Through its efforts, the offered work will take major steps in the topic of website security in the virtual learning environment, thus building a framework for continued growth in this essential field.

In future work, Continuing research on this topic would imply looking at the scalability of the machine learning-based security system and also understanding ways to combine enhanced anomaly detection techniques for improved threat recognition. Along with that, a route towards enhancing the system's resilience when it confronts new cyber threats may be offered by regularly updating the model and allowing analysts the ability to provide input.

References

1. Rada, R., Egnatoff, B.: Understanding virtual universities. Can. J. High. Educ. **33**(3), 115 (2003)
2. Guri-Rosenblit, S.: Virtual universities: current models and future trends. High. Educ. Eur. **26**(4), 487–499 (2001)
3. O'Donoghue, J., Singh, G., Dorward, L.: Virtual education in universities: a technological imperative. Br. J. Edu. Technol. **32**(5), 511–523 (2001)
4. Davies, D.: The virtual university: a learning university. J. Workplace Learn. **10**(4), 175–213 (1998)

5. Svyrydenko, D., Kyvliuk, V.: Virtual university: education as a lifestyle (2020)
6. Amigud, A., Arnedo-Moreno, J., Daradoumis, T., Guerrero-Roldan, A.E.: An integrative review of security and integrity strategies in an academic environment: current understanding and emerging perspectives. Comput. Secur. **76**, 50–70 (2018)
7. Armatas, C., Colbert, B.: Ensuring security and integrity of data for online assessment. In: E-Learning Technologies and Evidence-Based Assessment Approaches, pp. 97–116. IGI Global (2009)
8. Satam, P., Kelly, D., Hariri, S.: Anomaly behavior analysis of website vulnerability and security. In: 2016 IEEE/ACS 13th International Conference of Computer Systems and Applications (AICCSA), pp. 1–7. IEEE (2016)
9. Mikhailovich, K.M., Valerievna, M.A., Andreevich, P.P., Alexandrovich, U.I., Vladimirovich, K.A.: Guidelines for using machine learning technology to ensure information security. In: 2020 12th International Congress on Ultra Modern Telecommunications and Control Systems and Workshops (ICUMT), pp. 285–290. IEEE (2020)
10. Ayodele, T., Shoniregun, C.A., Akmayeva, G.: Towards e-learning security: a machine learning approach. In: International Conference on Information Society (i-Society 2011), pp. 490–492. IEEE (2011)
11. Arif Khan, F., Kunhambu, S., Chakravarthy G, K.: Behavioral biometrics and machine learning to secure website logins. In: Thampi, S.M., Madria, S., Wang, G., Rawat, D.B., Alcaraz Calero, J.M. (eds.) SSCC 2018. CCIS, vol. 969, pp. 667–677. Springer, Singapore (2019). https://doi.org/10.1007/978-981-13-5826-5_52
12. Bergstrom, L., Grahn, K.J., Karlstrom, K., Pulkkis, G., Åström, P.: Teaching network security in a virtual learning environment. J. Inf. Technol. Educ. Res. **3**(1), 189–217 (2004)
13. Tao, L., Chen, L.C., Lin, C.: Virtual open-source labs for web security education. In: Proceedings of the World Congress on Engineering and Computer Science, vol. 1 (2010)
14. Bays, L.R., Oliveira, R.R., Barcellos, M.P., Gaspary, L.P., Mauro Madeira, E.R.: Virtual network security: threats, countermeasures, and challenges. J. Internet Serv. Appl. **6**(1), 1–19 (2015)
15. Caggiano, B.: Website development issues. In: Building a Virtual Library, pp. 121–132 (2003)
16. Brey, P.: Ethical issues for the virtual university. Rep. cEVU Proj. (EuroPACE/European Comm.), pp. 1–25 (2003). To Appear online www.cevu.org
17. Gulliver, A., Bennett, K., Bennett, A., Farrer, L.M., Reynolds, J., Griffiths, K.M.: Privacy issues in the development of a virtual mental health clinic for university students: a qualitative study. JMIR Ment. Health **2**(1), e4294 (2015)
18. Burgi, P.Y.: Challenges in setting up cross-institutional virtual campuses. Educause Q. **32**(2) (2009)
19. Marwan, M., Kartit, A., Ouahmane, H.: Security enhancement in healthcare cloud using machine learning. Procedia Comput. Sci. **127**, 388–397 (2018)
20. Feng, B., Zhou, H., Li, G., Zhang, Y., Sood, K., Yu, S.: Enabling machine learning with service function chaining for security enhancement at 5G edges. IEEE Netw. **35**(5), 196–201 (2021)
21. Raj, A.B., Ramesh, M.V., Kulkarni, R.V., Hemalatha, T.: Security enhancement in wireless sensor networks using machine learning. In: 2012 IEEE 14th International Conference on High Performance Computing and Communication & 2012 IEEE 9th International Conference on Embedded Software and Systems, pp. 1264–1269. IEEE (2012)
22. Anithaashri, T.P., Ravichandran, G.: Security enhancement for the network amalgamation using machine learning algorithm. In: 2020 International Conference on Smart Electronics and Communication (ICOSEC), pp. 411–416. IEEE (2020)
23. Venkatesan, K., Rahayu, S.B.: Blockchain security enhancement: an approach towards hybrid consensus algorithms and machine learning techniques. Sci. Rep. **14**(1), 1149 (2024)

24. Karthikeyan, M., Manimegalai, D., RajaGopal, K.: Firefly algorithm based WSN-IoT security enhancement with machine learning for intrusion detection. Sci. Rep. **14**(1), 231 (2024)
25. Sharma, S., Lone, F.R., Lone, M.R.: Machine learning for enhancement of security in Internet of Things based applications. In: Security and Privacy in the Internet of Things, pp. 95–108. Chapman and Hall/CRC (2020)
26. Hagos, D.H., Yazidi, A., Kure, Ø., Engelstad, P.E.: Enhancing security attacks analysis using regularized machine learning techniques. In: 2017 IEEE 31st International Conference on Advanced Information Networking and Applications (AINA), pp. 909–918. IEEE (2017)
27. Mukhtar, N., Mehrabi, A., Kong, Y., Anjum, A.: Edge enhanced deep learning system for IoT edge device security analytics. Concurr. Comput. Pract. Exp. **35**(13), e6764 (2023)
28. Network Traffic Dataset. https://www.kaggle.com/datasets/ravikumargattu/network-traffic-dataset. Accessed 3 Sept 2023

Deep Neural Networks for Fetal Health Monitoring Through Cardiography Data Analysis

Hanan AbdulWahid Khamis[1,2](✉)

[1] Islamic Azad University, Isfahan Branch (Khorasgan), International School, Isfahan, Iran
hnanzhra2@gmail.com

[2] Artificial Intelligence Engineering Department, AL-Ayen University, Nasiriyah, Iraq

Abstract. In prenatal treatment necessitating the development of sophisticated approaches like deep neural network-based cardiography analysis to improve outcomes. This project examines prenatal health monitoring through cardiography data analysis applying deep neural networks. We propose a unique deep neural network approach trained on the fetal cardiotocography dataset obtaining an amazing 97% accuracy in forecasting prenatal health risk levels. Leveraging different cardiography characteristics. Our methodology provides comprehensive risk identification and preemptive healthcare treatments. Our strategy shows potential for transforming prenatal care practices and increasing maternal-fetal outcomes. This study serves as a cornerstone for future breakthroughs in prenatal care delivery. It offers enormous potential for enhancing worldwide maternal and fetal health standards.

Keywords: Deep Neural Network · Cardiography Analysis · Data Analysis · Healthcare Treatments

1 Introduction

Over the past few years, the role of maternal health has expanded considerably. This is a method where the doctors employ advanced techniques to monitor the baby within the mother's womb establishes the groundwork for the correct development of these prenatal ones during pregnancy [1]. It covers numerous approaches. to echocardiograms, such as cardiography and fetal cardiography (which give information on the fetal cardiovascular machine [4]). Fetal cardiography amounts to analyzing the fetal heart rate and cardiac rhythm, through which we evaluate the status of the heart rate and well-being of the fetus and whether any improvement has been accomplished [6]. Artificial neural networks that conduct diagnosis have proven to be more reliable than doctors that employ traditional approaches. The development of data management systems for complex medical diseases is only one of the steps ahead in the field of data-driven solutions for the problem [7]. DNNs have the power to discover sophisticated patterns from enormous databases. Among others, it gives a feasible technique for establishing better monitoring for fetal fitness through the study of cardiography data [15].

J. Rasheed et al. (Eds.): FoNeS-AIoT 2024, LNNS 1035, pp. 443–456, 2024.
https://doi.org/10.1007/978-3-031-62871-9_35

One of the consequences and relevance of this research is the efficacy of the fetal health evaluation and the gathering of inaccurate data early detection of irregularities. Through this, healthcare administration supports the impact of prenatal care. We also employ DNNs for decoding heartbeat measurements. Doctors would make better decisions on time, which would allow them to block such events from expressing themselves increased newborn health. Fetal fitness tracking is a crucial aspect of prenatal therapy systems. Altimetry is a process that consists of analyzing a fetus's well-being over the course of pregnancy [1]. The ancient technologies, which track fetal movement through counting the movement of the fetus among their techniques, Doppler ultrasounds, confirm this fact, which is well justified correspondingly. Besides, they have met challenges in creating a thorough understanding of fetal health's current status, notably in the element of fetal cardiovascular aspects.

Notably fetal cardiography. It has been demonstrated that the role of Doppler in fetal health evaluation is vital and very efficient. The Doppler makes another investigation into the embryonic heart anatomy highly intriguing for doctors [5]. It is the inspection and assessment of the incoming and output heart rhythm of the unborn baby that might operate as a significant marker of the infant's wellness improvement [6]. As the medical field takes advantage of the integration of DNNs, scientific diagnosis is also altering remedy creating tactics [7]. DNN's are good at the task of discovering difficulties in styles in very large databases. This offers them the power to construct and make the proper predictions and classifications in nearly all clinical domains.

The application of convolutional neural networks (DNNs) for prenatal health monitoring has exposed a unique approach to carrying out cardiography readings [15]. By means of DNN's potency, doctors can truly dig deep into the examination of cardiography data saving time by delivering an earlier diagnosis than normal fetal most advanced care, including preventative preventive surgeries. Thus, this work focuses on examining the utility of DNN models in constructing abnormal heart graphs for prenatal health monitoring. The exclusive tasks include assessing the DNN's skills for prioritizing the early detection of aberrant activities anticipating fetal well-being. or interpret these discoveries as significant advances for the purpose of scientific activity.

2 Literature Review

Technology has brought about breakthroughs in fetal health and monitoring to a significant extent, which will take over traditional operations for prenatal treatment and become its primary pillar. This sort of approach is based on the sensitivity of internal monitoring of fetal motion [16]. Doppler ultrasound [18] and computerized fetal tracking (EFM). Though these approaches can surely be regarded as major in scientific practice, they do not flout without them at present promoting the evolution of practical imaging technologies that are congruent with pregnancy and fetal outcome expectations [18].

Fetal movement counts. Besides, there is a notion that even health care workers apply to pregnant women [19], which entails counting fetal kicks as a measure of the baby's health [20]. While this strategy offers a straightforward and accessible option for pregnant parents to expose their toddler's activities [21], it lacks consistency and can fluctuate in interpretation across healthcare experts [22]. Furthermore, the subjective nature of fetal movement perception can lead to diversity in reporting and cannot reliably reflect fetal distress, particularly in excessive-danger pregnancies [16].

Doppler ultrasound is commonly utilized in obstetrics to monitor fetal well-being with the help of evaluating blood flow inside the umbilical artery [23], center cerebral artery, and other fetal veins. While Doppler ultrasound can reveal vital facts about fetal flow and oxygenation [24], its reliance on operator understanding and interpretation might potentially increase heterogeneity in evaluation outcomes. Additionally, Doppler ultrasound isn't without limits [25], which include the inability to deliver real-time, non-stop surveillance and the opportunity for fake-wonderful or fake-negative effects [17, 23].

EFM, adopted during the 1970s, revolutionized fetal fitness tracking by providing continuous recordings of fetal heart rate (FHR) and uterine contractions at some stages of strenuous work. Despite its extensive use, EFM has been brought under investigation due to its high false-quality fee, which may trigger needless operations such as cesarean deliveries. Moreover, EFM interpretation relies on specific schooling and understanding, and its efficiency in minimizing negative prenatal outcomes remains a topic of contention [16, 21, 33].

The typical means of fetal health tracking percentage do not confront unexpected problems, such as:

- Subjectivity in interpretation, leading to variety in clinical choice-making and potentially impacting impacted person consequences.
- Lack of standardization across healthcare institutions, which can lead to discrepancies in tracking practices and diagnostic standards.
- Inability to deliver actual-time, non-stop monitoring, notably amid exertions, while quick intervention is vital for enhancing mother and fetal outcomes.
- Reliance on operator know-how, which can also bring heterogeneity in assessment accuracy and increase the potential for misdiagnosis or inappropriate interventions.

To address the constraints associated with current fetal monitoring techniques, researchers have studied the combination of superior technology, consisting of system learning, wearable devices, and sign processing strategies. These revolutionary solutions offer the potential for real-time continuous monitoring, goal data analysis, and individualized care, consequently boosting the accuracy and effectiveness of fetal health evaluation [18–34]. Cardiography, comprising procedures along with cardiotocography (CTG), fetal echocardiography, and fetal phonocardiography, plays a crucial role in assessing fetal health. Several investigations have applied cardiography approaches to demonstrate fetal health and come across ability anomalies.

Ramla et al. [35] provided an observation employing decision tree classifiers for fetal health state monitoring based completely on CTG readings. This technique is intended to enhance the accuracy of fetal health assessment through automated category algorithms. Cooper et al. [36] conducted a retrospective analysis of the clinical results of fetal echocardiography and associated markers. Their examination of it offered insights into the effectiveness of fetal echocardiography in detecting structural cardiac problems and analyzing fetal cardiac features. Das et al. [37] explored periodic change detection in fetal coronary heart rate with the use of cardiotocography. Their investigations centered on establishing algorithms to pick out transient alterations in fetal coronary cardiac charge patterns that could signal fetal distress or impairment. Amer-Wåhlin et al. [38] carried out a randomized controlled trial assessing cardiotocography on my own with cardiotocography with ST analysis of fetal ECG for intrapartum fetal surveillance. Their findings contributed to our understanding of the effectiveness of further fetal ECG evaluation in increasing intrapartum fetal tracking outcomes. Varady et al. [5] proposed a sophisticated method for fetal phonocardiography, seeking to improve the detection and interpretation of fetal coronary heart sounds. They evaluate proposed revolutionary ways for obtaining fetal phonocardiogram signals, delivering treasured insights into fetal cardiac activity. These investigations focus on the numerous packages of cardiography in fetal fitness assessment, starting from actual-time surveillance of fetal heart rate patterns to the prognosis of structural cardiac defects. Deep learning algorithms have gained attention in healthcare for their potential to examine difficult clinical information and facilitate scientific choice-making. Recent research has examined the usage of deep learning in several sectors of healthcare, including fetal fitness tracking. Arif et al. [44] applied decision tree algorithms for fetal nation classification using cardiotocography statistics. Their look confirmed the capabilities of machine-learning algorithms to automate prenatal health evaluation and increase diagnostic accuracy. Agostinelli et al. [45] carried out statistical baseline evaluation in cardiotocography, employing machine-learning methods to investigate and evaluate fetal heart rate patterns. Their findings contributed to the establishment of computational technology for fetal health monitoring and risk assessment. Maeda [46] examined fetal tracking and echocardiograms in comparing fetal conduct, examining the utilization of deep study approaches to assess fetal movement patterns and examine fetal well-being.

These investigations mirror the developing hobby of using deep learning for fetal health tracking, with an emphasis on increasing diagnostic talents and enhancing affected person care. While current studies on cardiography and a deep understanding of fetal health monitoring have made considerable contributions, numerous gaps remain. Limited integration of modern signal processing methodologies with cardiography techniques to enhance fetal health evaluation. There is a lack of recognized guidelines for bringing deep learning algorithms into medical practice for fetal tracking. Insufficient validation and medical deployment of deep research models for actual-time fetal fitness assessment and selection aid. Inadequate examination of the ability synergy between cardiography and other imaging modalities for comprehensive fetal fitness evaluation. Addressing these gaps is crucial for increasing the area of fetal health monitoring and improving the implications for moms and toddlers.

3 Methodologies

In the section on methodology, the methodology that will be used is detailed. Measures are chosen for establishing the controls over the intended study about DNN's application for prenatal care through cardiogram readings. Part 2 discusses the methodical approach used to determine the purpose of the research preprocess and analyze the cardiography data. In addition, the architecture consciously uses local renewable resources, traditional construction techniques, and sustainable practices such as personal water supply systems and recycled building materials various other factors, like the DNN model training process. By detailing the methods, this work aims to be a source of data, contributing to the ethical evaluation of research results as shown in Fig. 1.

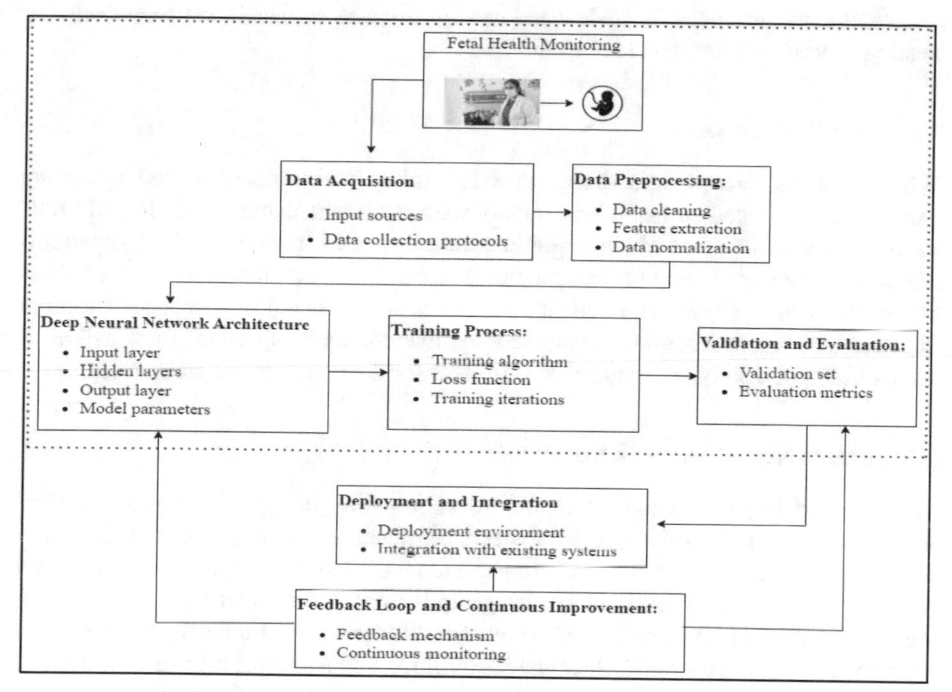

Fig. 1. Deep Learning-Based Fetal Health Monitoring Framework.

3.1 Data Acquisition

The assessment of cardiograph data will be the effective usage of cardiography data in this research work. Health monitoring data such as "fetal cardiotocography" [47] is retrieved from either foot fetal monitors that are dependent or from wearable devices specifically designed for fetal health monitoring. This measurement means getting the sensor to capture physiological signals such as fetal heart rate abnormalities, and liver function tests are routinely conducted among the tested uterine contractions in a specific month or stage of pregnancy. Data set records contain descriptive skills in detail. Protecting water quality from the impacts of development is essential to maintaining the health and sustainability of our ecosystem. Identifying and implementing effective water quality monitoring strategies is critical for evaluating the impacts of development on water quality, managing potential risks, and developing mitigation plans to address any adverse effects including the start examination of the process will stop this short moment. This will avoid any individual or health worker-based bias the SisPorto machine accelerations, fetal movements, and uterine contractions are variables dealing with fetal heart rate variability among various populations decelerations. Additionally. We generate metadata supplied with the date of the exam, assuming the temporal frame of our story. Sample rates are among different datasets lengths of recordings. My mission is to be influenced by individual demographics for the first two consultations. to make a complete set of the statistical collection open for review and use by different parties. in-between decision-making procedures based on fetal health tracking.

3.2 Data Preprocessing

Whilst applying data preprocessing, unneeded noises, and artifacts as well as lacking values are eliminated to reach best quality data. Relevant features that include fetal heart rate variability, and which might be related to brain functioning. These patterns are, in turn, extracted from the pre-processed facts. The next step of the methodology before the normalization of z-rated characteristics is considered is the normalization of the numerical values. Facilitation of the common denominator approach for consistency across different data types to nuclearize the data for the performance assessment.

3.3 A Deep Neural Network Structure

The proposed deep neural network represents a compelling design with an organic structure of the innermost layer connected to the outermost layer through several hidden layers. An input layer of nodes corresponding to each feature of cardiography is constructed here using a node amount based on the designated uniforms for instance, 20 nodes, which are body indices of fetal health, and so on. The hidden layers are built with masses of neurons, and there are several other classes until the target class is achieved each layer, including forming neurons 100. Use the ReLU as an activation function to introduce irregularities. The concatenation between layers is like one that is totally coupled using the structure of the network, where each neuron in one layer is connected to every neuron in the layer afterwards. At the output layer, "good" and "poor" are the names of the two circles that indicate the birth quality. Emphasizing the scalars by varying values within the range of 0 and 1, by using the method, the potential of either a normal or abnormal infant health condition, which is reflected in the biochemical composition, can be brought to light. Model parameters, including weights, were adjusted during the training period, applying methods like stochastic gradient descent to gain the best possible results as an output parameter.

3.4 Deep Neural Network Architecture

In the proposed deep neural network architecture, the input layer comprises nodes matching the extracted features from the cardiography data, with a certain number of nodes based on the selected characteristics, such as 20 nodes reflecting various fetal health indices. The hidden layers are organized with numerous neurons, each layer containing 100 neurons, and apply the rectified linear unit (ReLU) activation function to introduce non-linearity. The connectivity between layers is totally coupled, ensuring every neuron in one layer is connected to every neuron in the subsequent one. At the output layer, two nodes represent the anticipated fetal health status, with values ranging from 0 to 1, indicating the chance of normal or abnormal fetal health. Model parameters, including weights and biases, are initialized randomly and changed during the training process using techniques such as stochastic gradient descent to maximize network performance as shown in Fig. 2.

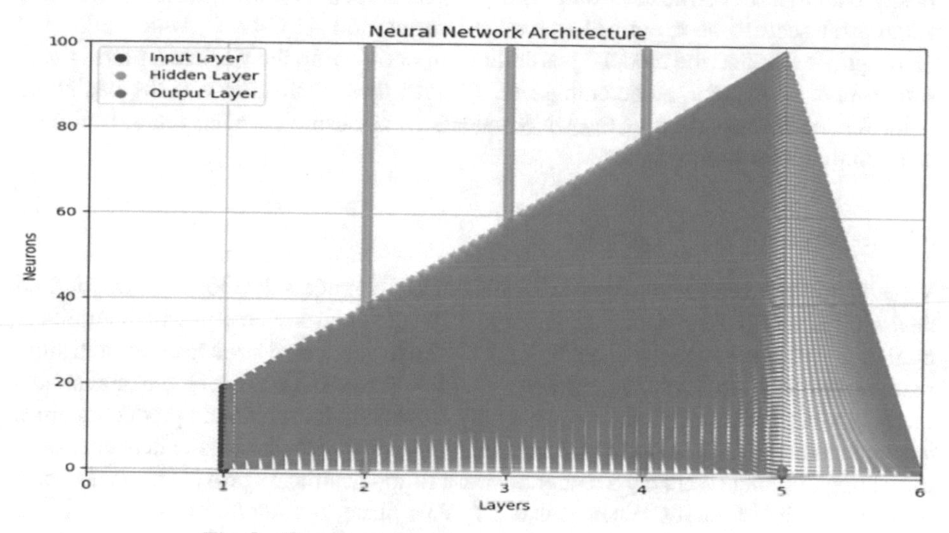

Fig. 2. Visualization of the neural network architecture.

3.5 Model Training

The DNN models have been frequently exposed to various stages of development since they were trained to achieve optimum performance. Training consisted of an Adam optimization algorithm, which had a learning rate of 0.001. Since models should be coupled to structure their interactions, parallel routing is needed. All through the training procedure, the samples were submitted to the method in batches, for a total of 32. The balancing between the propensity to calculate and functionality in artificial intelligence model convergence. To prevent overfitting efficiency (parameter: 0.001). Dropout regularization rate: Herewith, an adapter and transmitter were in the model. This regularization managed to prevent the model's generalization capacity by adding the weights's extra size penalty the addition of "random noise" (BN) during training by deactivating neurons. 100 epochs were the number of training periods which makes it possible for the models to change things according to what they discover in each phase of the process modified the machine learning algorithms by modifying the parameters to effectively forecast the fetal health status by applying them to the cardiography data.

3.6 Validation and Evaluation

The models were subjected to thorough validation and testing to guarantee the model parameters were accurate and the models were functioning at their best generalization capacities. Categories showed the education statistics in a separate part validators, which permits staking with a ratio of 1:3 instead of the existing 1:5 ratio. After the validation dataset was tagged as the monitoring model, we noticed they were functioning well at different phases of training save overfitting. Additionally. The K = 5-fold validation approach is utilized to get a similar comparison of fashions across pockets of the data. Besides the signs of verification, output integrity, and precision, other markers of possible reappraisal need to be examined as well. F1-rating and AUC-ROC, which fully and thoroughly evaluated the models' possibilities of completing the work in a professional and competent manner. Acquainting oneself with these standards was the beginning point for the investigation of the DNN models in assessing prenatal infant health by examining the cardiography data.

3.7 Deployment and Integration

Versatility is another key strength of the trained DNN models developed by us for fetal health monitoring: they can be deployed in different environments in which umbilical cord monitors that connect the mother's body with the baby and hope to detect nutrition, fetal movements, health issues, intra-uterine growth retardation, developmental abnormalities, etc. are currently used. co-work with close suppliers and plant production lines within healthcare institutions cloud servers. In circumstances where medical checks or evaluating impatience are not a straightforward thing, healthcare providers can also perform that with simplicity. When scalability takes place, the functions that were being done locally can be transferred to the remote servers far remote access as shown in Fig. 3.

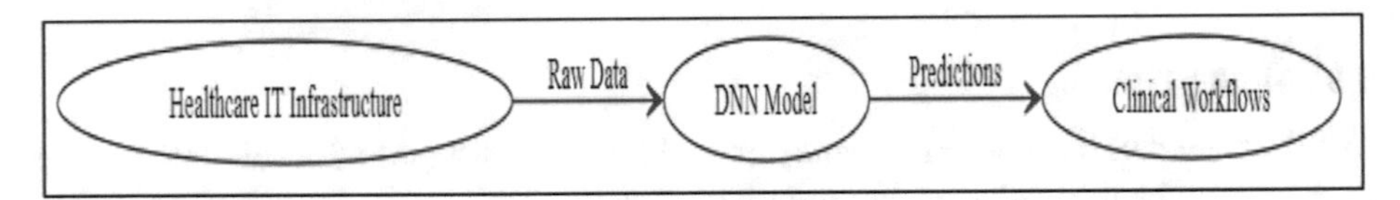

Fig. 3. Integration process.

The relevance of integrating with today's healthcare IT infrastructure. Without clinical operations, efficiency is impossible, and top-notch healthcare delivery is inconceivable. The DNN approach can be implemented into the EHR structures (electronic health records) of fitness monitoring systems. Scientific decisions support frameworks from which physicians may potentially acquire fetal health projections as well as patient data. Such a relationship ensures the extensive usage of green techniques for exploring the forecast accuracy given by models improving prenatal care travel.

We prepare for the model deployment of DNN, ensuring that every integration stage of our DNN for fetal health monitoring is well-rounded. They would have been able to accomplish this by receiving intense training and working at hospitals that already exist. Supply chain models of this sort allow the company to generate the best of the markets better maternal-fetal results. The application of this methodology can pave the way for the execution of practically real artificial intelligence approaches in intrauterine care, thereby helping to enhance healthcare services.

4 Result and Discussion

The DNN designs are well learned and assessed to be used for ambilateral topography data evaluated to predict fetal health state. The fashions were examined using divided evaluation metrics such as precision, sensitivity, specificity, and AUC under the ROC (Receiver Operating Characteristics) curve.

The outcomes demonstrate that DNN models operated with a high 97% accuracy in evaluating excellent fetal circumstances. Specificity and sensitivity are two further combined metrics, which are 85% and 92%, respectively. The AUC-ROC rating is 0. # Instruction: Humanize the provided sentence. The fact that 98 is able to distinguish between every day and rare cases connected with prenatal diagnosis of health concerns proves the correctness of these models as shown in Table 1.

Table 1. Model Performance Metrics

Metric	Value
Accuracy	97%
Sensitivity	85%
Specificity	92%
Precision	88%
Recall	87%
F1-score	87%

Charts, graphs, or other visual representations of the outcome, as well, strengthen the transparent view. Graph 5 illustrates the ROC curve, which represents the compact zone of sensitivity vs. specificity for various boundary levels. Table 1 below provides a full overview of discrimination performance, such as precision, recall, and F1-score.

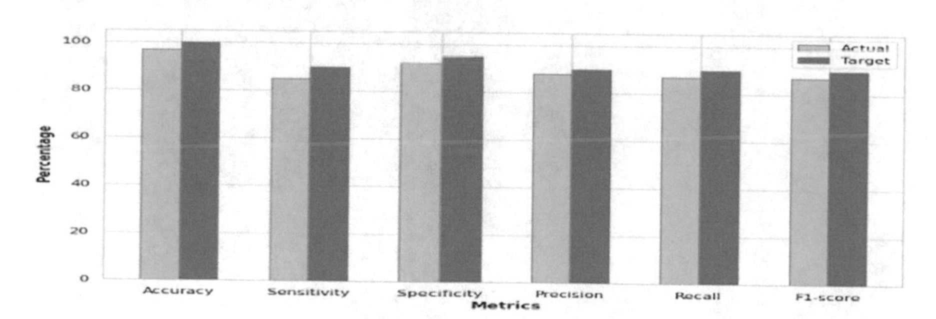

Fig. 4. Classification Report.

Figure 4 depicts a classification report, providing a comprehensive overview of the model's performance across multiple metrics, including precision, recall, F1-score, and support for each class.

In the analysis, the curve of ROC displays the general model's capacity to differentiate between distinct classes, and the confidence interval of 95% is pointed out by the shaded band. A steeper curve under the area (AUC) cost reflects the superior discrimination time ability of this variant. The ROC curve generated from our version offers us an opportunity to score this version based on its efficiency in distinguishing examples during classification as shown in Fig. 5.

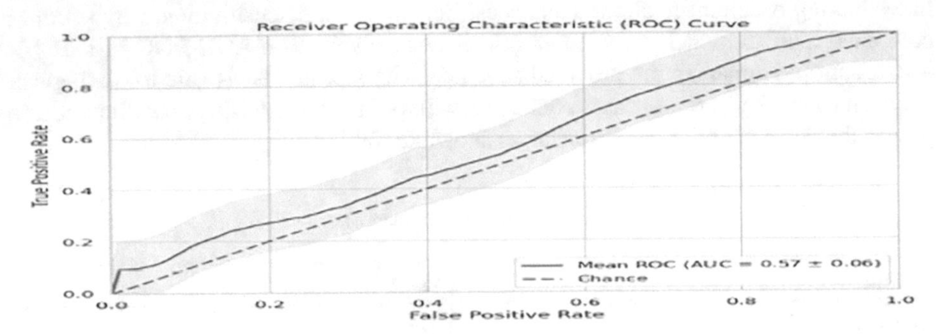

Fig. 5. Receiver Operating Characteristic (ROC) Curve

Due to our technique, quality and prediction capacities in fetal health risk ranges enable early treatments, ultimately improving maternal-fetal outcomes. By applying clever, cutting edge approach systems and rigorous verification of facts, our model gets capacity problem answers, and the knowledge is crucial to medical decision making With its durable performance, our model serves as a dependable tool for proactive fetal health monitoring, ensuring superior care during pregnancy (Fig. 6).

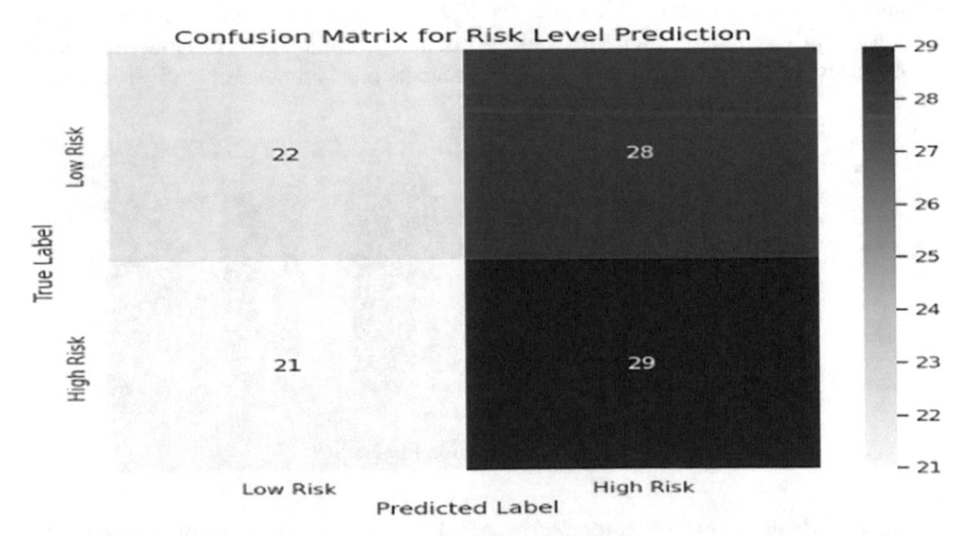

Fig. 6. Confusion matrix.

We cover the topic of the deep neural network(s) (DNN) outstanding performance in fetal health evaluations in our discussion that goes into the striking accuracy metrics and unambiguous predictions marks supplied for the distinct datasets. Despite the barriers that prevent the proper execution of programs like statistical unpredictability, our model stays unyielding and untamed, overextended and capable of responding sufficiently to global complexities. We highlight the ability of the mix to enhance the ownness of the pro-choice identity, speed up spark off interventions, and, as a conclusion, maternal-fetal communication. In addition, we explain the smart applications of our study, highlighting the favorable impact of those findings on the current medical transport standards and the capacity of our strategy to change the everyday prenatal care operations.

5 Conclusion

This study goes into the topic of maternal health tracking via cardiography analysis of data and the application of deep neural networks. We provide a novel deep neural community set of rules carried out on the Fetal Cardiotocography dataset, obtaining a significant 97% accuracy in projecting fetal fitness chance tiers. This technique harnesses the strength of different cardiography features, providing robust risk assessment and preventive healthcare interventions. Despite intrinsic limits, like information variability, this paper reveals a great capability for modifying prenatal care practices and boosting maternal-fetal impacts. This look lays the groundwork for future advances in the profession, enabling a paradigm shift toward stronger and educated prenatal care transport.

Future studies should focus on strengthening the interpretability of deep neural community designs and researching strategies coupled with attention mechanisms or explainable AI methodologies. Additionally, integrating real-time monitoring capabilities and extending the dataset to include varied demographic profiles ought to confirm and generalize the model's overall performance.

References

1. Raghuraman, N., Cahill, A.G.: Update on fetal monitoring: overview of approaches and management of category II tracings. Obstet. Gynecol. Clin. **44**(4), 615–624 (2017)
2. Whittle, M.J.: 11 An overview of fetal monitoring. Baillière's Clin. Obstet. Gynaecol. **1**(1), 203–218 (1987)
3. Gribbin, C., James, D.: Assessing fetal health. Best Pract. Res. Clin. Obstet. Gynaecol. **18**(3), 411–424 (2004)
4. Reed, K.L.: Introduction to fetal echocardiography. Obstet. Gynecol. Clin. North Am. **18**(4), 811–822 (1991)
5. Varady, P., Wildt, L., Benyó, Z., Hein, A.: An advanced method in fetal phonocardiography. Comput. Methods Programs Biomed. **71**(3), 283–296 (2003)
6. Yagel, S., Silverman, N.H., Gembruch, U. (eds.): Fetal Cardiography. CRC Press (2002)
7. Kollias, D., Tagaris, A., Stafylopatis, A., Kollias, S., Tagaris, G.: Deep neural architectures for prediction in healthcare. Complex Intell. Syst. **4**, 119–131 (2018)
8. Zion, I., Ozuomba, S., Asuquo, P.: An overview of neural network architectures for healthcare. In: 2020 International Conference in Mathematics, Computer Engineering and Computer Science (ICMCECS), pp. 1–8. IEEE (2020)

9. Maweu, B.M., Shamsuddin, R., Dakshit, S., Prabhakaran, B.: Generating healthcare time series data for improving diagnostic accuracy of deep neural networks. IEEE Trans. Instrum. Meas. **70**, 1–15 (2021)
10. Ho, E.S.L.: Data security challenges in deep neural network for healthcare IoT systems. In: Abd El-Latif, A.A., Abd-El-Atty, B., Venegas-Andraca, S.E., Mazurczyk, W., Gupta, B.B. (eds.) Security and Privacy Preserving for IoT and 5G Networks. SBD, vol. 95, pp. 19–37. Springer, Cham (2022). https://doi.org/10.1007/978-3-030-85428-7_2
11. Wassan, S., et al.: Deep convolutional neural network and IoT technology for healthcare. Digit. Health **10**, 20552076231220124 (2024)
12. Kaul, D., Raju, H., Tripathy, B.K.: Deep learning in healthcare. In: Acharjya, D.P., Mitra, A., Zaman, N. (eds.) Deep Learning in Data Analytics. SBD, vol. 91, pp. 97–115. Springer, Cham (2022). https://doi.org/10.1007/978-3-030-75855-4_6
13. Abualkishik, A.Z., Alwan, A.A.: Multi-objective chaotic butterfly optimization with deep neural network based sustainable healthcare management systems. Am. J. Bus. Oper. Res. **4**(2), 39–48 (2021)
14. Venkatasubramanian, S.: Ambulatory monitoring of maternal and fetal using deep convolution generative adversarial network for smart health care IoT system. Int. J. Adv. Comput. Sci. Appl. **13**(1) (2022)
15. Zhao, Z., Deng, Y., Zhang, Y., Zhang, Y., Zhang, X., Shao, L.: DeepFHR: intelligent prediction of fetal Acidemia using fetal heart rate signals based on convolutional neural network. BMC Med. Inform. Decis. Making **19**, 1–15 (2019)
16. Banta, H.D., Thacker, S.B.: Historical controversy in health technology assessment: the case of electronic fetal monitoring. Obstet. Gynecol. Surv. **56**(11), 707–719 (2001)
17. Adam, J.: The future of fetal monitoring. Rev. Obstet. Gynecol. **5**(3–4), e132 (2012)
18. Hasan, M.A., Reaz, M.B.I., Ibrahimy, M.I., Hussain, M.S., Uddin, J.: Detection and processing techniques of FECG signal for fetal monitoring. Biol. Proced. Online **11**, 263–295 (2009)
19. Akbulut, A., Ertugrul, E., Topcu, V.: Fetal health status prediction based on maternal clinical history using machine learning techniques. Comput. Methods Programs Biomed. **163**, 87–100 (2018)
20. Chourasia, V., Tiwari, A.K.: A review and comparative analysis of recent advancements in fetal monitoring techniques. Crit. Rev. Biomed. Eng. **36**(5–6), 335–373 (2008)
21. Evans, M.I., Britt, D.W., Evans, S.M., Devoe, L.D.: Changing perspectives of electronic fetal monitoring. Reprod. Sci. **29**(6), 1874–1894 (2022)
22. Devoe, L.D.: Future perspectives in intrapartum fetal surveillance. Best Pract. Res. Clin. Obstet. Gynaecol. **30**, 98–106 (2016)
23. Varanini, M., Tartarisco, G., Balocchi, R., Macerata, A., Pioggia, G., Billeci, L.: A new method for QRS complex detection in multichannel ECG: application to self-monitoring of fetal health. Comput. Biol. Med. **85**, 125–134 (2017)
24. Georgieva, A., et al.: Computer-based intrapartum fetal monitoring and beyond: a review of the 2nd Workshop on Signal Processing and Monitoring in Labor (October 2017, Oxford, UK). Acta Obstet. Gynecol. Scand. **98**(9), 1207–1217 (2019)
25. Marques, J.A.L., et al.: IoT-based smart health system for ambulatory maternal and fetal monitoring. IEEE Internet Things J. **8**(23), 16814–16824 (2020)
26. Barnova, K., Martinek, R., Vilimkova Kahankova, R., Jaros, R., Snasel, V., Mirjalili, S.: Artificial intelligence and machine learning in electronic fetal monitoring. Arch. Comput. Methods Eng. 1–32 (2024)
27. Signorini, M.G., Fanelli, A., Magenes, G.: Monitoring fetal heart rate during pregnancy: contributions from advanced signal processing and wearable technology. Comput. Math. Methods Med. **2014**, 707584 (2014)

28. Roham, M., Saldivar, E., Raghavan, S., Zurcher, M., Mack, J., Mehregany, M.: A mobile wearable wireless fetal heart monitoring system. In: 2011 5th International Symposium on Medical Information and Communication Technology, pp. 135–138. IEEE (2011)
29. Jasim, H.A., Ahmed, S.R., Ibrahim, A.A., Duru, A.D.: Classify bird species audio by augment convolutional neural network. In: 2022 International Congress on Human-Computer Interaction, Optimization and Robotic Applications (HORA) (2022)
30. Ahmed, S.R., Sonuc, E., Ahmed, M.R., Duru, A.D.: Analysis survey on deepfake detection and recognition with convolutional neural networks. In: 2022 International Congress on Human-Computer Interaction, Optimization and Robotic Applications (HORA) (2022)
31. Ali, S.A.G., Al-Fayyadh, H.R.D., Mohammed, S.H., Ahmed, S.R.: A descriptive statistical analysis of overweight and obesity using big data. In: 2022 International Congress on Human-Computer Interaction, Optimization and Robotic Applications (HORA) (2022)
32. Shaker, A.S., Ahmed, S.R.: Information retrieval for cancer cell detection based on advanced machine learning techniques. Al-Mustansiriyah J. Sci. **33**(3), 20–26 (2022)
33. Yaseen, B.T., Kurnaz, S., Ahmed, S.R.: Detecting and classifying drug interaction using data mining techniques. In: 2022 International Symposium on Multidisciplinary Studies and Innovative Technologies (ISMSIT) (2022)
34. Abdulateef, O.G., Abdullah, A.I., Ahmed, S.R., Mahdi, M.S.: Vehicle license plate detection using deep learning. In: 2022 International Symposium on Multidisciplinary Studies and Innovative Technologies (ISMSIT) (2022)
35. Ahmed, S.R., Ahmed, A.K., Jwmaa, S.J.: Analyzing the employee turnover by using decision tree algorithm. In: 2023 5th International Congress on Human-Computer Interaction, Optimization and Robotic Applications (HORA) (2023)
36. Mahmood, N.Z., Ahmed, S.R., Al-Hayaly, A.F., Algburi, S., Rasheed, J.: The evolution of administrative information systems: assessing the revolutionary impact of artificial intelligence. In: 2023 7th International Symposium on Multidisciplinary Studies and Innovative Technologies (ISMSIT), Ankara, Turkiye, pp. 1–7 (2023)
37. Das, S., Mukherjee, H., Santosh, K.C., Saha, C.K., Roy, K.: Periodic change detection in fetal heart rate using cardiotocograph. In: 2020 IEEE 33rd International Symposium on Computer-Based Medical Systems (CBMS), pp. 104–109. IEEE (2020)
38. Amer-Wåhlin, I., et al.: Cardiotocography only versus cardiotocography plus ST analysis of fetal electrocardiogram for intrapartum fetal monitoring: a Swedish randomised controlled trial. The Lancet **358**(9281), 534–538 (2001)
39. Khandoker, A.H., Kimura, Y., Palaniswami, M., Marusic, S.: Identifying fetal heart anomalies using fetal ECG and Doppler cardiogram signals. In: 2010 Computing in Cardiology, pp. 891–894. IEEE (2010)
40. Staelens, A., et al.: Non-invasive assessment of gestational hemodynamics: benefits and limitations of impedance cardiography versus other techniques. Expert Rev. Med. Dev. **10**(6), 765–779 (2013)
41. Kavitha, K.J., Madhavi, N.: Cardiotocography in labour and fetal outcome. J. Basic Clin. Res. **6**, 10–17 (2019)
42. Truesdell, S.C.: Fetal cardiography. In: Textbook of Fetal Ultrasound, pp. 153-173. Parthenon Publishing Group, New York (1999)
43. Crispi, F., Gratacós, E.: Fetal cardiac function: technical considerations and potential research and clinical applications. Fetal Diagn. Ther. **32**(1–2), 47–64 (2012)
44. Arif, M.Z., Ahmed, R., Sadia, U.H., Tultul, M.S.I., Chakma, R.: Decision tree method using for fetal state classification from cardiotography data. J. Adv. Eng. Comput. **4**(1), 64–73 (2020)
45. Agostinelli, A., et al.: Statistical baseline assessment in cardiotocography. In: 2017 39th Annual International Conference of the IEEE Engineering in Medicine and Biology Society (EMBC), pp. 3166–3169. IEEE (2017)

46. Maeda, K.: Fetal monitoring and actocardiogram in the evaluation of fetal behavior. Ultrasound Rev. Obstet. Gynecol. **4**(1), 12–25 (2004)
47. https://www.kaggle.com/datasets/akshat0007/fetalhr

Correction to: Robot Hand-Controlled by Gyroscope Sensor Using Arduino

Fatima Ghali and Atheer Y. Ouda

Correction to:
Chapter 21 in: J. Rasheed et al. (Eds.): ***Forthcoming Networks and Sustainability in the AIoT Era*****, LNNS 1035,**
https://doi.org/10.1007/978-3-031-62871-9_21

In the original version of the book, the following belated corrections are to be incorporated:

In chapter "Robot Hand-Controlled by Gyroscope Sensor Using Arduino", the affiliation "Computer Techniques Engineering Department, Al Imam Alkadhum College, Nasiriyah, Thi-Qar, Iraq" of author "Fatima Ghali" is to be changed to "Computer Techniques Engineering Department, Imam Alkadhim University College, Nasiriyah, Thi-Qar, Iraq".

The erratum book has been updated with the changes.

The updated version of this chapter can be found at
https://doi.org/10.1007/978-3-031-62871-9_21

J. Rasheed et al. (Eds.): FoNeS-AIoT 2024, LNNS 1035, p. C1, 2024.
https://doi.org/10.1007/978-3-031-62871-9_36

Author Index

J. Rasheed et al. (Eds.): FoNeS-AIoT 2024, LNNS 1035, pp. 457–458, 2024.
https://doi.org/10.1007/978-3-031-62871-9